Manipulation of Nanoscale Materials
An Introduction to Nanoarchitectonics

RSC Nanoscience & Nanotechnology

Series Editors:
Professor Paul O'Brien, *University of Manchester, UK*
Professor Sir Harry Kroto FRS, *University of Sussex, UK*
Professor Ralph Nuzzo, *University of Illinois at Urbana-Champaign, USA*

Titles in the Series:
1: Nanotubes and Nanowires
2: Fullerenes: Principles and Applications
3: Nanocharacterisation
4: Atom Resolved Surface Reactions: Nanocatalysis
5: Biomimetic Nanoceramics in Clinical Use: From Materials to Applications
6: Nanofluidics: Nanoscience and Nanotechnology
7: Bionanodesign: Following Nature's Touch
8: Nano-Society: Pushing the Boundaries of Technology
9: Polymer-based Nanostructures: Medical Applications
10: Metallic and Molecular Interactions in Nanometer Layers, Pores and Particles: New Findings at the Yoctolitre Level
11: Nanocasting: A Versatile Strategy for Creating Nanostructured Porous Materials
12: Titanate and Titania Nanotubes: Synthesis, Properties and Applications
13: Raman Spectroscopy, Fullerenes and Nanotechnology
14: Nanotechnologies in Food
15: Unravelling Single Cell Genomics: Micro and Nanotools
16: Polymer Nanocomposites by Emulsion and Suspension
17: Phage Nanobiotechnology
18: Nanotubes and Nanowires: 2nd Edition
19: Nanostructured Catalysts: Transition Metal Oxides
20: Fullerenes: Principles and Applications, 2nd Edition
21: Biological Interactions with Surface Charge Biomaterials
22: Nanoporous Gold: From an Ancient Technology to a High-Tech Material
23: Nanoparticles in Anti-Microbial Materials: Use and Characterisation
24: Manipulation of Nanoscale Materials: An Introduction to Nanoarchitectonics

How to obtain future titles on publication:
A standing order plan is available for this series. A standing order will bring delivery of each new volume immediately on publication.

For further information please contact:
Book Sales Department, Royal Society of Chemistry, Thomas Graham House, Science Park, Milton Road, Cambridge, CB4 0WF, UK
Telephone: +44 (0)1223 420066, Fax: +44 (0)1223 420247,
Email: booksales@rsc.org
Visit our website at http://www.rsc.org/Shop/Books/

Manipulation of Nanoscale Materials
An Introduction to Nanoarchitectonics

Edited by

Katsuhiko Ariga
National Institute for Materials Science (NIMS), Tsukuba, Japan
Email: ARIGA.Katsuhiko@nims.go.jp

RSCPublishing

RSC Nanoscience & Nanotechnology No. 24

ISBN: 978-1-84973-415-8
ISSN: 1757-7136

A catalogue record for this book is available from the British Library

© The Royal Society of Chemistry 2012

All rights reserved

Apart from fair dealing for the purposes of research for non-commercial purposes or for private study, criticism or review, as permitted under the Copyright, Designs and Patents Act 1988 and the Copyright and Related Rights Regulations 2003, this publication may not be reproduced, stored or transmitted, in any form or by any means, without the prior permission in writing of The Royal Society of Chemistry or the copyright owner, or in the case of reproduction in accordance with the terms of licences issued by the Copyright Licensing Agency in the UK, or in accordance with the terms of the licences issued by the appropriate Reproduction Rights Organization outside the UK. Enquiries concerning reproduction outside the terms stated here should be sent to The Royal Society of Chemistry at the address printed on this page.

The RSC is not responsible for individual opinions expressed in this work.

Published by The Royal Society of Chemistry,
Thomas Graham House, Science Park, Milton Road,
Cambridge CB4 0WF, UK

Registered Charity Number 207890

For further information see our web site at www.rsc.org

Printed and bound in Great Britain by CPI Group (UK) Ltd, Croydon, CR0 4YY, UK

Preface

Most of us as scientists and non-scientists know a term nanotechnology that seems to create many goods in our convenient life. This fact may not be true. Most useful materials, apparatuses, and machines can be fabricated by well-sophisticated micro-technology. Although nanotechnology has revealed various interesting phenomena, most of them still remain just as scientific topics. We have not yet utilised our knowledge of nanotechnology for materials science. What is necessary for the next steps? What comes after nanotechnology? Nanotechnology drastically develops ways to observe and analyze nanoscale worlds. We need to re-construct novel materials from the most basic material (atoms, molecules, and nano-units) based on knowledge from nanotechnology. A new paradigm for constructing materials from nanoworlds is nanoarchitectonics. Nanoarchtectonics must come after nanotechnology.

In the nanoscale worlds, differences between sciences such as physics, chemistry and biology are less, because phenomena and mechanisms would be fundamental and simple there. Therefore, nanoarchitectonics has to be accomplished by efforts of a wide range of research fields. This book collects excellent research examples from various fields for an introduction to nanoarchitectonics from the viewpoint of manipulation of nanoscale materials. I hope that readers can learn something about nanoarchitectonics from the chapters in this book.

Katsuhiko Ariga

Contents

Chapter 1	**Introduction: Nanoarchitechtonics for Materials Innovation**	**1**

Nanoarchitechtonics for Materials Development

Chapter 2	**Supramolecular materials nanoarchitechtonics**	**7**
2.1	Introduction	7
2.2	Molecular-level Complex and Arrangement	8
2.3	Self-assembly for Functional Structures	15
2.4	Intentional Assembly for Functional Structures	18
2.5	Bridging between Molecular and Macro	22
2.6	Summary	25
	Acknowledgements	26
	References	26
Chapter 3	**Controlled Multiscale Dewetting of Self Organized Block Copolymers**	**28**
3.1	Introduction	28
3.2	Background	30
	3.2.1 Fundamentals of Wetting and Dewetting Phenomena	30
	3.2.2 Stability of Block Copolymer Thin Films	36
3.3	Control of Dewetting of Polymer Films	43
	3.3.1 Overview	43
	3.3.2 Methods for Controlled Dewetting of Liquid Films	44
3.4	Control of Dewetting of Block Copolymer Films	50
	3.4.1 Overview	50

RSC Nanoscience & Nanotechnology No. 24
Manipulation of Nanoscale Materials: An Introduction to Nanoarchitectonics
Edited by Katsuhiko Ariga
© The Royal Society of Chemistry 2012
Published by the Royal Society of Chemistry, www.rsc.org

		3.4.2	Control of Block Copolymer Self Assembly	51
		3.4.3	Controlled Dewetting of Block Thin Film by Solvent Evaporation	54
		3.4.4	Controlled Dewetting of Block Thin Film by Chemically Patterned Substrates	56
		3.4.5	Controlled Dewetting of Block Thin Film by Topographically Patterned Substrates	59
	3.5	Control of Wetting of Block Copolymer Films		68
		3.5.1	Overview	68
		3.5.2	Controlled Wetting of Block Copolymer Droplets	69
		3.5.3	Controlled Anisotropic Wetting of Block Copolymer Droplets	74
	3.6	Concluding remarks		78
	References			78

Chapter 4 Nanoarchitectures Based on Clay Materials 87

	4.1	Introduction		87
	4.2	Layered Clays as Building Blocks for Nanoarchitectonics		88
		4.2.1	Pillared Clays (PILCs)	88
		4.2.2	Porous Clay Heterostructures (PCHs)	91
		4.2.3	Delaminated Porous Clay Heterostructures (DPCHs): Inorganic-Inorganic Nanocomposites	94
		4.2.4	Assembling NPs and Layered Clays	96
	4.3	Bottom-up Heteroarchitectures based on Fibrous and Tubular Clays		98
		4.3.1	Heterostructures based on Sepiolite and Palygorskita	98
		4.3.2	Heterostructures based on Halloysite and Imogolite	101
	4.4	Nanoarchitectonics based on Related Inorganic Solids		102
	4.5	Concluding Remarks		104
	Acknowledgements			104
	References			105

Contents ix

Chapter 5	Mesoporous Nanoarchtechtonics		112
	5.1	Introduction: The Nanoarchitectonics and Mesoporous Story	112
	5.2	The Carbon Nanocage and its Function	114
	5.3	Mesoporous Carbon Nitride and Mesoporous Boron Nitride	119
	5.4	Layered Hierarchic Structure	123
	5.5	Summary	127
	Acknowledgements		127
	References		128

Chapter 6	Nanoscale Oxides in Catalysis			129
	6.1	Introduction		129
		6.1.1	Overview	129
		6.1.2	Design and Synthesis of Nanomaterials	130
	6.2	Wet-Chemical and Low Temperature Routes for the Synthesis of Nanocrystalline Metal Oxides		131
		6.2.1	Hydrolysis/Chemical Precipitation	131
		6.2.2	Sol–Gel Synthesis	132
		6.2.3	Hydro-/Solvo-thermal	132
		6.2.4	Thermolysis	133
		6.2.5	Sonochemical	133
		6.2.6	Electrochemical	133
		6.2.7	Microwave Synthesis	134
		6.2.8	Biomimetic Mineralization	134
	6.3	Catalysis		134
		6.3.1	Effects of Nanostructuring and Morphology	135
	6.4	Some Typical Nanostructured Metal Oxides and their Catalytic Applications		135
		6.4.1	Nanocrystalline Magnesium Oxide (MgO)	135
		6.4.2	Copper Oxide (CuO)	144
		6.4.3	Titania (TiO_2)	147
		6.4.4	Zinc oxide	150
		6.4.5	Iron Oxides (Fe_2O_3, Fe_3O_4, Mixed Ferrites)	153
	6.5	Conclusions		154
	References			155

Chapter 7	Nanoarchitechtonics of Photocatalytic Materials			165
	7.1	Introduction		165
	7.2	Morphology Control		167
		7.2.1	One-dimensional Nanostructures	167
		7.2.2	Facet-controlled Nanostructures	169

	7.2.3 Hierarchical Composite Nanostructures	173
7.3	Nano-assembly	175
	7.3.1 Electronic Coupling Assembly	175
	7.3.2 Plasmon–Exciton Coupling Assembly	179
	7.3.3 Optical Coupling Assembly	181
7.4	Conclusion	183
	Acknowledgements	184
	References	184

Materials Nanoarchitechtonics for Bio-Conjugates and Bio-Applications

Chapter 8 Design, Synthesis and Application of Bio-conjugate Nanostructures 191

8.1	Introduction	191
8.2	Saccharides	191
	8.2.1 Monosaccharides	191
	8.2.2 Polysaccharides	192
8.3	Phospholipids	195
	8.3.1 Lipid	195
	8.3.2 PEG–Lipid	197
	8.3.3 Calcium Phosphate–Lipid	198
8.4	Proteins	199
	8.4.1 Polymer–Protein	200
	8.4.2 Gold–Protein	200
	8.4.3 Quantum Dot–Protein	201
	8.4.4 Iron Oxide–Protein	202
	8.4.5 Carbon–Protein	202
8.5	Nucleic Acids	203
	8.5.1 DNA	203
	8.5.2 Metal–DNA	204
	8.5.3 Graphene–DNA	205
8.6	Extension	206
8.7	Conclusions	206
	Acknowledgements	207
	References	207

Chapter 9 Architectonics of Active Sites: Life Processes at Nanodimensions 213

9.1	Enzymes, Active Sites and Vital Biological Reactions	213
9.2	Influence of the Architectonics of Active Sites on Enzymatic Reactions	216
9.3	Nanodimension of Active site, Confinement and Chiral Discrimination	219

Contents

	9.4 Example of a Life Process in a Nanospace: Aminoacylation Reaction in the Active Site of Aminoacyl tRNA Synthetase (aaRS)	223
	9.4.1 Architectonics of the Active Site of aaRS	224
	9.4.2 Influence of Active Site of aaRS on Aminoacylation Reaction	231
	9.5 Future Prospects	236
	Acknowledgements	237
	References	237

Chapter 10 Nanotechnology in Drug Delivery Systems — 242

10.1 Introduction	242
10.2 Nanocrystals	243
10.3 Surfactant/Polymer Micelles	244
10.4 Emulsions and Microemulsions	246
10.5 Liposomes	248
10.6 Polymer Nanogels	250
10.7 Molecular Conjugates	250
10.8 Dendrimers	251
10.9 Carbon Nanomaterials	251
10.10 Inorganic Nanomaterials	252
10.11 Inhalable particles	253
10.12 Summary	253
References	254

Chapter 11 Separation of Medically Useful Radionuclides: Role of Nano-sorbents — 259

11.1 Radionuclides for Use in Nuclear Medicine	259
11.2 Classifications of Radionuclides Based on Their Application: Diagnostic and Therapeutic	260
11.2.1 Diagnostic Radionuclides	260
11.2.2 Therapeutic Radionuclides	261
11.3 Production of Radioisotopes for Nuclear Medicine Applications	262
11.3.1 Cyclotron-produced Radioisotopes	262
11.3.2 Reactor-produced Radionuclides	262
11.3.3 Generator-produced Radionuclides	262
11.4 The Concept of the Radionuclide Generator and the Historical Perspective of its Development	263
11.5 The Mathematical Equations of Radioactive Decay and Growth in Radionuclide Generators	264

11.6	\multicolumn{2}{l}{The Available Options for the Preparation of Radionuclide Generators}	266

11.6 The Available Options for the Preparation of Radionuclide Generators — 266
  11.6.1 Column Chromatography — 266
  11.6.2 Solvent Extraction — 267
11.7 Essential Essential Components of a Column Chromatographic Radionuclide Generator — 268
11.8 Sorbent: The 'Heart' of Column Chromatographic Radionuclide Generator Systems — 269
11.9 Nanomaterials as New Generation Sorbents for the Preparation of Radionuclide Generators — 270
11.10 Procedures Involved in Evaluation of a Sorbent Material to Determine its Suitability for the Preparation of Radionuclidic Generators — 272
11.11 Quality Control of the Generator Produced Radioisotopes — 275
11.12 Shelf-life of a Radionuclide Generator — 277
11.13 Use of Nanomaterial-based Sorbents for the Preparation of ^{99}Mo/^{99m}Tc and ^{188}W/^{188}Re Generators — 278
  11.13.1 Current Status and Future Perspectives of ^{99}Mo/^{99m}Tc and ^{188}W/^{188}Re Generators — 278
  11.13.2 Polymer Embedded Nanocrystalline Titania for the Preparation of ^{99}Mo/^{99m}Tc and ^{188}W/^{188}Re Generators — 280
  11.13.3 Nanocrystalline Zirconia as Sorbent for the Preparation of ^{99}Mo/^{99m}Tc and ^{188}W/^{188}Re Generators — 288
11.14 Conclusions — 295
11.15 Terminology used in this chapter — 296
References — 297

Chapter 12 Sensing of Biomolecular Charges at Designer Nanointerfaces — 302

12.1 Introduction — 302
12.2 Label-free Biosensing with Semiconductor Devices — 303
  12.2.1 FET Designer Nanointerfaces — 303
  12.2.2 Nucleic Acids Sensing — 305
  12.2.3 Protein sensing — 307
12.3 Manipulation of Charges for Sensitive Biosensing — 308
  12.3.1 DNA Binders and Intercalators — 308
  12.3.2 Site-selective Charge Conversion of Proteins — 309
12.4 Summary — 311
Acknowledgements — 311
References — 311

Contents   xiii

Chapter 13   Nanostructured materials for biosensor applications:
             comparative review of preparation methods   318

    13.1 Introduction   318
    13.2 Nanostructures Manufactured Using Plasmonic
          Colloidal Nanoparticles   319
          13.2.1 Utilization of Noble Metal Nanostructures in
                  Plasmonic Biosensors   319
          13.2.2 Preparation of Colloidal Plasmonic
                  Nanostructures   320
          13.2.3 Core–Shell Nanoparticles   321
          13.2.4 Surface Nano-patterning of Random-fashion
                  Nanostructures Using Colloidal Au   324
          13.2.5 Surface Nano-patterning of Periodic Ordered
                  Nanostructures Using Colloidal Au   329
    13.3 Lithography-free Methods   331
          13.3.1 Oblique Angle Deposition Method   331
          13.3.2 Synthesis of Hybrid Nanostructures of Gold
                  Nanoparticles and Carbon Nanotubes   331
          13.3.3 Anodic Porous Alumina Membranes   334
    13.4 Lithographic Methods   337
          13.4.1 Scanning Beam Lithographies   337
          13.4.2 Colloidal Lithographies   339
          13.4.3 Nanoimprint Lithography   344
          References   347

Materials Nanoarchitechtonics for Advanced Devices

Chapter 14   Nanostructure Manipulation in Organic Solar Cells   359

    14.1 Introduction   359
    14.2 Nanostructure Manipulation In Conjugated Polymer–
          Fullerene Organic Solar Cells   361
          14.2.1 During Solution Preparation and Film
                  Formation   362
          14.2.2 Post-Film Formation Treatment   375
    14.3 Conclusion   385
          References   385

Chapter 15   Substrate alignment by surface probe lithography   392

    15.1 Introduction   392
    15.2 Calculation Using Single Molecules or Atoms   393
          15.2.1 Quantum Qubits   395
          15.2.2 Single Molecule Quantum Logic   396

15.3	Anchoring Molecules to Substrates		397
	15.3.1	Gold	397
	15.3.2	Oxides	399
	15.3.3	Silicon	400
15.4	Surface Patterning		402
	15.4.1	Scanning Tunneling Microscope	403
	15.4.2	AFM	404
	15.4.3	Vacuum Deposited Molecular Arrays	410
15.5	Scale Registration		412
	15.5.1	Direct Patterning	412
	15.5.2	Self-assembly	413
	15.5.3	Polymerization	416
	15.5.4	Alignment to Prefabricated Electrodes	418
15.6	Concluding Remarks		418
References			420

Chapter 16 Nanomechanical Sensors and Membrane-type Surface Stress Sensor (MSS) for Medical, Security and Environmental Applications   428

16.1	Introduction		428
	16.1.1	Demands for New Sensors	428
	16.1.2	Introduction to Nanomechanical Sensors	429
16.2	Cantilever Sensors		430
	16.2.1	A Brief History of Cantilever Sensors	430
	16.2.2	Operation Modes of Cantilever Sensors	431
	16.2.3	Surface Functionalization	435
	16.2.4	Read-out Methods	437
16.3	Membrane-type Surface Stress Sensor (MSS)		439
	16.3.1	Strategy for the Improvement in Sensitivity; Towards Membrane-type Surface Stress Sensor (MSS)	439
	16.3.2	Performance of MSS	443
16.4	Summary and Future Prospects		445
Acknowledgements			445
References			445

Subject Index   449

CHAPTER 1
Introduction: Nanoarchitechtonics for Materials Innovation

KATSUHIKO ARIGA[a,b]

[a] World Premier International (WPI) Research Center for Materials Nanoarchitectonics (MANA), National Institute for Materials Science (NIMS), 1-1 Namiki, Tsukuba 305-0044, Japan; [b] JST, CREST, 1-1 Namiki, Tsukuba 305-0044, Japan
E-mail: ARIGA.Katsuhiko@nims.go.jp

One of the most distinct differences between human beings as compared with other living creatures could be the fact that we can create various kinds of tools, apparatus, and machines. They can been seen commonly in our everyday activities. Tweezers, cloths, televisions, computers, and houses have been all developed by humans and they cannot be evolved in natural spontaneous processes. In their production, we have two key factors: (i) selection of materials (such as woods and metals) and (ii) method of making (construction and manufacturing). We have basically used raw materials, which can be obtained from surroundings, for manufacturing throughout most of our history. However, deep understanding of nanoscale materials in recent researches has changed this common situation drastically.

Rapid progresses in nanoscience and nanotechnology have created several fantastic ways to observe nanoscale objects such as atoms, molecules, clusters, and assemblies. Such progress will open new avenues in materials production, possibly by constructing even raw materials from nanoscale units. Therefore, we do not always have to select naturally-occurring substances that can be created with our 21st-century technologies. We can design functional materials through manipulation of the nanoscale, like carpenters construct houses and buildings. This novel methodology can be named nanoarchitectonics from nano + architecture + nics. This terminology was originally invented by Dr Masakazu Aono[1,2] at the National Institute for Materials Science (NIMS) and the World Premier International (WPI) Research Center for Materials Nanoarchitectonics (MANA).

It is necessary to differentiate two key terms, nanotechnology and nanoarchitectonics. Nanotechnology has become a popular word these days. Even non-scientists may frequently use this word in daily conversation. Nanotechnology (and/or nanoscience) deals with nanoscale of nano-related objects. On the basis of scale, it can be regarded as advanced form of microtechnology, which can be activated by considerable developments of microfabrications. However, nanotechnology and microtechnology are somewhat different. Most of the phenomena at the microscale (10^{-6} m range and the related scales) can be estimated from macroscopic phenomena of the same materials. In contrast, various unexpected properties becomes apparent for materials at the nanoscale (10^{-9} m range and the related scales). The latter features can be seen in various quantum phenomena and lots of abnormalities. Fullerene, carbon nanotubes, and graphene are zero-, one- and two-dimensional objects, respectively, where some dimensions are degenerated into nanoscales. Their properties cannot be deduced from those observed for macroscopic carbon materials such as carbon ash and diamonds. The unexpected phenomena found for nanocarbons actually create a novel scientific field.

With the emerging possibilities in nanotechnology, we anticipate lots of innovations in materials science. However, innovations, inventions, and dreams do not always come true. Most of the research in nanotechnology fields are oriented for observation and analyses on nanoscaled systems and materials. It may be said that major parts of nanotechnology still remain nanoscience. We need a breakthrough. It requires a novel concept, nanoarchitechtonics, which can bridge fundamental science in nanotechnology and practically useful materials through the application of knowledge and techniques in nanotechnology to explore useful materials, including atomic and molecular manipulation, chemical nanofabrication, self- and field-controlled organization, and theoretical modeling. In this methodology, controlled manipulation of atom/molecular-level nanostructures leads to creation of unexplored functional materials and systems.

Understanding novel concepts is always difficult. One of the best ways to promote the world of nanoarchitectonics would be highlighting the varied frontier research within this concept. From this point of view, this book includes current active research on nanoarchitectonics including (i) materials development in supramolecular chemistry, self organized polymers, in organic substances such as clay and mesoporous materials, catalysts design, and bio-related materials and (ii) systems innovations seen in biosensors, drug delivery systems, bio-related separations, lithographic techniques, and solar cell design.

Of course, these topics cannot cover all the targets in nanoarchitectonics, but they will surely stimulate reader's minds to realize nanoarchitectonics for material manipulations and material production.

References

1. This terminology was first proposed by Dr Masakazu Aono at the 1st International Symposium on Nanoarchitectonics Using Suprainteractions (NASI-1) at Tsukuba in 2000.
2. P. S. Weiss, A conversation with Dr. Masakazu Aono: leader in atomic-scale control and nanomanipulation, *ACS Nano*, 2007, **1**, 379.

Nanoarchitechtonics for Materials Development

CHAPTER 2

Supramolecular Materials Nanoarchitechtonics

KATSUHIKO ARIGA*[a,b], GARY J. RICHARDS[a], MASAAKI AKAMATSU[a], HIRONORI IZAWA[a,b] AND JONATHAN P. HILL[a,b]

[a] World Premier International (WPI) Research Center for Materials Nanoarchitectonics (MANA), National Institute for Materials Science (NIMS), 1-1 Namiki, Tsukuba 305-0044, Japan; [b] JST, CREST, 1-1 Namiki, Tsukuba 305-0044, Japan
*E-mail: ARIGA.Katsuhiko@nims.go.jp

2.1 Introduction

As seen in recent developments of portable computers and cellular phones, higher functions can be incorporated into very compact machines. The use of portable devices allows us to freely communicate information which can drastically change our lifestyles. Developments of portable, energy-efficient devices can impact on our societies in various ways. For example, the use of portable devices and ease of transfer of information and data between them can impart various freedoms to our lifestyles. As data and information communication between remote locations simplifies, problems such as overpopulation and traffic congestion within certain areas can be reduced. This in turn, can help reduce consumption of energy sources and the production of unnecessary environmental pollutants. Such innovations are based on the great success of microfabrication technologies which has led to the miniaturization of various systems.

Following on from the successes of microfabrication technologies, nanotechnology looks set to become the central subject of both scientific and technological research. Even though nanotechnology bears technology in its name, its technological aspect remains immature. Most results in nanotechnology research are observation-based and analysis-oriented. Although interesting scientific phenomena have been discovered, further development is required for more practical applications. For technological developments, production of useful materials and systems based on our knowledge of nanotechnology is necessary. We have to construct functional materials from nanostructures as architectures. This will be conducive to the creation of new functionalities that may be exhibited by nanoscale structural units through their mutual interactions even though these functionalities are not properties of the isolated units. Aono proposed a new term "nanoarchitectonics" to express this innovation of nanotechnology.[1] Materials nanoarchitectonics is a technology system aimed at arranging nanoscale structural units, which are groups of atoms, molecules or nanoscale functional components.

Application of materials nanoarchitectonics to organic soft matters and organic/inorganic hybrids requires control of molecular interactions. The field of control of molecular interaction and synthesis of functional molecular complexes is categorized as supramolecular chemistry. Therefore, two concepts: materials nanoarchitectonics and supramolecular chemistry, share a common interest. The combined concept, supramolecular materials nanoarchitectonics can be a useful methodology for the construction of organic and/or hybrid functional materials based on nanostructure control.

In this chapter, we introduce several topics of our research concerning supramolecular materials nanoarchitectonics. These topics are roughly categorized into (i) molecular-level complexes and arrangement, (ii) self-assembly for functional structures, (iii) intentional assembly for functional structures, and (iv) bridging between the molecular and macro scales.

2.2 Molecular-level Complex and Arrangement

An initial step for supramolecular materials nanoarchitectonics is to make molecular complexes in designed and controlled ways. In this process, a rational pairing of molecules becomes a key step and is generally called molecular recognition. Molecular recognition is the most fundamental and important concept in supramolecular materials nanoarchitectonics, because architecting any supramolecular materials involves selective molecular combination. The main concept associated with molecular recognition is based on the "lock and key" principle. Although this concept is well established, development of molecular recognition systems continues to this day.

Among the various types of molecular recognition, chiral recognition has been paid special attention because of the practical importance of such recognition for certain kinds of molecules including drugs and toxins is widely recognized. The enantiomeric excess (ee) is a critical parameter both as a determinant of the efficacy of chiral therapeutic agents, and as an indicator of success of organic asymmetric reactions. Recently, we have reported a technique for the determination of the ee in chiral carboxylic acids by using achiral porphine macrocycles as nonchiral–chiral solvating agents (Figure 2.1).[2] This system is unique in that chiral sensing is due to the fast exchange of analyte molecules at a protonated meso-substituted porphyrin dication rather than the formation of diastereomeric complexes with a chiral solvating reagent. This figure schematically illustrates the translation of chiral information from the analyte to the achiral host. A porphyrin dication binds two chiral guests in fast exchange equilibrium and each face of the macrocycle acts independently because of its saddle-shaped structure. The chiral guest bound between the opposing pyrrole groups induces non-equivalency in the adjacent pyrrolic proton signals, and anisochronicity of their resonances as a result of the preferred shielding direction. The (R)-/(S)- chiral environment induces different shielding in the place of these protons. The method is notable for the achirality of the reagent, its ease of implementation, and its application to pharmaceuticals because many drugs contain a carboxylic function in the vicinity of a chiral moiety. Exploitation of this phenomenon will lead to a new class of chiral solvating agents with unique applications.

Molecular complex systems can be used as mimics of information processing systems. Certain kinds of molecules are promising candidates for incorporation into electronic or optical devices because of their synthetic flexibility, processability and small size. One of the fundamental challenges in the development of molecular information processing systems is to prepare molecular memory and logic functions. In such systems, it is desirable for data to be stored in binary form, based on changes in optical or electronic properties, and toggled using an external stimulus, such as light, temperature, chemical concentration, voltage or a magnetic field. In particular, changes in optical absorbance and/or fluorescence can be detected as photonic output(s) and these may contain more information than simple electronic outputs. As illustrated in Figure 2.2, we have developed a nonvolatile-type memory system

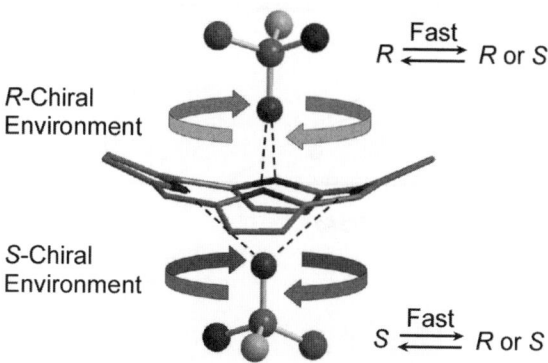

Figure 2.1 Achiral porphine macrocycle as nonchiral–chiral solvating agents.

based on fluorescence switching and based on subtle structural modifications of dye molecules.[3] The memory system can be operated primarily by using fluoride anions as specific writing signals. The nonvolatile memory system based on porphyrin derivatives can be achieved as follows. Fluoride anions induce the oxidative conversion from the porphyrin derivative with subsequent complexation of fluoride anion (memory writing). However, removal of the fluoride anion does not cause any change in the structure of the oxidized state, although a dissociation of the complex occurs (memory retention). Finally, the oxidized molecule is reduced to the original state by using ascorbic acid (memory erasing). These processes demonstrated fluoride-writable memory systems using simple porphyrin derivatives based on the relative oxidisability of these porphyrins in the presence of fluoride anions. In this case, OFF and ON states exhibited large differences in fluorescence emission intensity. This characteristic has great potential as a read-out signal for memory systems.

We also developed mechanical deformation of porphyrin macrocycles triggered by a photoredox reaction, which could be used as photo-driven molecular machines in future developments. As illustrated in Figure 2.3, 5,15-bis(3,5-di-*tert*-butyl-4-hydroxyphenyl)-substituted porphyrins can be operated as reversible photochemical molecular switches.[4] This porphyrin undergoes photoinduced aerial oxidation to porphodimethene with drastic variations of both electronic and conformational structures. The oxidized form undergoes a photoinduced reduction, reverting to the original form in the presence of a sacrificial electron donor such as triethylamine. Forward and reverse processes are governed by different chemical mechanisms, leading to better locking of the redox states. The oxidation process entails loss of aromaticity and an extension of conjugation over a greater number of atoms, and the oxidation reaction is coupled with a profound structural change due to steric and electronic requirements. This chimeric porphyrin-hydroquinone hybrid combines the

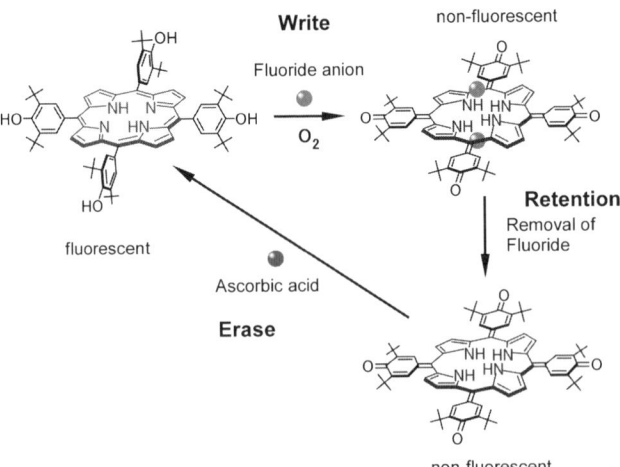

Figure 2.2 Nonvolatile-type memory system based on fluorescence switching.

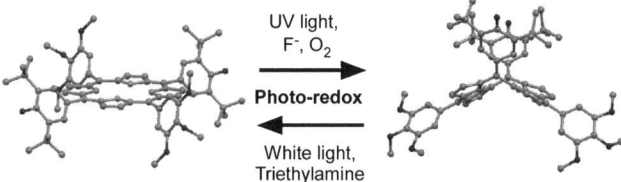

Figure 2.3 Photo-driven molecular machine.

photonic functionality of porphyrins with the redox activity of hydroquinones. Such molecular design indicates a simple and versatile method for producing photoredox macrocyclic compounds, which could lead to a new class of advanced functional materials suitable for bottom-up fabrication of molecular devices and machines.

Molecular complex formation can be used for regulation of physical events between chromophores. This heterotropic allosteric regulation can be a good mimic of certain processes in nature. For example, nature uses chloride anions as a cofactor of the oxygen-evolving complex. The supramolecular oligochromophoric model system shown in Figure 2.4 contains sites for binding of a reagent species and an anionic species.[5] The bis-porphyrin-substituted oxoporphyrinogen has two different binding sites, in which one site (composed of two porphinatozinc(II) units) is capable of binding bis(4-pyridyl)-substituted guests through coordination to the central zinc cations and the other site (composed of two pyrrole-type amine groups of the oxoporphyrinogen unit) interacts with anionic species through hydrogen bonding. This system demonstrated anion-complexation-induced enhancement of the charge separation in a supramolecular complex to produce singlet charge-separated states and also stabilization of the triplet charge-separated states in an oligochromophoric molecule possessing exclusive binding sites for both a guest electron

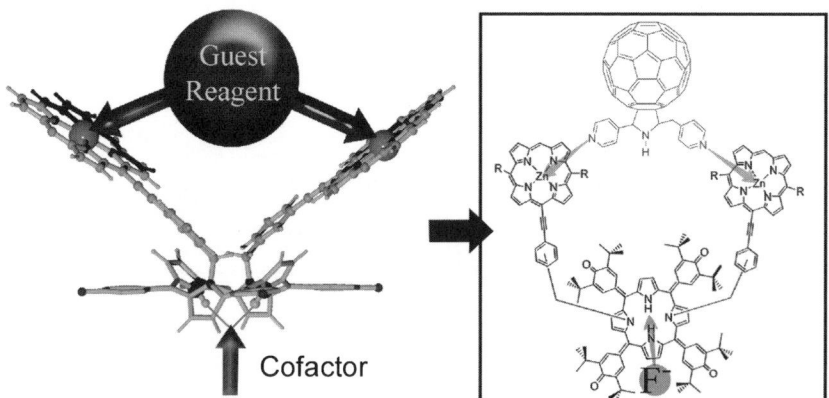

Figure 2.4 Bis-porphyrin-substituted oxoporphyrinogen with different binding sites, one for binding bis(4-pyridyl)-substituted guests and another for interaction with anionic species.

acceptor and an anionic cofactor species. The oxoporphyrinogen framework allows for the selective positioning of substituents and permits electrochemical control of its redox potentials in the presence of inorganic anions.

Control of molecular complex formation is not limited to complexes with a small numbers of component molecules. Formation of infinite molecular arrays is also a target of active research. In particular, molecular array formation at the surface has a high potential to make well-designed molecular networks on surfaces of devices. We have been recently developing systematic approaches to modulate structures of porphyrin molecular arrays using the

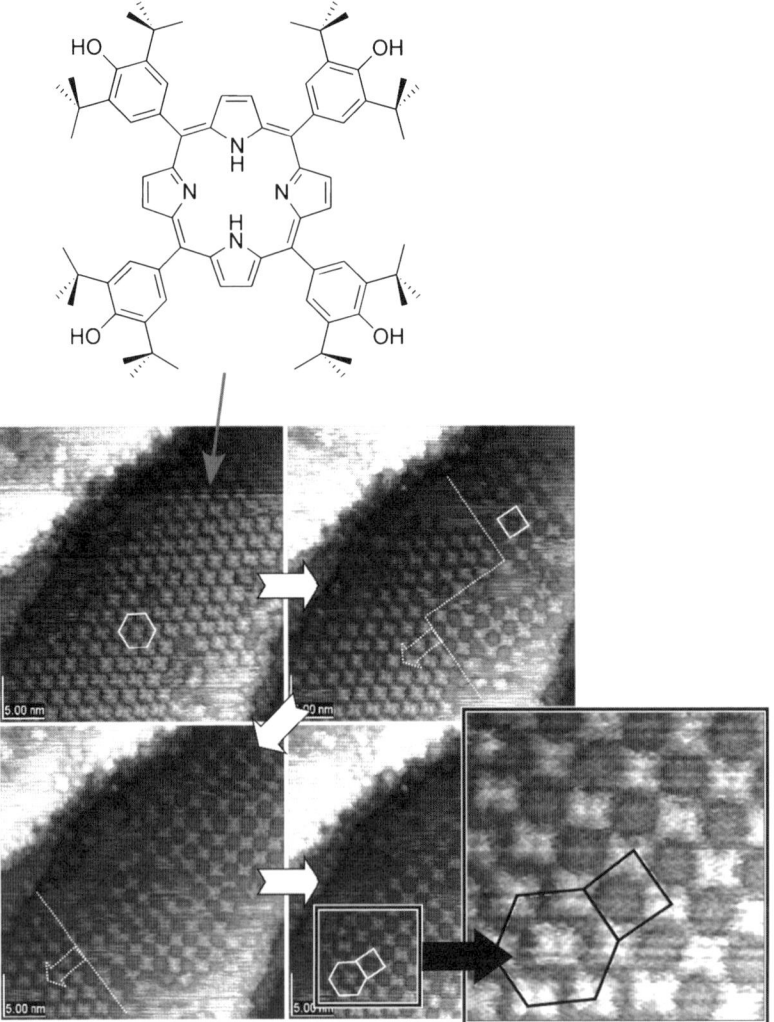

Figure 2.5 Phase transition of the two-dimensional molecular arrays of the oxoporphyrinogen.

oxoporphyrinogen unit. Phase transition of the two-dimensional molecular arrays of the oxoporphyrinogen in its reduced form (*tetrakis*(3,5-di-*tert*-butyl-4-hydroxyphenyl)porphyrin) are illustrated in Figure 2.5.[6] The reduced form at submonolayer coverage, which was deposited on Cu(111) under ultrahigh vacuum, comprises surface-mobile hexagonally-packed domain islands interspersed in a two-dimensional gas phase at lower temperature. Upon increasing the temperature to ambient, the hexagonally-packed structure undergoes a transition to the square packed grid motif. Because this phase transition occurs on a timescale of seconds, observation of the phase transition becomes possible by scanning tunneling microscopy (STM). Oxidation of these molecules drastically changes the situation, where hexagonal packings are maintained even at ambient temperature. The lack of mobility in the oxidized form leads to a much stronger molecule-surface interaction. The increased coplanarity of the oxidized porphyrin increases the multiplicity of contacts between *tert*-butyl substituents and the metal surface allowing formation of structures that are stable and immobile even at ambient temperatures. Detailed STM analyses revealed an interesting mechanism for molecular packing at the phase boundary between the two 2-dimensional patterns. We found a mismatch between lattice parameters of two-dimensional molecular arrays of the reduced form on the Cu(111) surface.[7] The molecules in the hexagonally-packed phase are composed of eight bright spots, corresponding to the *tert*-butyl groups. These molecules adopt a conformation in which the phenyl substituent groups and tetrapyrrole macrocycle approach coplanarity (Figure 2.6). Rotation of the phenyl substituents to a low dihedral angle, in response to surface adsorption, causes distortion of the tetrapyrrole core to a saddle shape. Surprising features were revealed by further inspection of the phase boundary structure. There is an apparent mixed conformation of the molecules at the

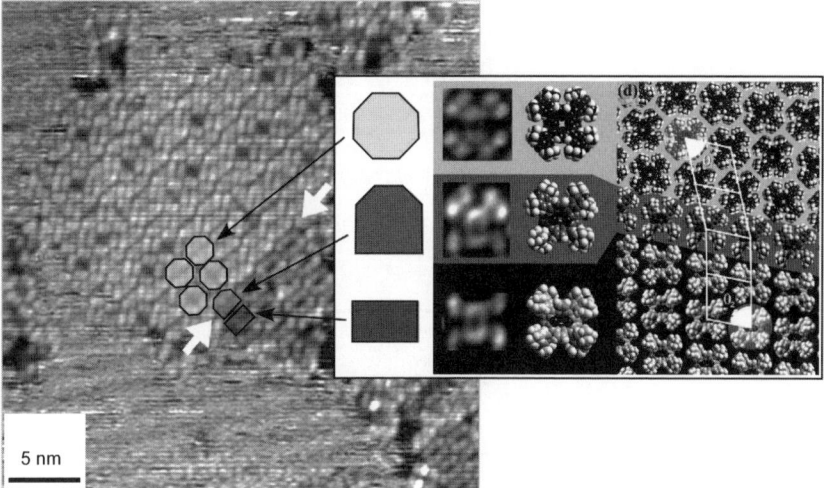

Figure 2.6 Adaptation of molecular conformation at the boundary region.

periphery of the assembly. For those molecules, phenyl substituents at the interior of the phase domain are likely to have similar dihedral angles (*i.e. ca.* 45°) while those at the exterior appear to have much smaller dihedral angles between them and the porphyrin macrocycle. One significant feature of the substituent mixed conformation of the porphyrin molecule is that it further permits its assemblies to exist in contact with other supramolecular assemblies. In the present case, a lattice match between two non-identical phases is not achieved by an interlock but by the creation of a conformationally unsymmetrical molecule, which exists at the interface between two differing domain structures.

The structural possibilities of allowing hydrogen bonding were also investigated using the porphyrin substituted with 3,5-dimethyl-4-hydroxyphenyl groups (5,10,15,20-*tetrakis*(3,5-dimethyl-4-hydroxyphenyl)porphyrin).[8] When the methyl-substituted analogue was adsorbed at the Cu(111) surface, intermolecular interactions occur at the meso substituents and are presumed to be hydrogen bonds because of the strength of these interactions presenting in the crystal structure. At submonolayer coverage (Figure 2.7), there is a pronounced tendency for the molecules to form a trimer unit through interactions at the meso substituents. The trimeric units represent a potential subunit of larger structures especially in macrocyclic form, which could be a

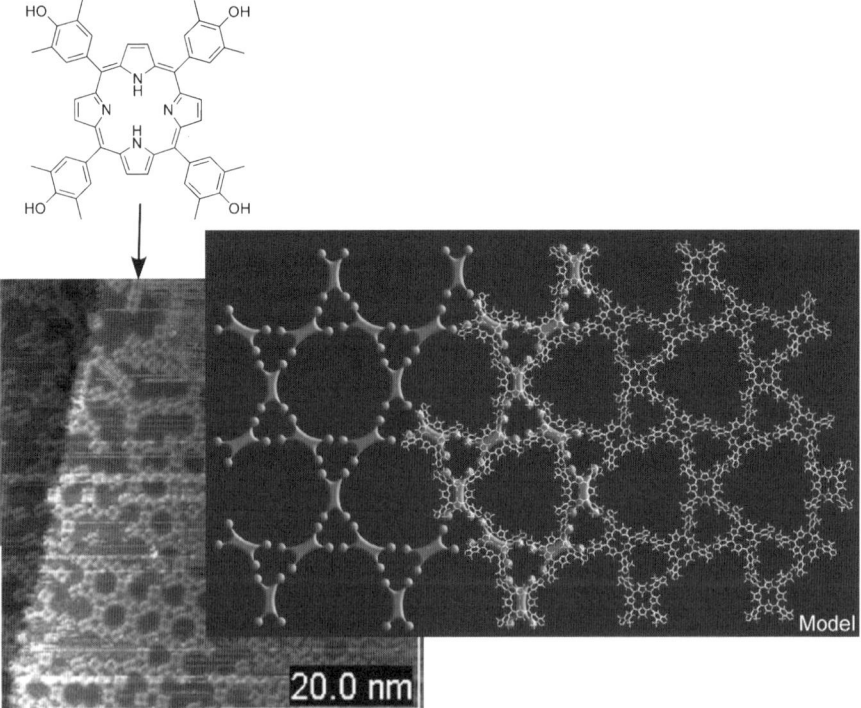

Figure 2.7 Kagomé lattice in two dimensions.

vertex of an extended hexagonal structure. In fact, a hexagonal structure could occasionally be observed. This represents the simplest case for a hexagonal structure built from a trimer having C_3 symmetry and is a porous honeycomb-type structure, where it can be seen as a Kagomé structure containing void areas within its lattice. An absence of gas-phase molecules in the enclosed areas of the Kagomé lattice was indicated by the high contrast of the STM image in those regions.

2.3 Self-assembly for Functional Structures

In order to use the concept of supramolecular nanoarchitechtonics in useful materials, construction of larger structures at the micron-scale or larger will be required. For such structures, assembling numerous molecules and/or unit materials to well-defined shapes becomes an important process. This can be achieved by spontaneous self-assembly.[9] In this section, our recent research on the self-assembly of well-controlled shapes, dynamic morphologies, and functional structures are exemplified.

Preparation of nano- and micro-structures with predetermined morphologies through self-assembly is one of the key processes in bottom-up approaches to nanotechnology. In particular, construction of shape-defined nanomaterials of functional nanocarbons such as fullerenes could lead to development of photo-electronic devices for ultrahigh information density processing. Recently, we reported controlled formation of two-dimensional (2D) objects including hexagons and rhombi and their selective shape-shifting (change or growth of crystals) into one-dimensional (1D) rods though solvent-dependent variations of the crystal lattice of C_{60} (Figure 2.8).[10] Interfacial precipitation of C_{60} by addition of appropriate solvents such as i-propyl alcohol (IPA) or *tert*-butyl alcohol (TBA) into saturated C_{60} solutions (benzene, toluene, CCl_4, CS_2, and *m*-xylene) resulted in 2D assemblies in various shapes. Uniformly shaped rhombi and hexagons were obtained at TBA/toluene and IPA/CCl_4 interfaces respectively, while polygon mixtures were collected from TBA/benzene interface. This research clearly demonstrates controlled formation of 2D nanosheets of various shapes (hexagons, rhombi, and mixed polygons) and selective shape-shifting (change or growth of crystals) to nanorods (short nanowhiskers) as C_{60} assemblies. This technically innovative concept based on our findings represents a new methodology in fullerene-based bottom-up nanotechnology.

We recently successfully realized highly dynamic micro-objects that self-assemble and reassemble spontaneously (Figure 2.9).[11] The component molecule is a trigeminal porphyrin with 4-*n*-dodecyloxyphenyl groups, oligoethyleneoxide substituents attached to one of the porphyrin phenyl groups through a 4-benzyloxyphenyl substituent, and bromine atoms at the meso position. These molecules initially present as vesicles in solution. However, once deposited on mica the vesicles gradually disappeared concurrently with nanowire growth (3.2 nm in height and ~8.5 nm in width).

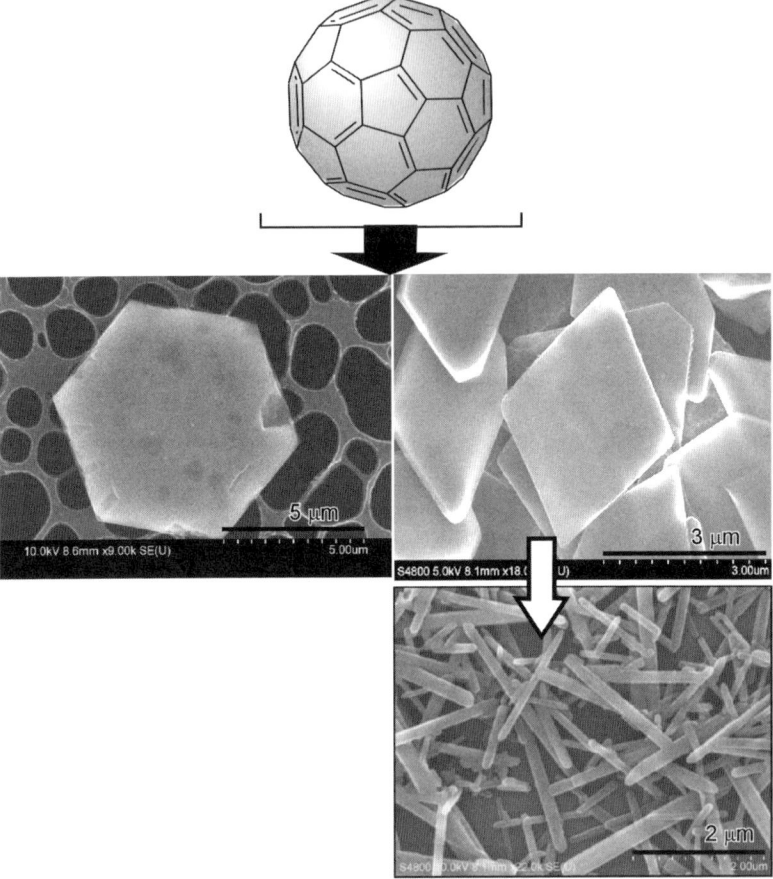

Figure 2.8 Shape-shifting C_{60} Assemblies.

A prominent feature of this study is the direct observation of the dynamic nanowire regrowth through scratching of existing nanowires by an atomic force microscope (AFM) tip. The trigeminal structural motif is a powerful method for assembling these molecules into a variety of nanostructures. This system can be used for controlled growth of surface-anchored nanoscale organic circuit structures in self-assembled molecular electronic devices.

Self-assembled structures with well-designed component units are often useful in the development of various functional devices. For example, Figure 2.10 shows that nanoarchitectures suitable for efficient organic photovoltaic devices can be realised by controlling the nanoscale morphology of electron donor/acceptor moieties based on self-assembly with molecular level phase segregation of block copolymers composed of the corresponding components.[12] In such a design, donor and acceptor domain sizes should be similar to the exciton diffusion length, usually in the order of 10 nm. We have prepared block-copolymer-based nanowires with nanosized donor/acceptor

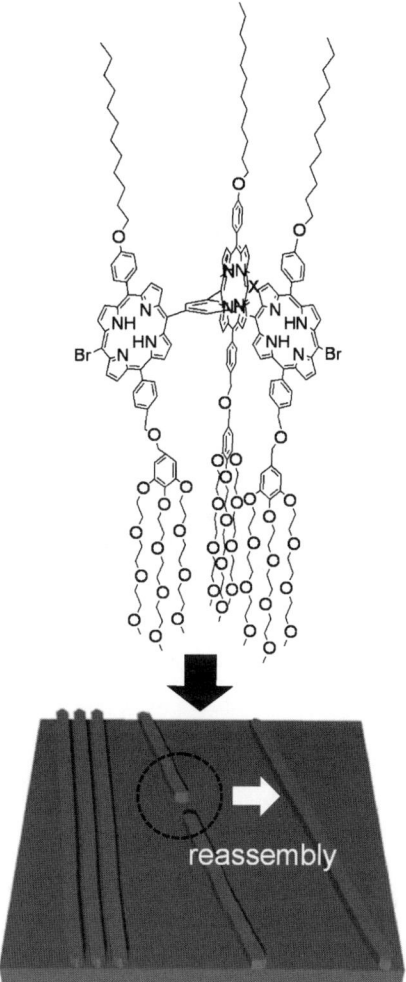

Figure 2.9 Dynamic self-assembly of trigeminal porphyrin molecule.

domain segregation with a domain width of *ca.* 5 nm. The nanowires consisted of regularly alternating domains arranged perpendicular to their long axes with a periodic domain width of around 5 nm. These nanowire structures were spontaneously formed by spin-coating or drop-casting their chloroform solutions onto several substrates (silicon, mica, glass, or highly ordered pyrolytic graphite (HOPG)) due to the amphiphilicity of the polymers. This donor/acceptor segregation appeared ideal as *p/n* heterojuction arrays for photovoltaics, and the resulting 1D nanowire thin films. Charge carriers generated at the nanosized interface between porphyrin and fullerene domains result from photoinduced electron transfer from porphyrins to fullerenes, leading to a photocurrent by their propagation through interwire carrier

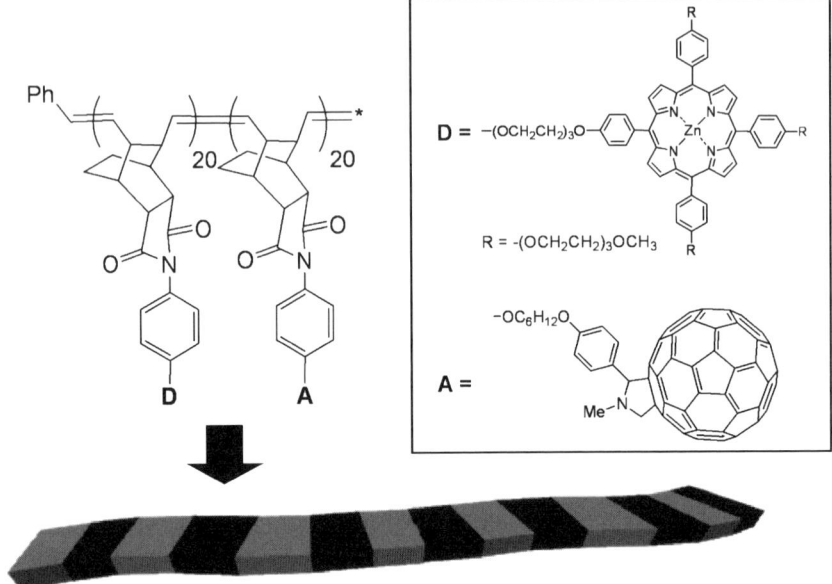

Figure 2.10 Self-assembled nanowires with molecular level phase segregation of block copolymers composed of donor and acceptor components.

transfer between domains. These self-assembled 1D nanostructured organic materials are interesting candidates for fabrication of bulk heterojunction solar cells or electronic devices.

2.4 Intentional Assembly for Functional Structures

Spontaneous self-assembly is not always a convenient method for the construction of desirable assembled structures. It is often necessary to have more direct control over the assembly process. Some techniques, such as the Langmuir–Blodgett (LB) method[13] and layer-by-layer (LbL) assembly,[14] provide controlled layered architectures. The LbL assembly method, in particular, has been embraced as a versatile method for thin film preparation and has rapidly developed as a method for self-assembly at interfaces. A typical method of electrostatic LbL assembly is illustrated in Figure 2.11. A cationic polyelectrolyte first adsorbs onto a negatively charged surface of a solid support, resulting in over-adsorption and surface charge reversal. The subsequent adsorption of an anionic particle again reverses the surface charge. Alternation of the surface charge permits continuous fabrication of the layered structure. The simplicity of the LbL assembly process allows a vast number of materials to be used with this technique. In this section, our recent research on functional microstructure preparation through LbL assembly is introduced, as well as a unique template synthesis on microcapsule structures.

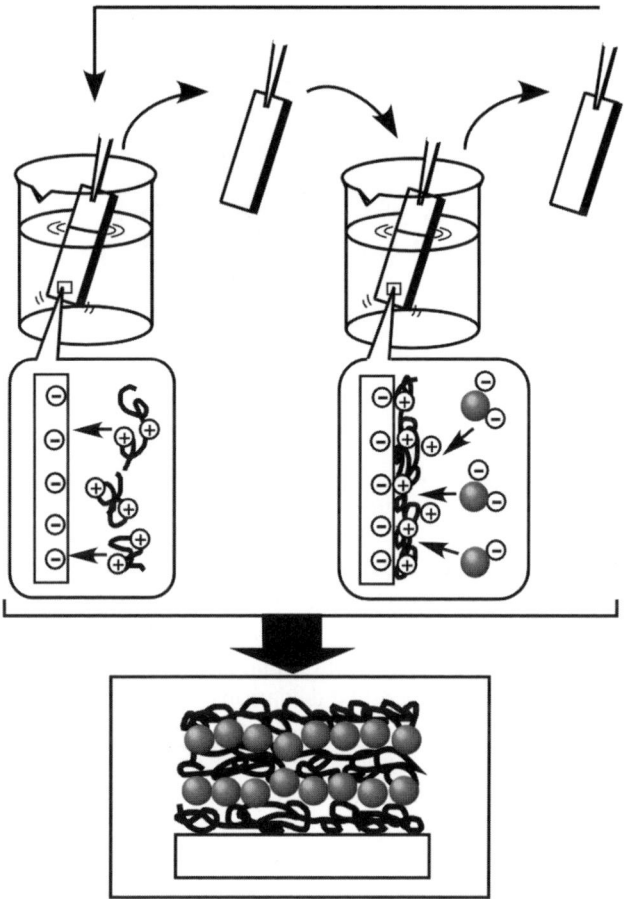

Figure 2.11 Typical method of electrostatic LbL assembly.

Recently, graphene has become one of the most attractive materials for research in nanoscience and nanotechnology. In our recent research, pieces of graphene can be disassembled from graphite then re-assembled into hierarchic structures through the LbL technique.[15] Graphene oxide sheet (GOS) was first prepared by oxidization of graphite under acidic conditions, followed by reduction to graphene sheet (GS) in the presence of ionic liquids in water. Composites of graphene sheet/ionic liquid (GS-IL) were assembled alternately with poly(sodium styrenesulfonate) (PSS) by LbL adsorption on appropriate solid supports, such as a QCM resonator (Figure 2.12). Exposure of the composite films to various saturated vapours under an ambient atmosphere caused an *in situ* decrease in frequency of QCM due to gas adsorption. The obtained results are a striking indication of the highly selective detection of aromatic guests within the well-defined π-electron-rich nanospace in the GS-IL films. Significantly higher selectivity (more than 10 times) for benzene vapour

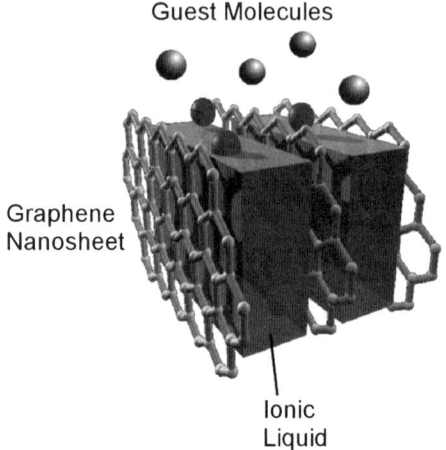

Figure 2.12 LbL assembly of graphene-sheet/ionic liquid (GS-IL) composite.

over cyclohexane was observed despite their similar molecular sizes, molecular weights, and vapour pressures. Detection of vapours can be repeated through alternate exposure and removal of the subject solvents. Responses to mixtures of benzene and cyclohexane at different molar ratios showed an approximately linear relation with small cooperative deviations, facilitating estimation of gas fractions in mixtures. Control of the electronic properties of the GS-IL films upon gas adsorption was also demonstrated. Adsorption of CO_2 vapours from a saturated sodium hydrocarbonate solution into the GS-IL films showed enhanced adsorption volume compared to the GS films without intercalated ionic liquids. These experimental results also suggest great potential for practical usage of GS-IL films.

We also developed a novel strategy for electrochemical coupling layer-by-layer (ECC-LbL) assembly based on a direct coupling reaction (Figure 2.13),[16] where *N*-alkyl carbazole dimerization was selected as a coupling reaction. This reagent-less clean process is especially useful for constructing covalently linked layer-controlled thin films on a sensitive device surface. In addition, unlike other electro-polymerizable precursors, such as aniline and thiophene, the resulting *N*-alkyl dicarbazole is transparent in the visible light region, and so does not impair the optical or electrical properties of the active moieties in the film. We synthesized several carbazole derivatives carrying distinctive donor and acceptor moieties. Both homo- and hetero-ECC-LbL assemblies were demonstrated. A prototype p/n heterojunction device with the structure ITO/donor/acceptor/Al was fabricated by the ECC-LbL assembly method and its photovoltaic properties were investigated. The photovoltaic switching response of this film, studied using pulsed single wavelength (500 nm) irradiation with 10 s intervals, demonstrates repetitive and stable device performance of the p/n heterojunction composed of a **PZ4C/C_{60}2C** ECC-LbL assembly. Appropriate variation of film structure and composition for broad optical absorption, low

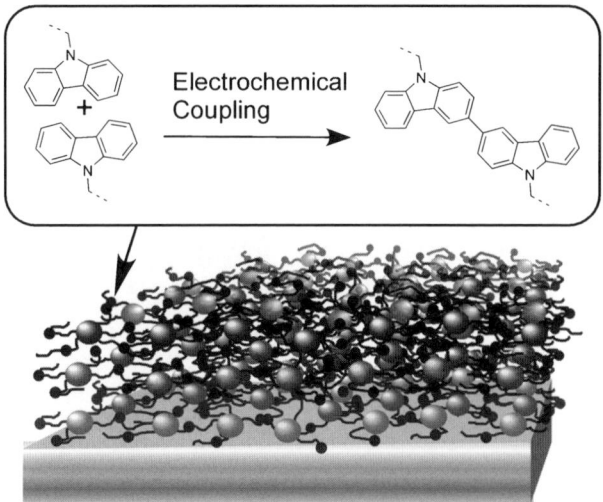

Figure 2.13 Electrochemical coupling layer-by-layer (ECC-LbL) assembly.

contact resistance and trapping density would improve device performance. Conventional preparation methods for thin film preparation such as vacuum deposition and spin-coating are limited in terms of which components can be applied depending on their molecular weights or solubilities, respectively. In contrast, use of electrochemical N-alkyl carbazole coupling has enabled us to fix cross-linked thin films at a substrate surface from solutions of the components. Functional moieties such as porphyrin and fullerene interact within the film indicating that the coupling reaction does not have a detrimental effect on the incorporated units. Although the preparation of this p/n heterojunction device with 500 nm thickness took about 30 min, the ECC-LbL processing time can easily be optimized to a few minutes under selected conditions.

Not limited to the LbL assembly, material nanoarchitectonics with guided structures is useful for creating designed structures appropriate for desired functions. We recently developed platinum microcapsules with open mouths, so called metallic cells, by template synthesis using polystyrene spheres (Figure 2.14).[17] A substantial increase of their electrode capability for methanol oxidation and catalytic activities for carbon monoxide (CO) oxidation were observed. Significant catalytic activity for CO oxidation was confirmed at an onset temperature (temperature at which 1% CO was converted) of 125 °C, which is much lower than values reported previously for bulk Pt or Pt-supported catalysts (>200 °C). The observed high CO oxidation activity of the metallic cells can be ascribed to the highly active and agglomeration-free interior surface, and demonstrates the advantages of using the interior surface of microcapsules for electrochemical and gaseous catalytic applications. The synthetic strategy of capsular catalysts could be widely

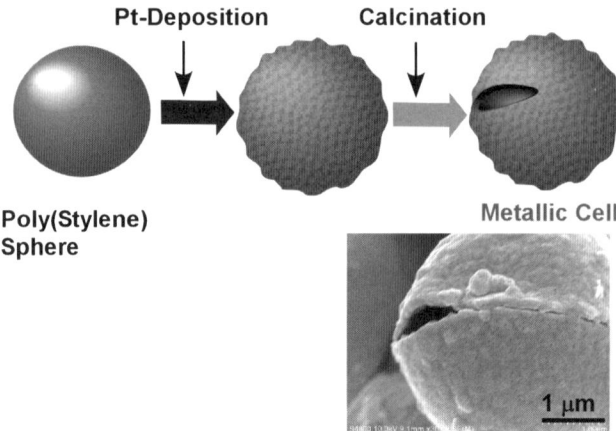

Figure 2.14 Preparation of platinum microcapsules with open mouths (metallic cells).

applicable to alloys of Pt, Pd and/or Rh to realize superior activities relative to the materials currently in use.

2.5 Bridging between Molecular and Macro

In the above sections, we discussed molecular-level architectonics and microscopic (or larger) architectonics. These researches are always considered separately because of significant differences in the sizes of the studied materials. However, bridging between molecular and macroscopic systems can create novel functional concepts, such as mechanical control of nano and molecular systems. This can be achieved through two-dimensional nanoarchitectonics on fluid surfaces. The most significant advantage of this concept is the possibility of dynamic conformational changes and spatial rearrangement. Demonstrations of two-dimensional nanoarchitectonics at dynamic interfaces are introduced below.

Some kinds of molecules are called molecular machines and respond to external stimuli with conformational or functional changes. They can be manipulated either in solution or on static surfaces. However, the lack of any direct connection between bulk stimuli and molecular functions often condemns molecular machines to be mere scientific curiosities. In order to incorporate them into operating machines, direct coupling of macroscopic motion and molecular motion is crucial. Molecular machines at a dynamic interface, *e.g.* air–water, could allow direct operation by bulk mechanical motion. Harmonized motion of molecular machines within the organized array can be coupled with macroscopic force or other stimuli. We selected a molecular machine otherwise known as a steroid cyclophane, that contains a 1,6,20,25-tetraaza[6.1.6.1]-paracyclophane cyclic core connected to four cholic acid moieties through a flexible L-lysine spacer.[18] As illustrated in Figure 2.15, the cholic acid moiety has both hydrophobic and hydrophilic faces so that, in a monolayer at low pressures, it forms an open conformation. Compression of the monolayer induces shrinkage of the steroid cyclophane molecule resulting in a cavity conformation. Molecular capture upon cavity formation of the

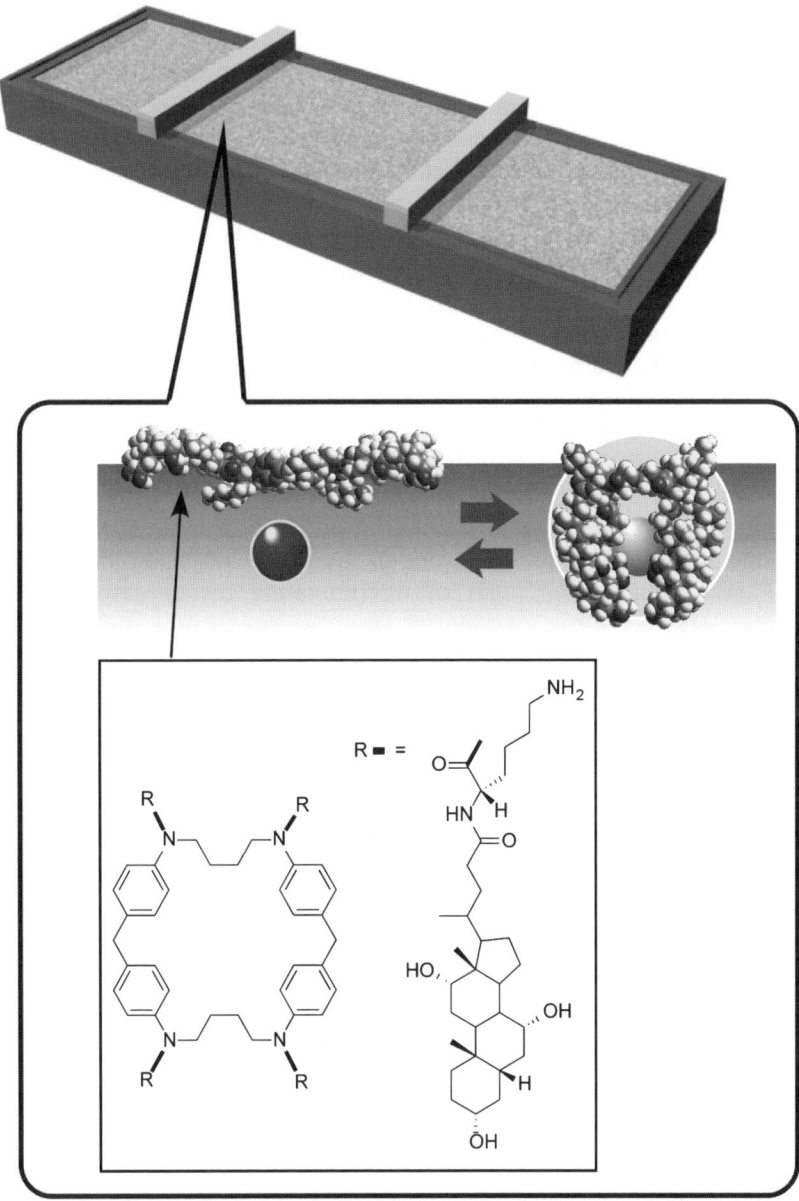

Figure 2.15 Molecular capture upon cavity formation of a steroid cyclophane.

steroid cyclophane was demonstrated using 6-(*p*-toluidino)naphthalene-2-sulfonate (TNS) as a model guest molecule through observation of *in situ* surface fluorescence spectroscopy. Repeated compression and expansion cycles of the steroid cyclophane monolayer by external mechanical force induced a periodic change in the fluorescence intensity, indicating that the capture and release of TNS occurs by the dynamic cavity formation. The results obtained clearly demonstrate that a molecular machine function (capture and release of guest molecule as a nanometre-level function) can be driven by application of macroscopic motion (monolayer compression and expansion over tens of cm).

Other modes of molecular motion can also be utilized for molecular machine functions. We utilised the twisting motion of a *N*-substituted cyclen, containing a 1,4,7,10-tetraazacyclododecane core with four cholesteric side arms, as a molecular machine (Figure 2.16).[19] Packing control of the *N*-substituted cyclen receptor through application of lateral mechanical force gives controlled modification of enatioselectivity in binding of aqueous amino acid guests. This resulted in an inversion of the magnitude of the binding constant between L- and D-enantiomers in the case of a valine guest. This system potentially allows discrimination of chiral substances by bulk motion even through hand operation. So, this concept can be named as hand-operated nanotechnology. This may be only the second successful chiral separation by hand since Prof. Pasteur's separation of crystals of tartaric acid.

This concept has been extended to a novel strategy for the fine tuning of the recognition capabilities of molecular machines. We have applied the concept of hand-operated nanotechnology to one of the most challenging biomolecular recognition problems, *i.e.*, that of discriminating thymine from uracil.[20]

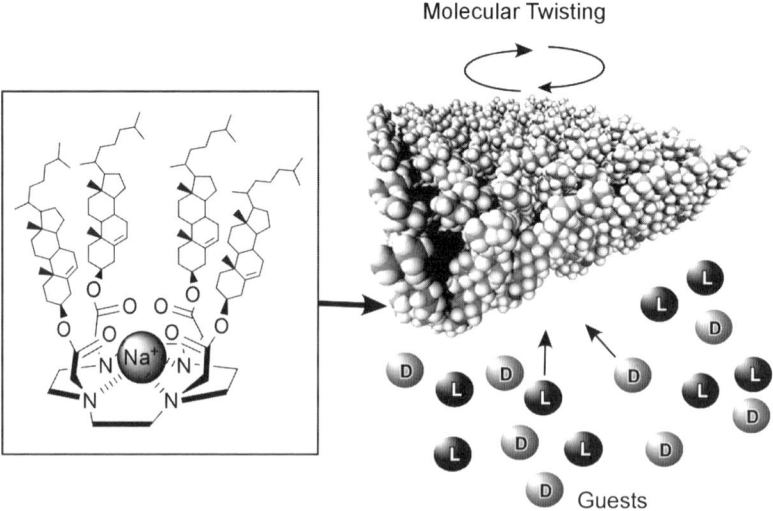

Figure 2.16 Chiral discrimination of amino acids through twisting motion of a *N*-substituted cyclen.

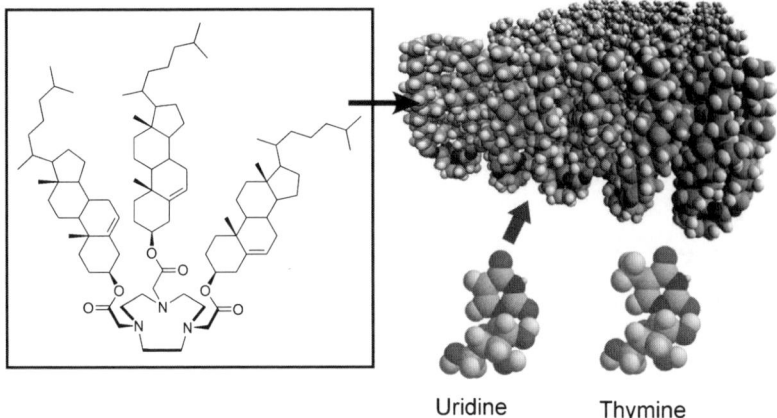

Figure 2.17 Cholesterol-substituted triazacyclononane used as a molecular machine to discriminate thymine from uracil.

Cholesterol-substituted triazacyclononane was used as a molecular machine to discriminate a subtle change in the structure of guest molecules (Figure 2.17). It was subjected to structural tuning by compression of its Langmuir monolayer in the absence and presence of Li^+ cations in the subphase. The monolayer of the triazacyclononane host selectively recognizes uracil over thymine (*ca.* 64 times) under optimized conditions ([LiCl] = 10 mM at a surface pressure of 35 mN m^{-1}). Optimization of molecular machine structures for the best recognition through mechanical tuning by an external force should become a novel methodology in host–guest chemistry as an alternative to the more traditional molecular design strategies.

2.6 Summary

In this chapter, we have briefly summarized aspects of supramolecular materials nanoarchitectonics using several examples, mainly from our recent accomplishments. Structural control over molecular arrays and materials arrays provides nanoscale structures, where molecular attributes such as morphology and functional group positioning require thoughtful design. For micro-sized objects obtained through self-assembly, careful design of component structures is indispensable for successful assembly. Intentional assembly such as LbL assembly is quite an effective means to construct highly desirable functional microstructures. Finally, some techniques such as dynamic manipulation of self-assembled molecules at the air–water interface can provide unique media for coupling of bulk mechanical motions with molecular function. These systems are mainly related to bottom-up processes in nanofabrication. Coupling of these exploratory researches with well-developed microlithography techniques will create meaningful and useful integrations of bottom-up and top-down nanotechnologies.

Acknowledgment

We thank World Premier International Research Center Initiative (WPI Initiative), MEXT, Japan and Core Research for Evolutional Science and Technology (CREST) program of Japan Science and Technology Agency (JST), Japan.

References

1. This terminology was first proposed by Dr Masakazu Aono at the 1st International Symposium on Nanoarchitectonics Using Suprainteractions (NASI-1) at Tsukuba in 2000.
2. J. Labuta, S. Ishihara, A. Shundo, S. Arai, S. Takeoka, K. Ariga and J. P. Hill, *Chem.–Eur. J.*, 2011, **17**, 3558.
3. A. Shundo, J. P. Hill and K. Ariga, *Chem.–Eur. J.*, 2009, **15**, 2486.
4. S. Ishihara, J. P. Hill, A. Shundo, G. J. Richards, J. Labuta, K. Ohkubo, S. Fukuzumi, A. Sato, M. R. J. Elsegood, S. J. Teat and K. Ariga, *J. Am. Chem. Soc.*, 2011, **133**, 16119.
5. F. D'Souza, N. K. Subbaiyan, Y. Xie, J. P. Hill, K. Ariga, K. Ohkubo and S. Fukuzumi, *J. Am. Chem. Soc.*, 2009, **131**, 16138.
6. J. P. Hill, Y. Wakayama, W. Schmitt, T. Tsuruoka, T. Nakanishi, M. L. Zandler, A. L. McCarty, F. D'Souza, L. R. Milgrom and K. Ariga, *Chem. Commun.*, 2006, 2320.
7. J. P. Hill, Y. Wakayama and K. Ariga, *Phys. Chem. Chem. Phys.*, 2006, **8**, 5034.
8. J. P. Hill, Y. Wakayama, M. Akada and K. Ariga, *J. Phys. Chem. C*, 2007, **111**, 16174.
9. K. Ariga, J. P. Hill, M. V. Lee, A. Vinu, R. Charvet and S. Acharya, *Sci. Technol. Adv. Mater.*, 2008, **9**, 014109.
10. M. Sathish, K. Miyazawa, J. P. Hill and K. Ariga, *J. Am. Chem. Soc.*, 2009, **131**, 6372.
11. Y. Xie, M. Akada, J. P. Hill, Q. Ji, R. Charvet and K. Ariga, *Chem. Commun.*, 2011, **47**, 2285.
12. R. Charvet, S. Acharya, J. P. Hill, M. Akada, M. Liao, S. Seki, Y. Honsho, A. Saeki and K. Ariga, *J. Am. Chem. Soc.*, 2009, **131**, 18030.
13. S. Acharya, J. P. Hill and K. Ariga, *Adv. Mater.*, 2009, **21**, 2959.
14. K. Ariga, J. P. Hill and Q. Ji, *Phys. Chem. Chem. Phys.*, 2007, **9**, 2319.
15. Q. Ji, I. Honma, S.-M. Paek, M. Akada, J. P. Hill, A. Vinu and K. Ariga, *Angew. Chem., Int. Ed.*, 2010, **49**, 9737.
16. M. Li, S. Ishihara, M. Akada, M. Liao, L. Sang, J. P. Hill, V. Krishnan, Y. Ma and K. Ariga, *J. Am. Chem. Soc.*, 2011, **133**, 7348.
17. S. Mandal, M. Sathish, G. Saravanan, K. K. R. Datta, Q. Ji, J. P. Hill, H. Abe, I. Honma and K. Ariga, *J. Am. Chem. Soc.*, 2010, **132**, 14415.
18. K. Ariga, Y. Terasaka, D. Sakai, H. Tsuji and J. Kikuchi, *J. Am. Chem. Soc.*, 2000, **122**, 7835.

19. T. Michinobu, S. Shinoda, T. Nakanishi, J. P. Hill, K. Fujii, T. N. Player, H. Tsukube and K. Ariga, *J. Am. Chem. Soc.*, 2006, **128**, 14478.
20. T. Mori, K. Okamoto, H. Endo, J. P. Hill, S. Shinoda, M. Matsukura, H. Tsukube, Y. Suzuki, Y. Kanekiyo and K. Ariga, *J. Am. Chem. Soc.*, 2010, **132**, 12868.

CHAPTER 3
Controlled Multiscale Dewetting of Self Organized Block Copolymers

JUNE HUH AND CHEOLMIN PARK*

Department of Materials Science and Engineering, Yonsei University, Seoul, Korea
*E-mail: cmpark@yonsei.ac.kr

3.1 Introduction

The 2 dimensionally ordered micro/nanostructures of polymer thin films have been of great importance not only for fundamental understanding of structure formation but also for the variety of their technological applications in the microelectronic industry *e.g.*, coatings, lubricants, cheap electronics, biosensors.[1–4] The polymeric nanopatterns currently available by the advanced top-down lithographic techniques, including EUV photo, e-beam and X-ray lithography and so on, are in general high cost and often time consuming. One of the alternatives for nanopatterning polymer thin films is soft lithography, in particular combined with various self organization principles of polymers such as macro-phase separation of blends, micro-phase separation of block copolymers and dewetting.[5–8] Control of the polymer self assembly in 2 dimensionally periodic space offers a great potential for fabricating a variety of unprecedented ordered structures. In spite of many methods proposed for patterned polymeric structures based on self assembly, most of the resulting structures were in the micron scale,[3,8,9] and the goal of nanofabrication still remains a difficult and elusive except block copolymer self assembly which undergoes spontaneous tens of nanometre scale phase separation with various structural motifs.

During the last two decades, block copolymers have gained enormous attention not only due to their abundant nano-sized structures but also due to their ability to undergo morphological reformation under various external fields.[10,11] Intriguing hierarchical structures evolved on the nanometre, micron to millimetre scale with periodic orders have been successfully produced by self assembly of block copolymers.[12–14] The simple and cost effective characteristics of the block copolymers have drawn many emerging applications in nanotechnology, such as nanolithography, photonic crystals, nano-templates for harvesting metals and semi-conductors and drug carriers.[12–15] In particular, thin films of block copolymers have emerged as a promising new route for technologically relevant nanopatterns on solid substrates.[11]

The structural stability of a thin viscoelastic polymer film strongly depends on the surfaces the film is in contact with, and often experiences film instability and subsequently dewets on either solid or liquid substrate due to its interaction with the substrate by long-range van der Waals forces and short range polar and molecular forces. In a typical dewetting process involving ruptures of an initially uniform film into droplets on a nonwetting substrate, the mechanism is elucidated by either spinodal or nucleation and growth dewetting. In spinodal dewetting, the unstable growth of periodic capillary waves of polymer concentration is amplified with time, followed by spontaneous rupture mainly due to the long-range molecular forces.[16] Nucleation and growth, on the other hand, involves the growth of holes nucleated mostly on the defect sites of the film and the subsequent formation of rims, polygons and droplets.[17] On the other hand, droplets of the polymer spontaneously spread on a favorable surface, resulting in a stable thin film. In fact, the spreading of non-volatile liquid droplets on a wettable substrate has been in fact a long standing issue attracting much interest due to its implications for understanding lubrication, molecular scale friction, and coating, and thus, for controlling hard surfaces.[18–22] The spreading frequently accompanies the formation of unique molecular and/or nanometre scale terraces characterized by distinct monomolecular layers in the direction of z normal to the surface.[23–28]

When a self-assembled material with an ordered structure, on a molecular or nanometre level, is employed for wetting or dewetting on a hard surface, the system becomes more complicated due to the interplay between the ordered structure and the surface energies. For instance, in the case of droplet wetting of a structured fluid on a hard surface, one should take into account not only the first layered terrace directly on the solid surface, frequently known as a frontier layer, but also multiple terraces regularly spaced in the z-direction arising from the ordered structure.[23–28] There have been several theoretical[23–25] and experimental studies[26–28] dealing with the dynamics of terraced droplets of structured fluids, such as smectic liquid crystals and block copolymers. For instance, block copolymers with lamellar microstructures showed very unique terraced hierarchical structures formed upon either dewetting a thin homogeneous film or wetting a distinct droplet.[29–32] Disk-like concentric rings with the characteristic periodicity of the block copolymer were, for instance, successively piled up in the droplet whose cross section resembles a triangle with a stepped side profile.[30,32] In this case, the monolayered first terrace was evenly formed on a substrate frequently called the brush monolayer of a block copolymer, in which one of the blocks is preferential to the surface with the other non-preferential block pointing toward the air.[30] Similarly, unique dewetting behaviors have been observed of a self assembled block copolymer thin films which involve hole and island formation and autophobic dewetting.[33]

To utilize the self organizing wetting and dewetting of a polymer thin film for technologically beneficial patterned structures, the control of those driving forces is essential and the controlled dewetting and wetting achieved by introducing surface heterogeneity of substrate is known to be an effective way for fabricating ordered micro structures of either a homopolymer or blend of two immiscible polymers.[34–37] The controlled mass transportation by certain periodic boundaries defined by either chemically or topographic patterned substrates in principle leads to the various ordered structures in ultrathin polymeric films. The controlled dewetting is additionally beneficial because it not only provides various ordered structures but also induces the size reduction of individually patterned domains due to the thermodynamic tendency of surface area minimization. The resulting ordered structures have dewet domains that are much smaller than that of an original pre-pattern. In fact, this approach has offered an efficient route for fabrication of nanopatterns of metallic thin films.

Control of wetting and dewetting in self assembled block copolymer thin films has been of great attention not only for scientific fundamental aspects of the interplay between multiscales in structural dimensions as well as between the multi-principles of self assembling driving forces, but also for technological advances of ordered hierarchical patterned structures from the molecular level to the macroscopic one. Considering that a block copolymer nanopattern can be further functionalized as a nano-reactor for developing various metallic, semi-conducting nanoobjects with ease, the hierarchically ordered block copolymer thin film combined with controlled wetting and dewetting can open numerous potential applications. The chapter begins with the fundamental understanding of dewetting and wetting of a thin polymer and a thin block copolymer film, followed by various principles of technological approaches for fabricating interesting ordered wetted and dewetted nanostructures.

3.2 Background

3.2.1 Fundamentals of Wetting and Dewetting Phenomena

The spreading of a liquid on a solid surface is one of the most fundamental and important properties of liquids. This phenomenon is of great practical importance in a number of industrial processes, such as lubricating films, active material components and protective coatings. In those practical situations, the film stability with respect to the uniformity of the film thickness is the most essential prerequisite and therefore has been one of the most important issues in the film science.[38]

The wetting statics of thick films in its basic form is very simple, whereas it becomes complicated for very thin films where van der Waals forces play an important role. A liquid will wet a surface if the total interfacial energy is lowered by doing so. This is the case for simple liquids such as water as well as more complex liquids such as polymers. The spreading coefficient (S) is a useful measure for the wettability of a liquid (*l*) on a solid surface (*s*) in a vapor medium (*v*) and is defined by

$$S = \gamma_{sv} - (\gamma_{sl} + \gamma_{lv}) \qquad (1)$$

where γ_{sv}, γ_{sl}, and γ_{lv} are the solid/vapor, solid/liquid, and liquid/vapor interfacial tensions in the solid/liquid/vapor system, respectively. The wetting occurs for $S > 0$ if other factors affecting the film stability (*e.g.*, dispersion force) are unimportant. When $S < 0$, the liquid does not spread and forms a droplet of spherical cap, or that of pancake on the surface if the gravity effect is not too large. The contact angle θ, at which the liquid/vapor interface meets the solid interface, is related to the three interfacial tensions:[39]

$$\gamma_{sv} = \gamma_{sl} + \gamma_{lv} \cos\theta + \frac{\tau}{R \sin\theta} \qquad (2)$$

where R is the radius of curvature of the droplet and τ is known as the line tension arising from the perimeter of the droplet (three phase contact line).[39] By ignoring the line tension effect that is negligible for a macroscopic droplet having large R, eqn (2) reduces to the famous Young's equation:[40]

$$\gamma_{sv} = \gamma_{sl} + \gamma_{lv} \cos\theta \qquad (3)$$

$$S = \gamma_{lv}(\cos\theta - 1) \qquad (4)$$

The equilibrium shape of the droplet is that of a spherical cap if the liquid drop is small enough to neglect the gravity effect. On the other hand, a large droplet is flattened by gravitational force balanced with lateral force due to interfacial tensions. The dewetting state, consisting of some areal fraction ψ of the flattened droplet and the remaining areal fraction of bare surface, has the mean free energy per unit area:

$$F_d(h) = \psi(h) F_w(h_c) + (1 - \psi(h))\gamma_{sv} \qquad (5)$$

with

$$F_w(h) = \gamma_{sl} + \gamma_{lv} + \frac{\rho g h^2}{2} \qquad (6)$$

where ρ is the mass density of the liquid, g is the gravitational acceleration, and h_c is the height of flattened droplet, and h is the average thickness satisfying the volume conservation of a non-volatile fluid ($\psi = h/h_c$). The minimization of F_d with respect to h_c gives the height of the pancake-like droplet,

$$h_c = \sqrt{\frac{-2S}{\rho g}} \qquad (7)$$

The minimized free energy of F_d should be compared with the free energy per unit area of the wetting state given by eqn (6). It can be shown from eqn (5)–(7) that the dewetting occurs ($F_d < F_w$) only if $h < h_c$. This film instability is initiated by hole formation created from impurities or processing stimuli such as strain in the film, *etc*. When a hole is nucleated, the hole expands until the lateral force due to the interfacial effect is balanced with the gravity force. On the other hand, for a thick film with $h > h_c$, the liquid wets the surface even on the nonwettable surface characterized by $S < 0$ due to the dominant gravity effect. The typical value of h_c is the in order of millimetres. Figure 3.1 presents the free energy per unit area *versus* the film thickness for the case $S < 0$, which depicts the wetting/dewetting in the thick films.

While the sign of S and the gravity control the wetting stability of thick films, the contribution of ubiquitous, long-range van der Waals attraction plays a critical role in the stability of thin films with a thickness down to tens of nanometres ($h < 100$ nm). The long-range van der Waals attraction includes three kinds of attraction energies, Keesom interaction between two permanent dipoles, Debye interaction between a permanent dipole and its induced dipole, and London dispersive interaction between two instantaneously induced dipoles, all of which vary as the inverse-sixth power of distance, $-c_1/r^6$; r and c_1 being the separation distance between two molecules and the positive constant, respectively. The pairwise summation of the intermolecular van der

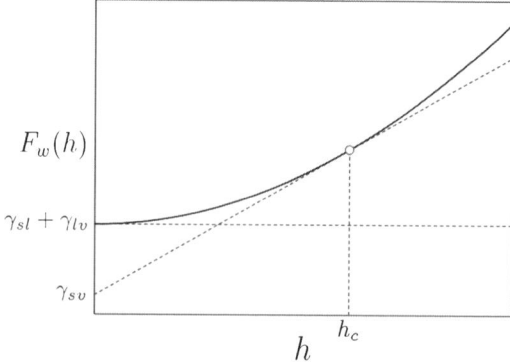

Figure 3.1 Free energy per unit area of a thick homopolymer film as a function of film thickness for $S < 0$. The blue dashed line represents the line by common tangent rule.

Waals attractions between molecules belonging to each of their macroscopic bodies, known as Hamaker theory,[41] gives an approximate estimation for the net van der Waals attractions present inside the liquid (V_{ll}) and those between the liquid and the substrate (V_{sl}) as a function of the film thickness h:

$$V_{ll}(h) = -\frac{v_l^2 c_{ll}}{2} \int_{0 \leq z < h} d\mathbf{r} \int_{0 \leq z' < h} d\mathbf{r}' \frac{1}{(\mathbf{r}-\mathbf{r}')^6} \tag{8}$$

$$V_{sl}(h) = -v_s v_l c_{sl} \int_{\infty < z < 0} d\mathbf{r} \int_{0 \leq z' < h} d\mathbf{r}' \frac{1}{(\mathbf{r}-\mathbf{r}')^6} \tag{9}$$

where v_α is the number density of macroscopic body α (liquid or solid substrate) and $c_{\alpha\beta}$ is the positive-valued coefficient proportional to the attraction between macroscopic body α and β. The integration $\int_{z_1 < z < z_2} d\mathbf{r}$ is the volume integration over the volume element $d\mathbf{r}$ belonging to the thickness range of $z_1 < z < z_2$. Eqn (8) and (9) lead to the net intermolecular potential energy per unit area of the film due to the long-range van der Waals attraction (Φ_{LR}):[42,43]

$$\Phi_{LR}(h) = \frac{A_{sl} - A_{ll}}{12\pi h^2} = \frac{A_{eff}}{12\pi h^2} \tag{10}$$

Here $A_{\alpha\beta} = \pi^2 v_\alpha v_\beta c_{\alpha\beta}$ is known as the Hamaker constant[41] between component α and β. The potential energy per unit area $\Phi_{LR}(h)$ can be interpreted as an effective interaction energy per unit area between two interfaces, s/l and l/v, being separated from each other by a separation distance h. The sign of the effective Hamaker constant, A_{eff} ($= A_{sl} - A_{ll}$), which characterizes the competition between the adhesive interactions between solid and liquid (A_{sl}) and the cohesive interactions holding the liquid together (A_{ll}), is of crucial importance in determining the wetting stability of thin films.[38,42] When the adhesion interaction between the liquid and the solid is stronger than the cohesive interaction of the liquid ($A_{eff} > 0$), there is an effective repulsion between s/l and l/v, implying that the formation of a liquid wetting layer between the two interfaces is favored. On the other hand, if $A_{eff} < 0$, the two interfaces attract to each other by the net effect from adhesive and cohesive forces, tending to eliminate the liquid layer between them. The effective force per unit area or the disjoining pressure,[44,45] $\Pi(h) = -d\Phi_{LR}(h)/dh$, can be measured by experiment (e.g., atomic force microscopy), which therefore allows one to estimate A_{eff} for a film system. It should be also pointed out that eqn (10) neglects the contribution of the molecular interactions with the upper vapor medium (corresponding to the case of the films in vacuum). Such contribution from the vapor medium becomes important in the case of vapor-assisted processes such as solvent annealing. In this case, the effective Hamaker constant is given as $A_{eff} = A_{sl} + A_{lv} - A_{ll} - A_{sv}$, which determines the sign of the effective potential between the two interfaces (repulsion or attraction).

For the complete description of intermolecular interactions, however, a proper reconciliation of eqn (10) when $h \to 0$ is necessary because the functional form of $-c_1/r^6$ for the long-range van der Waals attraction is no longer valid for two molecules at extremely short separation distances. It is well known that there exists an immensely large repulsion between the two molecules (or atoms) when they are very close together, known as the Born repulsion, originating from overlapping electron orbitals (*i.e.*, steric repulsion). The Born repulsion is described as the inverse-twelfth power of separation distance, c_2/r^{12}, in the 12-6 Lennard-Jones potential, or alternatively, $a\exp(-r/r_o)$ in the Buckingham potential, where c_2, a, and r_o are positive constants. Furthermore, there are also other types of short-range forces such as hydrogen bonding and hydration force, which decay exponentially with separation distance rather than follow an algebraic form. For a general description of the intermolecular force to some extent, Sharma and Jameel proposed an effective potential composed of the combination of short-range (Φ_{SR}) and long-range interactions (Φ_{LR}) by[46,47]

$$\Phi_m(h) = \Phi_{SR}(h) + \Phi_{LR}(h) = S_p(-h/\xi) + \frac{A_{eff}}{12\pi h^2} \quad (11)$$

Here the coefficient S_p, which can have either positive or negative value, represents the nature of short-range interactions (*e.g.* polar–polar, polar–apolar, apolar–apolar interactions), and ξ is the decay length characterizing the range of short-range interactions. The advantage of the exponential decay form as a short-range potential is that it can describe not only Born repulsion (by Buckingham-like potential with positive S_p) but also the attractive interactions between molecules where polar components are involved (with negative S_p).

The functional form in eqn (11) is an extended-range-description for the effective interaction potential between s/l and l/v covering very short large distance with a cutoff distance $d_o \sim 1.6$ Å.[46] The total potential per unit area $\Phi(h)$ and the free energy of wetting film per unit area $F_w(h)$ for $h > d_o$ are then expressed by

$$\Phi(h) = \Phi_m(h) + \frac{\rho g h}{2} \quad (12)$$

$$F_w(h) = \gamma_{sl} + \gamma_{lv} + \Phi(h) \quad (13)$$

with the asymptotic values of $\Phi(h \to 0) = S$ and $F_w(h \to 0) = \gamma_{sv}$ for $h < d_o$. By introducing short-range term, the potential energy per unit area $\Phi(h)$ can have a minimum at a finite thickness h_e that is often referred to as equilibrium thickness for a given film system. It can be shown that the equilibrium thickness is related to the spreading coefficient by[48]

$$\Phi(h_e) = \gamma_{lv}(1 - \cos\theta) = -S \quad (14)$$

The eqn (14) reveals the links of the intermolecular interaction to the macroscopic properties (contact angle, spreading coefficient).

It can be readily shown from the curve $F_w(h)$ (or $\Phi(h)$) that the films are unstable for a film thickness where the second derivative of $F_w(h)$ with respect to h is negative.[49-53] This instability is reminiscent of spinodal instability in the incompatible polymer mixture where the mixture components are spontaneously demixed from each other if the second derivative of the free energy of homogeneous phase with respect to the composition is negative. Likewise, for the film thickness where $F''_w(h) < 0$, the films are unstable to the capillary waves among which a certain wave is amplified the fastest. By this instability, the dewetting proceeds spontaneously via gradually amplifying the sinusoidal undulation of the film thickness. For a film thickness where $F''_w(h) > 0$, on the other hand, the films are stable or metastable. The metastable case is also in analogy with binodal instability in polymer mixture. In this case, a mean free energy, $\bar{F}_w(h) = \frac{h-h_2}{h_1-h_2} F_w(h_1) + \frac{h_2-h}{h_2-h_1} F_w(h_2)$, given by a combination of the potential energies at two different thicknesses (h_1 and h_2), becomes lower than $F_w(h)$, by overcoming an energy barrier in the course of reaching h_1 and h_2 from the initially uniform thickness h. The equilibrium values of h_1 and h_2 can be determined by the common tangent rule in the plot of $F_w(h)$. Analogous to the phase-separation in the polymer mixture driven by binodal instability, the dewetting of the metastable film is triggered either by the formation of nucleus from impurities (heterogeneous nucleation)[53,54] or by thermal fluctuation (homogenous nucleation)[55] overcoming the energy barrier. It should be also pointed out that for not-too-thin films, dewetting via a nucleation and growth mechanism is observed even in the spinodal regime because of the extremely slow process of spinodal rupture (with the rupture time $\tau \propto h^5$).[50]

Four typical cases for the curves of $F_w(h)$, classified by the sign of S_p and A_{eff}, are shown in Figure 3.2. For the film systems with $S_p > 0$ and $A_{eff} > 0$ (type I systems), the films are stable in the entire range of film thickness because $F''_w(h) > 0$ h. The systems of type II ($S_p < 0$ and $A_{eff} > 0$), type III ($S_p > 0$ and $A_{eff} < 0$), and type IV ($S_p < 0$ and $A_{eff} < 0$), which have a regime of $F''_w(h) < 0$, characterize the dewetting systems but exhibit different dewetting characteristics. The curve for type II, which is the case of $S_p < 0$ combined with $A_{eff} > 0$, has a finite value of global minimum at $h_e \cong (-A_{eff}\xi/6\pi S_p)^{1/3}$. In this case, the films thicker than h_e (but thinner than a certain critical thickness) undergo self-dewetting on a thin wetting layer with a thickness $\sim h_e$, which is referred to as "partial dewetting".[56-58] After partial dewetting, the fluid forms the dewetted droplets on the thin wetting layer of the fluid itself via spinodal dewetting or nucleation and growth mechanism. Spinodal dewetting occurs for $h_{sm} < h < h_{sg}$ where $h_{sm} \cong (-A_{eff}\xi^2/2\pi S_p)^{1/4}$ and $h_{sg} \cong (-6A_{eff}/\rho g)^{1/4}$. On the other hand, the dewetting proceeds via nucleation and growth mechanism in the range of $h_{cm}(\cong h_e) < h < h_{sm}$ and $h_{sg} < h < h_{cg}$ where h_{cm} and h_{cg} are determined by the common tangent rule. It is worth noting that the film stability in the regime of smaller film thickness ($h < h_{cm}$) is driven by the dominant contribution of intermolecular interactions, while the film stability at $h > h_{cg}$ is gravity-driven. In contrast, in the case of type III ($S_p > 0$ and $A_{eff} < 0$), the film exhibits the opposite behavior with respect to the dewetting

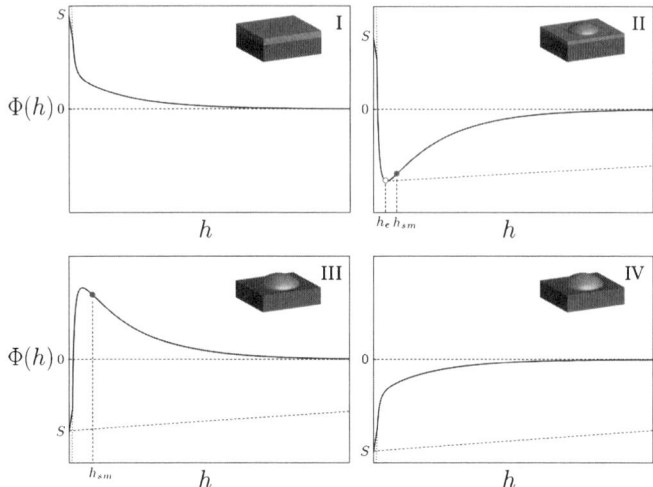

Figure 3.2 Free energy per unit area of homopolymer thin film as a function of film thickness for type I ($S_p > 0$, $A_{eff} > 0$), type II, ($S_p < 0$, $A_{eff} > 0$), type III ($S_p > 0$, $A_{eff} < 0$), and type IV ($S_p < 0$, $A_{eff} < 0$). The curves are shown in the range of very small film thickness where the gravity effect is negligible. The filled circles and the open circle represent the spinodal point and the minimum point, respectively. The blue dashed lines represent the lines by common tangent rule.

regimes. In the spinodal regime of $h < h_{sm}$ where $\Phi''(h) < 0$, the liquid dewets the (bare) substrate by spontaneously amplifying the capillary waves and the metastable regime is identified at $h > h_{sm}$ where the bare substrate is dewetted *via* nucleation and growth mechanism. Lastly, the films belonging to the type IV are unstable in the film thickness smaller than $h_{sg} \cong (-6A_{eff}/\rho g)^{1/4}$ and metastable at $h_{sg} < h < h_{cg}$.

3.2.2 Stability of Block Copolymer Thin Films

3.2.2.1 Theoretical Model

Molten block copolymer, which displays distinctive periodic nanostructure and ordering transition driven by incompatibility between two or more covalently linked polymer blocks, exhibits unique film behaviors owing to the crystal-like properties of the formed nanostructures that still maintain liquid-like mobility.[59,60] The best-known structures of ordered block copolymer are lamellae, hexagonally arranged cylinders, spheres arranged on a body-centered cubic structure, and bicontinuous gyroids. This periodic nanostructure, which have a characteristic interdomain spacing of λ_o, is a result of the tradeoff between minimal stretching energy associated with spring-like chain deformation and minimal interfacial energy between the phase-separated domains. The film geometry imposing a certain perturbation on this structured fluid can induce a film instability purely originated from a misfit between the film geometry and the unit cell of the crystal-like structure, which becomes even more complicated when coupled with the effect of short- and long-range intermolecular interactions with the medium bound to the film.

The simplest example is the thin films of symmetric an AB-diblock lamellar phase where the planar A/B interfaces of diblock lamellae are parallel to the substrate. In this case, the free energy per unit area for the film of parallel lamellae, F_w, can be expressed by the sum of three contributions:

$$F_w(h,n) = F_{w,\text{conf}}(h,n) + \Phi_{SR}(h,n) + \Phi_{LR}(h,n) + \Phi_g(h,n) \quad (15)$$

Here, the first term $F_{w,\text{conf}}$ is the free energy per unit area of parallel lamellae due only to the confinement of parallel lamellae, Φ_{SR} and Φ_{LR} accounts for the short- and long-range molecular interactions between parallel lamellae with the solid and the vapor medium (s, v) and those between lamellar layers, Φ_g ($=\rho g h^2/2$) is the contribution of gravity effect, and n is the number of A/B interfaces. Provided that the A- and B-blocks are strongly segregated, $F_{w,\text{conf}}$, can be formulated using strong segregation theory:[61–63]

$$F_{w,\text{conf}}(h,n) = \left(\frac{27\pi^2 \gamma_{AB}^2 \rho_o \kappa_B T}{32 N^2 a^2}\right)^{\frac{1}{3}} \left(\frac{4h^3}{3n^2 \lambda_o^2} + \frac{n\lambda_o}{3} - h\right) + \zeta(n) \quad (16)$$

where

$$\zeta(n) = \gamma_{As} + \gamma_{Av} + \frac{1-(-1)^n}{2}(\gamma_{Bv} - \gamma_{Av}) \quad (17)$$

Here $\gamma_{\alpha\beta}$ is α/β α/β interfacial tension, ρ_o is the mean monomer density, N is the number of monomers in a diblock copolymer, a is the monomer size, and λ_o is the intrinsic interlamellar spacing of diblock lamellae. In eqn (17), we postulate that the domain layer next to the solid substrate is A-domain without loss of generality. The number of A/B interfaces at equilibrium is obtained by minimizing Φ with respect to the quantized n. It can be easily shown from eqn (16) that $F_{w,\text{conf}}$ attains periodic minima at $h = n\lambda_o/2$. The minima at odd (even) numbers n are pronounced if the solid and the vapor medium favor asymmetric (symmetric) lamellae.

With the periodic elasticity arising from $F_{w,\text{conf}}$, the overall film stability is determined by $F_w = F_{w,\text{conf}} + \Phi_{SR} + \Phi_{LR} + \Phi_g$. Following the potential proposed by Sharma, the short-range effective potential has the form of

$$\Phi_{SR}(h,n) = S_{p,n} \exp(-h/\xi_n) \quad (18)$$

where $S_{p,n}$ and ξ_n are coefficients defined similarly in eqn (11) but for the stratified films of parallel lamellae with n A/B interfaces. Assuming additivity of long-range forces between layers, Φ_{LR} can be written as

$$\Phi_{LR}(h,n) = \frac{H_n}{12\pi h^2} = \frac{n^2}{3\pi h^2} \sum_{\mu=0}^{n} \sum_{v>\mu}^{n+1} \frac{A^{\alpha}_{\mu|v} - (1-\delta[\mu-v-1])A^{\alpha}_{v-1|v}}{(2(v-\mu) - \delta[\mu] - \delta[v-n-1])^2} \quad (19)$$

Here, H_n is an effective Hamaker constant between s/l and l/v in the stratified system having n A/B interfaces, the μ and v are the A/B interface indices between s/l interface ($\mu=0$) and l/v interface ($\mu=n+1$), $\delta[x]$ is the Kronecker delta, and $A^{\alpha}_{\mu|v}$ is the effective Hamaker constant in the single layer of α-component ($\alpha = $ A if v is odd and $\alpha = $ B otherwise) between the component below μ-th interface and the component above v-th interface. The eqn (16)–(19) are valid at $h < d_o$ and $F_w(h \to 0) = \gamma_{sv}$ $F_w(h \to 0) = \gamma_{sv}$ for $h < d_o$.

The computation of Φ_{LR} for multilayered structure such as lamellae, albeit feasible in principle, is a tedious task because of many combinations of layer–layer interactions. However, because the long-range force for large n is negligibly small compared to $F_{w,conf}$, it may be sufficient to take into account the layer–layer interactions for first few numbers of n. Given that the molecular constants of diblock copolymer and the bounded surface, the actual curve $F_w(h)$ is obtained by minimizing $F_w(h,n)$ with respect to the number of A/B interfaces n. Although not covering numerous cases of block copolymer films that can be classified in terms of the type of ordered structures, orientations, and the interfacial conditions, the concept used in eqn (15)–(19) allows one to compute the free energy of the film, which explains the stability of block copolymer film.

Figure 3.3 demonstrates some typical cases of the curve $F_w(h)$ for the films of parallel lamellae. Among a number of possibilities depending on $\gamma_{\alpha\beta}$ and $A_{\alpha\beta}$, the focus is restricted to the cases where asymmetric lamellae are favored over symmetric lamellae, i.e., $\gamma_{sA} < \gamma_{sB}$, $\gamma_{Av} < \gamma_{Bv}$. The curve for type V characterizes the film systems with the wettable substrate condition combined with favorable long-range interactions ($S_{p,1} > 0$, $S_{p,2} > 0$, $H_1 > 0$, $H_2 > 0$), which may correspond to the film systems of diblock copolymer on the wettable substrate that is strongly selective to A-block ($A_{sA} - A_{sB} > A_{AA} - A_{AB}$) in a strongly B-selective vapor medium ($A_{Bv} - A_{Av} > A_{BB} - A_{AB}$). In this case, the block copolymer fluid wets the entire area of the substrate but the film becomes metastable by non-vanishing elastic force whenever the initial film thickness is incommensurate with the intrinsic interlamellar spacing. At this incommensurate condition, the elasticity drives the film to have two different film thickness with a thickness difference λ_o. This film instability in the block copolymer film resulting in non-uniform film thickness is often referred to as "holes and islands",[64–66] exemplifying the elastic response of crystal-like block copolymer liquid under confinement. Type VI characterizes a partial dewetting system with the condition of a parameter combination of $S_{p,1} > 0$, $S_{p,2} > 0$ and $H_1 < 0$ and $H_2 < 0$. This film system can be found in the diblock films on the wettable and mildly A-selective substrate ($A_{sA} - A_{sB} < A_{AA} - A_{AB}$) in a weakly B-selective vapor medium ($A_{Bv} - A_{Av} < A_{BB} - A_{AB}$). Similar to ordinary autophobic fluid, the dewetted droplets of diblock copolymer fluid with height $h > n\lambda_o/2$ $h > n\lambda_o/2$ where $n>2$ are formed on a

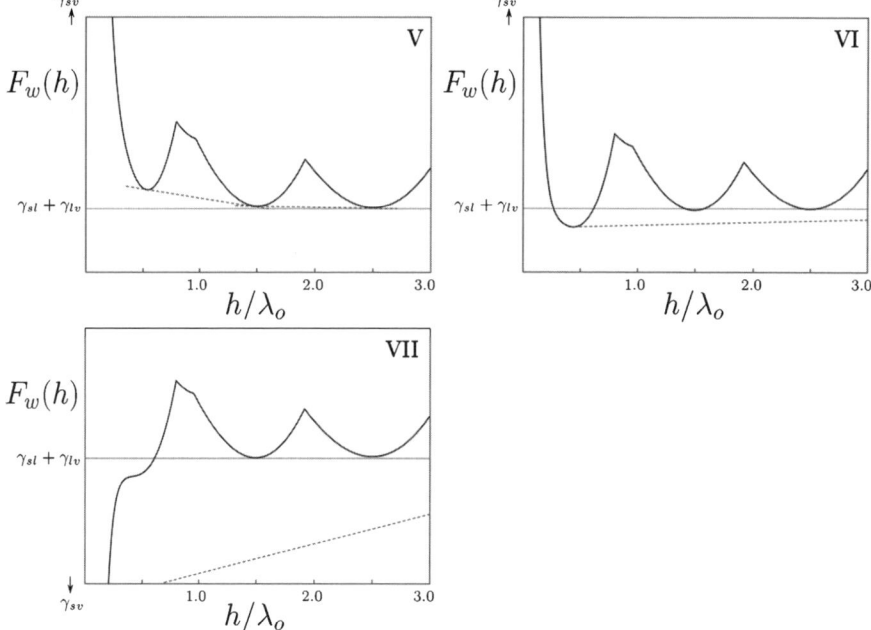

Figure 3.3 Free energy per unit area of diblock copolymer thin film as a function of film thickness for type V ($S_{p,1} > 0$, $S_{p,2} > 0$, $H_1 > 0$, $H_2 > 0$), type VI ($S_{p,1} > 0$, $S_{p,2} > 0$, $H_1 < 0$, $H_2 < 0$), and type VII ($S_{p,1} < 0$, $S_{p,2} < 0$, $H_1 < 0$, $H_2 < 0$). The blue dashed lines represent the lines by common tangent rule.

thin wetting layer of diblock copolymer itself with $h = \lambda_o/2$. A fundamental difference of this structured fluid from ordinary autophobic systems is that the dewetting always proceeds *via* the nucleation and growth mechanism because $F''_w(h) > 0$ h due to the periodically concave shape of $F_{w,\text{conf}}(h)$ despite $\Phi''_{LR}(h) < 0$ h. Lastly, the curves in the type VII, which have a parameter set of $S_{p,1} < 0$, $S_{p,2} < 0$ and $H_1 < 0$ and $H_2 < 0$, characterizes typical dewetting films that are driven by the nucleation and growth mechanism. In this case, the block copolymer liquid dewets the substrate to form droplets with their heights quantized the intrinsic interlamellar spacing λ_o.

In addition to these simple cases shown in Figure 3.3, the interplay between the commensurability-induced elasticity and short- and long-range interactions can possibly yield a number of quite complex shapes of $F_w(h)$ with many local minima, which is currently under investigation.

3.2.2.2 *Experiments on the Instability of Block Copolymer Thin Films*

Thin films of block copolymer have been intensively studied in recent years partly due to a rapidly growing interest in nanopatterned polymer structures that rely mainly upon thin film structures. While a large number of experimental studies on block copolymer thin films have focused on the formation of nanostructures and their orientations under wettable surface conditions, there have been also steady efforts to enlighten the dewetting in block copolymer films where the interdependence of surfaces or interfaces and nanostructure formation is quite complex.[29,59,67–75] Some representative works on the dewetting of block copolymer fluid are briefly reviewed in the following.

As an earlier effort to understand the dewetting of "structure-less" block copolymer fluid, Green *et al.* have examined the dewetting of a thin symmetric diblock copolymer film on a silicon substrate at temperatures higher than the bulk order-disorder transition temperature (ODT).[33,67] They found that the disordered diblock copolymer fluid dewets a single brush layer of diblock copolymer with a brush height h_L in the thickness regime of $h > h_L$, suggesting that the disordered block copolymer fluid undergoes partial dewetting (or autophobic dewetting) on its ordered layer driven by the preferential interaction of the substrate with one of the blocks. Limary *et al.* examined the dynamics of autophobic dewetting of a thin symmetric diblock copolymer film above the bulk ODT.[68] It was found that the dewetting is initiated by the formation of discrete holes without their characteristic peripheral rims. During this early stage, the hole radius R increased exponentially with time, which was followed by a narrow intermediate regime where the rim develops and $R \sim t$. When the rim became fully mature, R increased with $t^{2/3}$. At the final stage of the process, droplets of the block copolymer, a few microns in diameter and with heights on the order of tens of nanometres, existed on a dense brush of diblock copolymer anchored to the substrate, clearly indicating that this dewetting process is autophobic.

The dewetting of ordered block copolymers has attracted much interest not only for academic curiosity but also for its technological importance in fabricating well-organized nanopatterned structures in two-dimensions. Carvalho *et al.* have reported that thin films of an ordered symmetric diblock lamellae form a terraced droplet that are composed of aligned lamellae.[29] Croll *et al.* have investigated the detailed shape of terraced droplet of an ordered diblock copolymer.[71] They found from a measurement using atomic force microscopy that the droplet shape changes from a smooth spherical cap to a terraced hyperbolic profile as the liquid goes from a disordered state to an ordered layered structure. In the ordered state, the droplet has the morphology of a stack of concentric disks, each of which has a constant edge height corresponding to the thickness of a bilayer of diblock copolymer. Two competing interactions are suggested to be responsible for this non-classical, hyperbolic shape of terraced droplets: the edge tension which drives the droplet from a smaller disk to a large disk, and the edge repulsion which prevents two adjacent edges of stacked disks from coming too close from each other. They also extended the work to study the spreading dynamics of symmetric diblock copolymer droplets above and below the ODT.[72] It was found that above ODT the radius of droplet R grow as a power law in time with $R \sim t^{0.096}$ in

Table 3.1 Recent studies of dewetting of thin block copolymer films. Adapted and modified from T. H. Kim's PhD thesis **2010** (Yonsei University, Korea)

BCP	Structure	Driving force for dewetting	Characteristics	Ref.
PS-PMMA	sphere	Solventannealing	μGISAX study of dewetting induced roughness structures.	76
CBCP	ribbon	Solventannealing	The desetting process effects for the formation of the dispersed ribbonds.	77
PS-PFS	cylinder	Thermal/Solvent annealing	Study of dewetting-induced mesoscale structures.	78
PS-PAA	spherical micelle	Solvent evaporation	Nanorings formation is induced by the dewetting process of the solvent.	79
PS-PEO	cylinder	Thermal annealing	Study of thermal annealing-induced dewetting structures.	80
PEO-PBO	lamellar	Solvent evaporation	Study of controlled size-distributed dewetting.	81
PS-PEO	cylinder	Thermal annealing	Study of thermal annealing-induced dewetting.	82
PS-PI	cylinder	Spincoating	Surface morphologies of BCP/NP films.	83
PS-P2VP	cylinder	Solvent annealing	Study of morphological change of films.	84
PS-PtBA	cylinder	Solvent evaporation	Study of a mechanism of QD/polymer assemblies dewetting.	85
PS-PEO				
PI-PS-PEO	gyroid	Thermal annealing	Study of terraced dewetting on surface gradient substrate.	86
SEBS	cylinder	Thermal annealing	The surface morphology of the dewetting with brush layer.	87
EB	lamellar	Thermal annealing	Effect of substrate on dewetting of semicrystalline BCP.	88
PS-PMMA	cylinder	Thermal annealing	Study of BCP/polymer dewetting mechanism.	89
PS-PEO	cylinder	Solvent annealing	Study BCP dewetting at air/water interface.	90
PS-PMMA	lamellar	Solvent annealing	Morphology evolution of BCP dewetting	91
SEBS	cylinder	Solvent evaporation	Study of BCP dewetting after dipping.	98
SEBS	cylinder	Solvent annealing	Multiscale dewetting of BCP.	93
PS-PMMA	lamellar	Solvent annealing	Selective solvent effects on BCP dewetting.	94

Table 3.1 (Continued)

BCP	Structure	Driving force for dewetting	Characteristics	Ref.
PS-PpMS	lamellar	Solvent annealing	GISAX study of dewetting of BCP.	95
PS-PMMA	lamellar	Thermal annealing	Dewetting of BCP/homopolymers.	96
PS-PMMA	cylinder	Thermal annealing	Gradient hetrogeneous dewetting.	97
PS-PDMS-PtBA	cylinder	Thermal annealing	Film thickness effects on dewetting.	98
PS-P2VP	lamellar	Thermal annealing	BCP dewetting induced by residual solvent.	99
SEBS	cylinder	Thermal annealing	Dewetting of mechanically heterogenous films.	100
PS-PI	lamellar	Thermal annealing	Study of topological coarsensing of BCP films.	101
SVT	lamellar	Solvent annealing	Atuophobic dewetting of BCP films.	102
PS-PMMA	lamellar	Thermal annealing	Influence of surface ordering on BCP dwetting.	103
PS-PpMS	lamellar	Thermal annealing	Spinodal dewetting of BCP films.	69
PS-PI	lamellar	Thermal annealing	Multiscale dewetting of BCP films.	104
PS-PMMA	sphere	Thermal annealing	Effect of BCP adsorption on dewetting kinetics.	105
PS-P2VP	various structures	Thermal annealing	Surface induced dewetting of BCP films.	106
PS-PMMA	lamellar	Thermal annealing	Nucleation and growth of BCP dewetted films.	107
PS-PI	lamellar	Thermal annealing	Dewetting of neutral surface.	108
PS-PnBMA EB	various structures	Spincoating	Dewetting of BCP films at ambient conditions.	109

good agreement with the classical power law found for homopolymer droplets. Droplets of diblock copolymer in the ordered phase, however, have a much slower power law $R \sim t^{0.05}$ due to the frustration by the internal microstructures of the droplets.

Müller-Buschbaum *et al.* have investigated the nanostructured polymer surfaces resulting from the dewetting of ordered diblock copolymer fluid, observing that the destabilized film decayed into isolated small droplets dewetted on the bare substrate without an additional brush layer. Due to the extremely small height of these droplets, the periodicity of ordered structure inside each of the droplets increased when compared to the bulk.[69] Kim *et al.* have found from their study on the wetting/dewetting of asymmetric block copolymer under solvent annealing that terraced droplets comprised of ordered spherical microdomains are formed on the brush monolayer of the asymmetric block copolymers.[73] Ya *et a.l* have also investigated the dewetting of thin films of asymmetric block copolymers using Grazing-incidence small-angle X-ray scattering (GISAXS) and found that the terraces exhibit a facet-like wedge at the edge resulting from partial wetting on a monolayer of block copolymer brushes.[74] Yan *et al.* studied the coupling behavior of microphase separation and autophobic dewetting in asymmetric diblock copolymer in the weak segregation regime.[75] Unlike the dewetting in the ordinary fluid of homopolymer or disordered block copolymer where the dewetting velocity decreases as the annealing temperature decreases, they found the opposite dewetting kinetics, *i.e.* that the dewetting velocity increases with decreasing annealing temperature. As a mechanism of this opposite dewetting kinetics, it was suggested that the microphase separation between the incompatible blocks of diblock copolymers, which wants the film thickness to be commensurate with half of the intrinsic interdomain spacing, accelerates the dewetting speed as the dewetting proceeds from the rupture of thin films. In Table 3.1, some recent studies of dewetting of ordered block copolymer films are summarized.

3.3 Control of Dewetting of Polymer Films

3.3.1 Overview

Dewetting of a thin polymer film in general makes it difficult to apply the film to further applications requiring uniformity of the film over a large area. Seemingly, exploring a wetting condition of a polymeric fluid on a substrate poses an extremely difficult task involving proper selection of system components (substrate, polymer, temperature and solvent and so on), which may not be plausible in many practical situations. As an alternative approach, if the dewetting is unavoidable, it is highly desirable to control the dewetting of polymer fluid in such a way that the dewetting droplets register regularly with certain symmetry desired for a specific application. As shown in the previous section, the droplets resulting from dewetting of a thin polymer film all varied in size from microns to millimetres. Narrowly distributed micron- or submicron-scale droplets would be potentially beneficial. In addition, registry of the micron-sized droplets is important for potential micropatterning applications. It would be very desirable, therefore, to developing a way to control all x and y-directions whereby, one could imagine nearly mono-dispersed micro-droplets arrayed on the xy plane in a periodic order. Next a brief review of controlled dewetting of thin polymer films will be given. For more detailed and comprehensive discussion on the subject, a recent review article should be referred to.[110]

3.3.2 Methods for Controlled Dewetting of Liquid Films

Dewetting of a thin polymer film on a periodically modified substrate with different surface energy offers an efficient route for controlling the dewetting. For instance, periodic lines with alternating hydrophobic and hydrophobic surface can make a hydrophilic polymer film wetted or dewetted in different manners when spin coated and subsequently annealed at a temperature above its glass transition temperature, giving rise to a periodically patterned film in which severe dewetting of the polymer occurred selectively on the hydrophobic surface. In this case, however, the resulting pattern of the polymer consists of periodic regularity but the individual lines still contain dewetted droplets randomly organized with the size varied. Further control of the dewetting of a thin polymer film has been accomplished by spatial confinement of nucleation and growth of dewetting domains. In other words, nucleation of dewetting occurs at individual nucleating points intentionally designed in periodic order, followed by the growth of domains in confined geometry.

A variety of principles capable of producing either chemically or topographically periodic surface with different surface energy can be combined with polymer dewetting, including conventional photolithography, soft lithography, imprinting lithography, gradient surface, control of pH, mechanically driven surfaces and solvent evaporation. In the next section, diverse methods are introduced to control the dewetting of a polymer thin film and, in addition, we will briefly discuss the control of wetting of a thin polymer film based on electric field.

3.3.2.1 *Controlled Dewetting by Photolithography and Related Soft Lithography*

Dewetting of a polymer thin film was successfully controlled in combination with a topographic pre-pattern fabricated either directly by top-down lithographic techniques such as photolithography and electron beam lithography or by replication of a pre-pattern fabricated by photolithography.[111-119] For instance, low-duty-ratio micropatterns with no residual polymer were readily developed by the combination of dewetting of a thin polymer film and subsequent selective lift-off of the dewetted regions.[113] In addition, the dewetted micropatterns of polymers on a topographic pre-pattern were further shaped by a transfer bonding technique in which a non-circular dewetted structure was transformed into a well-defined one with sharp edges.[114] A similar controlled dewetting principle was applied with photo-switchable molecules, giving rise to a periodic nanodot patterns for optical memory.[115]

We also developed a new method for fabricating nanoscale patterned arrays of polymer thin film over a large area. The method is based on heat treatment of a polymer thin film spin coated directly on a topographic pre-pattern. We investigate the influence of pattern geometry on the final morphology of the dewetted polymer films using both mesa and indent patterned substrates. The position selective dewetting of the polymer film was initiated at the edges of either individual mesas or indents, leading to globally ordered spherical cap domains. The localized film thinning at the edge of the mesa observed was interpreted by the Gibbs–Thomson relation, $\Delta \mu = \kappa \gamma \Omega$ where $\Delta \mu$ is the excess chemical potential, κ is local curvature, γ is surface energy of mesa and Ω is the atomic volume of PS.[120] In the topographically patterned substrate with height variation, the positive chemical potential at the pattern boundary proportional to $\kappa_A = 1/R_A$ induces the selective mass flow, rendering the film thinner at the edges of mesas that now become initial places for dewetting of PS film on the mesas. Simultaneously, the polymer also migrates from the pattern edge to the canyon to reduce the local excess negative chemical potentials related to $\kappa_B = 1/|R_B|$, giving rise to the stabilization of the film on the canyon regions. Our method also suggests a convenient route for fabricating nanostructures by the significant pattern reduction of more than 200 % during the controlled dewetting. The arrays of PS domains of approximately 70 nm in diameter form successfully on those of 200 × 200 nm^2 mesas. We also demonstrated the versatility of our method by employing P4VP polymer with a fluorescent dye.

3.3.2.2 Controlled Dewetting by Chemically Heterogeneous Patterns

Topographically homogeneous but chemically patterned surfaces were prepared by various soft lithographic techniques such as micro-contact printing and micro-capillary printing and those patterned surfaces were effectively combined with dewetting of thin homopolymer films, polymer blends and composites with nanoobjects, leading to highly ordered arrays of dewetted polymer droplets.[121–134] In most cases, polymer films were cast on chemically periodic surfaces modified with self assembled monolayers (SAMs) containing different end-functional groups. Depending upon the interfacial energy between corresponding SAMs and polymer, dewetting of the polymer occurred on selective regions, leading to micropatterned dewetted structures. The preferential dewetting of a polymer thin film with SAMs was further applied for organic electronic devices such as transistors and light emitting diodes. For instance, the dewetted droplets of PMMA in a controlled manner was utilized as a template to pattern electrodes of organic transistors.[129] In addition, the controlled dewetting of a conducting polymer on pre-defined regions of the substrate offered an efficient route for fabrication of high performance self aligned organic transistors.[133] Dip-pen lithography also turned out to be an effective tool for chemically modified surface patterns in combination with dewetting of a polymer thin film. Various polymer blends were investigated on a nanometre scale pattern prepared by dip-pen lithography.

3.3.2.3 *Controlled Dewetting by Imprinting Lithography*

Micro- and nano-imprinting lithography has been a versatile and powerful technique for fabricating micro/nanopatterned structures at low cost and high throughput.[135–138] Imprinting can not only create resist patterns but also imprint functional device structures with various polymers, which can lead to a wide range of applications in optics, biomedical fields, electronics and data storage. Polymeric micro/nanopatterns imprinted with topographic master molds were further utilized as functional micro/nanometre scale structures, combined with dewetting of the film in the selected regions. Nanoimprinted polymer thin films were successfully combined with the dewetting of the polymer in the selected regions, called the molded dewetting process, giving rise to various high quality nanoscopic patterns from microscopic molds.[135]

We have also reported an efficient way for controlled dewetting of a polymer thin film based on micro-imprinting technique.[8] Ordered polymeric micropatterns over large areas were fabricated on various substrates by using dewetting and partial layer inversion of *topographically* patterned polymeric films *without* a pre-patterned substrate. Our method is based on utilizing microimprinting to induce the local thickness variation of an initially inverted bilayer which allows the controlled dewetting and partial layer inversion upon subsequent thermal annealing. As illustrated schematically in Figure 3.4, the microimprinting generates a topographically heterogeneous bilayer film, as the imprinted areas by a poly(dimethylsiloxane) (PDMS) mold are thinner than the other areas. The subsequent annealing above the glass transition temperatures of the constituent polymers induces the initiation of dewetting of the top layer selectively at thinner regions, leading to the localized dewetting of the top layer and the partial layer inversion of two layers. The kinetically driven, non-lithographical pattern structures were easily fabricated over large area by this approach. The method is applicable to a variety of substrates including Si, glass, Mica and even polymer substrate. The P4VP/PS bilayer prepared on an oxygen plasma treated polyimide film was successfully converted into an ordered micropattern by the dewetting and partial layer inversion. Furthermore, the micropatterned arrays of polymer droplets dewetted on a transparent glass substrate in a controlled manner were successfully useful for non-circular microlens arrays where the focal length of the lens was easily tunable by controlling thermal annealing.

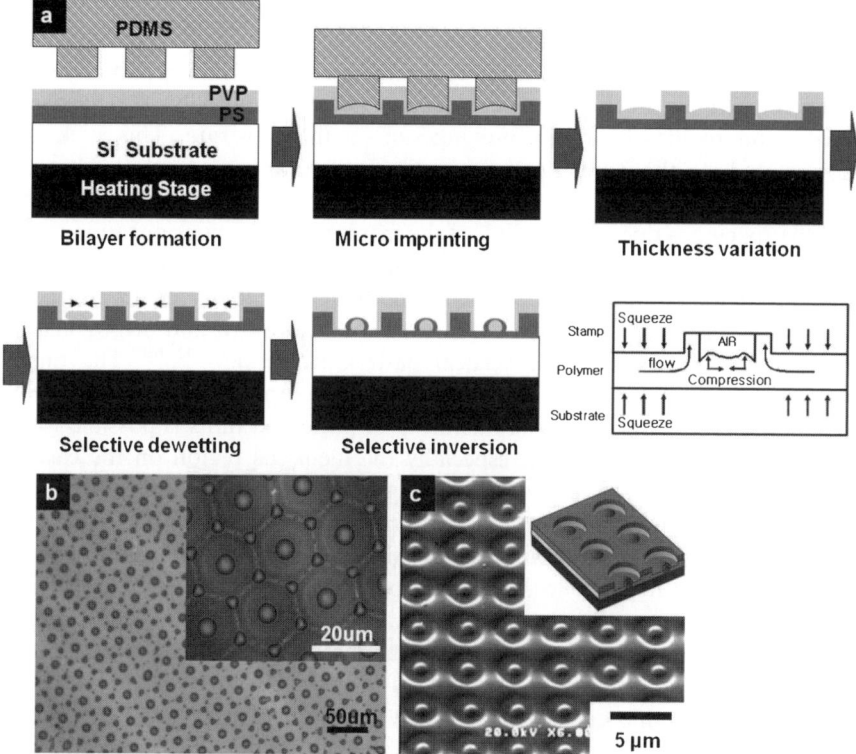

Figure 3.4 Schematic illustration of controlled dewetting and subsequent layer inversion of a poly(styrene) (PS)/poly(4vinyl pyridine) (PVP) bilayer combined with micro-imprinting. (b) and (c) A bright field OM image and a SEM image of examples of controlled dewetted patterns fabricated by the method in (a). Adapted and modified from ref. 8. (Reprinted with permission from *Macromolecules*, **39**, 901. Copyright 2006 American Chemical Society.)

3.3.2.4 Kinetically Driven Dewetting by Solvent Evaporation

Unique features may develop if the dewetting front is straight.[139–141] Small fluctuations of the rim width are spontaneously amplified since narrower sections of the rim move faster than wider ones due to frictional forces being proportional to the rim width. Instability leads eventually to an auto-control of the rim width by the continuous formation of droplets with a mean size proportional to the initial film thickness.[140] The same situation is expected if the movement of straight three-phase line shifts from the horizontal plane to a vertical plane. In dip-coating, the withdrawal of a substrate from a solution creates a downward moving three-phase line, which can be regarded as a straight line and can be used to pattern materials. During this process, an outward capillary flow of the solvent is necessary to compensate the loss of solvent at the moving front, which also carries the dispersed materials to the three phase contact line and leads to highly selective deposition along the moving front (the "coffee stain" effect[142]) forming a horizontal deposition.[143,144]

The upper-most line will eventually pin the meniscus which will be dragged and stretched as the substrate is pulled upward. Eventually, the meniscus breaks and recedes to a new pinning site on the substrate. This stick–slip motion leads to the deposition of one-dimensional arrays of solutes that are spatially resolved, where the solute lines are deposited during a "stick" event and the spacing is due to the "slip" of the meniscus. If the concentration is not so high, solutes tend to segregate into concentrated, periodically distributed domains at the contact line because of the finger instability of drying front. These domains would then cause a flow of solutes toward and orthogonal to the contact line forming the vertical pattern (fingers).[34,145–149] The finger patterns arise from a transverse instability of the evaporative dewetting front. Systematic production and manufacturing of patterned films is possible when the evaporating solution edge, especially the meniscus region on the casting substrate, is formed in a controlled casting process.

Stripe- and dot-shaped polymeric patterns were obtained by dewetting of a thin film of a polymer solution on a flat substrate, and the patterns are known to depend on both the concentration of polymer and the dewetting velocity. Similarly, periodic stripes of nanoparticles are recorded by dewetting of a thin film of a nanoparticle solution.[146,150] A hierarchic pattern was prepared by casting a toluene solution composed of PS and Ag nanoparticles (NP). Dewetting of the thin composite layer gave rise to the mesoscopic hexagonal arrays of PS-NP droplets on the mica substrate by control of humidity and solution concentration. The diameter of droplet and microstructure of PS/Au composite within a single droplet were controlled by the molecular weight of polymer in the dewetting solution. An extensive investigation of the effect of polymer concentration as well as casting speed on the formation of dewtted dot pattern of a polymer on a substrate was made. The work found that higher concentrations lead to the larger polymer droplets, whereas the faster roller speeds lead to the wider interdroplet spacing along the dewtting direction.[151] Proteins were also micropatterned by the dewetting of a non-polar polymer on a protein absorption-resistive surface. Polystyrene droplets achieved by the controlled dewetting during solvent evaporation effectively absorbed protein molecules, leading to their micropatterns.

3.3.2.5 *Controlled Dewetting and Wetting of a Polymer Film by Electric Field*

Dewetting and wetting of either thin polymer films or liquid droplets containing a polymer as a solute was controlled by direct electric field as well as electric wind driven by corona discharge. It is well known in the literature that a dielectric polymeric thin film–air interface sandwiched between two electrodes while heated above glass transition temperature and subjected to an externally applied electric field becomes unstable due to the electrostatic attraction engendered by the polarization of the dielectric polymer.[152–156] Experimental[152,155] as well as simulation[156] studies by different researchers reported the formation of ordered sub-micropatterns (columns/holes) induced by electrostatic instability and dewetting of dielectric polymer. On the other hand, use of an electric field, known as electro-wetting, has been utilized as an effective tool for controlling the wetting of liquid droplets. By applying a voltage between the droplet and an electrode submerged in the substrate, contact reductions in excess of 90° can be achieved with actuation speeds on the millisecond timescale. Recently, electro-wetting has become the most popular platform for so-called digital microfluidic systems that are based on the manipulation of discrete drops in a microfluidic chip.[157]

In the process of electrowetting, the contact angle of a liquid droplet is controlled when an external electric field is applied across a dielectric layer between a bottom electrode and the liquid droplet. The contact angle induced by electrowetting (θ_A) on a flat surface can generally be given by the Lippmann–Young equation, $cos\ \theta_A = cos\ \theta_o + (\varepsilon_r \varepsilon_o/2\gamma_{la}t)V_A^2$, where V_A is the applied potential across the dielectric layer, θ_o is the equilibrium contact angle at $V_A = 0$ V, ε_r is the relative permittivity of the dielectric layer, ε_o is the permittivity of vacuum, γ_{la} is the liquid–air interfacial tension, and t is the thickness of the dielectric layer. As we mentioned previously, the contact angle is one of the main effects to characterize wetting and dewttting behavior. Electro-wetting capable of manipulating contact angle of a droplet apparently offers a great potential for further liquid droplet based applications including biochips, sensors and mcirofabrication techniques.

We have developed a robust way to fabricate an ultra-thin polymer film by a so-called corona discharge coating technique based on utilizing directional electric flow, known as electric wind, of the charged uni-polar particles generated by corona discharge between a metallic needle and a bottom plate under high electric field ($\sim$5–10 kV cm^{-1}). Corona discharge, as a form of electrical breakdown in gases, occurs when the voltage difference applied between the sharp metal needle and the flat surface exceeds a threshold value such that the electric field strength in the vicinity of the needle becomes sufficiently large to ionize the gas molecules.[158–162] In the case of positive corona, negative ions are drawn toward the needle, while positive ions move outside and drift along the electric field lines toward the flat bottom electrode. The outside of an ionization region is so called a drift zone, where ions of one sign are dominant.[160] A unipolar charge current is established as ions in the drift zone are set in motion in response to the electric field. Collisions between drifting ions and electrically neutral air molecules give rise to momentum transfer that leads to the electro-hydrodynamic flow known as "corona wind".[160] The corona wind is physically driven by the Coulomb force from unipolar charge and electric field, and its direction follows the direction of ion drift along the electric field lines radiating from the needle toward the bottom electrode. The electric flow rapidly spreads out the polymer solution on the bottom plate and subsequently forms a smooth and flat thin film in large area within a few seconds. The method was found to be effective to fabricate uniform thin polymer films with the area of approximately larger than 30 mm^2. The effects of the voltage applied, tip-to-plate distance, and substrates were systematically investigated on the film formation as well as the resulting microstructure.

3.3.2.6 Other Methods

Many other principles have been proposed for controlling and directing dewetting of thin polymer films including surface gradient[163], pH[164] and mechanical fields.[165,166] Extensive study of dewetting of polystyrene films was made on surface energy gradient substrates prepared by a graded process of SAMs. The work revealed an interesting transition from a pattern-directed dewetting to an isotropic one at a critical surface energy difference between surface pattern domains. In addition, the gradient substrate provided a powerful approach for investigating the large number of parameters that governed the stability of polymer films and the physical factors that influenced the dewetted film morphology. pH sensitive dewetting of thin PVP films on a native oxide surface allowed the reversible wetting and dewetting in a liquid environment as a function of pH of the solution.

Dewetting of a thin polymer film was also successfully controlled by mechanically driven fields.[165,166] Anisotropic dewetting of a previously rubbed PS film happened when the rubbed film on a native oxide surface was heated above its glass transition temperature. In addition, non-spherically dewetted porous microdomains potentially applicable for polymer ferroelectrets were obtained when a thin polymer film was dewetted on a mechanically pre-stretched polyethylene substrate. Localized heating of a thin polymer film would be a robust way for controlling the film dewetting. The melting of a local area of a thin polymer on a non-favorable surface can induce confined dewetting of the polymer, and well defined arrays of the dewetted domains can be developed when the position of the melt zone is manipulated. In fact, the principle has been proposed in metallic thin films based on the local heating of the films by strong and localized pulsed lasers[167] and ion-beams.[168] Thin nickel films treated with a series of laser pulses were patterned into various shapes and dimensions. The edges and vertices of the patterned shapes acted as programmable instabilities, which enabled directed assembly *via* dewetting when the laser energy density was above the melting threshold. Ion-beam and Rayleigh instability was also utilized for fabricating metallic nanostructures with precisely controlled size, spacing and location.

3.4 Control of Dewetting of Block Copolymer Films

3.4.1 Overview

Dewetting of a self assembled block copolymer thin film has obviously been of great interest for understanding the origin and dynamics of the complicated phase separation and the resulting nanostructures; however, there are several issues which require consideration for utilizing it in further applications. Similar to the droplets resulting from dewetting of a thin polymer film, the droplets of a block copolymer containing various self organized nanostructures are still all varied in size from microns to millimeters. It is apparent that narrowly distributed micron- or submicron-scale droplets with functional nanostructures regularly organized on a substrate would potentially provide interesting micropatterning applications, such as self cleaning surfaces, biochemical sensors and microlens arrays. Furthermore, since a hemi-spherical or sphere capped block copolymer droplet dewetted on a substrate in general contains multiple layers of self assembled nanodomains, one should also consider the registry of the nanodomains on the xy-plane as well as the z-direction in an individual droplet. This suggests that the dewetting of a thin block copolymer film should be controlled in a way where not only dewetted micron or submicron scale droplets but also self assembled nanostructures in the droplets are simultaneously designed. The next section will focus on various technical approaches for fabricating arrays of the dewetted block copolymer droplets. The methods are again based on employment of chemical or topographical patterned surfaces which can control the nucleation and growth of block copolymer dewetting by spatial confinement similar to the control principle of dewetting of polymer thin films we discussed in the previous section. The section will begin with a brief survey of controlling block copolymer nanostructures, with particular emphasis on solvent vapor induced ordering.

3.4.2 Control of Block Copolymer Self Assembly

Orientation of block copolymer microdomains in a thin film is governed by preferential interactions at the surfaces when constituent blocks are mobile by either temperature or solvent. For example, the preferential interaction of one block with both the substrate and free surface leads to parallel orientation to the substrate with poorly ordered microdomains. Many approaches for creating ordered nanostructures with various external fields have been developed such as mechanical,[169] electro-magnetic,[170-172] pre-patterns[173,174] and neutral surface,[175,176] solvent[177-191] and so on. For nanometre scale patterned structures that need a 2 dimensional thin film geometry, only a few methods, such as graphoepitaxy[173,174], utilization of neutral surface[175] and solvent annealing,[188] turn out to be particularly beneficial because they are based on thin film casting without need for additional complicated apparatus. A detailed review of the various approaches for fabricating globally ordered nanopatterns of a block copolymer is beyond the scope of the chapter and several insightful review articles regarding the topic are recommended.[10-14] As mentioned earlier, to control both the dewetting of a block copolymer thin film and its self assembled nanostructure at the same time, utilization of solvent vapor would be effective which makes polymer molecules mobile and thus significantly affects dewetting and wetting of a block copolymer film.

Utilization of solvent for controlling block copolymer nanostructures can be categorized into two main approaches. One approach is to control the solvent evaporation rate when the block copolymer forms a film from a homogeneous solution[177–186] and the other approach is to use a solvent vapor or a mixture of solvent vapors to treat spin cast thin films for a certain time. The latter is usually referred to as solvent annealing.[187–191] Early work by Kim and Liebera[177] has shown that the orientation of cylindrical microdomains strongly depends on solvent evaporation rate. Several following works have also demonstrated well aligned block copolymer microdomains by controlling solvent evaporation combined with additional boundary fields, such as mechanical strain induced by directional solvent drying,[178] pinning of solution droplet,[180] meniscus of solution formed in a cylindrical tube,[181] addition of non-solvent droplet during spin casting[182] and directional crystallization of a solvent.[183–185]

Solvent annealing has been known as one of the most effective ways in particular for fabricating vertically ordered cylindrical microdomains with excellent hexagonal registry, although the mechanism of structural ordering still remains unclear. The original work by Kim et al.[188] revealed that a neutral environment created by benzene vapor successfully induced large area orientation of cylindrical poly(styrene-*block*-ethylene oxide) microdomains aligned perpendicular to the substrate. The solvent is typically a good solvent for both blocks, and the annealing process can be done at ambient temperature. Under the saturated solvent vapor, the films are highly swollen and are driven into the disordered phase. During solvent evaporation, it can be expected that the evaporation begins at the surface, and a gradient in solvent concentration will develop normal to the surface. As the solvent evaporates further, the top surface reaches the ordered phase and the ordering propagates

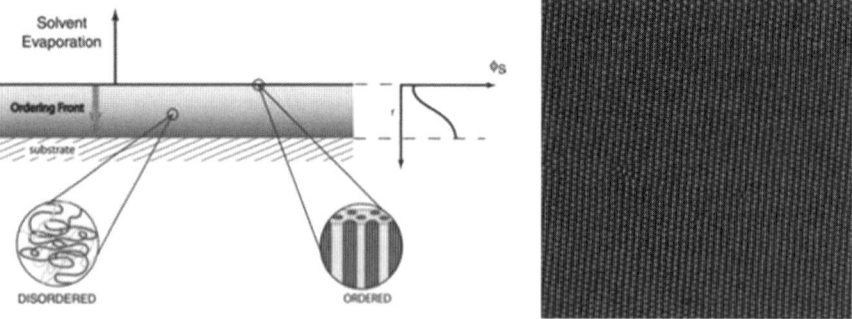

Figure 3.5 Schematics of structure evolution of a block copolymer upon solvent annealing (left) Adapted from ref. 188. An AFM image of height contrast of a poly(styrene-*block*-ethylene oxide) (PS-*b*-PEO) thin film spin-coated, followed by solvent annealing with benzene vapor (right). Highly ordered cylindrical PEO microdomains are apparent with the diameter of approximately 20 nm. (Reprinted with permission from *Advanced Materials*, **16**, 226. Copyright 2004 John Wiley and Sons.)

Table 3.2 Recent studies of thin block copolymer films developed by solvent annealing. Adapted and modified from T. H. Kim's PhD thesis (Yonsei University, Korea)

BCP/solvent vapor	Structure in bulk	Method for film formation	Characteristics	Ref.
SBS/toluene	cylinder	drop casting	Solvent evaporation rates effects	177
PS-P2VP-PtBMA/THF	lamellar	drop casting	Various solvent and solvent removal speed effects.	178
SBS/toluene	cylinder	drop casting	Solvent evaporation rates effects	177
PS-PEO/benzene	cylinder	spin coating	solvent concentration gradient effects throughout the film	179
PS-P2VP-PtBMA/ chloroform	cylinder	spin coating	Hexagonally perforated lamella structures	187
PS-PEO/benzene	cylinder	spin coating	Well ordered cylinders oriented normal to substrates.	199
PI-PLA/ benzene,chloroform	cylinder	spin coating	GISAXS study of solvent annealed films.	190
PEO-PMMA-PS/ benzene	cylinder	spin coating	Advantages of both lateral ordering and facile degradability	192
PS-PMMA/toluene	lamellar	spin coating	Selective solvents effects.	189
PS-PEO/ benzene,toluene,water	cylinder	spin coating	Fractal pattern formations.	91
PS-PEO+OS/ chloroform	lamellar	spin coating	Solvent annealing of BCP/inorganic hybrid.	194
PS-PEO+salt/benzene	cylinder	spin coating	Salt complexation effects on BCP orientation.	191
PS-P4VP/THF,toluene	cylinder	spin coating	Nanorings, nanodots formations.	199
PS-P2VP+gold/ dichloromethane	lamellar	spin coating	NP assembly into microdomains.	197
PS-PMMA/toluene	cylinder,lamellar	spin coating	Blending two BCP	196
PS-PI/MEK	cylinder	spin coating	Solvent annealing on SAMs	198
PEO-PMMA-PS/ benzene	cylinder	spin coating	Humidity effects on BCP morphologies.	186

Table 3.2 (*Continued*)

BCP/solvent vapor	Structure in bulk	Method for film formation	Characteristics	Ref.
PS-P4VP(PDP)/ chloroform	cylinder	spin coating	Solvent annealing of BCP/supramolecules.	200
PS-PEO/benzene	cylinder	spin coating	Solvent annealing of BCP on sawtoothed sapphire substrate.	201

from the top to bottom of the film, as shown in Figure 3.5. Many other groups have reported the controlled block copolymer structures under solvent vapor exposure with different block copolymers and solvents. Solvent vapor exposure time varied from a few hours to several tens of hours, depending on the particular system.

Solvent annealing has been also utilized for controlling the nanostructures of triblock copolymers,[192] supramolecular type block copolymer[193] and also various mixture type films including blends with homopolymers,[194] composites with inorganic precursors[195] and composites with nanoparticles.[196] In addition, solvent annealing was combined with metal deposition and subsequent thermal treatment, leading to novel nanostructures with metal decoration.[197] Co-assembly of a thin film with the mixture of two different block copolymers was successfully controlled by solvent vapor treatment.[198,199] Block copolymer nanostructures were also controlled by annealing in a compressible fluid, supercritical CO_2.[200] Recently, wafer-scale global orientation of a block copolymer was achieved on faceted surfaces of commercially available sapphire wafers in combination with solvent annealing.[201] Solvent annealing systems of block copolymer films are summarized in Table 3.2.

3.4.3 Controlled Dewetting of Block Thin Film by Solvent Evaporation

Controlled slip–stick motion of a polymer solution upon solvent evaporation was also combined with block copolymer self assembly.[202] The method successfully developed hierarchically ordered structures consisting of diblock copolymers using two consecutive self-assembly processes at different length scales. First, the evaporative self-assembly of a diblock copolymer solution induced fingering instabilities arising from the unfavorable interfacial interaction between one block and the substrate. In particular, controlled evaporation of the solution in a restricted geometry comprising a spherical lens on a flat substrate gave rise to intriguing concentric serpentines of a diblock copolymer at the microscopic scale. Subsequent solvent vapor annealing caused these serpentines to self-organize into a macroscopic web; at the same time, nanoscopic constituents of the diblock copolymer self-assemble into domains oriented vertically to the web surface. The resulting highly ordered structures exhibit two independent characteristic dimensions: global web-like macrostructures with local regular microporous mesh arrays by a top-down mechanism; and, by a bottom-up approach, vertical nanoscopic domains of self assembled diblock copolymer that span the entire web.

Directional solvent evaporation of a block copolymer solution also occurred when the solution confined in the gap between a slide glass and a silicon wafer was mechanically drawn at an appropriate velocity.[203] The hierarchical assembly was derived from two levels of spontaneous orderings, as shown in Figure 3.6. The large-scale ordering was defined by the periodic thickness

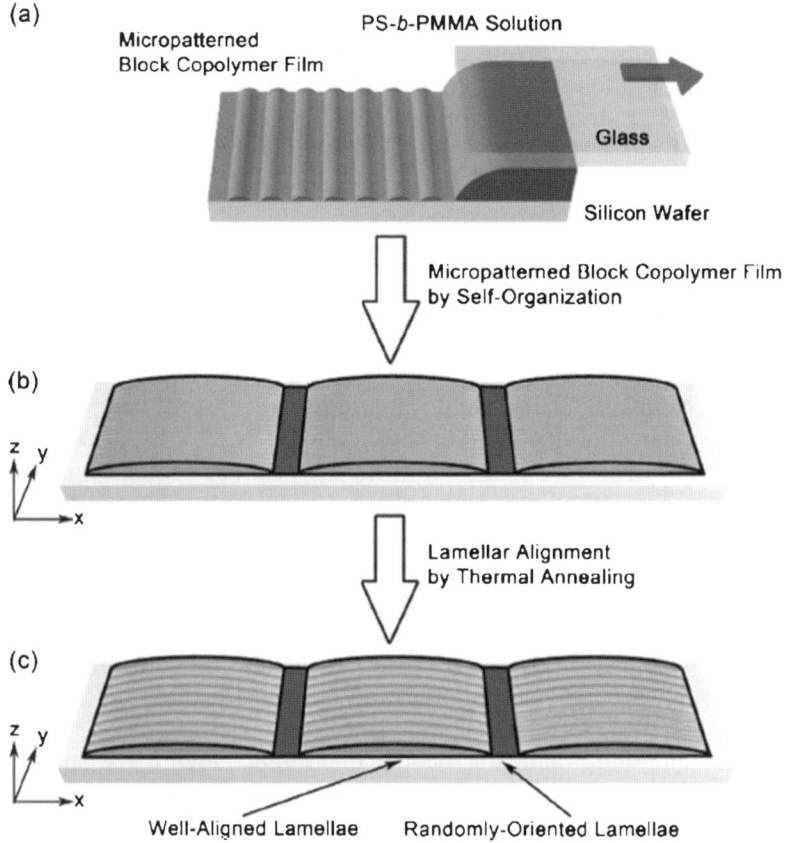

Figure 3.6 Schematics of micropatterned poly(styrene-*block*-methylmethacrylate) (PS-*b*-PMMA) copolymer film prepared by shear-driven casting. A highly ordered lamellae structure was obtained by subsequent thermal annealing along the thickness gradient. Adapted from ref. 203. (Reprinted with permission from *Advanced Materials*, **20**, 2303. Copyright 2008 John Wiley and Sons.)

modulation of a block-copolymer film, which was self-organized from the receding contact line of an evaporating block-copolymer solution upon a substrate. Its length scale was tunable over a scale of tens of micrometres by adjusting the processing parameters. The small-scale ordering corresponds to the self-assembled nanostructure of a block copolymer. Its length scale was on a scale of tens of nanometres, as determined by the molecular weight of the block copolymer. Unlike ordinary hierarchical assembly, where a small-scale structure determines the further organization into a larger structure, the large-scale structure directs the small-scale ordering in this approach.

3.4.4 Controlled Dewetting of Block Thin Film by Chemically Patterned Substrates

The dewetting of a self assembled amphiphilic block copolymer was controlled along chemically heterogeneous stripes as narrow as 70 nm fabricated by e-beam lithography, followed by selective gas phase silanation with different types of silanes, resulting in the formation of liquid ribbons of ~ 20 nm height and $300 \sim 800$ nm width. These liquid ribbons were stabilized due to the characteristic nanostructures of the block copolymer. The ordering of the block copolymer at the newly created interfaces provided the force required to prevent further dewetting. Under specific circumstances, the ordering even resulted in the formation of regular arrays of higher regions running parallel to the fluid ribbons. The work has demonstrated that the short range part of the effective interface potential of a surface can be tuned locally to generate a nanopattern of metastability for a thin polymer film. The higher dewetting rate of regions of higher metastability (more negative spreading coefficient) resulted in preferential dewetting of these regions.[204]

We have also envisioned that some of the ways for controlling the dewetting of block copolymer fluid could be utilized not only for achieving large area ordering of block copolymer thin film with solvent annealing but also for creating a fascinating structural hierarchy in which the microscopically self-assembled block copolymer droplets register macroscopically with a certain symmetry that could be controlled independently of the local microdomain shape and packing symmetries.[205] In order to achieve the control of both the dewetting of block copolymer thin film and its nanostructure, we localized the dewetting exclusively into micron scale patterned regions by the combination of solvent annealing and microcontact printing. Figure 3.7 shows the procedure of the method we developed for globally ordering PS-*b*-PEO copolymer with a hierarchical structure. We first micropatterned HDT SAMs, which is a resistive layer for spin casting block copolymer solution due to its strong hydrophobic nature, by conventional microcontact printing with the PDMS mold having hexagonal holes. Formation of the SAMs was confirmed by a water moisture test where the larger water droplets were condensed on the hydrophobic regions by SAMs, as reported previously.[206] The following spin casting of a PS-*b*-PEO solution forms a thin block copolymer film selectively in a hexagonal pattern corresponding to the bare Au regions. A large area micropatterned PS-*b*-PEO thin film is shown in the inset of Figure 3.7. Under our experimental conditions, the proper spin rate ranges from 1000 to 3000 rpm, leading to the block copolymer thin film micropatterned with relatively well defined pattern edge as shown in Figure 3.7.

The subsequent solvent annealing was performed on the micropatterned PS-*b*-PEO thin films with the solvent vapor mixture of benzene/water (the water fraction: 0.05). The dewetting during solvent treatment at each original thin hexagon film produced a convex lens-shaped spherical cap of the dewetting object nearly at the center of the corresponding hexagon, as shown in Figure 3.8. As noticed, significant reduction in domain size occurred from 20 μm to approximately 5 μm in diameter after the controlled dewetting. In general, the typical dewetting of a homopolymer thin film on a non-wettable solid substrate generates randomly distributed holes which subsequently grow with rim at their fronts. When the growing holes impinge with adjacent holes, polygons are formed. Finally, the scaffold of the polygons is broken up with time and eventually well-defined spherical droplets are formed. The nucleation of dewetting of a polymer thin film is in general restricted at the pattern boundary when the polymer thin film is formed on either chemically or topographically patterned surface, as described in the previous section.

Careful adjustment of the experimental conditions enabled us to fabricate a controlled dewetting pattern over large area (1 × 1 mm^2), as demonstrated in

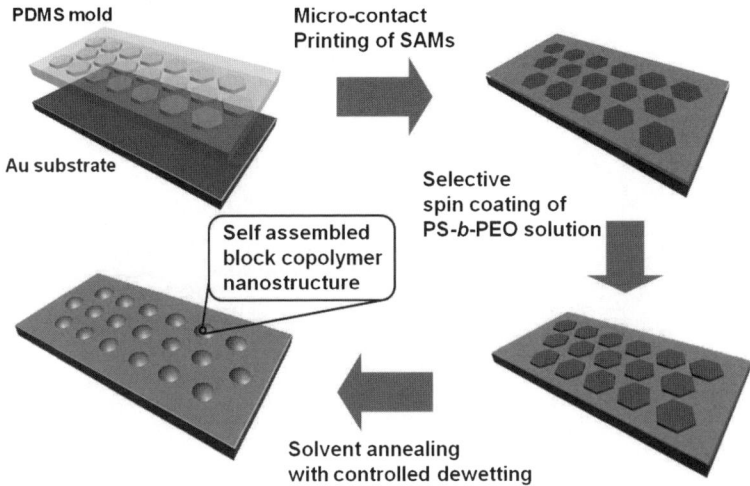

Figure 3.7 Schematics of experimental procedure of fabricating a micropatterned PS-*b*-PEO copolymer by controlled dewetting, combined with microcontact printing and subsequent solvent annealing. The dewetting of the film during solvent annealing is selectively confined onto micropatterned regions defined by microcontact printing. Adapted from ref. 205. (Reprinted with permission from *Advanced Materials*, **20**, 522. Copyright 2008 John Wiley and Sons.)

Figure 3.8a. The individual dewetting domains are approximately 4 μm in diameter with small variation in size. The dewetting polymer forms convex lens shaped spherical caps with the maximum height of approximately 700 nm, as shown in the inset of Figure 3.8a. The nanostructure of PS-*b*-PEO thin film prepared by the controlled dewetting upon solvent annealing was directly visualized by tapping mode AFM in phase contrast. Figure 3.8b is a top surface view of an isolated dewetting domain. A magnified structure of Figure 3.8b clearly displays highly ordered PEO microdomains with hexagonal symmetry (Figure 3.8c). The ordered area extends over all dewetting domain as a single grain except the boundary regions where some misaligned PEO microdomains are visible. The FFT pattern shown in the inset of Figure 3.8c also supports the high degree of orientation of PS-*b*-PEO with many sharp reflections from the hexagonal array. The ordered hierarchical structure was conveniently formed by the combination of block copolymer self assembly and controlled dewetting with the scale from nanometre, micron to millimetre.

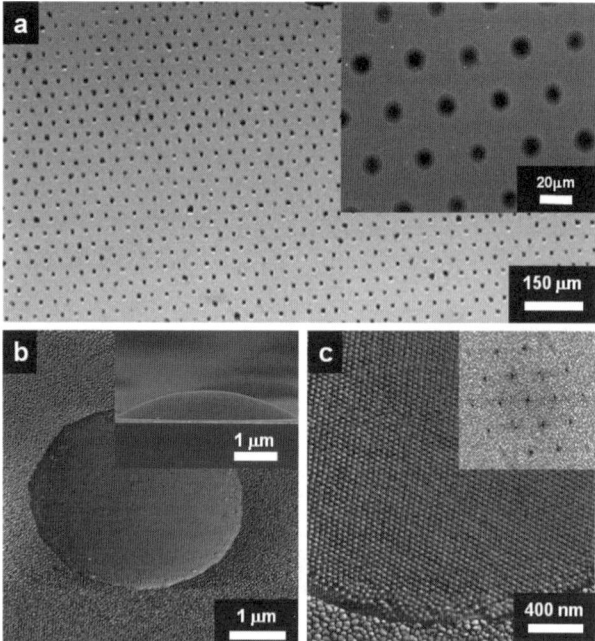

Figure 3.8 (a) Bright field OM image of hierarchically ordered PS-*b*-PEO copolymer arising from confined dewetting, followed by solvent annealing. Inset shows a SEM image magnified from (a). Individual droplet of PS-*b*-PEO with tomb-like side view is shown in (b) and inset, respectively. An AFM image in (c) clearly displays well-ordered PEO microdomains with hexagonal registry confirmed by FFT image in the inset of (c). Adapted and modified from the ref. 205. (Reprinted with permission from *Advanced Materials*, **20**, 522. Copyright 2008 John Wiley and Sons.)

3.4.5 Controlled Dewetting of Block Thin Film by Topographically Patterned Substrates

When a block copolymer film is prepared from its solution *via* various film casting methods, such as spin coating and dip coating, the film is frequently dewetted, depending on experimental conditions. The dewetting of a block copolymer can be controlled with a topographically pre-patterned substrate on which block copolymer molecules recognize the heterogeneous nature of the underlying pattern, resulting in various intriguing nanopatterned structures. Next, two examples of the controlled dewetting of a block copolymer are introduced.

3.4.5.1 Dewetting on Topographically Patterned Surface

Controlled dewetting of a symmetric diblock copolymer occurred through a combination of nanoimprint lithography and block copolymer self assembly, giving rise to a highly regular dewetted microdroplets in which hierarchically self assembled nanostructures were developed. The process was driven by the

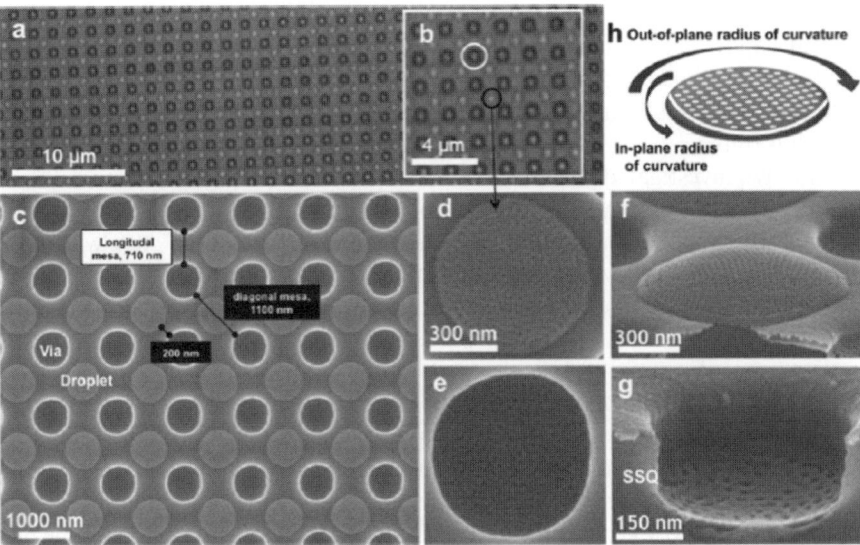

Figure 3.9 (a,b) Optical microscopy and (c) SEM images of a regular array of microdroplets formed on top of the diagonal mesas between the via-holes as a result of dewetting processes for a 1% wt PS-*b*-PMMA solution; (d,e) SEM images of microphase separation within the microdroplet on the mesa and within the via-hole (reservoir) taken from image (d); (f,g) Tilt SEM images of microphase separation within a microdroplet on mesas and via-holes; (h) 3D graphic of a droplet defining the in-plane and out-of-plane radii of curvature. Adapted from ref. 207. (Reprinted with permission from *ACS Nano*, **5**, 1073. Copyright 2011 American Chemical Society.)

unique chemical properties and geometrical layout of the underlying patterned silsesquioxane micrometre-sized templates. Given the presence of non-preferential substrate–polymer interactions, directed dewetting was utilized to produce uniform arrays of microsized droplets of microphase separated polystyrene-*block*-poly(methyl methylacrylate) (PS-*b*-PMMA), following thermal annealing at 180 °C, as shown in Figure 3.9. Directional, self-supporting (template-free), and self-aligned nanorings were produced when microdroplets are placed beyond a critical distance. The nanoring droplets adopt an aperiodic polydomain structure as the radius of curvature was pronounced at these dimensions. The cylinder reorientation process was driven by the high in-plane radius of curvature and was not governed by thickness effects. By utilizing both the trench and the mesa surface regions, the authors suggested further utilization of the process for high density magnetic nanoobjects.[207]

A thin block copolymer film prepared directly by spin coating on a topographically patterned substrate offered an efficient route for controlling the kinetically driven dewetting of the block copolymer. We have investigated the formation of *as-cast* thin films of a poly(styrene-*block*-4-hydroxystyrene) (PS-*b*-PHOST) copolymer *spin coated directly* on topographic pre-pattern substrates. Either wetting or dewetting of a polymer thin film on the elevated regions occurs in non-equilibrium state during spin coating with saturated solvent vapor and strongly depends on the dimensions of the pre-patterns.[208] We found that the ratio of the periodic unit area to the elevated one of a pre-pattern (β value) is found as one of the most important factors for wettability of a thin film.

First, we investigated the effect of line width of crest regions on thin PS-*b*-PHOST film formation using a series of pre-patterns denoted as Group I in **Table 3** which all have a ratio of periodicity of a pattern (λ) to crest width (d) (β value) of 1.95–2.92. A homogeneous flat film was formed both on the crest and trench with all the patterns investigated. Typical examples of the spin coated films are shown on the pre-patterns in Figure 3.10a–d. Even in a pre-pattern with the crest of 120 nm, a flat film was obtained in Figure 3.10d, which implies that absolute crest width is not very sensitive to de-stabilizing a film. Another set of pre-patterns (Group II) was examined on film formation with various β values in Table 3.3. As noted, the width of crests ranges from 140 nm to 2 µm, similar to that of the Group I patterns. The β values of Group II are from 4.10 to 5.86, as indicated in Table 3.3. Obviously, a film became de-stabilized on the pre-patterns and dewetting occurred, in particular at the elevated regions, as shown in Figure 3.10e–h. It is apparent that the dewetting was preferentially initiated at the sharp pattern edges of the crests. Directly compared with Figure 3.10a and c, the films in Figure 3.10e and f obviously dewet even on the pre-patterns with similar crest widths, respectively. In addition, a pattern with crest and pattern periodicities of 140 and 820 nm, respectively, induces the most severe dewetting of a PS-*b*-PHOST film, leading to an intriguing microstructure in which isolated dewet domains approximately 80 nm in size are regularly aligned along the line, as shown in

Table 3.3 Characteristics of various topographic patterned used in the study. Three groups are categorized, based on β values. Adapted and modified from ref. 208. (Reprinted with permission from *Macromolecules*, **41**, 9290. Copyright 2008 American Chemical Society.)

pre-patterns	Group I						Group II						Group III		
	1	2	3	4	5	6	7	8	9	10	11	12	13	14	15
crest width (d)/nm	120	370	630	860	1070	2100	140	400	600	840	1000	2100	600	1300	2000
periodicity (λ)/nm	350	800	1260	1720	2140	4100	820	1700	2600	3510	4160	8600	800	1650	2600
β (λ/d)	2.92	2.16	2.00	2.00	2.00	1.95	5.86	4.25	4.33	4.18	4.16	4.10	1.33	1.27	1.30

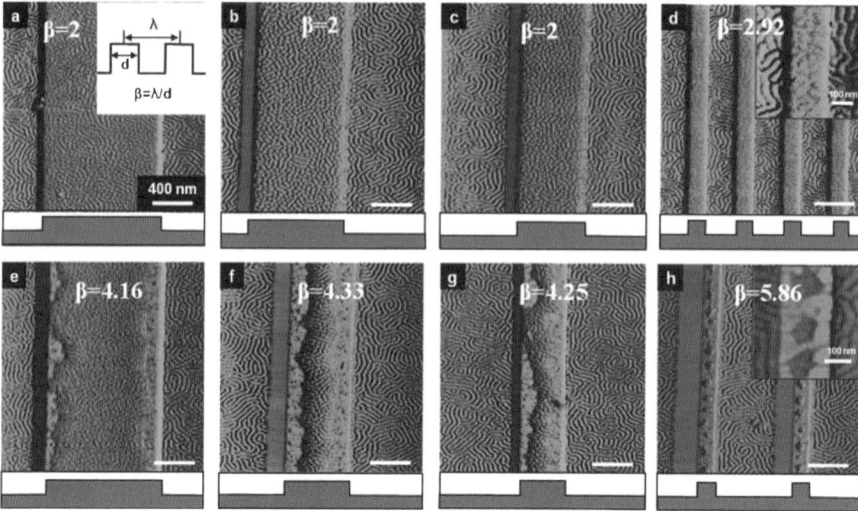

Figure 3.10 TM-AFM images in phase contrast of PS-*b*-PHOST thin films formed on 1D topographic line patterns. The schematic pattern profile is illustrated below each image. (a), (b), (c) and (d) correspond to the pre-patterns of 5, 4, 3, and 1 in Group I of Table 3.3, respectively. (e), (f), (g) and (h) are correspondent to the pre-patterns of 11, 9, 8 and 7 in Group II of Table 3.3 which show all the incomplete wetting of block copolymer films. A magnified image of (d) in the inset exhibits the homogeneous film formed on the crests. Isolated dewetted domains with the characteristic stretched arms are clearly visible in the inset of (h). Adapted and modified from ref. 208. (Reprinted with permission from *Macromolecules*, **41**, 9290. Copyright 2008 American Chemical Society.)

Figure 3.10h and the inset. This result at the same time suggests a fast and facile technique for fabricating an ordered nanostructure based on spontaneous dewetting of a polymer film without further thermal or solvent treatment.

The complicated submicron thick film formation over topography has been greatly investigated during spin coating, which involves the competition of three major driving fields of surface tension, viscosity of a solution and shrinking speed.[209] The planarization of a film on a topographic surface renders the film on crest regions thinner than on trenches, and is proportional to surface tension and inversely proportional to both viscosity and shrinking speed with a parameter corresponding to: [(surface tension/{(viscosity)x(shrinking speed)}]. Further planarization of a *ca.* 10 nm thick film on a topology with 45 nm height in our case gave rise to the incomplete wetting of the thinner film on the elevated crest regions, as shown in Figure 3.10e–h. The shape of the liquid solution surface during spin coating must be maintained if the crests are not to dewet. Dewetting is ultimately caused by the draining of fluid from the crests to the trenches. The driving force for this draining is capillarity. That is, the positive curvature on top of the crest drives flow into the trench. It is clear that curvature is initially large if both d and λ is small with low β values. The capillary pressure driven flow will rapidly tend to raise the thickness of the solution in the trenches, thus reducing the curvature difference and the driving capillary pressure. However, when β is large, the decrease in driving capillary pressure with flow of solution is much less, giving rise to such a thin solution film near the edge of the crest that it will eventually dewet.

The PS-b-PHOST film formation on pre-patterns with tetragonal arrays of square mesas also supports our argument about the influence of the β value. On the square mesas with size ranging from 2 µm to 340 nm, it is apparent that PS-b-PHOST films all dewetted, with the cylindrical microdomains perpendicular to the surface. The dewetting, again initiated at the mesa edges, becomes more dominant with the mesa of 340 nm in size. To correlate the results of the 2D mesa arrays with those from the previous 1D line patterns, we calculated the ratio of the periodic tetragonal unit cell area to the mesa one, which is the β values for 2D patterns, and examined the effect of the ratios on dewetting of PS-b-PHOST film. Consistent with the results with 1D line patterns, in which the patterns with β values greater than approximately 4 induced the preferential dewetting, the 2D patterns with the larger ratios also gave rise to the dewetting of the block copolymer.

We quantified the PS-b-PHOST film formation on the elevated topographic regions by calculating the averaged film coverage per unit of elevated area of a pre-pattern. Complete wetting of a film on either crest or mesa leads to a value of 1, which decreases in proportion with the degree of dewetting. Figure 3.11 shows a plot of the area fraction as a function of the β values of the pre-

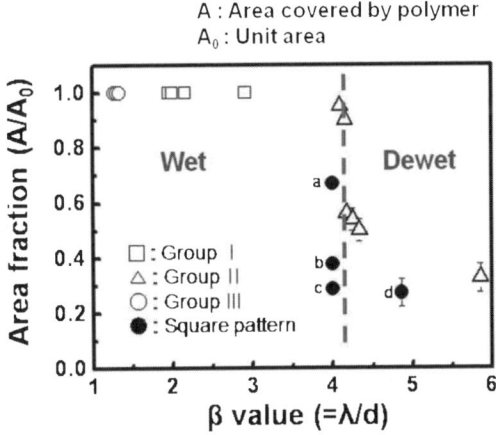

Figure 3.11 A plot of the area fraction covered with PS-b-PHOST film per unit cell area of a pre-pattern as a function of the β values. The dewetting of a film occurs with β value greater than approximately 4 for both line and square patterns, although the area fraction becomes significantly lower for the square patterns. Adapted and modified from ref. 208. (Reprinted with permission from *Macromolecules*, **41**, 9290. Copyright 2008 American Chemical Society.)

patterns. It is apparent that the dewetting of a PS-*b*-PHOST film was significantly promoted by β values greater than approximately 4 in our experimental conditions. The 1D pre-patterns of Group I and III in Table 3.3 with β values lower than 4 exhibit complete wetting of the elevated regions, while the pre-patterns of Group II with the β values greater than 4 clearly gave rise to the dewetting of a film. It should be noted that in our current experimental conditions, we did not observe dewetting behavior dependent upon the pattern dimensions with the homopolymers. For PS films, the dewetting always occurred regardless of the pre-patterns due to its large surface energy difference with Si substrate. On the other hand, homogeneous, flat PHOST films were obtained on all the pre-patterns used in the current work because of its good interaction with the substrate. We do still believe that the lateral aspect ratio dependent dewetting observed in the PS-*b*-PHOST copolymer is not unique but general to any thin block copolymer films. A system, however, requires more delicate adjustment of the experimental parameters, such as surface energies of solvent, substrate and polymers, dimension of pre-patterns.

3.4.5.2 *Smart adjustment of pinned micelles on topographic surface*

When a thin block copolymer film is prepared from a very dilute block copolymer solution, dewetting of the block copolymer is quite different from one conventionally observed in a thin film due to significantly insufficient block copolymer molecules on a substrate. In particular, when one of the blocks has a strong affinity to the substrate through hydrogen and ionic interactions, various dewetted nanostructures were obtained, from pinned spherical micelles, pancake shape micelles to holey layers, depending on the grafting density of the molecules to the substrate. We employed a PS-*b*-PHOST copolymer in which PHOST blocks had strong affinity to the substrate through the hydrogen bonding between hydroxyl side groups of PHOST and the oxide surface, and are expected to adopt a flat 2D conformation on the substrate, which leads to grafting of their covalently connected PS blocks onto the substrate. The PS blocks, on the other hand, agglomerate due to the unfavorable interactions with other components (*i.e.*, PHOST block, the substrate, and the air). This results in the surface morphology of pinned micelles where the PS blocks form a micelle core and the PHOST blocks pin the micelle to the substrate, as shown in Figure 3.12.

Each of theses micelles consists of many chains stretched as tethers to form a central globule or a "nucleus", *i.e.* a sphere with many legs stretching outward. On the other hand, in the regime of a high grafting density ($\rho N \gg 1$), the grafted chains become stretched due to the steric force from the neighboring chains, forming a uniform layer. In addition, the formation of worm-like micelles and holey layers has been reported in the intermediate grafting density

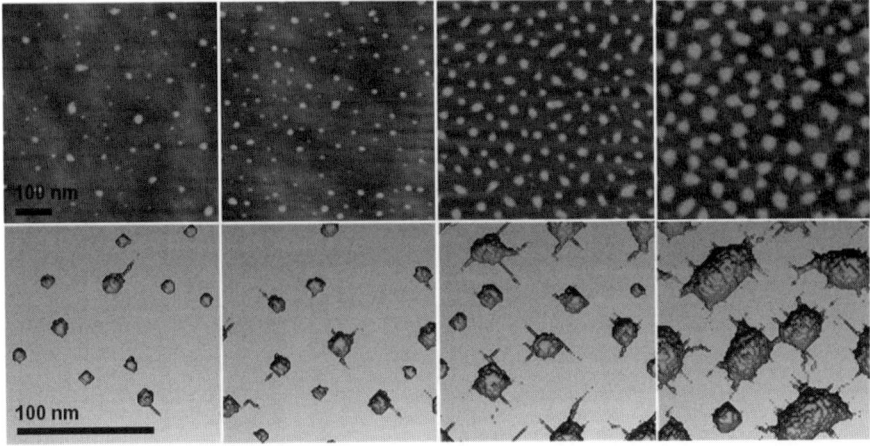

Figure 3.12 TM AFM images in height contrast (top row) of nanostructures of ultra thin PS-*b*-PHOST (73/27) films developed on the flat silicon oxide substrate by dip-coating with different immersion time. The immersion time increases from left to right. The captured images in the bottom row generated by computer simulation also show the increase of surface micelle coverage with the brush density.

between the octopus micelle and uniform layer, although their true thermodynamic stabilities still remain open to question.[210–212]

A topographic pre-pattern allowed us to control not only the size of the individual pinned micelles but also the domain-domain period (*i.e.* center-to-center distance, λ). We demonstrated how square shape mesas arrayed into *p4mm* symmetry with constant area fraction of mesa/canyon of 0.25 frustrate the conventional *p6mm* packing of the micelles and lead to the ability to control the area of micelles into an open lattice of *p4mm* symmetry having an unit–micelle distance of 400 nm, an order of magnitude larger than the closed packed *p6mm* lattice.[213]

Figure 3.13 displays a series of AFM images of the surface micelles formed on the topographic patterns comprised of square mesas 50 nm in height and side lengths ranging from 2 µm down to 400 nm arrayed with *p4mm* symmetry. Surface micelles are found randomly distributed on both the mesas and the canyons of the 2 × 2 µm² pattern, implying that the presence of the height-edge templates on this length scale does not affect the micellar arrangement. The random deposition of the surface micelles was maintained with a mesa size of down to 600 × 600 nm², as shown in Figures 3.13b, c, and d. However, on the mesas of 400 × 400 nm², the arrangement of the micelles was distinctly different; a single surface micelle is located approximately at the center of each mesa (Figure 3.13e). Considering that the averaged center-to-center distance between neighboring micelles is approximately 180 nm on a flat surface, one micelle per mesa observed on the pattern with 400 x 400 nm² is very intriguing. In addition, the outer diameter of the individual micelles on the 400 nm mesa is

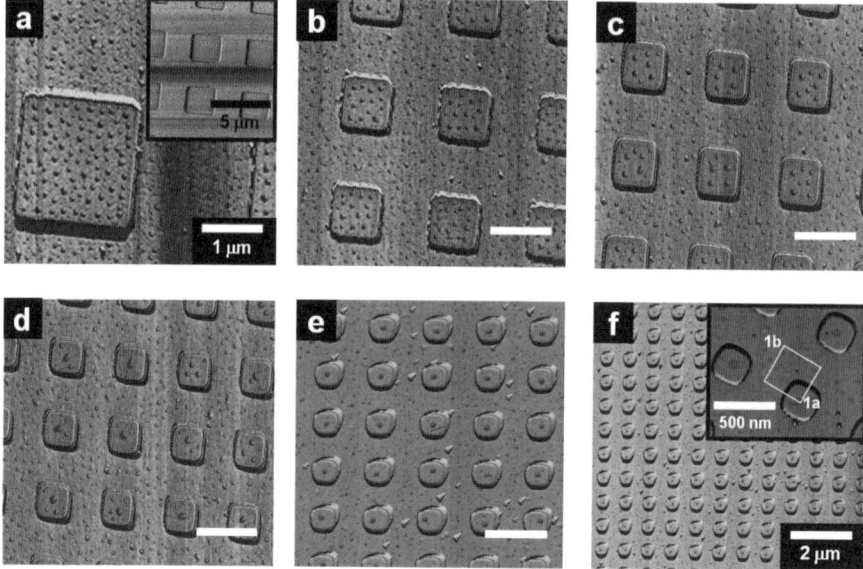

Figure 3.13 TM-AFM images of ultra thin PS-*b*-PHOST (73/27) films formed on topological patterns with square mesas arrayed into *p*4*mm* symmetry. (a) 2 × 2 μm², (b) 1 × 1 μm², (c) 800 × 800 nm², (d) 600 × 600 nm² and (e) 400 × 400 nm². Scale bars in (a)–(e) correspond to 1 μm. (f) Large area image of (e). Smart adjustment of the surface micelles is visualized in the inset of (f), magnified from (f). Wyckoff notation for *p*4*mm* symmetry is shown in the inset. Adapted from ref. 213. (Reprinted with permission from *Advanced Materials*, **19**, 3342. Copyright 2007 John Wiley and Sons.)

increased to approximately 70 nm, 20 nm larger than that of the micelles formed on a flat surface (inset of Figure 3.13f). The smart adjustment of the surface micelles to the center of each mesa occurred over a large area, as shown in Figure 3.13f.

The number density of the micelles (number of micelles per unit area) deposited on the surface as a function of the size of mesas was calculated. The density on 2 × 2 μm² mesas was similar to that on a flat surface, indicating no effect of the pattern on the micelle deposition. For smaller mesas, the change of the micelle packing density on a mesa can be ascribed to the boundary effect which forces the chains near the boundary to fuse into micelles away from the boundary. In the current situation where there are two sets of orthogonally oriented walls of each mesa, a repulsive wall–micelle interaction is propagated from each and the center of the mesa is the minimum energy location for a micelle. We denote the micelle origin as (0,0) which corresponds to site 1a in crystallographic Wyckoff notation in the *p*4*mm* unit cell, as indicated in the inset of Figure 3.13f. The micelles in the middle of the mesa therefore become fatter as the size of mesa decreases because they take up micelles near the periphery of the mesa. Ultimately, only a single micelle is centered when the mesa becomes small enough to strongly perturb the micellar properties (nucleus and corona size). It should be noted that the smart adjustment of the surface micelles results in a localized micellar array with a *p4mm* symmetry structure with a larger size.

Also of interest is that the formation of a single micelle in the central region of the mesa is induced by the strong boundary effect and the associated enlarged nucleus, as observed in Figures 3.13e and 3.13f. Obviously, the total tether penalty paid in forming a single micelle attains a minimum when the nucleus is positioned in the center. To visualize this in some detail, we also simulate the micelle formation of grafted chains on the mesa using a Monte Carlo simulation of lattice chains. Highly stretched chains converging toward the central region, which form an enlarged nucleus in the center, are observed when $L = 1.6R_o$ and $L = 2R_o$. A proper choice of the lateral dimension L can therefore lead to a technologically interesting micellar array where each micelle is centered on each of its respective mesas, as illustrated by the experimental AFM image (Figure 3.13e).

We also employed another two dimensional topographic pattern with circular indentations of 200 nm diameter and 50 nm depth, arrayed into *p4mm* symmetry with a lattice parameter of 400 nm. No PS-*b*-PHOST surface micelles were confined into the dents at the origin (0,0) position which is 1a in

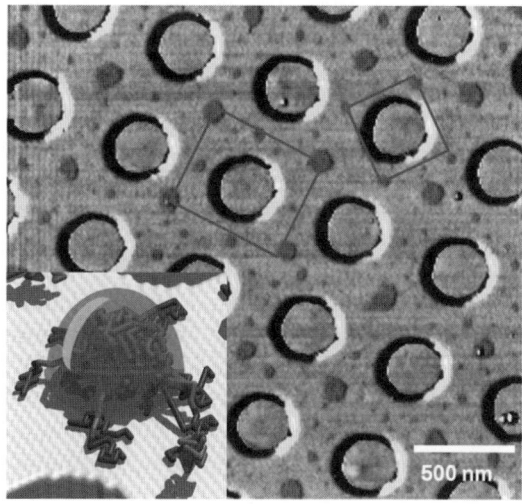

Figure 3.14 TM-AFM image in phase contrast of an ultra thin PS-*b*-PHOST (73/27) film formed on topological pattern with circular indentations of 200 nm in diameter arrayed into *p4mm* symmetry. The surface micelles formed with bimodal size distribution are assembled into two different tetragonal symmetries indicated by the thick solid and dot line, respectively. The inset exhibits a simulated PS-*b*-PHOST (73/27) pinned micelle. Adapted and modified from ref. 213. (Reprinted with permission from *Advanced Materials*, **19**, 3342. Copyright 2007 John Wiley and Sons.)

Wyckoff notation, as indicated in Figure 3.3.14c because of the solution depletion in the dents induced by a capillary interaction. In fact, the depletion of solution made individual surface micelles smaller than ones produced on a flat surface, as shown in Figure 3.13e. The micelles (dark objects in the AFM image) only assembled selectively onto the top of the thin boundary layer of the pre-pattern into an ordered structure occupying site 1b, being guided by the periodic surface boundary, leading to a tetragonal structure which is not the usual close-packed one self-assembly typically produces. It is also interesting to note that there is a bimodal size distribution of the surface micelles. As mentioned, relatively large micelles of approximately 70 nm diameter appear at site 1b surrounded by four neighboring indentations and a second set of smaller micelles of approximately 40 nm diameter formed in the middle of the corridor between two adjacent indentations at either (0, ½) or (½, 0) position with site symmetry 2*mm*, giving rise to another assembly with a square symmetry whose size is approximately 280 nm (Figure 3.14).

3.5 Control of Wetting of Block Copolymer Films

3.5.1 Overview

The spreading of non-volatile liquid droplets on a wettable substrate has been of great interest due to its scientific as well as technological implications.[18–22] Upon spreading, unique molecular and/or nanometre scale terraces characterized by distinct monomolecular layers were formed in the z direction normal to the surface.[23–28] When a self-assembled block copolymer with an ordered structure is employed for wetting on a hard surface, the system becomes more complicated, and therefore one should take into account not only the first layered terrace directly on the solid surface, frequently known as a frontier layer, but also multiple terraces regularly spaced in the z-direction arising from the ordered structure.[144–148] As described previously, block copolymers with lamellar microstructures exhibited very unique terraced hierarchical structures formed upon wetting a distinct droplet.[29–32] The droplet consisted of disk-like concentric rings with the characteristic periodicity of the block copolymer piled up.[30,32] The monolayered first terrace, called the brush monolayer of a block copolymer, was evenly formed on a substrate, in which one of the blocks is preferential to the surface, with the other non-preferential block pointing toward the air.[30]

The regularly spaced wetting of a structured fluid in the z-direction is no doubt useful for understanding the origin and dynamics of the structural formation; however, there are several issues which require consideration for utilizing it in further applications. The droplets that have been investigated so far are all varied in size from microns to millimetres. Narrowly distributed micron- or submicron-scale droplets would potentially provide interesting micropatterning applications, such as self cleaning surfaces,[214] biochemical sensors[215] and microlens arrays.[216,217] It would be very desirable, therefore, to develop a way that allows control in all x-, y-, and z-directions, whereby one could imagine nearly monodispersed micro-droplets terraced in a z-direction and at the same time arrayed on the xy plane in a periodic order. Such a system would also offer a more convenient route for investigating the growth of brush monolayers in a controlled manner. Next, some examples of periodic arrays of block copolymer droplets are introduced, based on controlled wetting of the block copolymer droplets.

3.5.2 Controlled Wetting of Block Copolymer Droplets

We have achieved the micropatterns of terraced block copolymer microdroplets wetted in a controlled manner, based on transfer-printing the microdroplets periodically dewetted on a topographic poly(dimethylsiloxane) (PDMS) pre-pattern onto a hard substrate, and subsequently solvent annealing them.[73] A poly(styrene-*block*-ethylene oxide) (PS-*b*-PEO) copolymer was employed for fabricating patterned arrays of the micro-droplets onto a topographic PDMS pre-pattern, as shown in the schematic of Figure 3.15. The dewetting occurred preferentially on the individual mesas when a PS-*b*-PEO solution was spin coated onto a PDMS pre-pattern. Dewetted domains, each of which forms a spherical cap, are located near the center regions of the 20 μm hexagonal mesas arrayed with *p6mm* hexagonal symmetry with the periodicity of 30 μm while trench regions are completely filled with the block copolymer film. The dewetting is initiated at the sharp edge of topographic patterns due to localized polymer flow at the pattern edges in order to spontaneously reduce the excess chemical potential induced by high curvature at the pattern edge, which causes draining of fluid from the mesas to the trenches, as explained in the previous sections.

The size of convex lens shaped, spherical caps of a dewetted droplet was varied by changing the size of the PDMS hexagons. As the size of the PDMS hexagon decreases, the size of the spherical cap does as well. The arrayed PS-*b*-PEO droplets dewetted on a PDMS pre-pattern were directly transferred onto a flat Si substrate with high transfer fidelity over 80%. The transfer was facilitated by conformal contact of PDMS mesas containing the dewetted droplets and a Si substrate for one minute without additional heat and pressure. The diameter and height of the transferred droplet domain are similar to those of the dewetted domains before transfer. On the surface of individually transferred droplets upside-down from the previous droplets on a PDMS, we also confirm that randomly ordered cylindrical PEO microdomains are embedded in the PS matrix with an average diameter of 15 nm and center-to-center spacing of 30 nm, respectively.

We have investigated how the transferred PS-*b*-PEO droplets are modified on a Si substrate during solvent annealing. It is apparent that the dewetted droplets transferred from the PDMS surface spread out on the Si substrate, making their diameter larger and larger with solvent annealing time. The captured morphologies of the sample after 3 minute solvent treatment exhibit

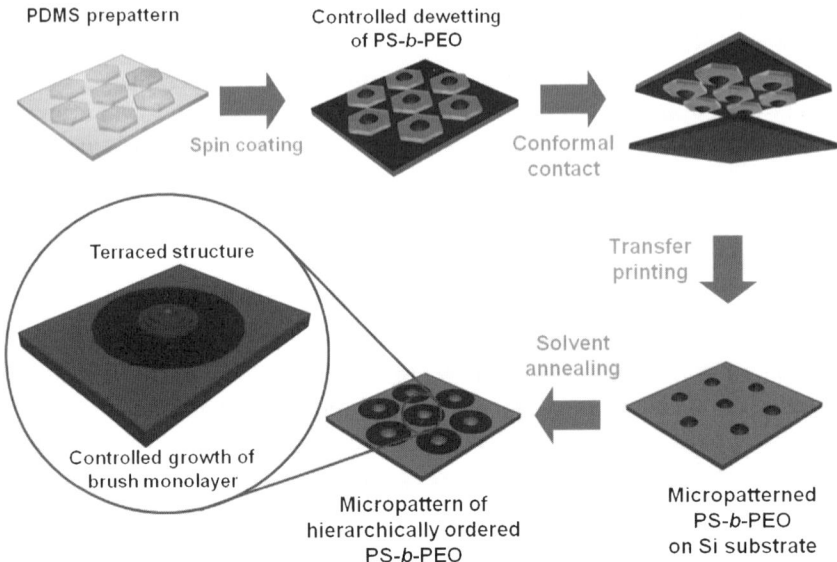

Figure 3.15 Procedure for controlling the growth of brush monolayers of micropatterned arrays of hierarchically ordered PS-*b*-PEO droplets on a solid substrate. The dewetting of the film on hexagonal patterned PDMS molds is selectively confined to the center of the mesa regions and subsequent transfer printing to Si substrate gives rise to patterned PS-*b*-PEO droplets. Solvent annealing provides sufficient mobility for the block copolymer molecules in convex-shaped patterned domains and allows control of the 2D circular spread of brush monolayers of the block copolymer, which leads to a hierarchical terraced structure with ordered microdomains. Adapted and modified from ref. 73. (Reprinted with permission from *Macromolecules*, **43**, 5352. Copyright 2010 American Chemical Society.)

two distinct features, as shown in Figure 3.16a and b. One is the formation of a unique terraced structure on each droplet and the other is the concentric growth of a thin sub-layer around each droplet.

Upon spreading of a dewetted droplet on the Si substrate, the characteristic terraced structure was developed, arising from the self assembled microstructure of PS-*b*-PEO copolymer. The similar terraced structures have been reported in lamellar forming poly(styrene-*block*-methylmethacrylate) copolymers when the droplets were thermally annealed at a temperature above the glass transition temperatures of the consistent blocks and below the order–disorder transition temperature (T_{ODT}) of the block copolymer;[26] whereas, droplets thermally annealed above T_{ODT} showed typical smooth surface profiles similar to those obtained from homopolymers. Step height of the terraced droplet, therefore, corresponds precisely to lamellae periodicity of the block copolymer. In our micropatterned, terraced droplets, the step is approximately 20 nm in height, as shown in the cross sectional profile of Figure 3.16b. The terraced structure was also developed in the arrays of the

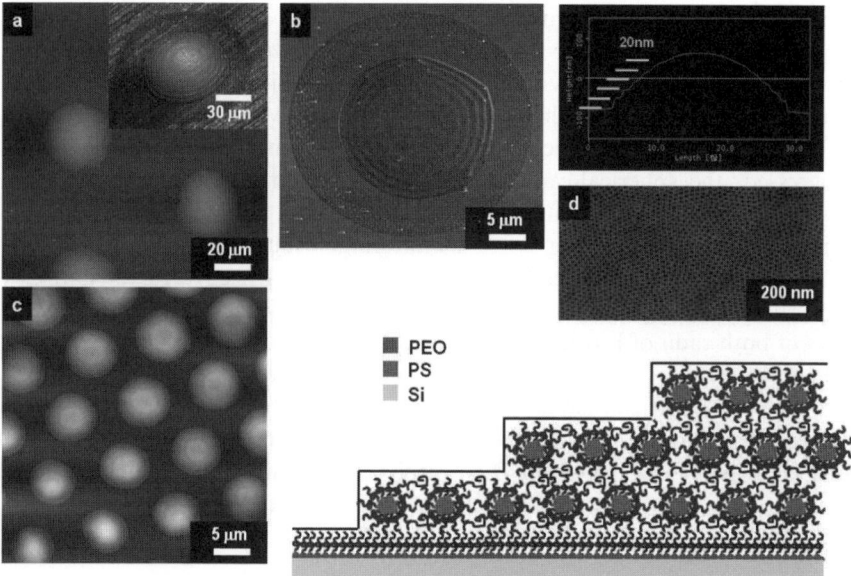

Figure 3.16 AFM images (a–d) of micropatterns of hierarchically terraced PS-*b*-PEO droplets with the concentric spreading of characteristic brush monolayers after solvent annealing. (a) Terraced PS-*b*-PEO droplets transferred with 40 μm hexagonal pre-patterns. The inset shows a magnified image of a terraced droplet in 3-D visualization. (b) (Left) Terraced structure of an individual droplet with the 2D concentric spread of a brush monolayer of the block copolymer. (Right) A height profile of the droplet of approximately 20 nm in step height, corresponding to the size of the block copolymer microdomains. (c) Microarrayed and terraced PS-*b*-PEO droplets transferred with 10 μm hexagonal pre-patterns. (d) Well-ordered PS-*b*-PEO nanostructures developed on the surface of droplets after solvent annealing. (e) A scheme of cross-sectional view of PS-*b*-PEO droplets with terraced structures after solvent annealing. Adapted and modified from ref. 73. (Reprinted with permission from *Macromolecules*, **43**, 5352. Copyright 2010 American Chemical Society.)

smaller droplets of 5 μm in diameter after solvent annealing, as shown in Figure 3.16c. The observation of the surface of terraced droplets revealed well developed circular PEO microdomains of approximately 20 nm in diameter, with improved ordering as compared to those observed in *as-transferred* droplets (Figure 3.16d).

The thin concentric layer grown around the terraced droplet is another characteristic feature of our system and corresponds to the brush monolayer of PS-*b*-PEO copolymer. The monolayered brush layer is approximately 7 nm in thickness in which polar PEO blocks are in contact with native oxide of Si substrate, as shown in the schematic of Figure 3.16e. The growth of the brush monolayer was monitored as a function of solvent annealing time. As *as-transferred* dewetted droplets were exposed to a benzene vapor for 20 seconds, concentric wetting of the brush monolayer began at each droplet, as shown in Figure 3.17a. At the same time, the size of the *as-transferred* PS-*b*-PEO droplets increased gradually with solvent exposure time (Figure 3.17b and c). The brush layer exhibited slow, continuous growth with the dewetted droplet acting as a reservoir which feeds block copolymer molecules. As time progressed, the spherical-cap droplets were transformed into terraced ones arising from the formation of the spherical PEO microdomains with the improved ordering. Further solvent annealing resulted in merging separate brush layers with each other (Figure 3.17e) and, finally, covered the entire Si substrate with the grown brush layer (Figure 3.17f). We examined the growth rates of both radii of brush monolayer and dewetted droplet, as defined in the scheme of Figure 3.16e. Interestingly, both radii increased linearly with $\sqrt{t}$ before the concentric brush layers were in contact with each other, as shown in Figure 3.16g, which indicates that both radii grow pseudo-diffusively with the relation of $R(t) \sim \sqrt{(Dt)}$ where D corresponds to a diffusion coefficient of the front. Spreading of symmetric block copolymer droplets above and below the order–disorder transition has recently been developed by Croll *et al.*[72] The

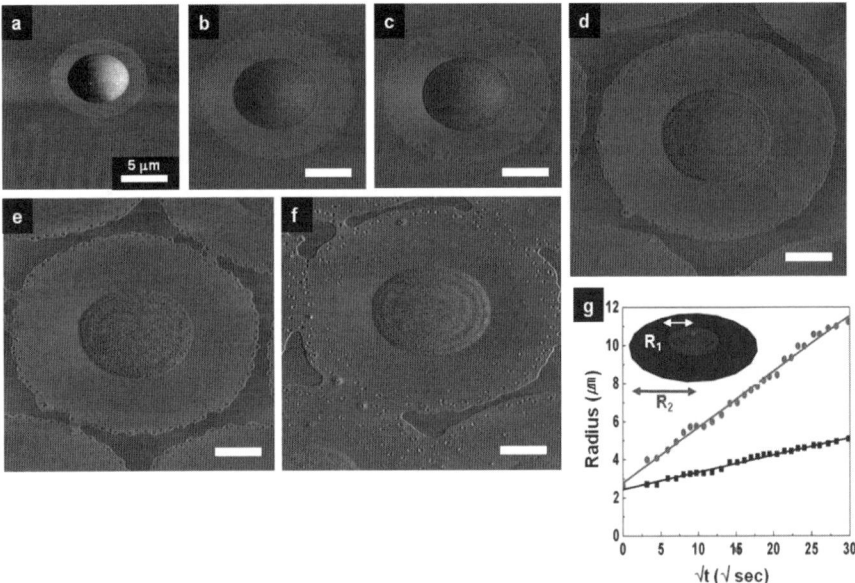

Figure 3.17 (a–f) AFM images of a dewetted PS-*b*-PEO droplet with the characteristic brush monolayer captured at different solvent annealing time: (a) 20 s. (b) 200 s, (c) 290 s (d) 890 s (e) 1130 s, (f) 1370 s. (g) Plots of the radii of the terraced droplet and the circular brush monolayer as a function of solvent annealing time. Both plots show the linear relationship between the root value of solvent annealing time and both the radii. Adapted and modified from ref. 73. (Reprinted with permission from *Macromolecules*, **43**, 5352. Copyright 2010 American Chemical Society.)

droplet in a disordered state grows as $R \sim t^m$, where t is time, with $m = 1/10$, consistent with Tanner's law. In contrast, a droplet in an ordered state spread much more slowly with $m \sim 0.05 \pm 0.01$. In our case, as noted, we observed a diffusive growth of both brush layer and droplet with an exponent, $m \sim 0.5$ much faster than that of droplets thermally induced. The diffusive motion of our system may be attributed to the solvent vapor, *i.e.* a solvent precursor wetting layer.

Our micropatterning technique of block copolymer droplets combined with solvent annealing provides an efficient way to tune the surface polarity of a substrate containing patterned droplets as a function of solvent annealing time. For instance, the contact angle of a water droplet on an *as-transferred* micropattern of approximately 55° almost linearly increases with solvent annealing time, as shown in Figure 3.18. In this particular micropattern containing approximately 2 µm droplets, solvent annealing for 100 seconds was long enough to cover the whole Si wafer with a brush monolayer grown from the individual droplets, resulting in a water contact angle of approximately 105°, which corresponds to that of pure PS, as shown in Figure 3.18. The controlled growth of a brush monolayer with the PS block pointing to the air by solvent annealing, therefore, allows us to tune the contact angle of a water droplet in the range over 50°.

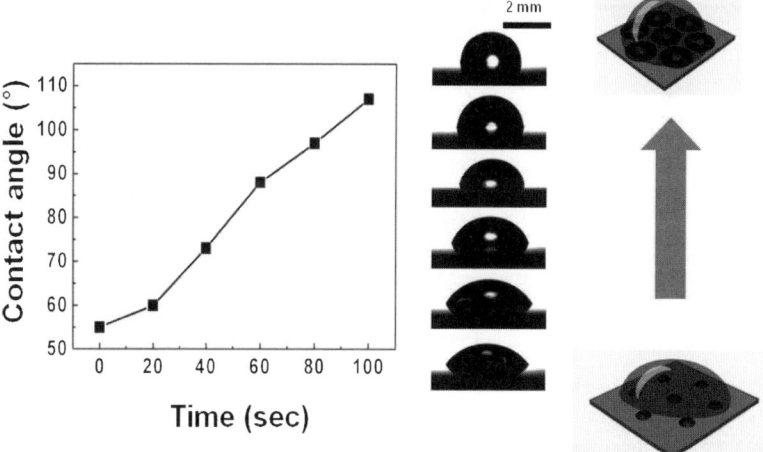

Figure 3.18 A plot of contact angle of a water droplet on micropatterned PS-*b*-PEO droplets transferred from 10 µm hexagonal PDMS pre-pattern as a function of solvent annealing time. The contact angle of approximately 55° on *as-transferred* pattern is varied to 105° after 100 seconds of solvent annealing, as shown in photographs from bottom to top. A schematic (not to scale) shows a water droplet on the controlled micropattern of PS-*b*-PEO. Adapted and modified from ref. 73. (Reprinted with permission from *Macromolecules*, **43**, 5352. Copyright 2010 American Chemical Society.)

3.5.3 Controlled Anisotropic Wetting of Block Copolymer Droplets

Although control of a block copolymer thin film has been of great interest upon wetting and dewetting of the film, until now block copolymer droplets with hierarchical structures have had a circular shape due to the surface area and surface energy minimization, which thus limits their applicability. Considering that the unique, more often anisotropic optical, electrical and mechanical properties of various non-spherical colloidal objects are potentially useful for plasmonics,[218] cosmetics,[219] biology,[220] pharmaceuticals,[221] and as structural materials,[222] it is interesting to develop a way for fabricating hierarchically ordered micropatterns of *non-circular* block copolymer droplets triggered by wetting and dewetting. For this purpose, we introduced two approaches to fabricate anisotropic hierarchical droplets of a block copolymer.

A dewetted poly(styrene-*block*-ethylene oxide) (PS-*b*-PEO) copolymer pattern of the micro-droplets onto a topographic PDMS was prepared, as shown in the schematic of Figure 3.19, as described previously. The sphere-

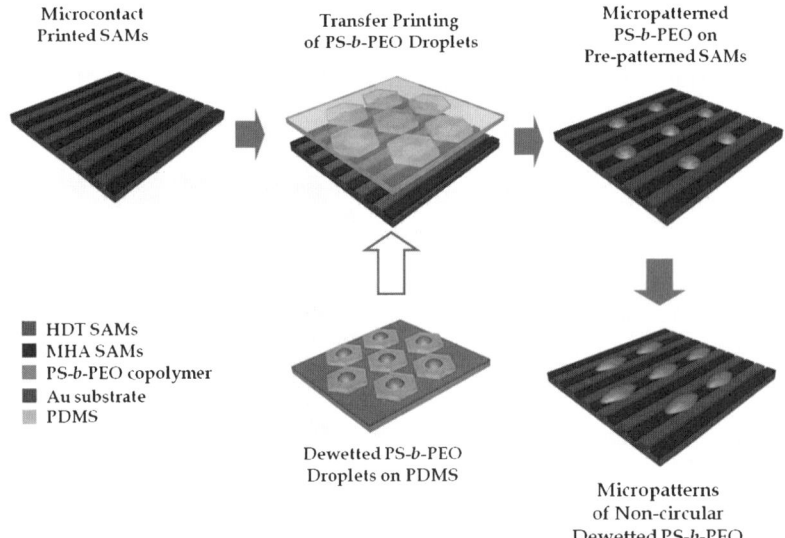

Figure 3.19 Procedure for fabricating micropatterns of non-circular droplets of a PS-*b*-PEO copolymer with hierarchically ordered nanostructures. The sphere-capped, dewetted droplets of the block copolymer selectively formed on the center of individual mesas of a hexagonally patterned PDMS mold are transferred to the periodic wettable MHA and non-wettable HDT SAMs line pattern. The chemically modified surface pattern controls the spreading of the transferred droplets upon solvent annealing, resulting in various shape non-circular micro-objects with hierarchical structures. Adapted and modified from the ref. 223. (Reprinted with permission from *Macromolecular Chemistry and Physics*, **213**, 431. Copyright 2012 John Wiley and Sons.)

capped PS-*b*-PEO microdroplets were obtained on the hexagonal posts by controlled dewetting, when a PS-*b*-PEO solution was spin coated onto a PDMS pre-pattern. After transfer printing of dewetted droplets on the periodically modified Au substrate with alternating stripes of hydrophobic and hydrophilic surfaces, the transferred droplets can be transformed to non-circular ones upon solvent annealing. The effective modulation of the spreading of the transferred droplets on the chemically periodic surface during solvent annealing led to micropatterns of non-circular droplets with various shapes.[223]

Before the investigation of the spreading behavior on micropatternd substrates, we examined the spreading of transferred droplets upon solvent annealing on Au substrates modified with either hexadecanethiol (HDT) or mercaptohexadecanonic acid (MHA). The arrays of dewetted PS-*b*-PEO droplets were transferred from PDMS to a chemically modified Au surface with either HDT or MHA. Both HDT and MHA have 16 carbon atoms in the alkane chain, and differ only in the end group (CH_3 and COOH, respectively) to produce SAMs that exhibit two opposing surface properties—hydrophobic and hydrophilic. After transfer-printing of dewetted PS-*b*-PEO droplets, we

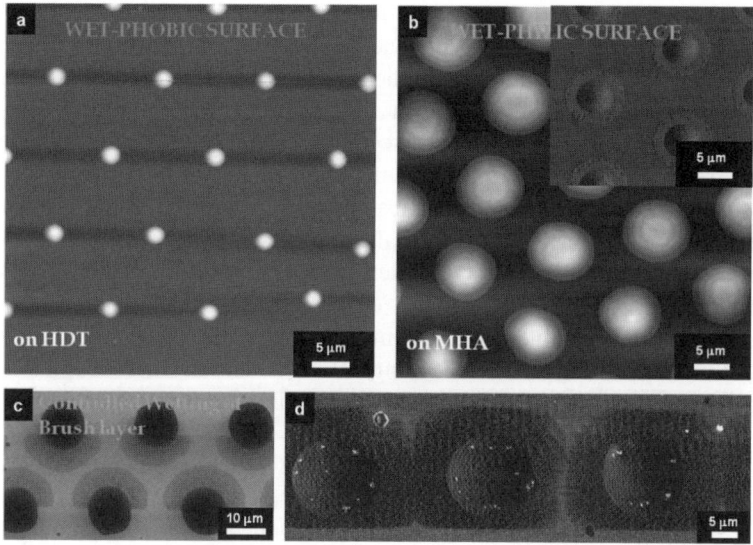

Figure 3.20 AFM images of the microarrayed PS-*b*-PEO droplets from a 5μm side wall hexagonal PDMS pattern (a) on wet-phobic HDT SAMs deposited Au substrate and (b) on wet-philic MHA SAMs deposited Au substrate after solvent annealing for one minute, respectively. A SEM image (c) and an AFM image (d) of shape controlled PS-*b*-PEO droplets on periodic lines of MHA and HDT with width andperiodicity of 20 and 40 μm, respectively, after solvent annealing for one minute. Adapted and modified from ref. 223. (Reprinted with permission from *Macromolecular Chemistry and Physics*, **213**, 431. Copyright 2012 John Wiley and Sons.)

monitored how the PS-b-PEO droplets were deformed in the progress of solvent annealing. As shown in Figure 3.20a, solvent annealing of *as-transferred* micropatterned droplets on a non-wettable HDT surface for 60 seconds resulted in almost no shape change from the originally dewetted domains.

In contrast, those on a wettable MHA surface showed the tendency to spread with a characteristic brush monolayer after solvent annealing for 60 s, as shown in Figure 3.20b. As shown in the inset of Figure 3.20b with a magnified phase image, the micro-droplets with approximately 8 μm in diameter have the characteristics of a terraced structure, arising from the self-assembled microstructure of PS-b-PEO copolymer in the droplet arrays after solvent annealing. In our micropatterned and terraced droplets, the step height is approximately 20 nm, which corresponds to the size of block copolymer microdomains after solvent annealing, consistent with our previous results.[185] The thin concentric brush layer around the droplet is approximately 7 nm in thickness in which the polar PEO blocks are in contact with the hydrophilic MHA surface and non-polar PS blocks are pointing towards the air. With these experiments and measurements, we have established a clear contrast in the shape-change of the transferred droplets on two different Au surface during solvent annealing.

To control the wetting of the droplets with non-circular shape, we introduced a chemical pattern with line periodicity comparable to the size of *as-transferred* droplets. Thus, we applied the micro-droplets from a 5 μm side wall hexagonal pattern of PDMS on wettable and non-wettable SAMs line pattern with line width of 5 μm. The size of *as-transferred* droplets (~3 μm) is comparable with the line width of 5 μm. The transfer of hexagonally arrayed droplets was performed on the chemical pattern in a way that the ^{1,1} direction of the hexagonal pattern is almost parallel to the line axis. Since the transferred hexagonal pattern of polymer droplets is incommensurate with the stripe pattern, the *as-transferred* droplets occupy areas with various ratios of wettable and non-wettable portions. Interestingly, in this particular condition, core sphere-capped droplets were changed to a variety of microdroplets with a different shape on the chemically patterned surface as shown in Figure 3.21a.

Careful investigation of the image suggests that 6 different shapes were repeated just like a unit cell, as marked in the white box of Figure 3.21a. An AFM image in the inset of Figure 3.21a shows the magnified view of the unit cell of shape-controlled droplets with the chemically modified line pattern underneath. It is apparent that the shape of the deformed droplets is closely related to how much of the area of the *as-transferred* droplets is initially in contact with the 5 μm chemical line pattern of periodic HDT, the MHA surface having a different extent of spreading of the droplets. The anisotropic, non-circular wetting of the droplets on the chemical line pattern involves two kinds of processes: the spreading of droplets and the growth of brush layers. The spreading of dewetted droplets occurred with the direction parallel to the MHA surface line, while was restricted near the region on the HDT surface, giving rise to the deformation of droplets. The growth of brush layers around the spread droplets was observed selectively on the wettable surface portion. Furthermore, on the surface of the resulting non-circular droplets, spherical PEO microdomains of approximately 20 nm in diameter developed, arising from microphase separation of PS-*b*-PEO during solvent annealing (Figure 3.21c). Our previous results have also demonstrated the cylinder-to-sphere transformation of PEO microdomains of the same block copolymer upon solvent annealing in a confined geometry.[172]

When the transferred droplets comparable in size with the chemical line width are deposited in the center of hydrophilic line pattern of MHA-covered surface, droplets with ellipsoidal shape were created after solvent annealing, as shown in

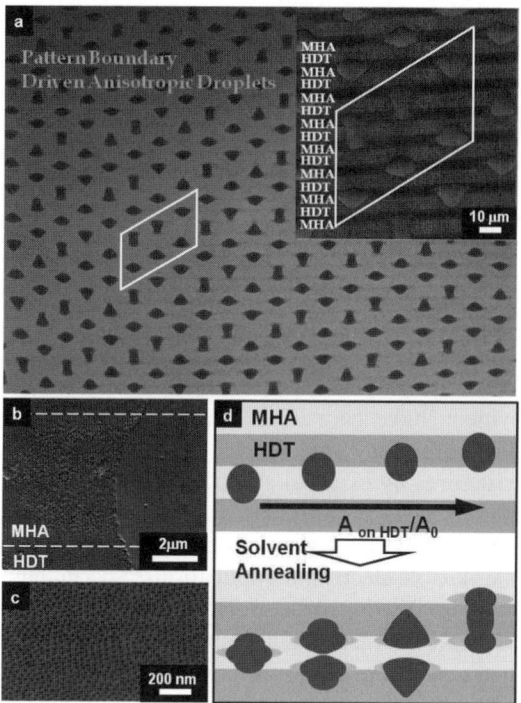

Figure 3.21 (a) A SEM and (b) an AFM image of the microarrayed PS-*b*-PEO droplets from a 5μm side wall hexagonal PDMS pattern on chemically periodic MHA and HDT lines with width and periodicity of 5 and 10 μm, respectively, after solvent annealing. Arrays of shape controlled, non-circular block copolymer droplets are formed with a repeating unit cell consisting of 6 different droplets, as marked with a white box. (c) Well-ordered PS-*b*-PEO nanostructures developed on the surface of non-circular droplets after solvent annealing. (d) A diagram showing the relationship between the shape of non-circular droplets and the ratio of area of an *as-transferred* droplet on HDT regions to the original droplet one (A_{onHDT}/A_0). Adapted and modified from ref. 223. (Reprinted with permission from *Macromolecular Chemistry and Physics*, **213**, 431. Copyright 2012 John Wiley and Sons.)

Figure 3.21b. The deformation of sphere-capped droplets into an ellipsoidal shape has been preceded from the droplet wetting and lateral growth of the brush layer on top of the MHA treated surface pattern starting from the three-phase-line of the polymer droplet on the MHA line pattern. The AFM image in Figure 3.21b shows the characteristic brush monolayer with the thickness of approximately 7 nm, which corresponds to the data previously obtained on a homogeneous MHA surface. Similarly, different non-circular droplets resulting from different initial registries of the droplets on the chemical pattern are understood by the schematic depicted in Figure 3.21d. The plot demonstrates the possible initial registries of droplets with respect to the chemical pattern varying the contact area ratio of A_{onHDT}/A_o, where A_o is the area of *as-transferred* dewetted droplets and A_{onHDT} is the droplets area on top of the HDT surface. Four different A_{onHDT}/A_o values, shown in the plot, produced the corresponding non-circular droplets with various shapes, as shown in the scheme. Two additional non-circular droplets can arise from two fold symmetry of the second and the third initial registries, as illustrated in Figure 3.21d.

3.6 Concluding remarks

Self-assembly driven by complicated but systematic hierarchical interactions provides a qualified alternative for fabricating functional micron- or nanometer-scale pattern structures that have been significantly useful for various organic and nanotechnological devices. Self-assembled nanostructures generated from synthetic polymer systems, such as controlled polymer blends and block copolymers, have gained great attention not only because of the variety of nanostructures they can evolve, but also because of the controllability of these structures by external stimuli, including electrical, mechanical, magnetic, solvent and surface confinement. This chapter introduced various methods to producing ordered micro/nanostructures of polymers potentially utilized for organic flexible electronic applications. Special emphasis was made on either dewetting or wetting of a polymer thin film on a hard surface, resulting from a delicate energetic balance between the polymer, substrate and environment. When dewetting and wetting were controlled *via* external stimuli, such as solvent, electric field, and chemical and topological periodic spatial boundaries, numerous micro- and nanostructures were developed in combination of the principle of self-assembly of functional polymers, *i.e.* block copolymers. The chapter extensively discussed a variety of the unprecedented periodic structures of hierarchically self-assembled synthetic block copolymer domains upon either wetting or dewetting, which we believe are extremely useful for various new emerging nanopattern and template applications.

References

1. G. Reiter, *Macromolecules.*, 1994, **27**, 3046.
2. Y. Xia and G. M. Whitesides, *Langmuir*, 1997, **13**, p2059.

3. M. Boltau, S. Welheim, J. Mlynek, G. Krausch and U. Steiner, *Nature*, 1998, **391**, 877.
4. S. Herminghaus, K. Jacobs, K. Mecke, J. Bischof, A. Fery, M. Ibn-Elhaj and S. Schlagowski, *Science*, 1998, **282**, p916.
5. J. L. Wilber, A. Kumar, E. Kim and G. M. Whitesides, *Adv. Mater.*, 1994, **6**, 600.
6. S. D. Evans, T. M. Flynn and A. Ulman, *Langmuir*, 1995, **11**, 3811.
7. T. Granlund, T. Nyberg, L. S. Roman, M. Svensson and O. Inganas, *Adv. Mater.*, 2000, **12**, 269.
8. B. K. Yoon, J. Huh, H. C. Kim, J. M. Hong and C. Park, *Macromolecules*, 2006, **39**, 901.
9. J. Peng, R. Xing, Y. Wu, B. Li, Y. Han, W. Knoll and D. H. Kim, *Langmuir*, 2007, **23**, 2326.
10. F. S. Bates and G. H. Fredrickson, *Phys. Today*, 1999, **32**, 52.
11. S. Forster and T. Planternberg, *Angew. Chem., Int. Ed.*, 2002, **41**, 688.
12. I. W. Hamley, *Angew. Chem., Int. Ed.*, 2003, **42**, 1692.
13. C. Park, J. Yoon and E. L. Thomas, *Polymer*, 2003, **44**, 6725.
14. C. J. Hawker and T. P. Russell, *MRS Bull.*, 2005, **30**, 952.
15. S. J. Jeong, J. E. Kim, H. S. Moon, B. H. Kim, S. M. Kim, J. B. Kim and S. O. Kim, *Nano Lett.*, 2009, **9**, 2300.
16. K. Jacobs, S. Herminghaus and K. R. Mecke, *Langmuir.*, 1998, **14**, 965.
17. Z. Zhang, Z. Wang, R. Xing and Y. Han, *Polymer.*, 2003, **44**, 3737.
18. P. G. de Gennes, *Rev. Mod. Phys.*, 1985, **57**, 827.
19. A. W.Adamson, *Physical Chemistry of Surfaces*, Wiley, New York, 4th edn, 1982.
20. A. M.Cazabat, *Contemp. Phys.*, 1987, **28**, 347.
21. L. Leger and J. F Joann, *Rep. Prog. Phys.*, 1992, **55**, 431.
22. X. Ma, J. Gui, L. Smoliar, K. Grannen, B. Marchon, C. L. Bauer and M. S. Jhon, *Phys. Rev. E: Stat. Phys., Plasmas, Fluids, Relat. Interdiscip. Top.*, 1999, **59**, 722.
23. J. De Coninck., U. D'Ortona., J. Koplik. and J.R. Banavar., *Phys. Rev. Lett.*, 1995, **74**, 928.
24. S. Betelú, B. M. Law and C. C. Huang., *Phys. Rev. E: Stat. Phys., Plasmas, Fluids, Relat. Interdiscip. Top.*, 1999, **59**, 6699.
25. J. De Coninck, N. Fraysse, M.P. Valignat and A.M. Cazabat, *Langmuir*, 1993, **9**, 1906.
26. R. Lucht and Ch. Bahr, *Phys. Rev. Lett.* 2000, **85**, 4080.
27. F. Heslot, N. Fraysse and A.M. Cazabat, *Nature*. 1989, **338**, 640.
28. L. Xu, M. Salmeron and S. Bardon, *Phys. Rev. Lett.* 2000, **84**, 1519.
29. B. L. Carvalho and E. L. Thomas, *Phys. Rev. Lett.* 1994, **73**, 3321.
30. A.B. Croll, M. V. Massa, M. V. Matsen and K. Dalnoki-Veress. *Phys. Rev. Lett.* 2006, **97**, 204502.
31. J. U. Kim and M.W. Matsen, *Soft Matter*, 2009, **5**, 2889.
32. T. H. Epps III, D. M. DeLongchamp, M. J .Fasolka, D. A. Fischer and E. L. Jablonski, *Langmuir* 2007, **23**, 3355.

33. P. F. J. Green, *J. Polym. Sci., Part B: Polym. Phys.* 2003, **41**, 2219.
34. J. Huang, F. Kim, A. R. Tao, S. Connor and P. Yang, *Nat. Mater.*, 2005, **4**, 896.
35. G. Nisato, B. D. Ermi, J. F. Douglas and A. Karim, *Macromoleculs*, 1999, **32**, 2356.
36. L. Rockford, Y. Liu, P. Mansky, T. P. Russel, M. Yoon and S. G. J. Mochrie, *Phys. Rev. Lett.*, 1999, **82**, 2602.
37. J. Lian, L. Wang, X. Sun, Q. Yu and E. C. Ewing, *Nano Lett.*, 2006, **6**, 1047.
38. M. Schick, *Liquids at Interfaces, Proceedings of the Les Houches Summer School, Session XLVIII*, ed. J. Charvolin, J.-F. Joanny, and J. Zinn-Justin, Elsevier, Amsterdam, 1990, p. 415.
39. R. D. Gretz, *J. Chem. Phys.*, 1966, **45**, 3169.
40. T. Young, *Phil. Trans. R. Soc.*, 1805, **57**, 827.
41. H. C. Hamaker, *Physica*, 1937, **4**, 1058.
42. J. Israelachvili, *Intermolecular and Surface Forces*, Academic, New York, 1992.
43. A. W. Adamson, *Physical Chemistry of Surfaces*, Wiley, New York, 1990.
44. J. A. de Feiter, *Thin Liquid Films*, ed. I. B. Ivanov, Dekker, New York, 1988, p. 1.
45. G. F. Teletzke, H. T. Davis and L. E. Scriven, *Rev. Phys. Appl.*, 1988, **23**, 989.
46. A. Sharma, *Langmuir*, 1993, **9**, 861.
47. A. Sharma, A. T. Jameel, *J. Colloid Interface Sci.*, 1993, **161**, 190.
48. A. N. Frumkin, *J. Phys. Chem. U. S. S. R.*, 1938, **12**, 337.
49. A. Vrij, *Discuss. Faraday Soc.*, 1966, **42**, 23.
50. E. Ruckenstein, R. K. Jain, *J. Chem Soc. Faraday Trans. 2*, 1974, **70**, 132.
51. M. B. Williams and S. H. Davis, *J. Colloid Interface Sci.* 1982, **90**, 1.
52. F. Brochard and J. Dailant, *Can. J. Phys.* 1990, **68**, 1084.
53. V. S. Miltin, *J. Colloid Interface Sci.*, 1993, **156**, 491.
54. V. S. Miltin, *Colloids Surf., A*, 1994, **89**, 97.
55. R. Blossey, *Int. J. Mod. Phys. B*, 1995, **9**, 3489.
56. M. W. J. Van der Wielen, M. A. Cohen Stuart and G. J. Fleer, *Langmuir*, 1998, **14**, 7065.
57. G. Reiter and J. U. Sommer, *Phys. Rev. Lett.*, 1998, **80**, 3771.
58. G. Reiter, A. Sharma, A. Casoli, M.-O. David, R. Khanna and P. Auroy, *Langmuir*, 1999, **15**, 2551.
59. M. J. Fasolka and M. J. Mayes, *Annu. Rev. Mater. Res.*, 2001, **31**, 323.
60. J. N. L. Julie and T. H. Epps III, *Mater. Today*, 2010, **13**, 24.
61. M. S. Turner, *Phys. Rev. Lett.*, 1992, **69**, 1788.
62. D. G. Walton, G. J. Kellog, A. M. Mayes, P. Lambooy and T. P. Russell, *Macromolecules*, 1994, **27**, 6225.
63. M. S. Turner, A. Johner and J. F. Joanny, *J. Phys.* I, 1995, **5**, 917.
64. G. Coulon, B. Collin, D. Ausserre, D. Chatenay and T. P. Russell, *J. Phys.* 1990, **51**, 2801.

65. B. Collin, D. Chatenay, G. Coulon, D. Ausserre and Y. Gallot, *Macromolecules* 1992, **25**, 1621.
66. A. M. Mayes, T. P. Russell, P. Bassereau, S. M. Baker and G. S. Smith, *Macromolecules*, 1994, **27**, 749.
67. R. Limary and P. F. Green, *Langmuir*, 1999, **15**, 5617.
68. P. F. Green and R. Limary, *Adv. Colliod Interface Sci.* 2001, **94**, 53.
69. P. Müller-Buschbaum, J. S. Gutmann, C. Lorenz-Haas, O. Wunnicke, M. Stamm and W. Petry, *Macromolecules*, 2002, **35**, 2017.
70. P. Busch, D. Posselt, D.-M. Smilgies, B. Rheinländer, F. Kremer and C. M. Papadakis, *Macromolecules*, 2003, **36**, 8717.
71. A. B. Croll, M. V. Massa, M. W. Matsen and K. Dalnoki-Veress, *Phys. Rev. Lett.*, 2006, **97**, 204502.
72. A. B. Croll and K. Dalnoki-Veress, *Eur. Phys. J. E: Soft Matter Biol. Phys.*, 2009, **29**, 234.
73. T. H. Kim, J. Huh and C. Park, *Macromolecules*, 2010, **43**, 5352.
74. Y.-S. Sun, S.-W. Chien and J.-Y. Liou, *Macromolecules*, 2010, **43**, 7250.
75. D. Yan, H. Huang, T. He and F. Zhang, *Langmuir*, 2011, **27**, 11973.
76. V. Körstgens, J. Wiedersich, R. Meier, J. Perlich, S.V. Roth, R. Gehrke and P. Müller-Buschbaum, *Anal. Bioanal. Chem.*, 2010, **396**, 139.
77. L. Zhao, M. D. Goodman, N. B. Bowden and Z. Lin, *Soft Matter*, 2009, **5**, 4698.
78. M. Ramanathan and S. B. Darling, *Soft Matter* 2009, **5**, 4665.
79. Y. Zhang, X. Xiao, J. J. Zhou, L. Wang, Z. B. Li, L. Li, L.Q. Shi and C.M. Chan, *Polymer*, 2009, **50**, 6166.
80. C. Neto, M. James, A. M. Telford, *Macromolecules*, 2009, **42**, 4801.
81. J. R. Howse, R. A. L. Jones, G. Battaglia, R. E. Ducker, G. J. Leggett and A. J. Ryan, *Nat. Mater.*, 2009, **8**, 507.
82. W. M. De Vos, A. De Keizer, J. Mieke Kleijn and M. A. Cohen Stuart, *Langmuir*, 2009, **25**, 4490.
83. M. M. A. Kashem, J. Perlich, L. Schulz, S. V.Roth, W. Petry and P. Müller-Buschbaum, *Macromolecules* 2007, **40**, 5075.
84. X. Li, J. Peng, J. Y. Wen, Y. D. H. Kim and W. Knoll, *Polymer*, 2007, **48**, 2434.
85. R. B. Cheyne and M. G. Moffitt, *Macromolecules*, 2007, **40**, 2046.
86. T. H. Epps III, D. M. DeLongchamp, M. J. Fasolka, D. A. Fischer and E. L. Jablonski, *Langmuir*, 2007, **23**, 3355.
87. L. Wang, S. Hong, H. Hu, J. Zhao and C. C. Han, *Langmuir*, 2007, **23**, 2304.
88. G. D. Liang, J. T. Xu, Z. Q. Fan, S. M. Mai and A. J. Ryan, *J. Phys. Chem. B* 2006, **110**, 24384.
89. B. Wei, P.G. Lam, M. B. Braunfeld, D. A. Agard, J. Genzer and R. J. Spontak, *Langmuir*, 2006, **22**, 8642.
90. G. G. Baralia, C. Filiâtre, B. Nysten and A. M. Jonas, *Adv. Mater.*, 2007, **19**, 4453.
91. J. Peng, D. H. Kim, W. Knoll, Y. Xuan, B. Li, Y. Han, *J. Chem. Phys.*, 2006, **125**, 064702.

92. A. J. F. Carvalho, M. A. Pereira-Da-Silva and R. M. Faria, *Eur. Phys. J. E: Soft Matter Biol. Phys.*, 2006, **20**, 309.
93. J. Zhu, J. Zhao, Y. Liao and W. Jiang, *J. Polym. Sci., Part B: Polym. Phys.*, 2005, **43**, 2874.
94. J. Peng, Y. Xuan, H. Wang, B. Li and Y. Han, *Polymer*, 2005, **46**, 5767.
95. P. Müller-Buschbaum, N. Hermsdorf, S. V. Roth, J. Wiedersich, S. Cunis and R. Gehrke, *Spectrochimica Acta, Part B*, 2004, **59**, 1789.
96. B. Wei, J. Genzer and R. J. Spontak, *Langmuir*, 2004, **20**, 8659.
97. I. Y. Tsai, M. Kimura and T. P. Russell, *Langmuir*, 2004, **20**, 5952.
98. J. T. Han and K. Cho, *Macromolecules*, 2003, **36**, 8902.
99. S. H. Lee, H. Kang, Y. S. Kim and K. Char, *Macromolecules*, 2003, **36**, 4907.
100. K. Swaminathan Iyer and I. Luzinov, *Langmuir*, 2003, **19**, 118.
101. D. N. Leonard, R. J. Spontak, S. D. Smith and P. E. Russell, *Polymer*, 2002, **43**, 6719.
102. K. Fukunaga, T. Hashimoto, H. Elbs and G. Krausch, *Macromolecules*, 2002, **35**, 4406.
103. R. Limary, P. R. Green, K. R. Shull, *Eur. Phys. J. E: Soft Matter Biol. Phys.*, 2002, **8**, 103.
104. D. N. Leonard, P.E. Russell, S. D. Smith and R. J. Spontak, *Macromol. Rapid Commun.*, 2002, **23**, 205.
105. R. Oslanec, A. C. Costa, R. J. Composto and P. Vlcek, *Macromolecules*, 2000, **33**, 5505.
106. J. P. Spatz, P. Eibeck, S. Mössmer, M. Möller, E. Yu. Kramarenko, P. G. Khalatur, I. I. Potemkin and P. Reineker, *Macromolecules*, 2000, **33**, 150.
107. R. Limary and P. F. Green, *Langmuir.*, 1999, **15**, 5617.
108. E. Huang, S. Pruzinsky, T. P. Russell, J. Mays and C. J. Hawker, *Macromolecules.*, 1999, **32**, 5299.
109. I. W. Hamley, E. L. Hiscutt, Y. W. Yang and C. Booth, *J. Colloid Interface Sci.*, 1999, **209**, 255.
110. L. Xue, Y. Han, *Prog. Polym. Sci.*, 2011, **36**, 269.
111. B. Yoon, H. Acharya, G. Lee, H. C. Kim, J. Huh and C. Park, *Soft Matter*, 2008, **4**, 1467.
112. R. Mukherjee, M. Gonuguntla and A. J. Sharma, *Nanosci. Nanotechnol.*, 2007, **7**, 2069.
113. D. H. Kim, M. J. Kim, J. Y. Park and H. H. Lee, *Adv. Funct. Mater.*, 2005, **15**, 1445.
114. S. Luan, Z. Cheng, R. Xing, Z. Wang, X. Yu and Y. J. Han, *Appl. Phys.*, 2005, **97**, 086102.
115. S. Harkema, E. Schäffer, M. D. Morariu and U. Steiner, *Langmuir.*, 2003, **19**, 9714.
116. S. Rath, M. Heilig, H. Port and J. Wrachtrup, *Nano Lett.*, 2007, **7**, 3845.
117. R. Mukherjee, D. Bandyopadhyay and A. Sharma, *Soft Matter*, 2008, **4**, 2086.
118. H. Celio, E. Barton and K. J. Stevenson, *Langmuir*, 2006, **22**, 11426.
119. A. Verma and A. Sharma, *Macromolecules*, 2011, **44**, 4928.

120. A. L. Giermann and C. V. Thompson, *Appl. Phys. Lett.*, 2005, **86**, 121903.
121. Z. Zhang, Z. Wang, R. Xing and Y. Han, *Langmuir*, 2003, **44**, 3737.
122. X. Li, R. Xing, Y. Zhang, Y. Han and L. An, *Polymer*, 2004, **45**, 1637.
123. Y. Nie, W. Li, L. An, D. Zhu, Z. Wang and B. Yang, *Colloid Surf., A*, 2006, **278**, 229.
124. H. Gau, S. Herminghaus, P. Lenz and R. Lipowsky, *Science*, 1999, **283**, 46.
125. N. Lu, X. Chen, D. Molenda, A. Naber, H. Fuchs, D.V. Talapin, H. Weller and L. Chi, *Nano Lett.*, 2004, **4**, 885.
126. A. Sehgal, V. Ferreiro, J. F. Douglas, E. J. Amis and A. Karim, *Langmuir*, 2002, **18**, 7041.
127. E. J. Tull and P. N. Bartlett, *Colloid Surf., A*, 2008, **327**, 71.
128. F. Fan and K. J. Stebe, *Langmuir*, 2004, **20**, 3062.
129. A. Benor and D. Knipp, *Org. Electronics.*, 2008, **9**, 209.
130. M. Boltau, S. Walheim, J. Mlynek, G. Krausch and U. Steiner, *Nature*, 1998, **391**, 877.
131. G. Nisato, B. D. Ermi, J. F. Douglas and A. Karim, *Macromolecules*, 1999, **32**, 2356.
132. J. H. Wei, D.C. Coffey and D. S. Ginger, *J. Phys. Chem. B*, 2006, **110**, 24324.
133. J. Z. Wang, Z. H. Zheng, H. W. Li, W. T. S. Huck and H. Sirringhaus, *Nat. Mater.*, 2004, **3**, 171.
134. Y. Cai and B. M. Zhang Newby, *Langmuir*, 2008, **24**, 5202.
135. H. L. Zhang, D. G. Bucknall and A. Dupuis, *Nano Lett.*, 2003, **4**, 1513.
136. I. T. Pai, I. C. Leu and M. H. Hon, *J. Micromech. Microeng.*, 2008, **18**, 105005.
137. R. Mukherjee, A. Sharma, G. Patil, D. Faruqui and P. S. G. Pattader, *Bull. Mater. Sci.*, 2008, **31**, 249.
138. R. Mukherjee, A. Sharma, M. Gonuguntla and G. K. Patil, *J. Nanosci. Nanotechnol.*, 2008, **8**, 3406.
139. A. Ghatak, M. K. Chaudhury, V. Shenoy and A. Sharma, *Phys. Rev Lett.*, 2000, **85**, 4329.
140. G. Reiter and A. Sharma, *Phys. Rev. Lett.*, 2001, **87**, 166103/1-4.
141. S. Gabriele, S. Sclavons, G. Reiter and P. Damman, *Phys. Rev. Lett.*, 2006, **96**, 156105/1-4.
142. R. D. Deegan, O. Bakajin, T. F. Dupont, G. Huber, S. R. Nagel and T. A. Witten, *Nature*, 1997, **389**, 827.
143. O. Karthaus, K. Ijiro and M. Shimomura, *Chem. Lett.*, 1996, **25**, 821.
144. J. Huang, R. Fan, S. Connor and P. Yang, *Angew. Chem., Int. Ed.*, 2007, **46**, 2414.
145. O. Karthaus, T. Koito and M. Shimomura, *Mater Sci. Eng.*, C, 1999, **8–9**, 523–6.
146. R. van Hameren, P. Schon, A. M. van Buul, J. Hoogboom, S. V. Lazarenko, J. W. Gerritsen, H. Engelkamp, P. C. M. Christianen, H. A. Heus, J. C. Maan, T. Rasing, S. Speller, A. E. Rowan, J. A. A. W. Elemans and R. J. M. Nolte, *Science*, 2006, **314**, 1433.

147. R. van Hameren, A. M. van Buul, M. A. Castriciano, V. Villari, N. Micali, P. Schon, S. Speller, L. M. Scolaro, A. E. Rowan, J. A. A. W. Elemans and R. J. M. Nolte, *Nano. Lett.*, 2008, **8**, 253.
148. Y. Tong, Q. Tang, H. T. Lemke, K. Moth-Poulsen, F. Westerlund, P. Hammershøj, K. Bechgaard, W. Hu and T. Bjørnholm, *Langmuir*, 2010, **26**, 1130.
149. A. J. Archer, M. J. Robbins and U. Thiele, *Phys. Rev. E: Stat., Nonlinear, Soft Matter Phys.*, 2010, **81**, 021602/1-5.
150. N.J. Suematsu, Y. Ogawa, Y. Yamamoto and T. Yamaguchi, *J. Colloid. Inter. Sci.*, 2007, **310**, 648
151. O. Karthaus, S. Mikami and Y. J. Hashimoto, *Colloid Interface Sci.*, 2006, **301**, 703.
152. E. Schaffer, T. Thurn-Albrecht, T.P. Russell and U. Steiner, *Nature*, 2000, **403**, 874.
153. M.D. Morariu, N.E. Voicu, E. Schaffer, Z. Lin, T.P. Russell and U. Steiner, *Nature Mater.*, 2003, **2**, 48.
154. S. Harkema and U. Steiner, *Adv. Funct. Mater.*, 2005, **15**, 2016.
155. Z. Lin, T. Kerle, T.P. Russell, E. Schaffer and U. Steiner, *Macromolecules*, 2002, **35**, 3971.
156. R. Verma, A. Sharma, K. Kargupta and J. Bhaumik, *Langmuir*, 2005, **8**, 3710.
157. F. Mugele, *Soft Matter*, 2009, **5**, 3377
158. J. D. Cobine, *Gaseous Conductors*, Dover, New York, 1958.
159. L. B. Loeb, *Electrical Corona*, University of California Press, Berkeley, 1965.
160. K. J. Nygaard, Rev. Sci. Instrum., 1965, **36**, 1320.
161. J. P. Boeuf and L. C. Pitchford, *J. Appl. Phys.*, 2005, **97**, 103307.
162. N. Sano and D. Yamamoto, *Ind. Eng. Chem. Res.*, 2005, **44**, 2982.
163. D. Julthongpiput, W. Zhang, J. F. Douglas, A. Karim and M. J. Fasolka, *Soft Matter*, 2007, **3**, 613.
164. R. Burtovyy and I. Luzinov, *Langmuir*, 2008, **24**, 5903.
165. X. Zhang, F. Xie and O. K. C. Tsui, *Polymer*, 2005, **46**, 8416.
166. E. Bormashenko, A. Musin, R. Pogreb, Y. Bormashenko and O. Gendelman, *Colloids Surf., A*, 2007, **303**, 253.
167. P. D. Rack, Y. Guan, J. D. Fowlkes, A. V. Melechko, M. L. Simpson, *Appl. Phys. Lett.* 2008, **92**, 223108.
168. J. Lian, L. Wang, X. Sun, Q. Yu and R. C. Ewing, *Nano Lett.*, 2006, **6**, 1047.
169. C. Park, C. De Rosa, L. J. Fetters and E. L. Thomas, *Macromolecules*, 2000, **33**, 7931.
170. T. Thurn-Albrecht, J. Schotter, G. A. Kastle, N. Emley, T. Shibauchi, L. Krusin-Elbaum, K. Guarini, C. T. Black, M. T. Tuominen and T. P. Russell, *Science*, 2000, **290**, 2126.
171. C. Osuji, P. J. Ferreira, G. P. Mao, C. K. Ober, J. B. Vander Sande and E. L. Thomas, *Macromolecules*, 2004, **37**, 9903.

172. T. Xu, A. V. Zvelindovsky, G. J. A. Sevink, K. S. Lyakhova, H. Jinnai and T. P. Russell, *Macromolecules*, 2005, **38**, 10788.
173. R. A. Segalman, H. Yokoyama and E. J. Kramer, *Adv. Mater.*, 2001, **13**, 1152.
174. J. Y. Cheng, A. M. Mayes and C. A. Ross, *Nature Mater.*, 2004, **3**, 823.
175. E. Huang, L. Rockford, T. P. Russell and C. J. Hawker, *Nature*, 1998, **395**, 757.
176. D. Y. Ryu, K. Shin, E. Drockenmuller, C. J. Hawker and T. P. Russell, *Science*, 2005, **308**, 236.
177. G. Kim and M. Libera, *Macromolecules*, 1998, **31**, 2569.
178. K. Fukunaga, H. Elbs, R. Maerle and G. Krausch, *Macromolecules*, 2000, **33**, 947.
179. Z. Lin, D. H. Kim, X. Wu, L. Boosahda, D. Stone, L. LaRose and T. P. Russell, *Adv. Mater.*, 2002, **14**, 1373.
180. M. Kimura, M. J. Misner, T. Xu, S. H. Kim and T. P. Russell, *Langmuir*, 2003, **19**, 9910.
181. J. Hwang, J. Huh, B. Jung, J. M. Hong, M. Park and C. Park, *Polymer*, 2005, **46**, 9133.
182. J. Hahm and S. J. Sibener, *Langmuir*, 2000, **16**, 4766.
183. C. De Rosa, C. Park, E. L. Thomas and B. Lotz, *Nature*, 2000, **405**, 433.
184. C. Park, C. De Rosa and E. L. Thomas, *Macromolecules*, 2001, **34**, 2602.
185. C. Park, J. Y. Cheng, C. De Rosa, M. J. Fasolka, A. M. Mayes, C. A. Ross and E. L. Thomas, *Appl. Phys. Lett.*, 2001, **79**, 848.
186. A. W. Harant and C. N. Bowman, *J. Vac. Sci. Technol., B*, 2005, **23**, 1615.
187. S. Ludwigs, A. Boker, A. Voronov, N. Rehse, R. Magerle and G. Krausch, *Nat. Mater.*, 2003, **2**, 744.
188. S. H. Kim, M. Minsner, T. Xu, M. Kimura and T. P. Russell, *Adv. Mater.*, 2004, **16**, 226.
189. J. Zhao, S. Jiang, X. Ji, L. An and B. Jiang, *Polymer*, 2005, **46**, 6521.
190. K. A. Cavicchi, K. J. Berthiaume and T. P. Russell, *Polymer*, 2005, **46**, 11635.
191. E. M. Freer, L. E. Krupp, W. D. Hinsberg, P. M. Rice, J. L. Hedrick, J. N. Cha, R. D. Miller and H.-C. Kim, *Nano Lett.*, 2005, **5**, 2014.
192. J. Bang, S. H. Kim, E. Drockenmuller, M. J. Misner, T. P. Russell and C. J. Hawker, *J. Am. Chem. Soc.*, 2006, **128**, 7622.
193. W. van Zoelen, T. Asumaa, J. Ruokolainen, O. Ikkala and G. ten Brinke, *Macromolecules.*, 2008, **41**, 3199.
194. S. H. Kim, M. J. Misner and T. P. Russell, *Adv. Mater.*, 2004, **16**, 2119.
195. E. M. Freer, L. E. Krupp, W. D. Hinsberg, P. M. Rice, J. L. Hedrick, J. N. Cha, R. D. Miller and H.-C. Kim, *Nano Lett.*, 2005, **5**, 2014.
196. Q. Li, J. He, E. Glogowski, X. Li, J. Wang, T. Emrick and T. P. Russell, *Adv. Mater.*, 2008, **20**, 1462.
197. S. Park, J. Y. Wang, B. Kim and T. P. Russell, *Nano Lett.*, 2008, **8**, 1667.

198. Y. Chen, Z. Wang, Y. Gong, Y. H. Huang and T. J. He, *Phys. Chem. B*, 2006, **110**, 1647.
199. S. H. Kim, M. J. Misner, L. Yang, O. Gang, B. M. Ocko and T. P. Russell, *Macromolecules*, 2006, **39**, 8473.
200. Y. Li, X. Wang, I. C. Sanchez, K. P. Johnston and P. F. Green, *J. Phys. Chem. B*, 2007, **111**, 16.
201. S. Park, H. L. Dong, J. Xu, B. Kim, W. H. Sung, U. Jeong, T. Xu and T. P. Russell, *Science*, 2009, **323**, 1030.
202. S. W. Hong, J. Wang and Z. Lin, *Angew. Chem., Int. Ed.*, 2009, **48**, 8356.
203. B. H. Kim, D. O. Shin, S. Jeong, C. M. Koo, S. C. Jeon, W. J. Hwang, S. Lee, M. G. Lee and S. O. Kim, *Adv. Mater.*, 2008, **20**, 2303.
204. G. G. Baralia, C. Filiatre, B. Nysten and A. M. Jonas Adv. *Mater.*, 2007, **19**, 4453.
205. T. H. Kim, J. Hwang, W. S. Hwang, J. Huh, H.-C. Kim, S.H. Kim, J. M. Hong, E. L. Thomas and C. Park, *Adv. Mater.*, 2008, **20**, 522.
206. S. H. Lee, P. J. Yoo, S. J. Kwon and H. H. Lee, *J. Chem. Phys.*, 2004, **121**, 4346,
207. R. A. Farrell, N. Kehagias, M. T. Shaw, V. Reboud, M. Zelsmann, J. D. Holmes, C. M. S. Torres and M. A. Morris, *ACS Nano*, 2011, **5**, 1073.
208. G. Lee, P. S. Jo, B. Yoon, T. H. Kim, H. Acharya, H. Ito, H.-C. Kim, J. Huh and C. Park, *Macromolecules*, 2008, **41**, 9290.
209. S. Hirasawa, Y. Saito, H. Nezu, N. Ohashi and H. Maruyama, *IEEE Trans. Semicond. Manuf.*, 1997, **10**, 438.
210. J. Huh, C.-H. Ahn, W. H. Jo, J. N. Bright and D. R. M. Williams, *Macromolecules*, 2005, **38**, 2974.
211. G. S. Grest and M. Murat, *Macromolecules*, 1993, **26**, 3108.
212. P.-Y. Lai and K. Binder, *J. Chem. Phys.*, 1992, **91**, 586.
213. B. Yoon, J. Huh, H. Ito, J. Frommer, B.-H. Sohn, J. H. Kim, E. L. Thomas, C. Park and H.-C. Kim, *Adv. Mater.*, 2007, **19**, 3342.
214. M. Tang, M. H. Hong and Y. S. Choo, *2008 IEEE Photonics Global at Singapore, IPGC*, 2008, 4781512.
215. L. O. Péres and J. Gruber, *Mater. Sci. Eng., C*, 2007, **27**, 67.
216. M.-H. Wu, C. Park and G. M. Whitesides, *Langmuir*, 2002, **18**, 9312.
217. Y. Lu, Y. Yin and Y. Xia, *Adv. Mater.*, 2001, **13**, 34.
218. S. T. Gentry and M. W. Bezpalko, *J. Phys. Chem. C*, 2010, **114**, 6989.
219. M. J. Murray and M. J. Snowden, *Adv. Colloid Interface Sci.*, 1995, **54**, 73.
220. M. Yoshida, K. H. Roh and J. Lahann, *Biomaterials*, 2007, **28**, 2446.
221. B. Y. Shekunov, P. Chattopadhyay, H. H. Y. Tong and A. H. L. Chow, *Pharm. Res*, 2007, **24**, 203.
222. L. J. Bonderer, A. R. Studart and L. J. Gauckler, *Science*, 2008, **319**, 1069.
223. T. H. Kim, J. Huh and C. Park, *Macromol. Chem. Phys.*, 2012, **213**, 431.

CHAPTER 4
Nanoarchitectures Based on Clay Materials

E. RUIZ-HITZKY*[a], P. ARANDA[a] AND C. BELVER[a,b]

[a] Instituto de Ciencia de Materiales de Madrid, CSIC, Madrid E-28049, Spain;
[b] Sección Ingeniería Química, Universidad Autónoma de Madrid, Madrid E-28049, Spain
*E-mail: eduardo@icmm.csic.es

4.1 Introduction

Nanoarchitectonics[1] is a term coined at the National Institute for Materials Science (NIMS) in Japan to define the preparation of materials by arranging structural units at the nanoscale using different experimental approaches, such as physical and/or chemical manipulation of atoms and molecules, field-induced manipulation and self-assembly synthesis. The concept of nanoarchitectonics applied to silicates belonging to the clay minerals family was probably pioneering, as it was used to design and synthesize the so-called "pillared" clays (PILCs). In these solids, as described below, metal-oxide NPs act as pillars that permanently separate the silicate layers, developing nanogalleries in a similar way to those of conventional buildings and mines but at the nanometrre scale.

Typical clay minerals are smectites (montmorillonite, hectorite, saponite, beidellite, *etc.*) that are charged alumino-phyllosilicates showing plate-like morphology. Their structure is based on layers built by cations in tetrahedral and octahedral coordination to oxygen and hydroxyl groups, leading to tetrahedral and octahedral sheets (Figure 4.1A). Isomorphic substitutions of silicon, aluminium and other metal atoms in the tetrahedral and/or octahedral

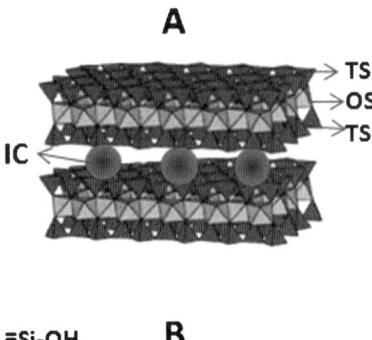

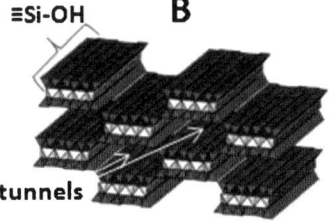

Figure 4.1 Schematic representation of: (A) a charged layered clay (montmorillonite) and (B) a fibrous clay (sepiolite). TS: tetrahedral sheet; OS: octahedral sheet; IC: interlayer exchangeable cations; Si–OH: surface silanol groups.

position by others with lower charge result in a net negative charge in the layers compensated for by interlayer cations, and therefore the clays show the ability to act as ion-exchangers.[2] Other clays show microfibrous morphologies, such as palygorskite and sepiolite (Figure 4.1B), or tubular morphologies, such as halloysite and imogolite.[2]

The present chapter will introduce an overview on the diverse types of, and main approaches developed to prepare, inorganic heterostructures based on clay minerals, from the classical PILCs to the most recent systems prepared by assembling of clay minerals and other inorganic solids. The state-of-the-art related to strategies of synthesis, and new trends and applications beyond catalysis, will be described to demonstrate the many opportunities still open in this research area.

4.2 Layered Clays as Building Blocks for Nanoarchitectonics

4.2.1 Pillared Clays (PILCs)

Pillared clays (PILCs) are nanoarchitectures with high surface areas and a microporous network created from smectites and other layered clays with cation exchange capacity. The synthesis of these materials is based on intercalation processes.[3] Actually, the majority of PILCs are porous clay-based solids formed in two step processes in which polyoxocations have to be firstly intercalated by an ion-exchange reaction by replacing the interlayer cations of the pristine clay, and then further transformed in oxide particles by an adequate thermal treatment.[4] Intercalated species are anchored to the clay layers due to the thermal treatment and their distribution yields channels and bidimensional galleries providing a microporous structure that can be tailored by the nature and size of the intercalated species. Hence, PILCs may be regarded as nanocomposites usually formed by oxyhydroxide particles of nano- and subnano-metre sizes whose aggregation to give large oxide particles is limited. Whichever the case, an interesting two-dimensional network is obtained in the interlayer space of the clays showing molecular dimensions and acidity comparable to that of zeolites.[5]

The intercalation of clays with the aim of developing porous solids was reported in the 1950s by Barrer *et al.* using organic species.[6] However, this modification yielded unstable solids because the intercalated species decomposed by temperature resulting in the clay layers collapse. Pillared clays received much attention in 1973 during the fuel crisis. Due to the fuel price, the petrochemistry industry looked for new and improved heterogeneous catalysts to be used for cracking reactions. Thus, the design of materials with thermal and hydrothermal stability, and optimum pore sizes to adsorb hydrocarbons received special attention. In this context, inorganic polycations were chosen as intercalated agents yielding to stable materials with high surface areas. The pillaring process by polycations is detailed in Figure 4.2, including two steps. In the first one (a), the cations constituting the cation exchange capacity (CEC) of the clay are replaced with polyoxocations usually derived from the partial hydrolysis of multivalent cations. The resulting solids have higher interlayer distances and are called intercalated clay (*pre*-PILC). In another step (b), the polyoxocations yield, upon calcination, thermally-stable oxide pillars that prop apart the clay layers. The properties and the synthesis methodology of PILCs depend on several factors.[3] For instance, (i) the nature of the pristine clay, including the chemical composition, the CEC or the presence of accompanying

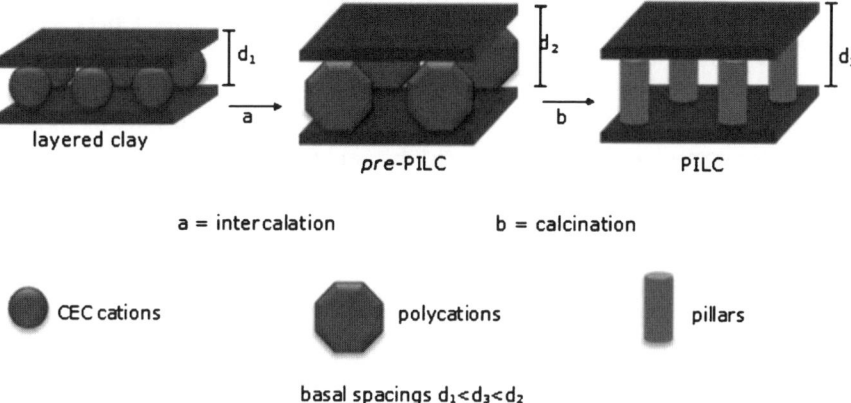

Figure 4.2 Schematic representation of the synthesis methodology of PILCs.

oxides and silicates impurities; (ii) the nature of the polycation; (iii) the synthesis parameters, such as pH, ratio polycation/clay, *etc.*; and (iv) the temperature of the thermal treatment, *i.e.* the final calcination step.

Authors as Brindley *et al.*,[7] Lahav *et al.*[8] and Vaughan *et al.*[9] were pioneers, independently reporting their results on the intercalation of aluminium polycations between the layers of smectites (montmorillonites, saponites, beidellites, stevensites, *etc.*). Since these works, many researches have been focussed on the intercalation of aluminium polyoxycations, mainly the so-called "Al_{13}" Keggin cage ($[Al_{13}O_4(OH)_{24}(H_2O)_{12}]^{7+}$), yielding PILCs with basal spacing in the 1.7–1.8 nm range and thermal stability up to 500 °C. Microwave irradiation has been also used in the consolidation step.[10] The mechanism of alumina pillar creation has been followed by NMR.[11] The thermal decomposition of the Al_{13} generated H^+ that attacked the tetrahedral sheet of the layer on the proximal OH of the polycation, thus the pillar kept anchored to the layer; this union caused the AlO_4 tetrahedra inversion and generated new Brönsted acid centres.[12]

Polycations derived from the hydrolysis of other metals, such as Ti(IV), Zr(IV), Cr(III), Fe(III) or Ga(III), have also been considered.[13] As regards the clay materials, several smectites (saponite, laponite, hectorite, *etc.*) have been studied, but to a lesser extent than montmorillonite.[14] The fact that Al-PILCs were the most studied resulting in stable pillared clays determined that they were frequently used to prepare mixed metal polycations. Thus, trends in PILCs are related to the optimum preparation of pillars with two elements, such as Fe/Al, Ga/Al, Si/Al, Zr/Al, or La/Al, amongst others[15,16,17], and sometimes even with more elements, *e.g.* Al-Ce-Fe PILCs.[18] Recently, the synthesis of mixed pillars has been also applied to other metals instead of Al, for instance Zr/Ti or Zr/Fe.[19] The inclusion of several species on pillars and the utilization of pillaring technology to different clays have led to the use of novel technologies of synthesis. For example, microwave irradiation (MW) has been used in the different steps of the pillaring process, such as the synthesis of the polycation, the preparation of the clay suspension or even for the drying process.[20] The literature in this area reports the development of PILCs with different polycations, from pure Al to multimetallic ones.[10,21,22] This technology guaranteed short times of preparation, providing ordered materials with different pillar distributions than the conventional synthesis methodology. Nevertheless, special attention must be paid to the synthesis parameters to obtain optimum materials, for instance high irradiation times hurt the structural and textural properties of PILCs.

The relevance of PILCs is evident from a review of the scientific literature, being very extensive in the last years. Some recent review articles and book chapters are included in the bibliography of this chapter.[3,5,13–17,23,24] An important fact for their relevance is that PILCs can be produced on a large scale,[25,26] providing commercial pillared clays with a low price. Besides, many natural clays from different deposits in the world are often evaluated to develop PILCs, thus the natural resources of the country can be valorised yielding to high added value materials.[15b,27,28] The major interest of PILCs has been addressed to application as heterogeneous catalysts,[23a,29] but other applications have been reported in the fields of environmental uses, thermal insulators, pigments, electrodes and membranes.[3,30]

Catalytic applications of pillared clays are mainly related to the nature and distribution of pillars, in addition to their surface properties.[5,24,31] Several approaches have been described to optimise and to improve the catalytic properties of PILCs.[24] The creation of pillars with a multimetallic composition is one of them. The main aim is to incorporate the catalytic active phase inside the pillars, but at the same time the methodology permits the creation of a porous network adequate for the catalytic reaction. Another application is the use of PILCs as supports for catalytic active phases, providing heterogeneous catalysts for different reactions. The number of scientific articles in this field is too extensive to be reviewed here. They include the incorporation of transition metals (Co, Cr, Cu, Fe, Mn, Ni, Pd, Pt, Rh, Au or V) in the pillars, or the creation of complex catalysts including both an active phase in the pillars and one or more active species at surface. Both approaches have mainly been applied to design catalysts, especially directed nowadays towards environmentally friendly processes, such as photocatalysis,[32] Fenton and photo-Fenton,[33] wet air catalytic oxidation (WACO), wet hydrogen peroxide catalytic oxidation (WHPCO),[19] VOCs deep oxidation[34] and DeNOx processes,[35] and also for green chemistry reactions, in which case organic species can be included in the pores of PILCs to enhance their acid character.[17a,24]

Other applications have been also explored. The adsorption capacity common on clays has been extended to PILCs, thus they have been tested as adsorbent materials for the removal of VOCs,[36] organic wastewaters,[37] natural and biogas components, such as carbon dioxide, methane, ethane, and nitrogen,[30a,38] heavy metal ions[39] and even as promising materials for H_2 storage.[40] For these purposes, the nature of the pillars, the pore network and also the acidity control the adsorption properties of PILCs, which can be enhanced by different means. For instance, the adsorption of non-ionic surfactants to Al-PILCs achieved higher adsorption values of chlorinated phenols from aqueous solutions.[37] Al-PILCs have even been used as porous templates to generate nanostructured carbonaceous materials,[41] which resulted in optimum anodes for rechargeable lithium batteries. In conclusion, this section describes the relevance of PILCs as nanoarchitectures based on clays. This extensive research field attracts the interest of researchers, especially regarding their application in environmental processes for removal of pollutants.

4.2.2 Porous Clay Heterostructures (PCHs)

Layered clays have been described as pristine materials for creating functional nanoarchitectures, such as the PILCs described in the previous section, and also the so-called porous clay heterostructures (PCHs).[4,42] The synthetic approach was proposed by Pinnavaia's group based on templated synthesis usually employed to prepare mesoporous silica.[43,44] The method is schematised in Figure 4.3. The initial step was the intercalation of a surfactant (usually long-chain alkylammonium cations) and a co-surfactant (typically a long-chain alkylamine) between the clay layers by an ion-exchange process, organizing micelles in the interlayer space of the clay. Then, the silica source (such as, tetraethoxysilane, TEOS) was incorporated and its *in situ* hydrolysis and polymerisation occurred around the micelles in a controlled way. Thus, the silica formed was templated by the micelle generating silica pillars in a very well ordered pattern. The final removal of the organic templates by calcination or extraction (with a suitable solvent) drove to materials with high surface area and porosity. The recent study of laponite-derived PCHs by FTIR has revealed that this synthetic method yields new SiOH groups of different environments, which are much more accessible than the structural OH groups of the parent laponite.[45] PCHs are characterized by ion-exchange properties, high thermal and hydrothermal stability, significant surface acidity, and a combined micro- and mesoporous structure, with BET surface areas up to 790 $m^2 g^{-1}$, although these properties depend on several factors. For instance, some PCHs have been synthesized with surface area values close to 1100 $m^2 g^{-1}$ when an extraction procedure instead of calcination was employed to remove the organic templates.[46] The nature of the pristine clays is also relevant, with it even being possible to enhance the acid properties by using acid-activated clays as host for PCHs.[47] Of course, these properties, and particularly the porous structure, are also controlled by the surfactant and cosurfactant as well as the synthesis conditions.[48] In this way, treatment of montmorillonite in alkaline medium with aqueous mixtures of cetyltrimethyl ammonium and polyethylene glycol 200 prior to the incorporation of TEOS results in mesoporous materials, probably due to partial exfoliation of the clay.[49]

This approach has been extended to the preparation of PCHs involving atoms other than silicon, which introduces novel synthesis aims. The

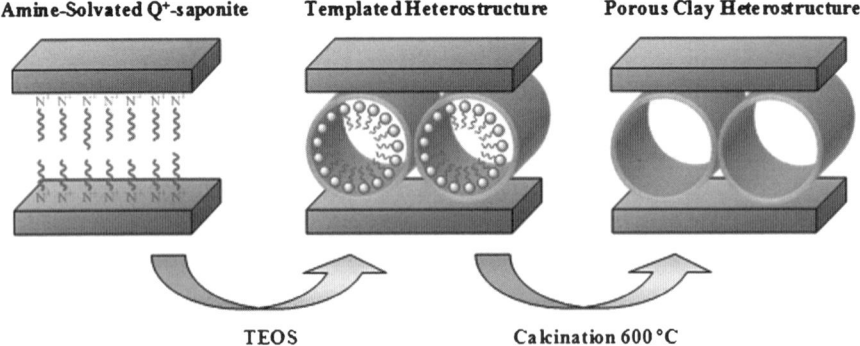

Figure 4.3 Schematic representation of the synthesis of PCHs using the process reported by Pinnavaia's group.[42]

generation of alumina/silica-PCHs has been achieved by Ahenach et al. using a molecular designed dispersion method.[50] The irreversible absorption of the aluminium acetylacetonate complex, followed by its thermal decomposition, yielded the aluminium oxide PCHs. In this way, AlO_x-species became grafted onto the surface of a previously formed PCH, leading to the formation of Si–(OH)–Al bonds by anchoring with surface silanol groups of the silica-PCH. These nanoarchitectures described stable and strong Brönsted acid sites that remained even after thermal treatment at 300 °C. The post-synthesis modification of silica-PCH precursors by aluminium grafting has been also described using other aluminium reagents ($AlCl_3$ or $NaAlO_2$).[51] The properties of the resulting PCHs depend on the choice of the aluminium reagent and subsequently to the synthesis conditions. For instance, the alumination with $AlCl_3$ treatment does not produce changes to the textural properties of the silica-PCHs, resulting in pores in the microporous to small mesoporous region with BET values close to 900 m^2 g^{-1}. ^{27}Al NMR results revealed that aluminium was located in tetrahedral positions within the gallery silica framework, resulting in protonated Al/Si-PCHs with increased acidity compared to the initial silica-PCH. Another way to get Al/Si-PCHs has been reported by Zhou et al., achieving a silica/alumina framework between clay layers by the in situ assembly of TEOS and aluminium isopropoxide precursors (ratio Si/Al equal to 100/1).[52] Although the aluminium was thus incorporated within pillars by Si–O–Al bonds, providing Lewis acid sites, the textural properties were not so improved compared to the other Al-incorporation methods described (BET values ca. 500 m^2 g^{-1}), due to the creation of a dense silica/alumina phase that completely filled the interlayer region of the pristine clay.

Porous clay heterostructures intercalated with silica–titania pillars have been prepared using a mixture of tetraethylorthosilicate (TEOS) and titanium *tetra*-isopropoxide (TIPOTI) in several molar ratios. These PCHs were thermally stable but the calcination process (at 600 °C) resulted in the formation of a disordered pillared structure (delaminated structure) for the Ti-PCHs with higher Ti amounts. The titanium incorporated was present as both hexa- and tetracoordinated Ti^{4+} cations, as well as partially polymerised Ti species, with no evidence for the presence of anatase clusters.[53] The introduction of titanium also resulted in a significant increase of the Brönsted acidity, in which acid centres were located on the surface of the clay layers and also on the silica-titania pillars.[54] These Ti-PCHs showed surface areas up to 600 m^2 g^{-1} and were tested as both catalytic supports and active catalysts for DeNOx reactions. Although these heterostructures were less active than similar Ti-PILCs, the incorporation of Cu or Fe as an active phase yields effective catalysts for high temperature DeNOx processes.[53,55]

An alternative approach to that reported by Pinnavaia's group has been more recently described by Zhu et al. for PCHs synthesis. A mixed sol of silicon and titanium hydroxides was employed to create a pillared bentonite, which after subsequent treatment with quaternary ammonium surfactants gave rise to micelles that act as a template.[56] The surfactant permitted control of the pore structure of the synthesized PCHs, with different mechanisms being involved to control the size of the framework pores.[57] This approach was also employed by the authors to intercalate alumina between laponite clay layers, yielding porous nanoarchitectures consisting of clay layers intercalated with alumina nanoparticles.[58]

The development of PCHs is focused on their functional applications, mainly related to catalysis[43,59] and adsorption,[38,60,61] as occurred with PILCs described in previous section. Recent advances are to develop more efficient and functional materials exploring different type of clays, modifying the synthetic methodology described by Pinnavaia's group and/or using the known PCHs as precursors. For instance, PCHs prepared from several smectites (saponite, montmorillonite or vermiculite) have been modified by ion-exchange to incorporate metal active phases (Fe, Cu, *etc.*) to develop optimum catalysts for DeNOx reactions with ammonia.[35a,59a,62] Other promising research trends are attempts to modify the silica-PCHs by grafting reactions with organosilanes. Aminopropylsilane (APS) has been employed to functionalise PCHs for further immobilization of vanadium and copper complexes, creating effective catalysts for epoxidation reactions.[59b] In the same way, PCHs modified with APS were used for immobilizing chiral Mn(III) complexes to develop heterogeneous catalysts for asymmetric epoxidation of alkenes.[63] PCHs modified with APS can be also used as a substrate for improving adsorption of CO_2, allowing its potential use for carbon dioxide capture as well as for catalytic purposes based on the activation of CO_2.[64] 3-Mercaptopropyltrimethoxysilane (MPTMS) has also been chosen to obtain PCHs provided of thiol functions, used as heavy metal (Cd, Cu, Mn, Ni, Pb and Hg) adsorbents useful for water treatment[65] and modified electrodes.[66]

Recently, PCHs have been also employed as matrices to prepare and tailor mesoporous carbon materials. Using silica-PCHs as matrix and furfuryl alcohol as a carbon precursor, the authors created carbon materials with surface areas up to 1400 $m^2\ g^{-1}$ and two domains of pores with sizes close to 9 and 17 nm; it was important to control the carbonisation temperature and the nature of the amine used to prepare the PCHs.[67] Sucrose has been also studied as a carbon precursor using PCHs as matrixes, although the textural properties were not very enhanced.[68]

4.2.3 Delaminated Porous Clay Heterostructures (DPCHs): Inorganic–Inorganic Nanocomposites

Another new type of nanoarchitecture related to PILCs and PCHs consists of the formation of oxide NPs in between exfoliated clay sheets, which has been named as delaminated porous clay heterostructures (DPCHs).[4] In this way, firstly Letaïef and Ruiz-Hitzky[69] described the formation of such delaminated silica-clay heterostructures using organoclays as intermediates. Their incorporation to polar solvents such as alcohols (*e.g.*, *n*-butanol) allows the expansion of the organoclay and the incorporation of alkoxides (*e.g.*, tetramethoxysilane, TMOS). At this stage, by adding small amounts of water a heterocoagulation (sol–gel transition) of the expanded organoclay suspension is provoked, while the alkoxide is hydrolysed, giving rise to NPs that remain assembled to the clay derivative network (Figure 4.4).[4,69,70,71] The consolidation of NPs-clay nanoarchitectures takes place during a further thermal treatment that produces the removal of the organic matter, the condensation of OH groups and the anchorage of NPs to the clay substrate, leading to porous delaminated heterostructures.

The first studies refer to silica-clay DPCHs that were prepared from two types of 2 : 1 charged phyllosilicates of small particle size (*ca.* 1μm) and present as larger crystals (montmorillonite and vermiculite, respectively).[69,70a] The characteristics of both the sol–gel process and the resulting materials depend on the type of the organoclay as well as on the nature of the silicon alkoxide precursor.[70a] A significant increase in the specific surface area values and porosity are observed in the formation of silica-based DCPHs. For instance those materials prepared from vermiculite showed *ca.* 500 m^2 g^{-1} and total volume pore of 0.43 cm^3 g^{-1}, representing an increase by more than 20 times compared to the pristine vermiculite. Probably the most interesting feature of these silica-based DPCHs is their ability to be modified by grafting reactions using functional organosilanes such as APS. The introduced amino groups are covalently bonded to these nanoarchitectonics, and after protonation confer anion-exchange properties (*ca.* 60 meq per 100 g), which together the CEC of the DCPH afforded by the clay moiety (*ca.* 55 meq per 100 g) gives rise to unique materials with simultaneous cationic and anionic exchange capacity.[69,70a]

The precedent strategy to prepare silica-based DCPHs has also been applied to prepare anatase-based DPCHs.[70b] Using titanium alkoxide precursors, the DPCHs obtained after thermal treatment at 500 °C originates TiO$_2$ NPs stabilized as an anatase phase, whose crystal size is in the order of 10 nm.[70b] The use of mixtures of silicon and titanium alkoxides as precursors results in the formation of DPCHs containing titanium–silicon mixed oxide NPs. One potential application of these last heterostructures is their use in photocatalysis based on the presence of photoactive anatase NPs homogenously distributed in a highly porous matrix. Preliminary studies indicate a high efficiency of these

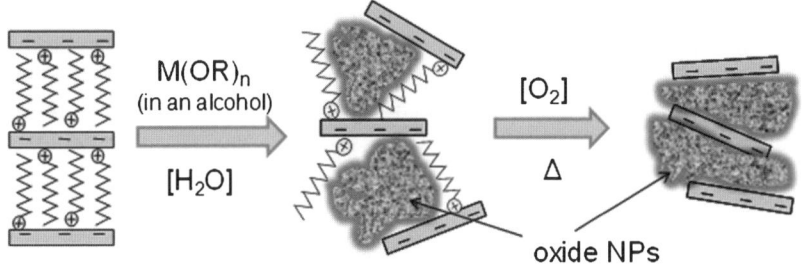

Figure 4.4 General procedure for the preparation of DPCHs introduced by Ruiz-Hitzky's group.[4,69–71]

TiO$_2$ and TiO$_2$–SiO$_2$ DPCHs systems tested for the photodecomposition of 2,4-dichlorophenol, compared to the commercial P25® catalyst, with the additional advantage of being easily recovered from the aqueous dispersions by filtration.[70b]

Other alternatives to develop DPCHs have been developed. For instance, using supercritical CO$_2$ in the treatment of organoclays with alkoxytitanates results in the formation of mesoporous materials.[72] Here, the alkylammonium ions also assist the intercalation of the alkoxide, giving rise to the formation of TiO$_2$ NPs inside the layered structure.

More recently DPCHs based on the incorporation of Al$_2$O$_3$ and SiO$_2$–Al$_2$O$_3$ have been attempted by Belver and co-workers using the same heterocoagulation approach and a commercial organoclay, avoiding the previous organoclay synthesis step.[71] The insertion of pure alumina was not able to carry out the exfoliation of the clay, because of the fast hydrolysis of the aluminium alkoxide that provokes the generation of an oxyhydroxide matrix at the exterior of the silicate particles. Nevertheless, mixed silica–alumina heterostructures with delamination of the clay were successfully prepared. The characterization of these materials revealed the creation of Si–O–Al linkages in the silica/alumina network and even the formation of linkages between this network and the clay layers. The resulting nanoarchitectures had high surface areas (up to 500 m^2 g^{-1}) with pores in the micro and small mesopore region and Brönsted acid sites at surface. These silica–alumina DPCHs exhibited efficient behaviour as catalysts for different green chemistry reactions, such as conversion of limonene to *p*-cymene[71] and production of glycerol-derived chemicals with high added value.[73]

4.2.4 Assembling NPs and Layered Clays

Layer-by-Layer (LbL) techniques[74] have been useful for preparation of nanoarchitectures based on the sequentially assembling of metal-oxide NPs and lamellar clays to form multilayer films. The negative charge of the clays determines the driving force for this assembling, as many colloidal metal-oxide NPs are positively charged (see for instance Ras *et al.*[75]). Contributions in this line are numerous making it difficult to consider all of them. An example is the assembling of ZnO NPs to hectorite, forming ultrathin films of about 400 nm, useful for photocatalytic applications.[76] The preparation of multilayers requires a multistep process, as illustrated in Figure 4.5, in which Zn(OH)$_2$ NPs and hectorite sheets are assembled and further consolidated by thermal treatment at 400 °C for 3 hours to stabilize the resulting ZnO–clay nanoarchitecture.

Another interesting case is the assembling of magnetic NPs to layered clays. For instance, using Pinnavaia's approach for PCHs, Shimizu and co-workers treated Na-4-mica and Na-3-mica synthetic clays with Fe^{3+} silica sol prepared from TEOS and iron nitrate.[77] After calcination at 500 °C, the resulting nanoarchitectonic materials show a relatively high specific surface area

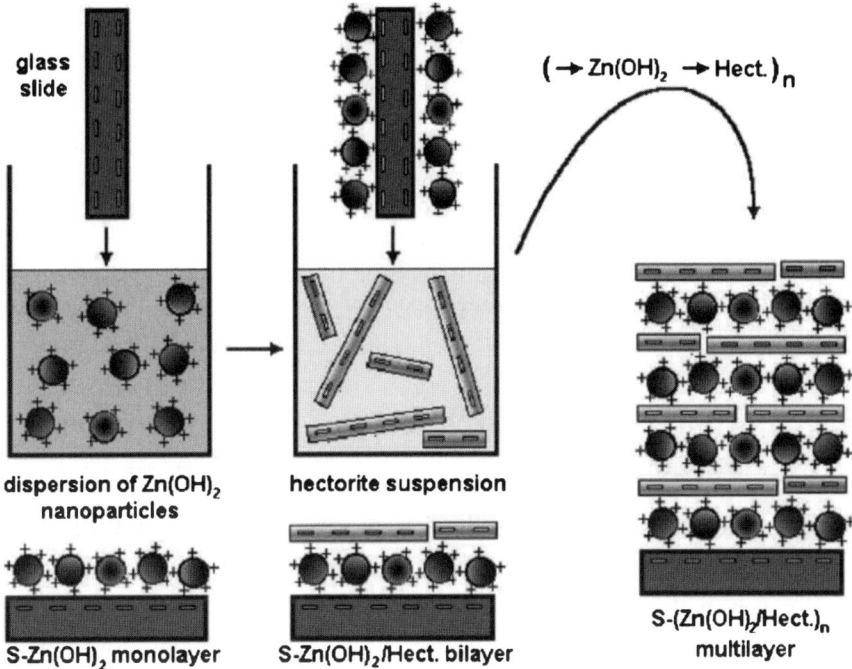

Figure 4.5 Scheme of the self-assembly preparation procedure for ZnO NPs-clay multilayer films, after Szabó et al.[76] Reproduced with permission from Elsevier.

(57–130 m^2 g^{-1}) due to the "house-of-cards" structural arrangement adopted by the clay derivative.

Attempts to assembly iron oxide NPs to clays are an interesting subject due to the potential utility for adsorption/magnetic separation. A contribution by Son et al.[78] shows the formation of α-Fe$_2$O$_3$/clay heterostructures from Fe-polycations, where the iron oxide NPs are intercalated in the silicate, increasing their saturated magnetization per Fe atom with nearly zero coercitive field. Petridis et al.[79] prepared iron oxide/laponite heterostructures containing superparamagnetic γ-Fe$_2$O$_3$ NPs that are highly porous and possess significantly higher surface area and saturation magnetization. In this case, the assembling method implies the *in situ* formation of iron oxide NPs from hydrated FeCl$_3$ in an alkaline medium. Alternatively, Fe$_3$O$_4$ NPs have been assembled to kaolinite clay mineral (a non-charged 1 : 1 aluminosilicate)^2 using a solvothermal method, and further functionalised with APS for immobilization of glucoamylase enzyme.[80] Ruiz-Hitzky et al. reported an innovative procedure to develop advanced multifunctional materials based on the assembly of magnetite NPs present in a ferrofluid to diverse porous solids, including different types of clays, which confers superparamagnetic properties to the final system allowing diverse advanced applications.[81] This procedure, applied to montmorillonite, provides solids with superparamagnetic behaviour while maintaining its ion-exchange capacity and adsorption ability,[82] and therefore these nanoarchitectures can be used to remove heavy and radioactive cations as well as organic pollutants.

Semiconducting NPs assembled to montmorillonite have been recently prepared using clay exchanged by Zn^{2+} and Pb^{2+} in water dispersions purged with a H_2S flux.[83] The authors of this work claim that nanosized clusters are formed in the interlayer space of the montmorillonite, although some of them remain agglomerated at the external surface of the clay according to SEM results. This strategy opens the way to develop functional nanostructures by *in situ* formation of NPs in the confined space of layered clays with optical and electronic transport properties.

4.3 Bottom-up Heteroarchitectures based on Fibrous and Tubular Clays

4.3.1 Heterostructures Based on Sepiolite and Palygorskite

Sepiolite and palygorskite are excellent candidates for nanoarchitectonic development with different types of NPs (metal and metal-oxides, hydroxides and oxyhydroxides, silica, nanosilicates, *etc.*)[4,84] Conventional strategies of synthesis of such heterostructures are mainly based on co-precipitation and impregnation processes, using salts as precursors that were further treated to produce oxidation or reduction. Other methods are based on the grafting of functional organosilanes, assembly of metal complexes ions *via* ion-exchange reactions, incorporation of metal or oxide nanoparticles in acid-treated sepiolite, and template synthesis in the presence of organoclays.

Sepiolite–metal oxide NPs heterostructures prepared by precipitation of metal salts as hydroxides and oxyhydroxides and further thermally treated to form the corresponding oxides or reduced to obtain metal nanoparticles, have been applied, for instance, to form NiO_x NPs.[85] These last NPs remain homogeneously distributed on the large surface area afforded by the sepiolite, therefore these materials are mainly applied in the heterogeneous catalysis field.[85,86,87] Iron cations absorbed on the palygorskite surface and treated with ammonia to precipitate Fe_3O_4 NPs give rise to heterostructures that show magnetic properties of interest in magnetic separation technologies.[88]

Monodispersed metallic NPs have been synthesized following a wet chemical route using aqueous suspension of acid treated sepiolite to which salts of transition-elements (Fe, Ni, Co, Cu, Ag, *etc.*) are added for precipitation of the corresponding oxyhydroxydes metal precursors.[89] After consolidation by thermal treatment under reducing conditions, NPs appear disperse and their size distributions are remarkably narrow, with an average particle size ranging from 3 to 6 nm[84]. Also, oxide phases, such as Fe_2O_3 (hematite), can be assembled to sepiolite fibers by the abovementioned procedure, the resulting heterostructures being used as active phases of magnetic sensors for relative humidity measurements.[89b]

Heterostructures based on sepiolite can also be prepared by controlled hydrolysis of metal alkoxides in the presence of sepiolite modified with cetyltrimethylammonium bromide (CTAB) and other surfactants (organo-sepiolite). In this way, silica,[90] titania, silica–titania[91] and silica–alumina[92] NPs have been formed *in situ* on sepiolite following a methodology related to the colloidal route reported above for the preparation of DPCHs based on layered clays (Figure 4.6). In the case of silicon alkoxides used as precursors (*e.g.*, TMOS), heterostructures are obtained in which silica NPs remain anchored to sepiolite microfibres through covalent bonds, giving rise to materials with elevated specific surface area (*ca.* 350 m^2g^{-1}). Using titanium alkoxides as precursors, it is possible to prepare titania and titania–silica/sepiolite heterostructures, both useful as efficient photocatalysts of interest for the removal of pollutants in water. In the adopted procedure, the anatase phase, which is the mainly responsible for the observed photoactivity, is stabilized by S-doping, incorporating thiourea during the synthesis. In this nanoarchitecture, the formation of Si-O-Ti bonds has been postulated,[91] assuring the stability of the microparticulated silicate and the anatase NPs. By replacing sepiolite by palygorskite, the formation of highly homogeneous distribution of titania NPs on the silicate surface are also observed.[93] Similar procedures avoiding the use of the surfactant modified clay result in TiO_2–sepiolite where the NPs appear as aggregates with an inhomogeneous distribution in the

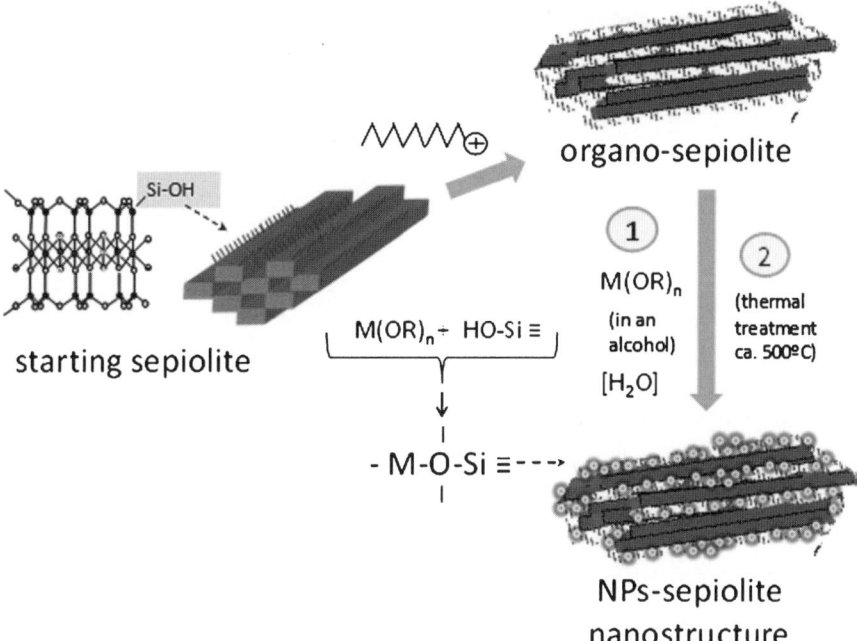

Figure 4.6 General procedure for assembling metal-oxide NPs to sepiolite *via* the colloidal route firstly described in ref. 91.

resulting material.[94] Similarly, the assembly of palygorskite and SnO_2–TiO_2 NPs has been reported as a very active nanoarchitecture for photocatalysis.[95] Also recently, by using Zr-alkoxides as precursors, Letaïef *et al.* prepared ZrO_2-sepiolite, useful as catalysts for the Knoevenagel condensation reactions.[96]

Using sepiolites modified by grafting of functional organic groups by reaction with organosilanes is an alternative for the anchorage of NPs of diverse natures. The first contribution refers to the immobilization of OsO_4 clusters on a vinyl-derivative of sepiolite,[97] which is useful as a photocatalyst for water splitting.[98] Recently, Au[99] and Keggin phosphotungsten NPs[100] have been anchored to sepiolite and palygorskite, respectively, previously modified by grafting of organo-silanes containing amino functions.

Nanoarchitectures based on fibrous clays and more complex inorganic solids, such as layered double hydroxides (LDHs),[81,90b] hydroxyapatite (HA),[101] zeolites[90b,102,103] and ferrites[104] have been recently reported. The

Figure 4.7 FE-SEM image of a nanoarchitecture based on zeolite ZSM-5 nanocrystals assembled to sepiolite fibers, prepared by the method described in ref. 90b.

adopted experimental conditions are decisive to accomplish the assembly between the involved micro/nano-particulated inorganic entities that have to be formed *in situ* in the presence of the fibrous clay. This is the case for zeolite–sepiolite architectures that can be formed as big zeolite crystals, where the clay fibers are assembled to them,[90b,103] or as nano-sized zeolites that remain assembled as monodispersed NPs on the sepiolite surface (Figure 4.7).[90b,102]

Another strategy for nanoarchitectonics based on fibrous clays and inorganic solids is the assembly of the silicate, mainly through its surface silanol groups, to already formed NPs. A recent example refers to the assembly of Fe_3O_4 NPs from a non-aqueous ferrofluid to the surface of sepiolite, palygorskite and other clay minerals.[82] The resulting heterostructures preserved both the superparamagnetic properties of the magnetite nanoparticles and the adsorption properties of the clay, being of interest for a large number of applications, including as adsorbent, for removal of pollutants in wastewaters or as a magnetic nanofiller of polymers.[81] More complex architectures based on NPs assembled to sepiolite consist of the production of scaffolds using directional freeze-drying techniques, as in the case of TiO_2 P25 NPs anchored to sepiolite in the presence of TEOS, useful as a photocatalyst.[105]

4.3.2 Heterostructures Based on Halloysite and Imogolite

Halloysite and imogolite are aluminosilicates of tubular morphologies, the former being regarded as a hydrated kaolinite and the latter related to allophane.[2] In halloysite the presence of water molecules curve and determine the formation of multilayer tubes with inner diameters of 10–150 nm, in which the outer surface of the nanotube is the SiO_2 sheet and inner cylinder core is the Al_2O_3 sheet.[106] Imogolite presents an opposite arrangement with the SiO_2 sheet at the interior, creating nanotubes of 0.65 nm inner diameter.[107] In both cases, it is possible to develop strategies for nanoarchitectonics based on the assembly of NPs on the external surface of the nanotubes as well as the interior, however just a few examples are reported in the literature.

In general, NPs have been assembled at the exterior of halloysite because of the negative surface potential provided by the silica sheet. In this way, Ag NPs have been assembled both by precipitation from an $AgNO_3$ solution added dropwise to a suspension of halloysite pre-treated with mercaptoacetic acid,[108] or by using LbL techniques and Ag NPs capped with a cationic polymer.[94a] An illustrative example of nanoarchitectonics based on halloysite in which NPs are incorporated on to both the external surface and within the tubular cavity has been reported by Fu and Zhang.[109] These authors used an electroless deposition technique for the assembly of Ni NPs at the external surface but also at the inner cavity. The incorporation of Ni at the interior of the nanotubes requires the elimination of the water by a freeze-drying treatment before use of a plating bath in order to grow Ni nanowires in the cavity. More recently, Papoulis and co-workers have prepared TiO_2-halloysite heterostructures where TiO_2 NPs in the range 10–30 nm are formed *in situ* at the exterior of the nanotube and also inside the cavity.[94a] Although no special driving force to incorporate the TiO_2 NPs precursors has been proposed, the decrease of average pore size confirms that NPs are also at least partially incorporated inside the nanotubes, which is supported by TEM.

In the case of imogolite the reactive silica sheet is at the interior of the narrow nanotube, making the access of reagents for the NPs assembly difficult. One of the few reported examples refers to the immobilization of OsO_4 NPs to imogolite modified by reaction with APS.[110] The incorporation of amine functions in the inner part of the nanotubes tends to stabilize and improve activity and selectivity of the catalyst, although the resulting material loses activity after the first cycle in the dehydroxylation of olefins. Another possibility to built nanoarchitectures from imogolite profits from the fact that the external surface shows a positive potential. In this way, imogolite nanotubes have been intercalated between the negatively charged layers of montmorillonite.[111] The resulting imogolite–smectite intercalated materials are porous heterostructures in which imogolite acts as pillar, showing interesting properties for selective absorption and catalysis.

4.4 Nanoarchitectonics Based on Related Inorganic Solids

Diverse alkaline layered silicates have been modified in a similar way to layered clays, giving rise to analogous nanoarchitectures. Pinnavaia described some time ago[112] the preparation of silica-pillared magadiite by the same approach described above for PCHs (described in section 4.2.2). More recently, the silica pillaring of magadiite and kenyaite has been reported by Park et al.,[113,114] who prepared a dodecylamine/TEOS co-intercalated silicate gel that acts as precursor for further hydrolysis of TEOS, giving rise to silica pillars. The physical properties of the resulting materials were similar to those described by MCM-41, with a regular gallery height of 3.0–3.4 nm, a narrow pore size of 2.5–3.0 nm, and large specific surface areas of 877 and 1029 $m^2\,g^{-1}$. Ni and Pd were successfully dispersed in the gallery of these materials, resulting in optimum catalysts for the partial oxidation of methane (POM).[114] The incorporation of silica pillars between magadiite layers was also described by Kwon et al.[115] by simultaneous intercalation of an amine and TEOS. The amine expands the silicate, acting simultaneously as catalyst and template for the hydrolysis and condensation of TEOS precursor. The silica-pillared magadiites showed lower surface areas than the previous methodology, 500–770 $m^2\,g^{-1}$, but the synthesis is here simplified.

Other nanoarchitectures based on magadiite have been also prepared in other ways. For instance, aluminium-pillared magadiite was prepared by intercalation of aluminium Keggin ions, although the magadiite was a previously modified hexylamine-intercalated magadiite.[116] The presence of alumina-pillars, which was confirmed by ^{27}Al NMR, is responsible for the creation of strong Al–OH acid sites. Other oxides and metal NPs have been also intercalated between magadiite layers. Ozawa et al. successfully intercalated 4.5 nm ZnO NPs by ion-exchange, which held up the plate-like structure of the magadiite after thermal treatment at 600 °C, although the amount of ZnO was quite low.[117] The same approach was applied to intercalate Ag NPs (3–5 nm) created by heating the [Ag(NO$_3$)$_2$]$^+$ precursor at 120 °C, whereas heterogeneous enlargement of these NPs was detected above 180 °C.[118] Ozawa *et al.* have even recently described the synthesis of heterostructures based on magadiite and Si-ZSM-11. The one-step hydrothermal synthesis obtained crystalline magadiite/Si-ZSM-11 composites, the properties of which can be controlled by using surfactants as templates.[119]

Layered titanates can be also considered as clay-related inorganic solids. The literature related to nanoarchitectures based on titanates is quite large and some examples are collected in Table 1. Among them, layered titanates have been pillared by several synthetic approaches incorporating diverse metal-oxides (TiO$_2$, ZnO, SnO$_2$, and Fe$_2$O$_3$). These pillared titanates were successfully prepared through an exfoliation–restacking route.[120,121] The mixture between the colloidal suspension of an exfoliated titanate and a MO$_x$ aqueous sol resulted in a restacking of the layers, entrapping oxide NPs. This approach permits the development of, for instance, mesoporous ZnO-pillared titanates showing interesting photocatalytic properties.[120b] Similar results have been recently reported for Fe$_2$O$_3$-pillared titanates as visible-light-driven photocatalysts.[120c] Other examples related to TiO$_2$-pillared titanates with photocatalytic applications are collected in Table 4.1, although their activity is limited when using visible light. The creation of Al$_2$O$_3$ and SiO$_2$ pillars in layered titanates has been also described some time ago, mainly using similar approaches to those abovementioned for clays, *i.e.*, performing the pre-intercalation of an amine that acts as precursor for the ulterior intercalation of the alumina or silica source.[122]

Nanoarchitectures based on layered titanates can also be prepared by their assembling to cationic species. For instance, [CeO]$^{2+}$ ions have been successfully intercalated by ion-exchange and further thermally treated to yield optimum photocatalysts.[132] Cu^{2+} ions have been chosen to create Cu-titanate films for amperometric sensors.[133a] Zr-complexes have also been

Table 4.1 Nanoarchitectures created by modification of layered titanates.

Layered titanate modification	Nanoarchitecture	References
Pillaring	ZnO-pillar/Fe-titanate	120b,123
	SnO$_2$-pillar/titanate	120a
	Fe$_2$O$_3$-pillar/titanate	120c,124
	TiO$_2$-pillar/titanate	121,125,126
	SiO$_2$-pillar/titanate	122b,127,128
	Al$_2$O$_3$-pillar/titanate	122,129,130,131
Intercalation	[CeO]$^{2+}$/titanate	132
	Cu^{2+}/titanate	133
	Zr^{4+} ions /titanate	134

intercalated by an exfoliation–restacking approach, giving rise to mesoporous nanohybrids with functionalities for CO_2 gas adsorption and for photocatalytic applications.[134]

The strategies used to development hybrid materials based on clays have been frequently applied to layered double hydroxides (LDHs) because they show a layered structure and a chemistry closely related to that of smectite clay minerals, but with anion exchange properties. In this way, the preparation of pillared structures based on LDHs implies the intercalation metal polyoxoanions (e.g., $V_{10}O_{28}^{6-}$, $\alpha\text{-}[H_2W_{12}O_{40}]^{6-}$, $SiW_9V_3O_{40}^{7-}$), in place of polyoxycations, typically in Mg–Al and Zn–Al LDHs (see for instance examples in Table 13.1.5 of ref. 135). Recent strategies aim to use environmentally friendly approaches and reagents, such as Ti peroxide, that avoid the use of organic or chlorine-containing hazards in the pillaring process and allow the preparation of high active oxidation catalysts.[136] Other nanoarchitectures based on LDHs that use synthetic strategies introduced in this Chapter refer to their assembly to zeolite Y,[137] or to carbon nanofibres for developing base catalysts useful, for instance, in the synthesis of methyl-isobutyl ketone (MIBK).[138]

Other related solids are layered phosphates. Intercalation of Cr(III) polyhydroxyacetate into $\alpha\text{-}Zr(HPO_4)_2$ allows the formation of pillars, resulting in a significant increase of the surface area (from 1 $m^2\ g^{-1}$ in the starting phosphate to 350 $m^2\ g^{-1}$ in the pillared structure).[139] On the other hand, treatment of γ-zirconium phosphate with water–acetone mixtures that contain tetrapropylammonium hydroxide allows the phosphate exfoliation and the incorporation of Pt/Pd NPs, creating in this case materials with surface area values close to 600 $m^2\ g^{-1}$.[140]

4.5 Concluding Remarks

As clays are particulate solids that offer diverse topologies, such as layers, fibers and tubules, many of them at the nanometre size, they are an attractive target for being assembled with many types of NPs. Therefore, nanoarchitectonics based on the assembling of clays and related inorganic solids is a research area that offers a large number of possibilities to develop new functional porous materials for adsorption, catalytic purposes, sensor devices and other diverse applications. In fact, many opportunities related to both the preparation of new heterostructures with functional properties and the exploration of applications are still open.

Acknowledgments

The authors acknowledge the financial support from CICYT (Spain; project MAT2009-09960). C.B. is indebted to the MICINN for a Ramon y Cajal postdoctoral contract.

References

1. http://en.wikipedia.org/wiki/Nanoarchitectonics, accesed on May 25, 2012.
2. *Handbook of Clay Science*, ed. F. Bergaya, B. K. G. Theng and G. Lagaly, Elsevier Science Ltd., Amsterdam, 2006.
3. J. T. Kloprogge, *J. Porous Mater.*, 1998, **5**, 5.
4. P. Aranda, C. Belver and E. Ruiz-Hitzky, in *Clays and Materials*, ed. P. Aranda, M. Ogawa and L. F. Drummy, The Clay Minerals Society vol. 18, Chantilly (VA), USA, in press.
5. A. Gil, S. A. Korili and M. A. Vicente, *Catal. Rev.*, 2008, **50**, 153.
6. R. M. Barrer and D. M. MacLeod, *Trans. Faraday Soc.*, 1955, **51**, 1290.
7. G. M. Brindley and R. E. Sempels, *Clay Miner.*, 1977, **12**, 229.
8. N. Lahav, U. Shani and J. Shabtai, *Clays Clay Miner.*, 1978, **26**, 107.
9. (a) D. E. W. Vaughan, R. J. Lussier and J. S. Magee, *U. S. Pat.* 4176090, 1979; (b) D. E. W. Vaughan, R. J. Lussier and J. S. Magee, *U. S. Pat.* 4271043, 1981.
10. A. M. De Andrés, J. Merino, J. C. Galván and E. Ruiz-Hitzky, *Mater. Res. Bull.*, 1999, **34**, 641.
11. (a) D. Plee, F. Borg, L. Gatineau and J. J. Fripiat, *J. Am. Chem. Soc.*, 1985, **107**, 2362; (b) J. F. Lambert, S. Chevalier, R. Franck, H. Suquet and D. Barthomeuf, *J. Chem. Soc. Faraday Trans.*, 1994, **90**, 675.
12. J. C. Galván, A. Jiménez-Morales, R. Jiménez, J. Merino, A. Villanueva, M. Crespin, P. Aranda and E. Ruiz-Hitzky, *Chem. Mater.*, 1998, **10**, 3379.
13. (a) T. J. Pinnavaia, *Science*, 1983, **220**, 365; (b) F. Figueras, *Catal. Rev.*, 1988, **30**, 457.
14. R. Burch, *Catal. Today*, 1988, **2**, 185.
15. (a) A. Gil, L. M. Gandía and M. A. Vicente, *Catal. Rev. Sci. Eng.*, 2000, **42**, 145; (b) S. Letaïef, B. Casal, P. Aranda, M. A. Martín-Luengo and E. Ruiz-Hitky, *Appl. Clay Sci.*, 2003, **22**, 263.
16. F. Bergaya, A. Aouad and T. Mandalia, in *Handbook of Clay Science. Developments in Clay Science.*, ed. F. Bergaya, B. K. G. Theng and G. Lagaly, Elsevier Science, Amsterdam, 2006, vol. 1, p. 393.
17. (a) M. A. Vicente, C. Belver, M. Sychev, R. Prihodkoc and A. Gil, *Ind. Eng. Chem. Res.*, 2009, **48**, 406; (b) C. Belver, M. A. Bañares-Muñoz and M. A. Vicente, *Appl. Catal., B*, 2004, **50**, 101.
18. J. Carriazo, E. Guelou, J. Barrault, J. M. Tatibouet, R. Molina and S. Moreno, *Water Res.*, 2005, **39**, 3891.
19. C. B. Molina, J. A. Casas, J. A. Zazo and J. J. Rodríguez, *Chem. Eng. J.*, 2006, **118**, 29.
20. G. Fetter and P. Bosch, in *Pillared Clays and Related Catalysts*, ed. S. K. A. Gil, R. Trujillano and M.A. Vicente, Springer, New York, 2010, p. 1.
21. F. J. Berry, K. K. Rao and G. Oates, *Hyperfine Interact.*, 1994, **83**, 343.
22. (a) G. Fetter, G. Heredia, A. M. Maubert and P. Bosch, *J. Mater. Chem.*, 1996, **6**, 1857; (b) G. Fetter, V. Hernández, V. Rodríguez, M. A. Valenzuela, V. H. Lara and P. Bosch, *Mater. Lett.*, 2003, **57**, 1220.

23. (a) Z. Ding, J. T. Kloprogge, R. L. Frost, G. Q. Lu and H. Y. Zhu, *J. Porous Mater.*, 2001, **8**, 273; (b) A. Gil, M. A. Vicente and L. M. Gandía, *Microporous Mesoporous Mater.*, 2000, **34**, 115.
24. *Pillared Clays and Related Catalysts*, ed. A. Gil, S. A. Korili, R. Trujillano and M. A. Vicente, Springer, New York, 2010.
25. A. Aouad, T. Mandalia and F. Bergaya, *Appl. Clay Sci.*, 2005, **28**, 175.
26. W. Jones, G. Poncelet, E. Ruiz-Hitzky, J. C. Galván, P. Pomonis, H. Van Damme, F. Bergaya, N. Papayanakos and N. Gangas, Synthesis Report for publication. Date 26-10-97. Contract No. BR2-CT-94-0629 EU BRITE/EURAM-II Project No. 8211, 1997.
27. L. S. Belaroui, J. M. M. Millet and A. Bengueddach, *Catal. Today*, 2004, **89**, 279.
28. (a) B. Casal, J. Merino, E. Ruiz-Hitzky, E. Gutierrez and A. Alvarez, *Clay Miner.*, 1997, **32**, 41; (b) S. Letaïef, B. Casal, N. Kbir-Ariguib, M. Trabelsi-Ayadi and E. Ruiz-Hitzky, *Clay Miner.*, 2002, **37**, 517.
29. (a) S. Cheng, *Catal. Today*, 1999, **49**, 303; (b) G. Centi and S. Perathoner, *Microporous Mesoporous Mater.*, 2008, **107**, 3.
30. (a) J. Pires and M. L. Pinto, in *Pillared Clays and Related Catalysts*, ed. A. Gil, S. A. Korili, R. Trujillano and M. A. Vicente, Springer, New York, 2010; (b) K. G. Bhattacharyya and S. S. Gupta, *Adv. Colloid Interface Sci.*, 2008, **140**, 114.
31. A. Vaccari, *Appl. Clay Sci.*, 1999, **14**, 161.
32. M. Lim, Y. Zhou, B. Wood, L. Z. Wang, V. Rudolph and G. Q. Lu, *Environ. Sci. Technol.*, 2009, **43**, 538.
33. (a) Q. Chen, P. Wu, Y. Li, N. Zhu and Z. Dang, *J. Hazard. Mater.*, 2009, **168**, 901; (b) E. G. Garrido-Ramírez, B. K. G. Theng and M. L. Mora, *Appl. Clay Sci.*, 2010, **47**, 182.
34. (a) S. F. Zuo, Q. Q. Huang, J. Li and R. X. Zhou, *Appl. Catal., B*, 2009, **91**, 204; (b) Q. G. Huang, S. F. Zuo and R. X. Zhou, *Appl. Catal., B*, 2010, **95**, 3.
35. (a) C. Belver, in *Pillared Clays and Related Catalysts*, ed. A. Gil, S. A. Korili, R. Trujillano and M. A. Vicente, Springer, New York, 2010, p. 255; (b) C. Belver, M. A. Vicente, A. Martínez-Arias and M. Fernández-García, *Appl. Catal., B*, 2004, **50**, 227.
36. (a) C. Volzone, *Appl. Clay Sci.*, 2007, **36**, 191; (b) J. Pires, A. Carvalho and M. B. de Carvalho, *Microporous Mesoporous Mater.*, 2001, **43**, 277.
37. L. J. Michot, O. Barrds, E. L. Hegg and T. J. Pinnavaia, *Langmuir*, 1993, **9**, 1794.
38. M. L. Pinto, J. Pires and J. Rocha, *J. Phys. Chem. C*, 2008, **112**, 14394.
39. T. S. Anirudhan, C. D. Bringle and S. Rijith, *J. Environ. Radioact.*, 2010, **101**, 267.
40. A. Gil, R. Trujillano, M. A. Vicente and S. A. Korili, *Int. J. Hydrogen Energy*, 2009, **34**, 8611.
41. J. Sandí, *J. New Mater. Electrochem. Syst.*, 2001, **4**, 259.
42. A. Galarneau, Barodawalla A., Pinnavaia, T. J., *Nature*, 1995, **374**, 529.

43. A. Galarneau, A. Barodawalla and T. J. Pinnavaia, *Chem. Commun.*, 1997, 1661.
44. M. Polverejan, T. R. Pauly and T. J. Pinnavaia, *Chem. Mater.*, 2000, **12**, 2698.
45. H. Pálková, J. Madejová, M. Zimowska and E. M. Serwicka, *Microporous Mesoporous Mater.*, 2009, **127**, 237.
46. (a) M. Benjelloun, P. Cool, T. Linssen and E. F. Vansant, *Microporous Mesoporous Mater.*, 2001, **49**, 83; (b) M. Benjelloun, P. Cool, P. Van Der Voort and E. F. Vansant, *Phys. Chem. Chem. Phys.*, 2002, **4**, 2818.
47. M. Pichowicz and R. Mokaya, *Chem. Commun.*, 2001, 2100.
48. H. Pálková, J. Madejová, M. Zimowska, E. Bielanska, Z. Olejniczak, L. Litynska-Dobrzynska and E. M. Serwicka, *Microporous Mesoporous Mater.*, 2009, **127**, 228.
49. F. Li, Y. Jiang, M. Xia, M. Sun, B. Xue and X. Ren, *J. Porous Mater.*, 2010, **17**, 217.
50. (a) J. Ahenach, P. Cool and E. F. Vansant, *Phys. Chem. Chem. Phys.*, 2000, **2**, 5750; (b) J. Ahenach, E. F. Vansant and P. Cool, *Proceedings of the Second Pacific Basin Conference on Adsorption Science and Technology*, Brisbane, Australia, 2000.
51. M. Polverejan, Y. Liu and T. J. Pinnavaia, *Chem. Mater.*, 2002, **14**, 2283.
52. C. Zhou, X. Li, Z. Ge, Q. Li and D. Tong, *Catal. Today*, 2004, **93–95**, 607.
53. L. Chmielarz, Z. Piwowarska, P. Kustrowski, B. Gil, A. Adamski, B. Dudek and M. Michalik, *Appl. Catal., B*, 2009, **91**, 449.
54. L. Chmielarz, B. Gil, P. Kustrowski, Z. Piwowarska, B. Dudek and M. Michalik, *J. Solid State Chem.*, 2009, **182**, 1094.
55. L. Chmielarz, Z. Piwowarska, P. Kustrowski, A. Wegrzyn, B. Gil, A. Kowalczyk, B. Dudek, R. Dziembaj and M. Michalik, *Appl. Clay Sci.*, 2011, In Press.
56. H. Y. Zhu, Z. Ding and J. C. Barry, *J. Phys. Chem. B*, 2002, **106**, 11420.
57. H. Y. Zhu, Z. Ding, C. Q. Lu and G. Q. Lu, *Appl. Clay Sci.*, 2002, **20**, 165.
58. H. Y. Zhu and G. Q. Lu, *Langmuir*, 2001, **17**, 588.
59. (a) L. Chmielarz, P. Kustrowski, R. Dziembaj, P. Cool and E. F. Vansant, *Catal. Today*, 2007, **119**, 181; (b) C. Pereira, K. Biernacki, S. L. H. Rebelo, A. L. Magalhães, A. P. Carvalho, J. Pires and C. Freire, *J. Mol. Catal. A: Chem.*, 2009, **312**, 53.
60. L. Mercier and T. J. Pinnavaia, *Microporous Mesoporous Mater.*, 1998, **20**, 101.
61. (a) F. Qu, L. Zhu and K. Yang, *J. Hazard. Mater.*, 2009, **170**, 7; (b) S. Arellano-Cárdenas, T. Gallardo-Velázquez, G. Osorio-Revilla and M. D. S. López-Cortez, *Water Environ. Res*, 2008, **80**, 60.
62. L. Chmielarz, P. Kustrowski, Z. Piwowarska, B. Dudek, B. Gil and M. Michalik, *Appl. Catal., B*, 2009, **88**, 331.
63. (a) I. Kuźniarska-Biernacka, A. R. Silva, A. P. Carvalho, J. Pires and C. Freire, *Catal. Lett.*, 2010, **134**, 63; (b) I. Kuźniarska-Biernacka, C. Pereira, A. P. Carvalho, J. Pires and C. Freire, *Appl. Clay Sci.*, 2011, **53**, 195.

64. M. L. Pinto, L. Mafra, J. M. Guil, J. Pires and J. Rocha, *Chem. Mater.*, 2011, **23**, 1387.
65. R. Tassanapayak, R. Magaraphan and H. Manuspiya, *Adv. Mater. Res.*, 2008, **55-57**, 617.
66. A. J. Tchinda, E. Ngameni, I. T. Kenfack and A. Walkarius, *Chem. Mater.*, 2009, **21**, 4111.
67. C. Santos, M. Andrade, A. L. Vieira, A. Martins, J. Pires, C. Freire and A. P. Carvalho, *Carbon*, 2010, **48**, 4049.
68. D. Nguyen-Thanh and T. J. Bandosz, *Microporous Mesoporous Mater.*, 2006, **92**, 47.
69. S. Letaief and E. Ruiz-Hitzky, *Chem. Commun.*, 2003, 2996.
70. (a) S. Letaïef, M. A. Martín-Luengo, P. Aranda and E. Ruiz-Hitzky, *Adv. Funct. Mater.*, 2006, **16**, 401; (b) E. Manova, P. Aranda, M. A. Martín-Luengo, S. Letaief and E. Ruiz-Hitzky, *Microporous Mesoporous Mater.*, 2010, **131**, 252.
71. C. Belver, P. Aranda, M. A. Martín-Luengo and E. Ruiz-Hitzky, *Microporous Mesoporous Mater.*, 2012, **147**, 157.
72. S. Yoda, Y. Sakurai, A. Endo, T. Miyata, K. Otake, H. Yanagishita and T. Tsuchiya, *Chem. Commun.*, 2002, 1526.
73. C. Belver, J. Esteban, M. Ladero, P. Aranda and E. Ruiz-Hitzky, Proceedings of EuropaCat X Conference, 28 August – 2 September, 2011, Glasgow, U.K., (pTh. 121).
74. R. K. Iler, *J. Colloid Interface Sci.*, 1966, **21**, 569.
75. R. H. A. Ras, J. Németh, C. T. Johnston, E. DiMasi, I. Dékány and R. A. Schoonheydt, *Phys. Chem. Chem. Phys.*, 2004, **6**, 4174.
76. T. Szabó, J. Németh and I. Dékány, *Colloids Surf. Physicochem. Eng. Aspects*, 2004, **230**, 23.
77. K.-I. Shimizu, Y. Nakamuro, R. Yamanaka, T. Hatamachi and T. Kodama, *Microporous Mesoporous Mater.*, 2006, **95**, 135.
78. Y.-H. Son, J.-K. Lee, Y. Soong, D. Martello and M. Chyu, *Chem. Mater.*, 2010, **22**, 2226.
79. T. Szabó, A. Bakandritsos, V. Tzitzios, S. Papp, L. Kőrösi, G. Galbács, K. Musabekov, D. Bolatova, D. Petridis and I. Dékány, *Nanotechnology*, 2007, **18**, 285602.
80. G. Zhao, J. Wang, Y. Li, X. Chen and Y. Liu, *J. Phys. Chem. C*, 2011, **115**, 6350.
81. E. Ruiz-Hitzky, P. Aranda and Y. González-Alfaro, *Spain Pat.* 201030333, 2010.
82. Y. González-Alfaro, P. Aranda, F. M. Fernandes, B. Wicklein, M. Darder and E. Ruiz-Hitzky, *Adv. Mater.*, 2011, **23**, 5224.
83. L. Jankovic, K. Dimos, J. Bujdák, I. Koutselas, J. Madejová, D. Gournis, M. A. Karakassides and P. Komadel, *Phys. Chem. Chem. Phys.*, 2010, **12**, 14236.
84. E. Ruiz-Hitzky, P. Aranda, A. Alvarez, J. Santaren and A. Esteban-Cubillo, in *Developments in Palygorskite-Sepiolite Research. A New*

Outlook on these Nanomaterials, ed. E. Galán and A. Singer, Elsevier, Oxford, 2011, p. 393.
85. R. Salvador, B. Casal, M. Yates, M. A. Martín-Luengo and E. Ruiz-Hitzky, Appl. Clay Sci., 2002, 22, 103.
86. (a) X.-L. Zhai, M.-Y. Jia and Y.-S. Shen, Chin. J. Chem., 2005, 23, 557; (b) N. Güngör, S. Işçi, E. Günister, W. Miśta, H. Teterycz and R. Klimkiewicz, Appl. Clay Sci., 2006, 32, 291.
87. (a) M. R. Sun Kou, S. Mendioroz, J. L. García-Fierro, I. Rodríguez-Ramos, J. M. Palacios, A. Guerrero-Ruiz and A. M. de Andrés, Clays Clay Miner., 1992, 40, 167; (b) K.-I. Shimizu, R. Maruyama, S.-I. Komai, T. Kodama and Y. Kitayama, J. Catal., 2004, 227, 202.
88. Y. Liu, P. Liu, Z. Su, F. Li and F. Wen, Appl. Surf. Sci., 2008, 255, 2020.
89. (a) C. Pecharromán, A. Esteban-Cubillo, I. Montero and J. S. Moya, J. Am. Ceram. Soc., 2006, 89, 3043; (b) A. Esteban-Cubillo, J.-M. Tulliani, C. Pecharromán and J. S. Moya, J. Eur. Ceram. Soc., 2007, 27, 1983; (c) A. Esteban-Cubillo, R. Pina-Zapardiel, J. S. Moya and C. Pecharromán, J. Nanopart. Res., 2010, 12, 1221.
90. (a) A. Gómez-Avilés, P. Aranda, F. M. Fernandes, C. Belver and E. Ruiz-Hitzky, at the General Meeting of the 2010TMC, Seville, Spain, 2010; (b) A. Gómez-Avilés, PhD Thesis, Autonomous University of Madrid, 2010.
91. P. Aranda, R. Kun, M. A. Martín-Luengo, S. Letaïef, I. Dékány and E. Ruiz-Hitzky, Chem. Mater., 2008, 20, 84.
92. C. Belver, P. Aranda and E. Ruiz-Hitzky, 5th International FEZA Conference : abstracts, ed. T. Blasco and E. Sastre, Editorial Universitat Politècnica de València, Valencia, Spain, 2011, p. 349.
93. L. Bouna, B. Rhouta, M. Amjoud, F. Maury, M.-C. Lafont, A. Jada, F. Senocq and L. Daoudi, Appl. Clay Sci., 2011, 52, 301.
94. (a) D. Papoulis, S. Komarneni, A. Nikolopoulou, P. Tsolis-Katagas, D. Panagiotaras, H. G. Kacandes, P. Zhang, S. Yin, T. Sato and H. Katsuki, Appl. Clay Sci., 2010, 50, 118; (b) D. Karamanis, A. N. Ökte, E. Vardoulakis and T. Vaimakis, Appl. Clay Sci., 2011, 53, 181; (c) M. Ugurlu and M. H. Karaoglu, Chem. Eng. J., 2011, 166, 859.
95. L. Zhang, J. Liu, C. Thang, J. Lv, H. Zong, Y. Zhao and X. Huang, Appl. Clay Sci., 2011, 51, 68.
96. S. Letaief, Y. Liu and C. Detellier, Can. J. Chem., 2011, 89, 280.
97. J. Barrios-Neira, L. Rodrique and E. Ruiz-Hitzky, J. Microsc. Spectrosc. Electron., 1974, 20, 295.
98. B. Casal, F. Bergaya, D. Challal, J. J. Fripiat, E. Ruiz-Hitzky and H. Van Damme, J. Mol. Catal., 1985, 33, 83.
99. L. Zhu, S. Letaïef, Y. Liu, F. Gervais and C. Detellier, Appl. Clay Sci., 2009, 43, 439.
100. (a) L. Zhang, Q. Jin, L. Shan, Y. Liu, X. Wang and J. Huang, Appl. Clay Sci., 2010, 47, 229; (b) L. Zhang, Q. Jin, J. Huang, Y. Liu, L. Shan and X. Wang, Appl. Surf. Sci., 2010, 256, 5911.
101. C. Wan and B. Chen, Nanoscale, 2011, 3, 693.

102. A. Gómez-Avilés, M. A. Camblor, P. Aranda and E. Ruiz-Hitzky, unpublished results.
103. Z. Shuqin, H. Yong, H. Shia, Q. Dong, T. Kewen and Y. Jianmin, *Acta Chim. Sinica*, 2010, **68**, 329.
104. V. Blanco-Gutiérrez, E. Urones-Garrote, M. J. Torralvo-Fernández and R. Sáez-Puche, *Chem. Mater.*, 2010, **22**, 6130.
105. M. Nieto-Suárez, G. Palmisano, M. L. Ferrer, M. C. Gutiérrez, S. Yudakal, V. Agugliaro, M. Pagliaro and F. del Monte, *J. Mater. Chem.*, 2009, **19**, 2070.
106. Y. M. Lvov and R. R. Price, in *Bio-inorganic Hybrid Materials. Strategies, Synthesis, Characterization and Application*, ed. E. Ruiz-Hitzky, K. Ariga and Y. Lvov, Wiley-VCH, Weinheim, 2008, p. 419.
107. M. F. Brigatti, E. Galán and B. K. G. Theng, in *Handbook of Clay Science*, ed. F. Bergaya, B. K. G. Theng and G. Lagaly, Elsevier Science Ltd., Amsterdam, 2006, vol. 1, p. 48.
108. P. Liu and M. Zhao, *Appl. Surf. Sci.*, 2009, **255**, 3989.
109. Y. Fu and L. Zhang, *J. Solid State Chem.*, 2005, **178**, 3595.
110. X. Qi, H. Yoon, S.-H. Lee, J. Yoon and S.-J. Kim, *J. Ind. Eng. Chem.*, 2008, **14**, 136.
111. I. J. Johnson, T. A. Werpy and T. J. Pinnavaia, *J. Am. Chem. Soc.*, 1988, **110**, 8545.
112. J. S. Dailey and T. J. Pinnavaia, *Chem. Mater.*, 1992, **4**, 855.
113. K. W. Park, J. H. Jung, S. Y. Jeong and O. Y. Kwon, *J. Nanosci. Nanotechnol.*, 2009, **9**, 3160.
114. K.-W. Park, J. H. Jung, H.-J. Seo and O. Y. Kwon, *Microporous Mesoporous Mater.*, 2009, **121**, 219.
115. (a) O. Y. Kwon, *J. Ind. Eng. Chem.*, 1999, **5**, 314; (b) O. Y. Kwon, K. W. Park and U. H. Pack, *J. Ind. Eng. Chem.*, 1999, **5**, 93; (c) O. Y. Kwon, H. S. Shin and S. W. Choi, *Chem. Mater.*, 2000, **12**, 1273.
116. S. T. Wong and S. Cheng, *Chem. Mater.*, 1993, **5**, 770.
117. K. Ozawa, Y. Nakao, Z. Cheng, D. Wang, M. Osada, R. Okada, K. Saeki, H. Itoh and F. Iso, *Mater. Lett.*, 2009, **63**, 366.
118. K. Ozawa, F. Iso, Y. Nakao, Z. Cheng, H. Fujii, M. Hase and H. Yamaguchi, *J. Eur. Ceram. Soc.*, 2007, **27**, 2665.
119. K. Ozawa, R. Okada, Y. Nakao, T. Ogiwara, H. Itoh and F. Iso, *J. Am. Ceram. Soc.*, 2010, **93**, 4022.
120. (a) J. Eun Ko, B. Jin Kwon and H. Jung, *J. Phys. Chem. Solids*, 2010, **71**, 658; (b) K.-Z. Zhang, B.-Z. Lin, Y.-L. Chen, B.-H. Xu, X.-T. Pian, J.-D. Kuang and B. Li, *J. Colloid Interface Sci.*, 2011, **358**, 360; (c) Z.-J. Chen, B.-Z. Lin, Y.-L. Chen, K.-Z. Zhang, B. Li and H. Zhu, *J. Phys. Chem. Solids*, 2010, **71**, 841.
121. (a) J. H. Choy, H. C. Lee, H. Jung and S. J. Hwang, *J. Mater. Chem.*, 2001, **11**, 2232; (b) J. H. Choy, H. C. Lee, H. Jung, H. Kim and H. Boo, *Chem. Mater.*, 2002, **14**, 2486.
122. (a) J. Yang and J. Ding, *Mater. Lett.*, 2004, **58**, 3872; (b) S. Udomsak, R. Nge, D. C. Dufner, S. E. Lott and R. G. Anthony, in *Stud. Surf. Sci.*

Catal., ed. G. Poncelet, J. Martens, B. Delmon, P. A. Jacobs and P. Grange, Elsevier, 1995, vol. Volume 91, p. 391.
123. (a) T. W. Kim, S. J. Hwang, Y. Park, W. Choi and J. H. Choy, *J. Phys. Chem. C*, 2007, **111**, 1658; (b) T. W. Kim, S. G. Hur, S. J. Hwang and J. H. Choy, *Chem. Commun.*, 2006, 220.
124. T. W. Kim, H. W. Ha, M. J. Paek, S. H. Hyun, I. H. Baek, J. H. Choy and S. J. Hwang, *J. Phys. Chem. C*, 2008, **112**, 14853.
125. (a) J. H. Choy, *J. Phys. Chem. Solids*, 2004, **65**, 373; (b) J.-H. Choy, S.-M. Paek, J.-M. Oh and E.-S. Jang, *Curr. Appl. Phys.*, 2002, **2**, 489.
126. Z. J. Chen, B. Z. Lin, B. H. Xu, X. L. Li, Q. Q. Wang, K. Z. Zhang and M. C. Zhu, *J. Porous Mater.*, 2011, **18**, 185.
127. (a) X. Chen, H. Wang, W. H. Hou and Q. J. Yan, *Chin. J. Inorg. Chem.*, 2002, **18**, 550; (b) F. Jiang and F. Shao, *J. Inorg. Mater.*, 2008, **23**, 1263.
128. (a) T. Sumida, R. Abe, M. Hara, J. N. Kondo and K. Domen, *J. Mater. Res.*, 2000, **15**, 2587; (b) S. Udomsak and R. G. Anthony, *Catal. Today*, 1994, **21**, 197.
129. (a) J. Yang, J. F. Ding, L. L. Zhang, L. D. Lu and X. Wang, *Chin. J. Inorg. Chem.*, 2004, **20**, 1459; (b) M. W. Anderson and J. Klinowski, *Inorg. Chem.*, 1990, **29**, 3260.
130. (a) F. Kooli, T. Sasaki, V. Rives and M. Watanabe, *J. Mater. Chem.*, 2000, **10**, 497; (b) F. Kooli, T. Sasaki, V. Rives and M. Watanabe, *Mater. Res. Soc. Symp. Proc.*, 1999, **549**, 79.
131. (a) A. Dakskobler and T. Kosmač, *J. Mater. Res.*, 2006, **21**, 448; (b) S. Cheng and T. C. Wang, *Inorg. Chem.*, 1989, **28**, 1283.
132. X. Gu, F. Chen, B. Zhao and J. Zhang, *Superlattices Microstruct.*, 2011, **50**, 107.
133. (a) S. Tong, H. Jin, D. Zheng, W. Wang, X. Li, Y. Xu and W. Song, *Biosens. Bioelectron.*, 2009, **24**, 2404; (b) S. Tong, W. Wang, X. Li, Y. Xu and W. Song, *J. Phys. Chem. C*, 2009, **113**, 6832.
134. I. Y. Kim, K. Y. Lee, T. W. Kim and S.-J. Hwang, *Mater. Lett.*, 2011, **65**, 894.
135. C. Forano, T. Hibino, F. Leroux and C. Taviot-Guého, in *Handbook of Clay Science*, ed. F. Bergaya, B. K. G. Theng and G. Lagaly, Elsevier Science Ltd., Amsterdam, 2006, vol. 1, p. 1021.
136. J. He, H. Shi, X. Shu and M. Li, *AIChE J.*, 2010, **56**, 1352.
137. H. Han, G. Seo, Y. Yoo and G. Park, *Worldwide Pat.* WO 2008/066275-A1, 2008.
138. F. Winter, V. Koot, A. Jos van Dillen, J. W. Geus and K. P. de Jong, *J. Catal.*, 2005, **236**, 91.
139. (a) P. Maireles-Torres, P. Olivera-Pastor, E. Rodríguez-Castellón, A. Jiménez-López and A. A. G. Tomlinson, *J. Mater. Chem.*, 1991, **1**, 739; (b) F. J. Pérez-Reina, P. Olivera-Pastor, P. Maireles-Torres, E. Rodríguez-Castellón and A. Jiménez-López, *Langmuir*, 1998, **14**, 4017.
140. B. Pawelec, S. Murcia-Mascaros and J. L. G. Fierro, *Langmuir*, 2002, **18**, 7953.

CHAPTER 5
Mesoporous Nanoarchitectonics

AJAYAN VINU*[a], QINGMIN JI[b], JONATHAN P. HILL[b,c] AND KATSUHIKO ARIGA*[b,c]

[a] Australian Institute for Bioengineering and Nanotechnology (AIBN), Corner College and Cooper Rds (Bldg 75), The University of Queensland, Brisbane Qld 4072, Australia; [b] World Premier International (WPI) Research Center for Materials Nanoarchitectonics (MANA), National Institute for Materials Science (NIMS), 1-1 Namiki, Tsukuba 305-0044, Japan; [c] JST, CREST, 1-1 Namiki, Tsukuba 305-0044, Japan
*E-mail: a.vinu@uq.edu.au

5.1 Introduction: The Nanoarchitectonics and Mesoporous Story

Syntheses of nanomaterials and control of nanostructures are undoubtedly important research subjects in current science and technology. Although various methods exist for preparation of attractive nanostructures of nanomaterials, the control of nanomaterials' compositions, structures and orientations are still regarded as great challenges. Some novel concepts for integration of the methodologies of different fields are required as a breakthrough. Aono proposed the novel concept of nanoarchitectonics,[1] which is a technology system for arranging nanoscale structural units in a required configuration. Nanoarchitectonics contributes to materials innovation not only by the creation of individual nanostructures and the elucidation of their functions, but also by the creation of macroscopic materials based on a profound understanding of mutual interactions between the individual nanostructures and their arbitrary arrangements, such as by atom/molecule manipulation, chemical nanomanipulation, field-induced material control, controlled self-assembly and organization.[2]

This concept should be reflected in materials syntheses, ideally in bulk amounts. As bulk quantity materials with precise internal nanostructures, mesoporous materials have been recently attracted much attention, demonstrating their great potential in practical applications such as catalysis, adsorption, separation, sensing, medical usage, ecology, and nanotechnology.[3] These materials are defined as porous materials with pore diameters in the size range of 0.2–2.0 nm, according to IUPAC classification. Although research of mesoporous materials was initiated before the proposal of the nanoarchitectonics concept, these attractive materials should be considered within the concept of nanoarchitectonics for further development.

The science and technology of mesoporous materials began in 1990. Kuroda and coworkers first reported the preparation of mesoporous silica with uniform pore size distribution from layered polysilicate kanemite (FSM-16, Folded Sheet Materials).[4] This was followed by a significant breakthrough in mesoporous materials research by Mobil scientists who disclosed the M41S family of materials,[5] which have large uniform pore structures, high specific surface areas and specific pore volumes; the family includes hexagonal-MCM-41, cubic-MCM-48 and lamellar-MCM-50. A typical method of mesoporous material synthesis is illustrated in Figure 5.1. Mesoporous materials are basically prepared through silica formation around template micelle assemblies followed by template removal by appropriate methods such as calcination. These pioneering findings were followed by various kinds of mesoporous materials. For example, highly ordered large-pore mesoporous silica SBA-15, with thicker pore walls and a two dimensional hexagonal structure, was made by using an amphiphilic triblock-copolymer of poly(ethylene oxide) and poly(propylene oxide) (Pluronic P123) as the structure directing reagent in highly acidic media.[6] Their pore diameters are highly tunable in the range of 5 to 30 nm, and these materials exhibit high hydrothermal stability.

One of the most unique approaches for organic–inorganic hybrid materials is synthesis of periodic mesoporous organosilicates (PMO).[7] This method uses

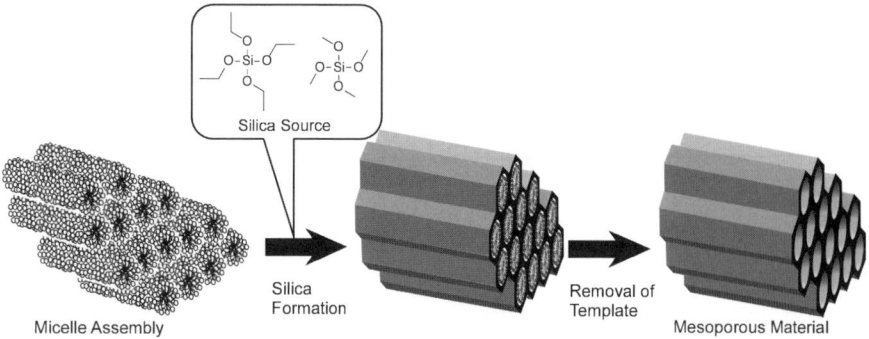

Figure 5.1 A general scheme of preparation of mesoporous silica.

organic molecules having multiple alkoxysilane groups as a silica source and introduces various organic components into the framework. Several examples of syntheses of the mesoporous materials with synthetically designed surfactants have been also reported.[8] For example, chirality was first introduced in mesoporous materials by using a chirally-defined surfactant template.[9] A newly-synthesized surfactant, N-acetyl-type alanine-based surfactant, was used as a structure-directing reagent, resulting in mesoporous silica materials with regularly twisted rod-like structures with diameters of 130–180 nm and lengths of 1–6 mm. In the obtained structure, hexagonally aligned mesoscopic channels with a diameter of 2.2 nm wound together in a particular direction. Preparation of non-oxide materials with mesoporous structures has also been paid attention. This research trend was virtually initiated from invention of mesoporous carbon materials.[10] As seen in these histories, various kinds of mesoporous materials have been created, like the construction of new architectures. In this chapter, we introduce recent our innovations, including carbon nanocages, mesoporous carbon nitrides, mesoporous boron nitrides, and their hierarchic assemblies.

5.2 The Carbon Nanocage and its Functions

Carbon materials with nanoscale pore sizes prepared from periodic inorganic silica templates have been receiving much attention because of their versatility and shape-selectivity, which are useful for various applications such as chromatographic separation systems, catalysts, nanoreactors, battery electrodes, capacitors, energy-storage devices and biomedical devices. Wise selection of the mesoporous silica template provides regular carbon materials with superior textural characteristics. Vinu and coworkers very recently have reported synthesis of a novel nanocarbon, a "carbon nanocage" (Figure 5.2),[11] through replica synthesis using three dimensional large cage-type face centered cubic mesoporous silica materials (KIT-5) as inorganic templates. An appropriate water to silica template weight ratio (*ca.* 2.5) in the syntheses was found to be a crucial factor by systematic investigation of synthetic conditions. With this initial knowledge, carbon nanocage materials with different pore diameter were prepared by using different KIT-5 mesoporous silica synthesized at different temperatures (from 100 to 150 °C) as templates and sucrose as the carbon source. Independently, the samples were prepared at different sucrose to silica weight ratios from 0.45 to 2.0. High-resolution transmission electron microscopy (HRTEM) images were recorded along two different crystallographic directions, and both of them confirm that the mesoporous carbon possesses a highly ordered structure with a uniform pore size distribution.

The textural characteristics of the prepared carbon nanocage materials were more deeply investigated by nitrogen adsorption–desorption measurement. The specific surface area and specific pore volume reached 1600 $m^2 g^{-1}$ and 2.1 $cm^3 g^{-1}$, respectively, in the case of the carbon nanocage at the lowest sucrose

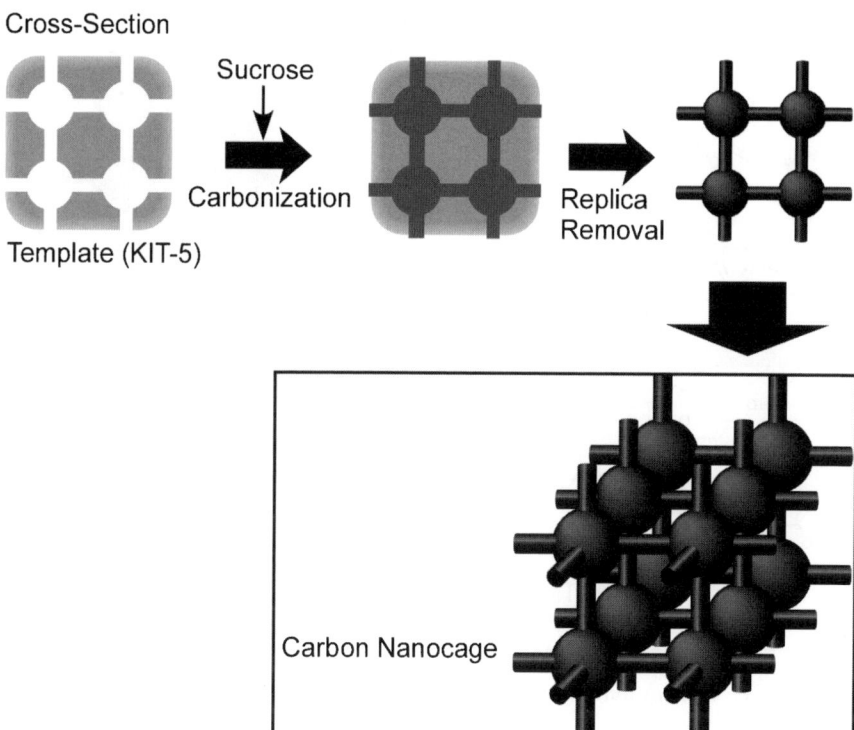

Figure 5.2 Preparation of carbon nanocage.

to silica ratio. These values are apparently larger than those reported for conventional mesoporous carbon, CMK-3 (surface area, 1260 m^2 g^{-1}; pore volume, 1.1 cm^3 g^{-1}). A further analysis with the method proposed by Ravikovitch et al.[12] provided a cage diameter of 15 nm for the corresponding carbon nanocage, which has a pore diameter of 5.2 nm. Integrated structures with a large difference between pore size and cage size would result in huge values of the surface area and pore volume.

The synthesized carbon nanocages were used for quantitative experiments on adsorption of the bioactive tea components, caffeine, catechin, and tannic acid (Figure 5.3).[13] The adsorbent was dispersed in an aqueous solution of the tea components with shaking for 24 hours at 20 °C. All three carbon adsorbents had similar adsorption capacity for caffeine, although caffeine was hardly adsorbed by hydrophilic SBA-15. Superior capacity of carbon adsorbents relative to SBA-15 was also evident for adsorption of catechin. Interestingly, only the carbon nanocage showed two-step adsorption and greater adsorption capacity at higher concentration. Variation in the behaviors of the adsorbents was much more obvious in isotherms for tannic acid adsorption. The carbon nanocage material exhibited larger adsorption capacity for tannic acid with a two-step adsorption, while CMK-3 showed

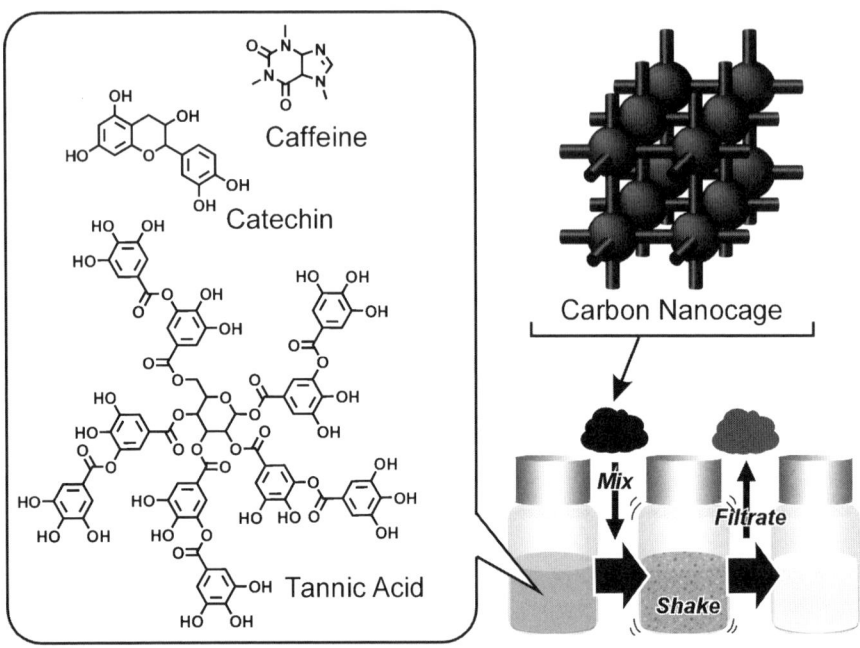

Figure 5.3 Adsorption of tea components to carbon nanocage.

lower capacity in single-step mode. Only poor adsorption capacity was detected for activated carbon and SBA-15.

The molecular dimensions of catechin (0.8 × 1.3 nm) and of tannic acid (approximate circular diameter 3 nm) were estimated using molecular models. Catechin can be contained in pores of the carbon nanocage (pore diameter 5.2 nm and cage diameter 15 nm), CMK-3 (3.0 nm), and activated carbon (< 2 nm), but the larger tannic acid is not permitted to enter activated carbon micropores and may have difficulty diffusing within small CMK-3 mesopores. The larger cage-type pores of the the carbon nanocage adsorbent promote its adsorption of both catechin and tannic acid in multilayer mode, and this may originate from interaction between adsorbed guests. Adsorption strengths of catechin and tannic acid on the carbon nanocage were estimated as binding constants of 18 000 and 56 000 M^{-1} for catechin and tannic acid, respectively. Competitive adsorption (guest selection) of catechin and tannic acid on the carbon nanocage adsorbent using solutions containing equal weights of the guests unexpectedly gave adsorption behaviors which departed significantly from those of the individual guests because of competitive adsorption. Catechin adsorption was suppressed drastically by the presence of tannic acid, especially at lower concentrations. Diminished catechin adsorption is caused by preferential adsorption of tannic acid on the carbon nanocage. Surprisingly, use of a carbon nanocage as adsorbent provided a highly selective adsorption of tannic acid (*ca.* 95%) in a simple one-pot process. This process cannot be

Mesoporous Nanoarchitectonics

achieved by activated carbon or conventional mesoporous carbon, CMK-3. Very high selectivity for adsorption of tea components (catechin and tannic acid) was achieved through a simple one-pot process using the novel nanocarbon carbon nanocage.

Carbon nanocage adsorbents could be used for efficient removal and extraction of biomaterials and will have a great impact on biomedical fields. Adsorption isotherms of the three nucleosides, guanosine, adenosine and thymidine onto various mesoporous materials, carbon nanocage (Figure 5.4), mesoporous carbon CMK-3, activated carbon and mesoporous silica KIT-5, were investigated.[14] Mesoporous silica KIT-5 is inert towards all the nucleosides, possessing very poor adsorption capacities. This result suggests the importance of hydrophobic interactions and π–π interactions, which are in any case not expected in the case of mesoporous silica upon adsorption of nucleosides to mesoporous materials. For the nanoporous carbons, adsorption amounts of nucleosides on activated carbon are apparently smaller than those

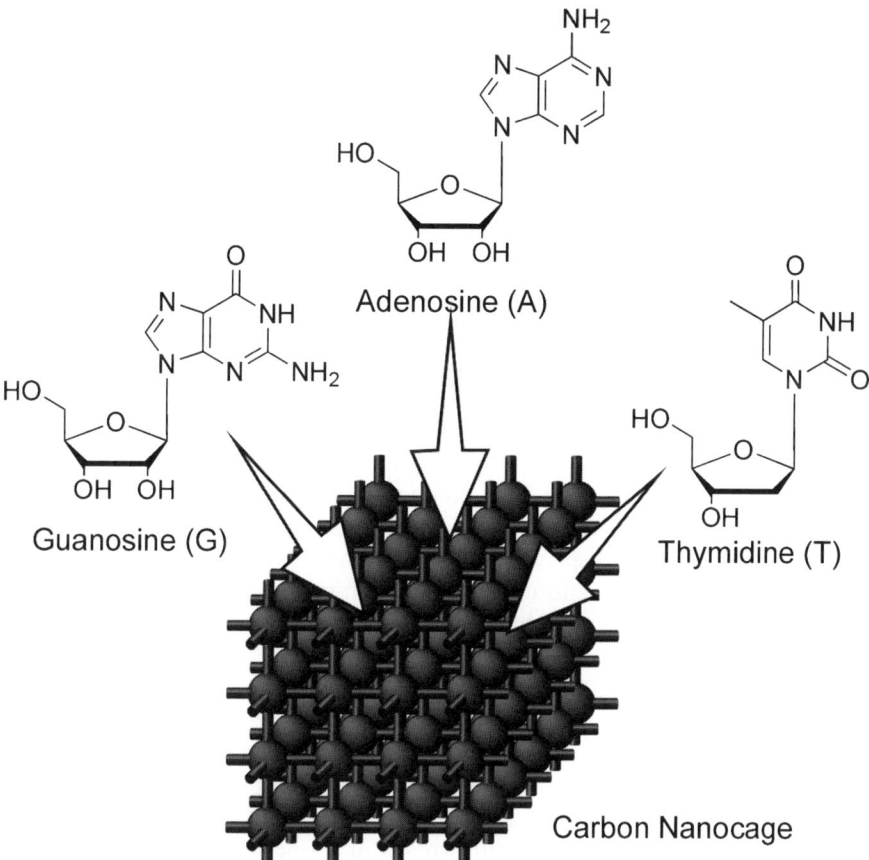

Figure 5.4 Adsorption of nucleosides to the carbon nanocage.

observed for the carbon nanocage and CMK-3. The molecular size of the nucleosides and the pore size of the activated carbon (pore size on the basis of non-local density functional theory) are less than 1 nm (specific surface area, 1629 $m^2 g^{-1}$; specific pore volume, 0.70 $cm^3 g^{-1}$), so that diffusion for efficient adsorption should be suppressed although the activated carbon material has a highly specific surface area.

Interestingly, the detailed adsorption profiles of nucleosides to the three carbon materials (carbon nanocage, CMK-3 and activated carbon) exhibit some dependency on nucleoside identity. The carbon nanocage showed the highest adsorption capacity to guanosine, while the difference between the carbon nanocage and CMK-3 is not so obvious. These mesoporous carbon materials exhibit superior adsorption capacity to microporous activated carbon and hydrophilic mesoporous silica KIT-5. Overall, guanosine adsorbed with high selectivity in all cases while thymidine always showed the minimum adsorption. Guanosine has a larger hydrophobic aromatic system than thymidine, which causes greater hydrophobic interactions and π–π interactions. Although activated carbon and CMK-3 exhibited an adsorption order of guanosine > adenosine > thymidine, the carbon nanocage clearly discriminates between purines (guanosine and adenosine) and pyrimidines (thymidine). At the same time, the binding affinity of adenosine over thymidine is pronounced especially in the case of carbon nanocage. These results indicate that the carbon nanocage discriminates between binding of purine and pyrimidine nucleosides and is an especially good material for differentiating adenosine and thymidine. Purine bases generally have a large aromatic π-electronic system, which is advantageous for their adsorption on carbon through hydrophobic interaction and π–π stacking. The larger pore volume of the carbon nanocage should provide sufficient freedom in motion of guest nucleosides to attain more favorable contact with the carbon surface even in the nanospace. Such effects should result in selective binding to purine-base nucleosides.

Adsorption of dyes (3,6-diaminoacridine hydrochloride, methyl violet and cytosine) were also investigated.[15] Some of these are suspected to be toxic intercalators. Generally, these guest compounds strongly adsorb to the carbon nanocage and CMK-3. Adsorption of 3,6-diaminoacridine hydrochloride onto various porous hosts displayed a two-step adsorption mode and greater adsorption at higher concentration. The carbon nanocage exhibited maximum adsorption capacity, followed by CMK-3, which showed a similar trend with slightly lower adsorption capacity. Activated carbon and SBA-15 similarly showed two-step adsorption behavior, wherein the plateau region was retained until higher concentration, as compared to that of the carbon nanocage and CMK-3 adsorbents. Adsorption isotherms of methyl violet onto different porous supports showed two-step adsorption and greater adsorption capacity at higher concentration in the cases of the carbon nanocage, CMK-3 and SBA-15, while activated carbon exhibited poor adsorption in a one-step mode. The carbon nanocage again exhibited the largest adsorption capacity, while adsorption to CMK-3 was 28.8 % less as compared to that of the carbon nanocage.

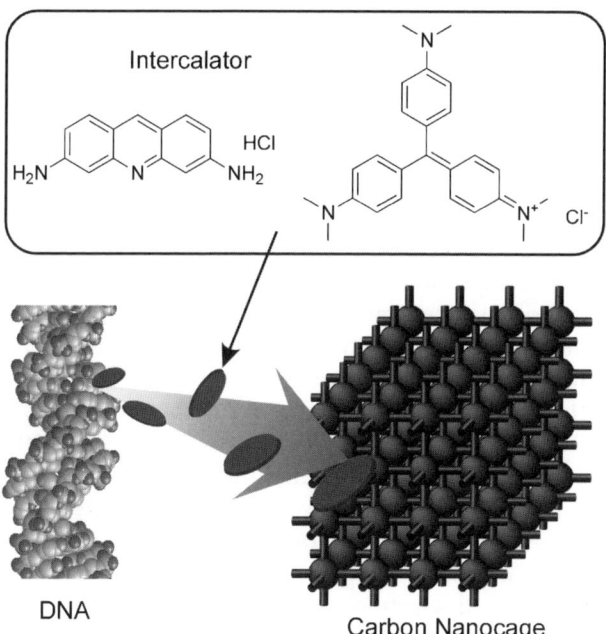

Figure 5.5 Removal of intercalators to the carbon nanocage.

We also examined competitive adsorption of methyl violet to porous adsorbents and DNA (Figure 5.5). Mixing DNA with methyl violet leads to DNA intercalation, wherein dye molecules slip in between the stacked bases of DNA. A spectral shift due to electronic stacking interactions of the dye molecules with the base pairs of the DNA helix characteristic of intercalation was observed. We expect that addition of adsorbents to intercalated DNA selectively adsorbs methyl violet molecules retaining DNA spectra. In the presence of the carbon nanocage or CMK-3, the original DNA spectrum was retained due to removal of the dye molecules, which were initially bound to DNA. In contrast, activated carbon and SBA-15 are not capable of inhibiting intercalant molecules. These results indicate the superior capability of the carbon nanocage and CMK-3 in inhibition of intercalation of dyes to DNA. DNA intercalants can be removed or altered by using nanoporous materials, such as the carbon nanocage that has hydrophobic characteristics as well as high surface area, large pore volume and pore size.

5.3 Mesoporous Carbon Nitride and Mesoporous Boron Nitride

The replica synthesis is not limited to synthesis of carbon materials with mesoporous structure and can be utilized for synthesis of novel mesoporous materials with elements other than carbon. Instead of using a single carbon source, use of sources with other elements and/or their mixtures could provide mesoporous structures with different materials as framework components. As one of the successful examples along this strategy, synthesis of mesoporous carbon nitride is described here.[16]

For synthesis of mesoporous carbon nitride materials (Figure 5.6), a co-material source, a mixture of carbon source and nitrogen source, was impregnated into an appropriate silica template. Mesoporous silica SBA-15 template was added to a mixture of ethylenediamine and carbon tetrachloride and the obtained composite was then heat-treated in a nitrogen flow. The mesoporous carbon nitride was recovered after dissolution of the silica framework in hydrofluoric acid. Thermogravimetric analysis under an oxygen atmosphere revealed that the maximum silica residue was confirmed to be less than 1 wt%. The structure of the mesoporous carbon nitride was investigated by powder XRD, giving three clear peaks assignable to the (100), (110) and (200) diffractions of a two-dimensional hexagonal lattice with a lattice constant of $a_{100} = 9.52$ nm. This powder diffraction pattern of mesoporous carbon nitride showed a single broad diffraction peak near 25.8°, corresponding to an interlayer d spacing of 0.342 nm.

Nitrogen adsorption–desorption measurement provided detailed data on the pore structures of the mesoporous carbon nitride materials. As compared with SBA-15, the pose size distribution of the mesoporous carbon nitride materials is somewhat broadened but still exhibits a narrow pore size distribution. Pore structures of the mesoporous carbon nitride were directly investigated by HRTEM. Only a stripe pattern could be detected when viewed down the [100] direction. Bright contrast strips on the under-focused image represent images of the pore walls, and empty channels appear as dark contrast cores. The cross-sectional HRTEM image clearly exhibits a hexagonal arrangement of the

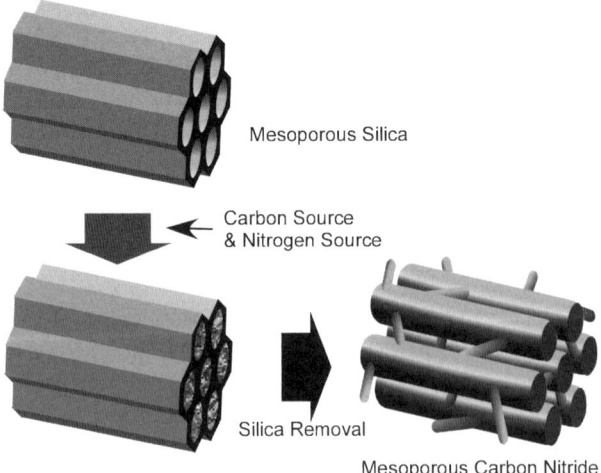

Figure 5.6 Preparation of mesoporous carbon nitride.

mesopores. In addition to the pore size analyses, elemental composition of the mesoporous carbon nitride materials was investigated by several methods. Nanoscopic distribution of carbon and nitrogen clearly confirm that both of the elements are homogeneously dispersed over the whole of the mesoporous structure. The X-ray photoelectron spectroscopy (XPS) survey spectrum of mesoporous carbon nitride did not show any peaks for elements other than carbon, nitrogen and oxygen. The XPS C_{1s} spectrum was deconvoluted into four peaks with binding energies of 289.3, 287.5, 285.7, and 284.1 eV. The lowest energy contribution fitted for C_{1s} is assigned to pure graphitic sites in the amorphous CN matrix, and the peak at 285.7 eV is attributed to the sp^2 carbon bonded to nitrogen inside the aromatic structure. The energy contributions at 287.5 and 289.3 eV are assigned to the sp^2 carbon and the sp^2 carbon in the aromatic ring attached to NH_2 groups, respectively. The XPS N_{1s} peak at higher binding energy (400.2 eV) corresponds to nitrogen atoms trigonally bonded to all sp^2 carbons, or to two sp^2 carbon atoms and one sp^3 carbon atom in an amorphous CN network, while another peak at 397.8 eV is attributed to nitrogen sp^2-bonded to carbon.

As described above, successful preparation of mesoporous carbon nitride was confirmed both from structural and elemental analyses. Upon wise selection of material sources and design of template structures, various kinds of materials with regular mesoporous structures could be created *via* replica syntheses. The synthesized mesoporous carbon nitride can also be a good medium for production of nanomaterials. For example, gold nanoparticles can be encapsulated inside the nanochannels without any stabilizing or size-controlling chemical agents (Figure 5.7).[17] Mesoporous carbon nitride materials have three different functions, namely stabilizer, size controller and reducing agent. Ultrasmall gold nanoparticles inside the confined nanoporous matrix were found to be highly active, selective and recyclable for the synthesis of fine chemicals such as propargylamines through the coupling reaction of benzaldehyde, piperidine and phenyl acetylene. Propargylamines are intermediates for the construction of nitrogen containing biologically active molecules and for the synthesis of polyfunctional amino derivatives.

Mesoporous materials have so far been prepared by two major routes, soft template synthesis using micelle assemblies and replica synthesis using hard mesostructured template. Vinu and coworkers has proposed a third method for synthesis of mesoporous materials.[18] This method can be called the "elemental substitution method", where component elements are substituted by other elements with retaining the mesoporous structure. For example, mesoporous boron nitride and mesoporous boron carbon nitride with a very high surface area and pore volume can be fabricated *via* substitution reactions at high temperatures, using a well-ordered hexagonal mesoporous carbon (CMK-3) as a template, boron trioxide as a boron source and nitrogen gas as a nitrogen source (Figure 5.8).[19] Selective preparation of these materials can be done by careful tuning of the synthetic temperature. The mesoporous boron nitride is usually synthesized at higher temperatures, such as 1750 °C, while the

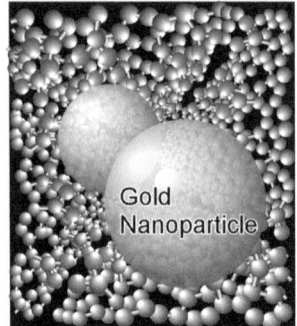

Figure 5.7 Immobilization of gold nanoparticles into mesopore channels of mesoporous carbon nitride.

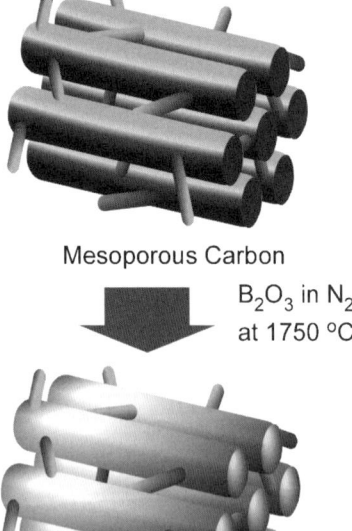

Figure 5.8 Preparation of mesoporous boron nitride.

mesoporous boron carbon nitride is obtained *via* the substitution reaction at lower temperature conditions (1450–1550 °C) with rather short reaction times. The mesoporous boron carbon nitride synthesized at 1450 °C and 1550 °C possess relatively high surface areas of 740 and 650 $m^2\ g^{-1}$ and large pore volumes of 0.69 and 0.60 $cm^3\ g^{-1}$, respectively. The specific surface area and the specific pore volume are decreased to 565 $m^2\ g^{-1}$ and 0.53 $cm^3\ g^{-1}$, respectively, for mesoporous boron nitride synthesized at 1750 °C.

HRTEM images suggest that the mesoporous boron nitride materials possess a less ordered mesoporous structure with a disordered boron nitride framework with highly crystalline layers. The image shows the well-ordered mesoporous structures with local interlinking of crystalline boron carbon nitride layers. In order to confirm distribution of component elements, elemental mapping was carried out in the same scope of the TEM image of mesoporous boron carbon nitride. The EEL spectrum of the mesoporous boron nitride revealed that the boron-to-nitrogen ratio in the corresponding material was 1.0 and that carbon signal was virtually absent. The EEL spectrum of mesoporous boron carbon nitride showed an increase of the carbon content (8.0 and 20.1% for the corresponding materials synthesized at 1550 and 1450 °C, respectively). The overall boron-to-nitrogen ratio of the mesoporous boron nitride obtained from the XPS analysis is 0.94, which is again very consistent with the ideal composition of boron nitride. The XPS measurement provided the elemental composition of mesoporous boron carbon nitride synthesized at 1450 °C as 21.1% of boron, 37.7% of carbon, and 23.0% of nitrogen, with 18.2% of oxygen as an unavoidable contaminant.

The above-mentioned results confirm successful preparation of mesoporous boron nitride and mesoporous boron carbon nitride from regularly structured mesoporous carbon, *via* the elemental substitution method. This preparative concept is new in fields of mesoporous materials and can be regarded as the third-generation method after template synthesis and replica synthesis.

5.4 Layered Hierarchic Structure

Biological systems have highly hierarchic structures, as seen in organelles, cells, tissues and organs. In biological systems and processes, hierarchic structures are constructed through spontaneous self-assembly that is achieved through molecular design and the appropriate selection of components. It would be very difficult to mimic all available natural processes by using non-biochemical approaches. One possible approach for constructing functional hierarchic structures would be a multi-step process where first nanostructured materials are synthesized followed by their subsequent further assembly into organized structures of higher order. As one of the most versatile techniques to construct organized structures, alternate layer-by-layer (LbL) assembly has been paid much attention.[20] The most important feature of the LbL technology is the wide freedom in the resulting layered structures, with layer thickness (the number of layers) and layer sequences being easily tuned. The LbL technique can be used in the fabrication of functional hierarchic structures using pre-synthesized nanostructured objects such as mesoporous materials.

As one example of LbL assemblies of mesoporous materials, we have also demonstrated sensor applications of the LbL structures prepared from mesoporous carbon materials and polyelectrolytes. Surface oxidation of mesoporous carbon (CMK-3) using ammonium persulfate enabled us to introduce negative carboxylate groups to mesoporous carbon (CMK-3). LbL assembly of oxidized CMK-3 was performed using polycation PDDA on a QCM plate (Figure 5.9).[21] Sensing performances were investigated in aqueous solution, where a QCM plate covered with the LbL film of CMK-3 was immersed and the sensing target was injected. A frequency shift upon adsorption of tannic acid to the CMK-3 LbL films was observed immediately after injection of tannic acid. Frequency shifts upon adsorption of tannic acid greatly exceed those for catechin and caffeine. The resulting sensitivity ratios of tannic acid to catechin or caffeine are *ca.* 3.9 and 13.6, respectively. The superior adsorption capacity for tannic acid likely originates in its molecular structure, *i.e.*, the multiple phenyl rings of the tannic acid molecule can interact with the carbon surface through π–π interactions and hydrophobic effects. In addition, size fitting of tannic acid (a roughly circular molecule with approximate diameter 3 nm) to the CMK-3 nanochannel may result in enhanced interactions between the guests themselves and/or the guest and carbon surface.

Adsorption quantities of tannic acid to the CMK LbL film at equilibrium exhibited a sigmoidal profile at low concentrations. This cooperative binding profile was absent for adsorption of tannic acid to the SAM surface of octadecanethiol. Highly cooperative behavior might result from confinement effects during adsorption. The observed behavior may be similarly explained by enhanced guest–guest interaction, since the adsorbed tannic acid can have effective π–π and/or hydrophobic interactions when confined. These observations will also promote our understanding of molecular interactions within nanospaces, especially non-specific interactions in aqueous media, a full exploration of which might clarify important phenomena, including those of biological systems.

Carbon capsules were also synthesized using zeolite crystals as templates. Surfactant-covering of the capsules enabled us to assemble non-charged substances in the LbL process with the aid of counterionic polyelectrolyte. Adsorption of various volatile substances onto the carbon capsule LbL films in

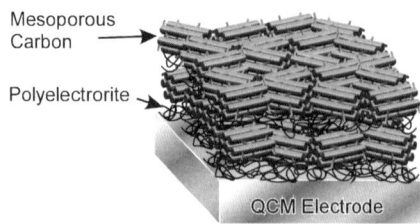

Figure 5.9 Layer-by-layer assembly of mesoporous carbon for sensing application.

Mesoporous Nanoarchitectonics

vapor-saturated atmospheres was investigated by *in situ* frequency decrease of the QCM resonator used as the film support (Figure 5.10).[22] Aromatic hydrocarbons, such as benzene and toluene, are better detected in this sensing system than aliphatic hydrocarbons such as cyclohexane. For example, the amount of benzene adsorbed at equilibrium is *ca.* 5 times larger than that of cyclohexane, despite their very similar vapor pressures, molecular weights and structures, indicating the crucial role of $\pi-\pi$ interactions on the adsorption of volatiles in the carbon capsule film. Selectivity could be easily tuned by impregnation with additional recognition components, that can be introduced after film preparation. The carbon capsule film impregnated with lauric acid showed the greatest affinities for non-aromatic amines and the second highest affinity for acetic acid. FT-IR spectra suggested strong entrapment of amines through acid–base interactions. In contrast, impregnation of dodecylamine into the carbon capsule films resulted in a strong preference for acetic acid. The prepared hierarchic layer-by-layer films with dual pore carbon capsules exhibit excellent adsorption capabilities for volatile guests such as aromatic hydrocarbons. In addition, selectivity of gas adsorption can be controlled flexibly by impregnation with second recognition sites. Such designed materials will find widespread applications as sensors or filters because of their designable guest selectivity. As the carbon materials used are stable in water, this system could also be used for removal of toxic materials from water.

We used mesoporous silica capsules for LbL assembly. Anionic silica capsules were deposited on a QCM resonator using LbL assembly (Figure 5.11).[23,24] The LbL assembly between the hollow capsules and PDDA was performed with the aid of anionic silica nanoparticles as a co-adsorber of the capsule component. QCM frequency changes were recorded during a LbL process involving

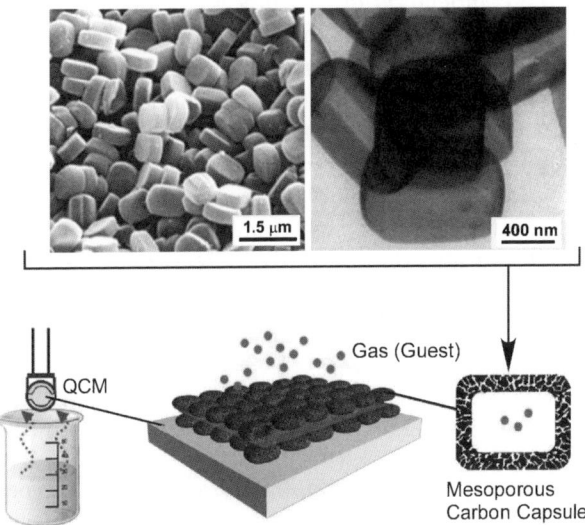

Figure 5.10 Layer-by-layer assembly of mesoporous carbon capsules for gas sensing applications.

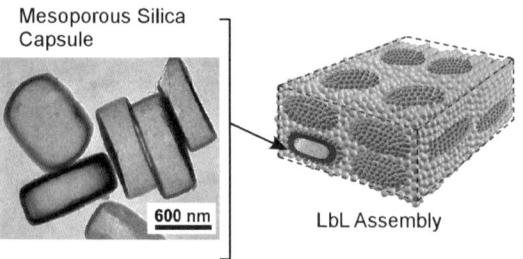

Figure 5.11 Layer-by-layer assembly of mesoporous silica capsules.

alternate immersion of the QCM resonator in a solution of a mixture of silica particles and silica capsules (w/w, 10 : 1) and a solution of PDDA, which resulted in the required structure. Scanning electron microscopic (SEM) images of the silica particle/capsule compartment films reveal that the silica capsules are dispersed among the silica particles. The top and cross-sectional views of the nanocompartment film illustrate that the silica capsules are embedded within the smooth film with retention of their capsular morphology. After immersing the compartment film on the QCM resonator into water and drying under nitrogen flow, the net change in the weight of the film after each cycle was measured in air by using QCM. An increase in QCM frequency corresponds to a decrease in mass and in this case is symptomatic of water release from the interior of the silica capsule within the compartment film. Surprisingly, the frequency shifts upon water evaporation from the mesoporous nanocompartment films possess a stepwise profile even though no external stimulus was applied.

In the proposed mechanism (Figure 5.12), stepwise release is assumed to originate from the combination of two processes: water evaporation from the pores and capillary penetration into the pores. Judging from the release profiles observed under different encapsulation conditions, the number of release steps seems to be related to the ratio of water volume to mesopore volume in the capsule wall. The interior volume of a silica capsule is four times as large as that of its mesoporous wall. If the first step of water release originates in water release from the mesopore region, then the number of steps becomes one, two, three or four, depending on the initial water content. Initially, water entrapped in mesopore channels evaporates to the exterior, which is observed as the first step of water release. After most of the water has evaporated from the mesopore channels, water enters that region from the capsule interior, probably through rapid capillary penetration. Subsequently, water again evaporates from the mesopore to the exterior and is apparent as the second evaporation step. This mechanism explains why the quantity of each step is almost identical and the number of steps is determined by the ratio between entrapped water amount and mesopore volume. In addition, the water evaporation rate at each step can be controlled by several factors such as temperature and the co-adduct materials (silica particle and polymer). Most of the currently available controlled-release systems perform modulation in release using some external stimulus. However, we have presented

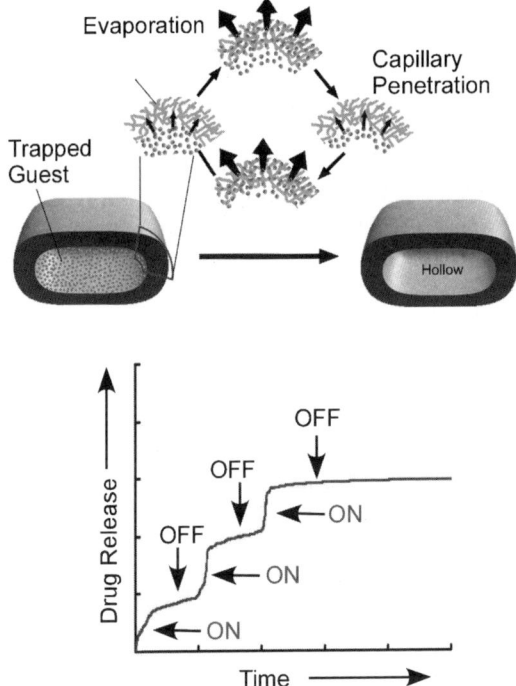

Figure 5.12 Stimulus-free controlled periodic materials release from silica capsules.

here a rare example of *a stimulus-free controlled-release medium*, which operates in a stepwise manner with prolonged release efficiency, a feature useful for controlled-release drug delivery. This new system has been shown to possess features of controlled loading/release, which are of great utility for development of energy-less and clean *stimulus-free* controlled drug release applications.

5.5 Summary

In this chapter, various methodologies for syntheses of mesoporous materials are described, including recent advances in this field. It is certainly a kind of architectonics for mesoporous materials. This feature can be especially pronounced in hierarchic design of mesoporous materials. These related approaches can be categorized as *mesoporous nanoarchitectonics*.

Acknowledgment

We thank the World Premier International Research Center Initiative (WPI Initiative), MEXT, Japan, and the Core Research for Evolutional Science and Technology (CREST) program of the Japan Science and Technology Agency (JST), Japan. A. Vinu thanks the Australian Research Council for the Future Fellowship and AIBN for the start-up grant.

References

1. This terminology was first proposed by Dr Masakazu Aono at the 1st International Symposium on Nanoarchitectonics Using Suprainteractions (NASI-1) at Tsukuba in 2000.
2. K. Ariga, J. P. Hill, M. V. Lee, A. Vinu, R. Charvet and S. Acharya, *Sci. Technol. Adv. Mater.*, 2008, **9**, 014109.
3. K. Ariga, A. Vinu, J. P. Hill and T. Mori, *Coord. Chem. Rev.*, 2007, **251**, 2562.
4. T. Yanagisawa, T. Shimizu, K. Kuroda and C. Kato, *Bull. Chem. Soc. Jpn.*, 1990, **63**, 988.
5. C. T. Kresge, M. E. Leonowicz, W. J. Roth, J. C. Vartuli, J. S. Beck, *Nature*, 1992, **359**, 710.
6. D. Zhao, J. Feng, Q. Huo, N. Melosh, G. H. Fredickson, B. F. Chmelka, G. D. Stucky, *Science*, 1998, **279**, 548.
7. S. Inagaki, S. Guan, T. Ohsuna and O. Terasaki, *Nature*, 2002, **416**, 304.
8. Q. Zhang, K. Ariga, A. Okabe and T. Aida, *J. Am. Chem. Soc.*, 2004, **126**, 988.
9. S. Che, Z. Liu, T. Ohsuna, K. Sakamoto, O. Terasaki and T. Tatsumi, *Nature*, 2004, **429**, 281.
10. R. Ryoo, S. H. Joo and S. Jun, *J. Phys. Chem. B*, 1999, **103**, 7743.
11. A. Vinu, M. Miyahara, V. Sivamurugan, T. Mori and K. Ariga, *J. Mater. Chem.*, 2005, **15**, 5122.
12. P. I. Ravikovitch and A.V. Neimark, *Langmuir*, 2002, **18**, 1550.
13. K. Ariga, A. Vinu, M. Miyahara, J. P. Hill and T. Mori, *J. Am. Chem. Soc.*, 2007, **129**, 11022.
14. K. K. R. Datta, A. Vinu, S. Mandal, S. Al-deyab, J. P. Hill and K. Ariga, *J. Nanosci. Nanotechnol.*, 2011, **11**, 3959.
15. K. K. R. Datta, A. Vinu, S. Mandal, S. Al-deyab, J. P. Hill and K. Ariga, *J. Nanosci. Nanotechnol.*, 2011, **11**, 3084.
16. A. Vinu, K. Ariga, T. Mori, T. Nakanishi, S. Hishita, D. Golberg and Y. Bando, *Adv. Mater.*, 2005, **17**, 1648.
17. K. K. R. Datta, B. V. Subba Reddy, K. Ariga and A. Vinu, *Angew. Chem., Int. Ed.*, 2010, **49**, 5961.
18. A. Vinu, T. Mori and K. Ariga, *Sci. Technol. Adv. Mater.*, 2006, **7**, 753.
19. A. Vinu, M. Terrones, D. Golberg, S. Hishita, K. Ariga and T. Mori, *Chem. Mater.*, 2005, **17**, 5887.
20. K. Ariga, J. P. Hill and Q. Ji, *Phys. Chem. Chem. Phys.*, 2007, **9**, 2319.
21. K. Ariga, A. Vinu, Q. Ji, O. Ohmori, J. P. Hill, S. Acharya, J. Koike and S. Shiratori, *Angew. Chem., Int. Ed.*, 2008, **47**, 7254.
22. Q. Ji, S. B. Yoon, J. P. Hill, A. Vinu, J.-S. Yu and K. Ariga, *J. Am. Chem. Soc.*, 2009, **131**, 4220.
23. Q. Ji, M. Miyahara, J. P. Hill, S. Acharya, A. Vinu, S. B. Yoon, J.-S. Yu, K. Sakamoto and K. Ariga, *J. Am. Chem. Soc.*, 2008, **130**, 2376.
24. Q. Ji, S. Acharya, J. P. Hill, A. Vinu, S. B. Yoon, J.-S. Yu, K. Sakamoto and K. Ariga, *Adv. Funct. Mater.*, 2009, **19**, 1792.

CHAPTER 6
Nanoscale Oxides in Catalysis

M. LAKSHMI KANTAM[a], SUNKARA V. MANORAMA[a], PRATYAY BASAK[a], VENKAT REDDY CHINTAREDDY[b] AND SURESH K. BHARGAVA[c]

[a] Inorganic and Physical Chemistry Division, Indian Institute of Chemical Technology, Hyderabad – 500607, Andhra Pradesh, India; [b] 1311 Gilman Hall, Department of Chemistry, Iowa State University, Ames, IA-50011, USA; [c] School of Applied sciences, RMIT University, Australia
*E-mail: mlakshmi@iict.res.in

6.1 Introduction

6.1.1 Overview

Catalysis provides a means of changing the rates and of controlling the yields of chemical reactions to increase the amounts of desirable products from these reactions and reduce the amounts of undesirable ones. The triumph of industrial production relies largely on the strong foundation of heterogeneous catalysis.[1–3] Virtually all nitrogen fertilizers are produced from ammonia made utilizing the *Haber–Bosch process*.[4] Since the realization of this ground-breaking discovery, aptly coined as the "invention of the last century", wherein the introduction of an iron catalyst accelerated the production of ammonia from nitrogen and hydrogen gases,[5] there has been an explosion in the area of heterogeneous catalysts research. Recent estimates reveal that more than one-third of production and manufacturing processes in all industries worldwide utilize catalysis in one form or another.[6,7] According to the Food and Agriculture Organization of the United Nations, the worldwide demand for nitrogen-based fertilizer will be 154.2 million tons in 2011, with an expected increase of 7.3

million tons annually. Today, the world faces a variety of challenges in reducing dependence on petroleum reserves, reducing harmful by-products in manufacturing and transportation, remediation of environmental issues, preventing pollution and creating safe pharmaceuticals.[8] Catalysis has been the core technology of the chemical industry, oil refineries, conversion of sustainable energy sources and environmental remediation for several decades now. The petroleum, chemical and pharmaceutical industries, which contribute around $500 billion to the gross national product of the United States, rely on catalysts to produce everything from fuels and "wonder drugs" to paints and cosmetics.[9,10] The reduced emissions of automobiles, the abundance of fresh food and affordable medicines are all made possible by chemical reactions using catalysts, and thus they are a major contributor to a healthy economy.

In 1959, the groundbreaking talk by Richard Feynman described the implications and possibility of "arranging atoms the way we want", in what was to become a nano-revolution.[11] Nanocatalysis, especially heterogeneous catalysis, is one of the primary beneficiaries from the explosively growing contributions of materials science research. The essence of nanostructuring is to work at the molecular level, atom by atom, to create hierarchical structures and assemblies with exciting properties. The fundamental size range defines the nanoscale materials and their dimensionality (nanoparticles/rods/tubes, nanoclusters and nanocrystallites). A high surface to volume ratio leading to increased surface reactivity, unique morphology and low dimensionality combined with quantum confinement effects manifests as exceptional physical, chemical, electrical, optical and magnetic properties. Attempts to exploit these distinctive properties to develop new products, devices and technologies are motivating the global research initiative.

Metal oxide nanostructures, in particular, represent the most diverse class of materials because of their unique structural, physical and chemical characteristics suitable for a variety of applications. Nanosized metal oxides such as titania, silica, zirconia, tin dioxide, ceria, iron oxide, zinc oxide and alumina, *etc.*, are under intensive study for their use as catalysts/photocatalysts, hydrogen storage materials, semiconductors, sensors, MRI contrast agents and biomaterials.[12–20]

A comprehensive outlook on some of the important metal oxides, the various strategies practiced to synthesize them, their application and performance in various catalytic reactions, along with a bird's eye view of the results, important breakthroughs and achievements in the last decade are briefly discussed. In this chapter, the highlights of our research efforts on nanoscale oxide materials employed in catalysis research are discussed briefly.

6.1.2 Design and Synthesis of Nanomaterials

Of crucial importance in heterogeneous catalysis to achieve excellent performance is the design and fabrication of nanomaterials. The main theme of size-dependent catalytic chemistry is to reduce the size of particles thereby increasing their activity. In recent years, rapid developments in novel morphological and structural nanomaterials have enabled the fabrication of catalytic materials with preferred orientation of reactive crystal planes and tailor designed active sites.

Over the years several approaches have been used by researchers to prepare nanocrystalline metal oxides, to study their material properties and explore their feasibility for a variety of applications. In physical techniques (such as gas phase condensation, mechanical alloying, thermal crystallization, molecular beam epitaxy), molecular precursors are allowed to react at rather high temperatures in the gas phase or as molecular solids. Control over the particle size distribution, shape and crystalline orientation is inherently difficult to achieve. In wet chemical approaches, a solvent is used as the reaction medium or vehicle, and the reaction is performed at relatively lower temperatures.

The traditional solid-state routes (*top-down* approaches) to obtain metal oxide nanoparticles with well-defined shape, size, crystallinity and also surface functionality, is not only energy intensive but also extremely difficult. Wet-chemistry routes (*bottom-up* approaches) provide an attractive alternative. The processes, however, need judicious choice of the molecular precursor, reaction medium and good kinetic control of the reaction parameters to form nanomaterials with high purity and compositional homogeneity at low processing temperatures. The morphology of the final product depends strongly on the precursor and solvent used, *i.e.*, metal oxides with the same composition and crystal structure but obtained from different precursors and/or solvents may yield different particle sizes and shapes. The reaction parameters for most of these processes, once successfully tuned, are highly reproducible, easy to scale up to gram quantities and applicable to a broad family of metal oxides. Some of the soft-chemical routes for synthesis of metal oxide nanoparticles at relatively lower temperatures are discussed in the following subsections.

6.2 Wet-Chemical and Low Temperature Routes for the Synthesis of Nanocrystalline Metal Oxides

6.2.1 Hydrolysis/Chemical Precipitation

The wet-chemical method is suitable for the rapid production of large quantities of materials. The precipitation technique is probably the simplest and most efficient chemical pathway to obtain nanoparticles. The main advantage of the precipitation process is that it can be easily scaled up. However, the control of particle size distribution is limited, because only kinetic factors control the growth of crystal formation. In a typical precipitation process, two critical stages are involved: (i) as the concentration of the species reaches critical super-saturation a short burst of nucleation occurs, and (ii) this is followed by a slow growth, mediated by diffusion of the solutes onto the surface of the crystal. To obtain mono-dispersed nanoparticles, these two stages should ideally be separate; *i.e.*, nucleation should be avoided during the period of growth.[21] Use of stabilizers, additives and soft-templating agents to obtain microemulsions are some of the popular approaches to control the particle size distributions. The co-precipitation technique offers the possibility of incorporating and dispersing another additive/catalytic component on to a particular precursor during synthesis. Hence, co-precipitation[22–26] and microemulsion[27–29] based techniques go hand-in-hand. The process typically involves many controllable parameters due to the complexity underlying the chemical processing. These include temperature, pH, time, concentration of the precursor, the hydrolyzing agent used, structure directing agents and stabilizers. The main drawbacks of the synthetic procedure are the use of large amounts of solvent, potentially expensive chemical precursors and the need for heat treatment post-synthesis, not only to remove volatile materials and impurities but also to help form crystallites from otherwise amorphous products.

6.2.2 Sol–Gel Synthesis

Sol–gel processing has been widely used for synthesizing high surface area materials and is a popular route towards the synthesis of colloidal dispersions of inorganic and organic-inorganic hybrid materials. A sol can be defined as a colloid suspended in a liquid medium, whereas a gel is as interconnected sol network, forming a rigid framework. Since the colloids that make up the sol are between 1–100 nm, the gel they make up can have pore szes in the nanometre to micrometre scale.[30] The sol–gel technique has proven extremely versatile since it enables the preparation of a large variety of metal oxides at relatively low temperatures *via* the processing of metal salt or metal alkoxide precursors. The alkoxide sol–gel route *via* hydrolysis and poly-condensation of a metal oxide in the presence of an acid/base catalyst is one of the most widely practiced. The structure and composition of nano-oxides formed by this method, however, depends on several parameters, such as the preparation condition, the nature of the precursors, the kinetics, growth reactions, hydrolysis, condensation reactions, the ion source, pH, and consequently the structure and properties of the gel.[31,32] In addition to being a controllable and practical method to make glasses and ceramics, the mild reaction conditions of sol–gel syntheses allow for the preparation of other unique material geometries such as films, fibres, coatings and, of particular interest, nanoparticles.[33] Generally, the sol–gel process is classified as either inorganic-based or alkoxide-based.[30,33–40] Although it offers advantages such as: (i) good process control, (ii) ease of formation of homogeneous multicomponent systems due to mixing in a liquid medium and (iii) low temperature for materials processing, similarly to the hydrolysis route, the procedure often needs heat treatment post-synthesis.

6.2.3 Hydro-/Solvo-thermal

The hydro-/solvo-thermal synthesis is a process of crystallizing substances from high temperature aqueous solutions or solvents of choice at high vapour pressures.[41] It can be defined as a method for the synthesis of single crystals, which are dependent on the solubility of minerals in hot water/solvents under high pressure. The crystal growth is performed in an apparatus consisting of a steel pressure vessel called an autoclave, in which a precursor is dispersed in a suitable medium. The reaction parameters that can be controlled are concentration, pH, time, temperature and pressure. The use of additives, mineralizers and structure directing agents provide further avenues for the control of the reaction pathways and crystal morphologies. Further, materials having a high vapour pressure near their melting points can be synthesized by the hydrothermal method. Possible advantages of the hydrothermal method over other types of crystal growth include the ability to create crystalline phases without the need for post-synthesis calcination. This method is also suitable for the growth of large higher quality crystals while maintaining good control over their composition.[41–45]

The solvo-thermal method is similar to the hydro-thermal method except that the solvent used is non-aqueous. In this case, much higher reaction temperatures can be used, since a variety of organic solvents with high boiling points can be employed. The solvo-thermal method generally has greater control on size than the hydrothermal method and has been found to be a versatile method for the synthesis of a variety of nanoparticles with narrow size distribution and dispersity.[46] The solvothermal method has been employed to synthesize, for example, TiO_2 nanoparticles and nanorods with and without the aid of surfactants.[45–48]

6.2.4 Thermolysis

In this method, the basic strategy is to synthesize nanoparticles using coordinating metal precursors in a non-hydrolytic process and then use the resulting nanoparticle as a seed for further growth. Depending on the size of the seed nanoparticle, monodispersed nanocrystals up to 20 nm in diameter have been synthesized.[26] Since the seed mediated growth takes place above the thermal decomposition temperature, the rate at which the growth occurs is the key factor in determining the resulting shape of the nanoparticle.[26,49]

6.2.5 Sonochemical

Sonochemistry uses ultrasound to assist or enhance a chemical reaction. Nanoparticles can be produced using sonochemical methods that involve the process of cavitation.[50,51] Cavitation is best described as the formation, growth and collapse of bubbles.[51] Using sonochemical methods, different titania materials have been produced.[52,53] In most cases, the end product is amorphous and highly aggregated, making it difficult to characterize as individual particles.[54]

6.2.6 Electrochemical

Electrodeposition is commonly employed to produce a coating, usually metallic, on a surface by the reduction of a species at the cathode. The substrate to be coated acts as the cathode and is immersed into a solution containing a salt of the metal to be deposited. The metallic ions are attracted to the cathode and reduced to metallic form. With the use of a template, TiO_2 nanowires can be obtained by electrodeposition.[55,56]

6.2.7 Microwave Synthesis

A dielectric material can be processed with energy in the form of high-frequency electromagnetic waves. The principal frequencies of microwave heating are between 900 and 2450 MHz. At lower microwave frequencies, conductive currents flowing within the material due to the movement of ionic constituents can transfer energy from the microwave field to the material. At higher frequencies, the energy absorption is primarily due to the molecules with a permanent dipole, which tend to reorient under the influence of a microwave electric field. This reorientation loss mechanism originates from the inability of the polarization to follow extremely rapid reversals of the electric field, so the polarization phasor lags the applied electric field. This ensures that the resulting current density has a component in phase with the field, and therefore power is dissipated in the dielectric material. The major advantages of using microwaves for industrial processing are the rapid heat transfer, and volumetric and selective heating. Microwave radiation has been used to prepare various metal oxide nanomaterials, especially TiO_2.[57-59]

6.2.8 Biomimetic Mineralization

Biomimetic mineralization within protein cages provides an attractive alternative approach for synthesizing monodispersed nanosized particles. Since maximum particle sizes are limited by the cage inner diameter, reactions carried out in cages can lead to highly monodispersed size distributions, particularly if nucleation rather than growth is the rate-limiting stage of the synthesis. Control of the size and shape of the particles can be obtained through the use of various distinct protein cages in an expanding library of structures.[60-63] Finally, the protein shells surrounding the particles synthesized by this method provide a framework of amino acid side-chains for further synthetic processing, useful, for example, for forming uniform spatial arrays.[64]

6.3 Catalysis

Catalysis in chemical reaction pathways is a technique employed to increase the rate at which equilibrium is achieved through the addition of a substance or combination of substances (catalyst, co-catalyst) such that, once the reaction is at equilibrium, the material is indistinguishable from its original form. Industry favors catalytic processes induced by a heterogeneous catalyst over the homogeneous one in view of its ease of handling, simple workup and regenerability. Heterogeneous catalysis involves at least one of the reactants being in a distinct phase from the catalyst. In the coming decades, the focus in this area would particularly revolve around: improvement of atom economy

Nanoscale Oxides in Catalysis 135

via catalysis, development of catalysts approaching 100% selectivity and catalysts as devices or integral components of devices.

6.3.1 Effects of Nanostructuring and Morphology

Nanostructured metal oxide surfaces have metal cations separated by oxygen anions that are intrinsically analogous to mononuclear metal complexes in solution. Thus, they offer possibilities of uniformity and site isolation that supported metal catalysts prepared by conventional means cannot. Additionally, surface atoms in nanomaterials possess fewer nearest neighbors and therefore have unsatisfied bonds exposed on the surface. Due to these unsatisfied bonds, they are under an inwardly directed force that leads to the shortening of bond distances between the surface atoms and the bulk atoms. This bond shortening becomes more significant as the ratio of surface atoms to interior atoms increases. The extra energy possessed by the surface atoms alters the properties of the bulk materials, including morphology, band gap, reactivity and catalytic potential. Thus, as the radius of the particle decreases, the chemical potential increases and leads to a more reactive surface. This is one of the most significant ways in which nanomaterials differ from bulk materials. Engineering of catalysts is thus typically focused on the shape, size, defects and components of the catalysts.

6.4 Some Typical Nanostructured Metal Oxides and their Catalytic Applications

6.4.1 Nanocrystalline Magnesium Oxide (MgO)

The synthesis of nanocrystalline MgO is well-established in the literature and various types are now commercially available from NanoScale Materials Inc., Manhattan, Kansas, USA. Abbreviations and specific properties for various types of MgOs are listed in Table 6.1.

These metals oxides possess very high lattice energies and melting points due to their high ionicity. Furthermore, these solids can exist with numerous surface sites with enhanced surface reactivity, such as crystal corners, edges or ion vacancies (Figure 6.1).[65–70] Nanocrystalline NAP-MgO has a three dimensional polyhedral structure, with high surface concentrations of edge/corner and various exposed crystal planes (such as 002, 001, and 111), leading to an inherently high surface reactivity per unit area. The NAP-MgO has Lewis acid

Table 6.1 Some physical properties of different crystallites of MgO.

Entry	Catalyst	Abbreviation	Specific surface area
1	Commercial MgO	CM-MgO	10–30 $m^2 g^{-1}$
2	Conventionally prepared MgO	NA-MgO	130–250 $m^2 g^{-1}$
3	Aerogel prepared MgO	NAP-MgO	300–590 $m^2 g^{-1}$

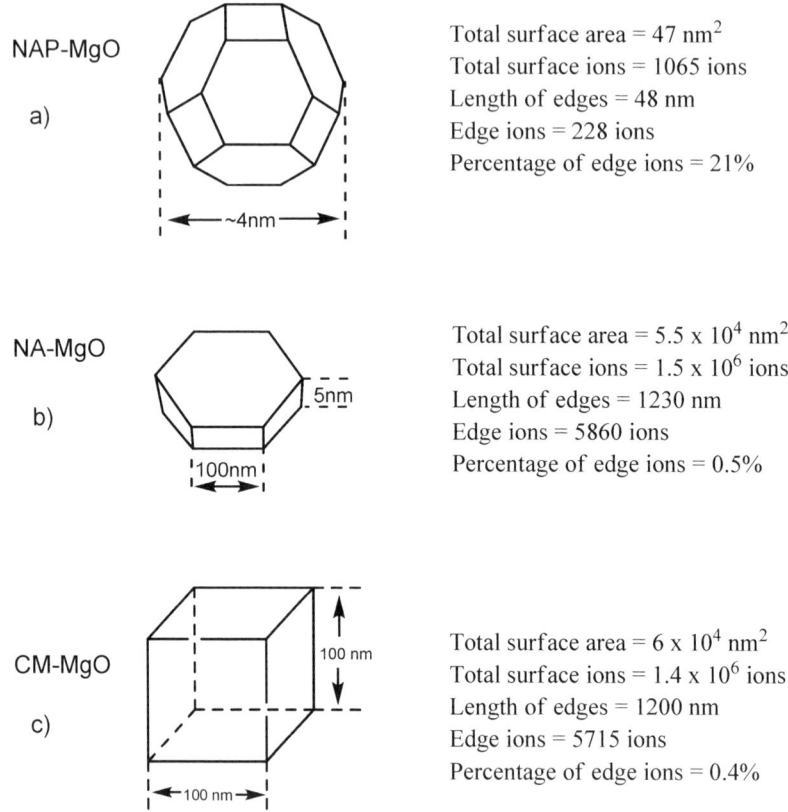

Figure 6.1 (a) Model of MgO nanocrystal of polyhedral shape (NAP-MgO); (b) model of MgO hexagonal microcrystal (NA-MgO); (c) model of cube shaped microcrystal (CM-MgO). (WILEY-VCH Verlag GmbH. Reprinted with permission.)

site Mg^{2+}, Lewis basic sites O^{2-} and O^-, lattice bound and isolated Brønsted hydroxyls, and anionic and cationic vacancies (see Figure 6.1 and 6.2).[65,66,71–77]

In general, residual surface hydroxides add to the rich surface chemistry exhibited by metal oxides such as MgO, CaO, SrO, BaO, Al_2O_3, TiO_2, Fe_2O_3, ZnO and others, and this chemistry is generally attributable to Lewis acid, Lewis base, and Brønsted acid sites of varying coordination; that is, respectively, metal cations, oxide anions and surface hydroxyls which can be isolated or lattice bound (Figure 6.2).

Preparation of (NA-MgO):[72,74,78] Several grams of commercially available MgO was refluxed with rapid stirring in 500 mL of distilled water overnight. After cooling, the slurry was filtered, and the filter cake was dried in an oven at 120 °C. The dried powder was broken into pieces and heat treated to 500 °C under vacuum (1×10^{-3} Torr) in a Pyrex reaction tube that fits into a cylindrical furnace. Heating took about 12 h, and the sample was maintained at 500 °C for several hours, usually overnight.

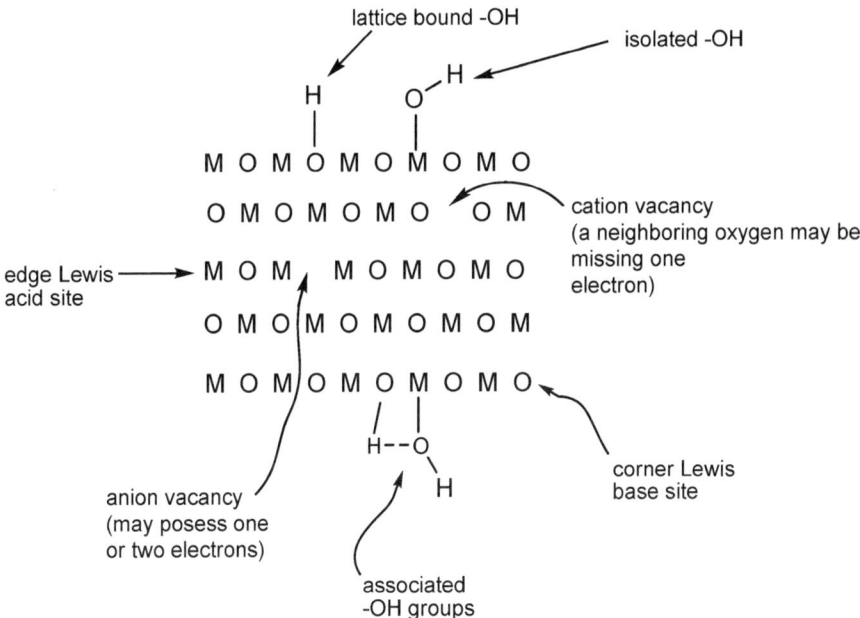

Figure 6.2 Illustration of reactive surface sites on highly ionic metal oxides. (WILEY-VCH Verlag GmbH. Reprinted with permission.)

Preparation of (NAP-MgO):[71,72,74,78] In a three-necked 2 L round bottom flask equipped with a mechanical stirrer, water cooled condenser and argon inlet with a three-way stopcock was placed 300 mL of toluene. In another flask, 2.4 g (0.1 mol) of Mg turnings was allowed to react with 100 mL of anhydrous CH_3OH under argon. The resulting 1 M solution of $Mg(OCH_3)_2$ was added dropwise to the toluene with vigorous stirring under argon. Then 4 mL (0.22 mol) of distilled water was added dropwise by a syringe over a 30 min period, and the solution was stirred at room temperature under argon overnight. The resulting slightly milky solution (gel-like) was placed in an autoclave, slowly heated to 265 °C, vented, and thus converted to a $Mg(OH)_2$ aerogel.[72] After cooling, the slurry was filtered, and the filter cake was dried in an oven at 120 °C. The dried powder was broken into pieces and heat treated to 500 °C under high vacuum in a Pyrex reaction tube that fits into a cylindrical furnace. Heating took about 12 h, and the sample was maintained at 500 °C for several hours, usually overnight.

Preparation of Sil-NAP-MgO:[77,78] A mixture of 0.5 g of NAP-MgO and 0.3 g of methoxytrimethylsilane in 20 mL of toluene was refluxed for 7 h. The reaction mixture was allowed to cool and centrifuged to obtain silylated NAP-MgO, which was washed several times with *n*-pentane.

Over the last two decades, Klabunde and co-workers have reported many uses of nanocrystalline magnesium oxide (Nano MgO), such as efficient destructive chemisorbents for toxic gases, NO_2, SO_2, SO_3, and HCl, as well as chlorinated and phosphorous containing compounds, and for the dehydrohalogenation of chlorohydrocarbons, and chlorination of alkanes.[65,79,80] Choudary and Lakshmi Kantam et al. have extensively researched in this area to further the applications of nanocrystalline magnesium oxide in organic synthesis. They have demonstrated for the first time the transfer of molecular chemistry to surface metal-organic chemistry by fabricating a single site catalyst for successful induction of asymmetric centre in a prochiral substrate[81] since nanomaterials are expected to have a well defined shape and size. Recently, there have been a number of reports on nanocrystalline magnesium oxide (Nano MgO) in many areas of chemistry, the catalytic applications in synthetic organic chemistry are summarized in Figure 6.3.

Lakshmi Kantam et al. have demonstrated the other uses of nano MgO as a catalyst in several organic transformations (Scheme 6.1), such as cyanosilylation of aldehydes and ketones,[82] Baylis–Hillman reaction,[83] Strecker reaction,[78] Wittig reaction,[84] synthesis of organic carbonates,[85] synthesis of α-diazo-β-hydroxy esters,[86] and flavanones.[87] Nano MgO has also been shown to be promising support for many metal-assisted reactions, including nanocrystalline magnesium oxide stabilized palladium-catalyzed Heck coupling-hydrogenation,[88] reduction of nitro compounds[89] and Heck and Sonogashira reactions,[90] nanocrystalline magnesium oxide stabilized molybdenum catalyzed aerobic oxidation of alcohols,[91] ruthenium nanoparticles stabilized on nanocrystalline magnesium oxide by ionic liquids for transfer hydrogenation,[92] and ruthenium species stabilized on nanocrystlaline magnesium oxide by basic ionic liquids for aerobic alcohol oxidation[93] and achiral dihydroxylation of olefins by osmate.[94] Nano MgO has also been employed for the the synthesis of biodiesel,[95,96] substituted 2-amino-2-chromenes,[97] the decomposition of ammonium perchlorate,[98] in the Claisen–Schmidt condensation[99] and for environmental remediation[100] etc.

The versatile Wadsworth–Emmons (WE) reaction has numerous applications in the elegant synthesis of intermediates for fine chemicals, such as perfumes, fragrances, analgesics, insecticides, carotenoids, pheromones,

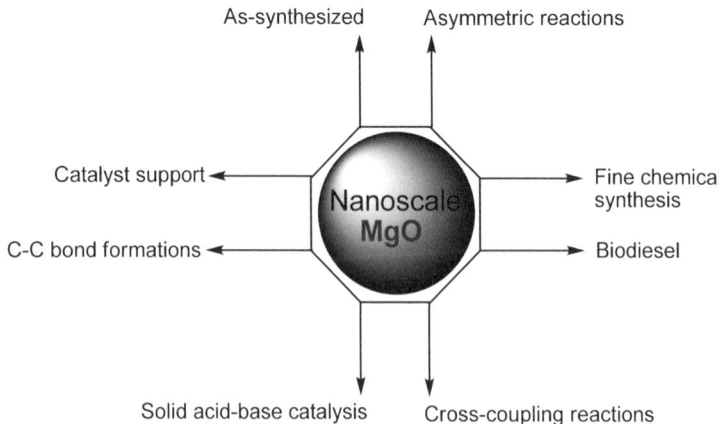

Figure 6.3 Uses of nano MgO in heterogeneous catalysis

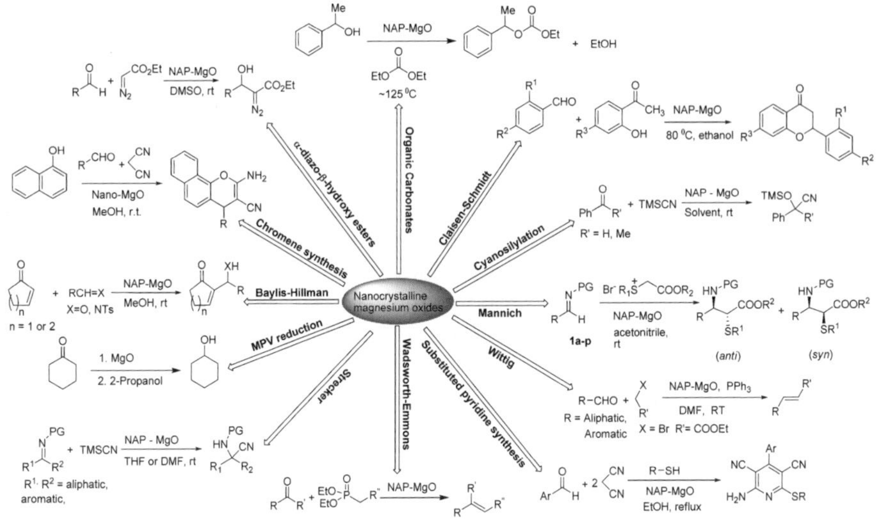

Scheme 6.1 Schematic representation of organic reactions catalyzed by nanocrystalline aerogel prepared magnesium oxide (NAP-MgO). (Springer, reprinted with permission.)

pharmaceuticals and prostaglandins, and is classically induced by bases under homogeneous[101–106] or heterogeneous conditions.[107,108]

Choudary et al. reported[109] the use of NAP-MgO as a catalyst for the Wadsworth–Emmons (WE) reactions in toluene at reflux temperature in good to excellent yields of α,β-unsaturated esters (Scheme 6.2). Various magnesium oxide crystals of CM-MgO, NA-MgO, NAP-MgO were evaluated in the WE reaction between benzaldehyde and triethylphosphonoacetate to understand the relation between structure and reactivity. All types of magnesium oxides catalyze the WE reaction with excellent yields. However, NAP-MgO shows higher reactivity over NA-MgO and CM-MgO. The scope of the methodology was further extended using NAP-MgO in the reaction of aromatic, aliphatic, cyclic and heterocyclic aldehydes with phosphonates.

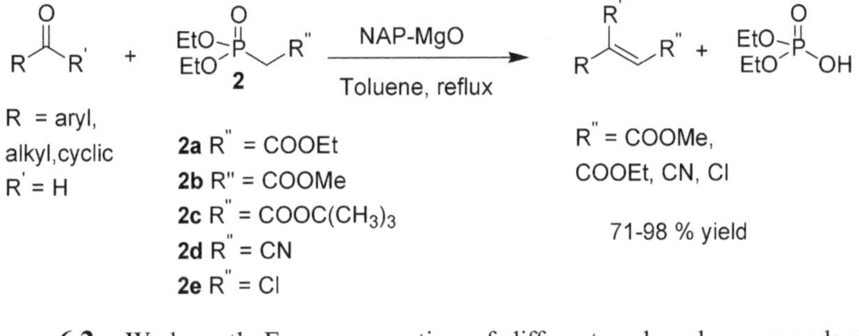

Scheme 6.2 Wadsworth–Emmons reaction of different carbonyl compounds with various phosphonates catalyzed by NAP-MgO

The Wittig reaction and its variants have been acknowledged as powerful and versatile tools in organic synthesis for the formation of C–C double bonds.[110–112] The most important intermediates for several biologically active molecules,[113,114] fluoro compounds,[115] have been synthesized through Wittig reactions.

The most impressive variant in the Wittig reaction is the replacement of the three step-process, involving the preparation of phosphonium salt, followed by base treatment to give ylide and subsequent reaction with carbonyl compounds to give olefinic products with the one pot synthesis. Choudary and co-workers reported[84] the nanocrystalline magnesium oxide catalyst (NAP-MgO) catalyzed Wittig reaction to afford α,β-unsaturated esters and nitriles in high E-stereoselectivity. Nanocrystalline MgO (NAP-MgO) was found to be more active than other MgO catalysts in Wittig reaction (Scheme 6.3).

Flavanone and its derivatives are important intermediates in the synthesis of anti-cancer, anti-inflammatory, anti-bacterial and anti-AIDS drugs.[116,117] Choudary and co-workers reported[87] NAP-MgO catalyzed direct synthesis of flavanones in a single pot with high selectivity and conversions under mild recyclable conditions (Scheme 6.4). All these MgO samples catalyzed both Claisen–Schmidt condensation of benzaldehyde with 2-hydroxyacetophenone followed by cyclization to obtain flavanones with moderate to good yield in a single-pot. However, the nanocrystalline MgO (NAP-MgO) was found to be more active than CM-MgO and NA-MgO.

Organic carbonates are important intermediates for the synthesis of fine chemicals, pharmaceuticals, plasticizers and synthetic lubricants.[118–121] In 2007, Kantam *et al.* reported[85] a method for the synthesis of unsymmetrical organic carbonates in quantitative yields *via* direct condensation of various alcohols with diethyl carbonate in the presence of nanocrystalline magnesium oxide, NAP-MgO (Scheme 6.5). Several magnesium oxide derivatives and other nano metal oxides were screened for the synthesis of organic carbonates. Among these metal oxides, NAP-MgO provided best results. NAP-MgO displayed the highest activity and selectivity compared to that of NA-MgO and CM-MgO. Besides this, NAP-MgO has a Lewis acid site Mg^{2+}, Lewis basic sites O^{2-} and O^-, lattice bound and isolated Brønsted hydroxyls, and anionic and cationic vacancies. The direct condensation of various alcohols with DEC is known to be driven by base catalysts, and accordingly, the surface –OH and O^{2-} of these oxide crystals are expected to trigger these reactions.

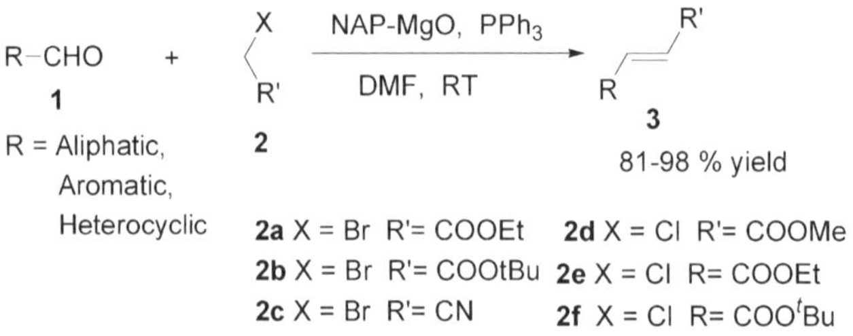

R–CHO + X–CHR' (2) → (NAP-MgO, PPh₃, DMF, RT) → R'CH=CHR (3)

1

R = Aliphatic, Aromatic, Heterocyclic

81-98 % yield

2a X = Br R'= COOEt
2b X = Br R'= COOtBu
2c X = Br R'= CN
2d X = Cl R'= COOMe
2e X = Cl R = COOEt
2f X = Cl R= COO^{t}Bu

Scheme 6.3 Wittig reaction of different aldehydes with various α-halo esters and bromo-acetonitrile catalyzed by NAP-MgO.

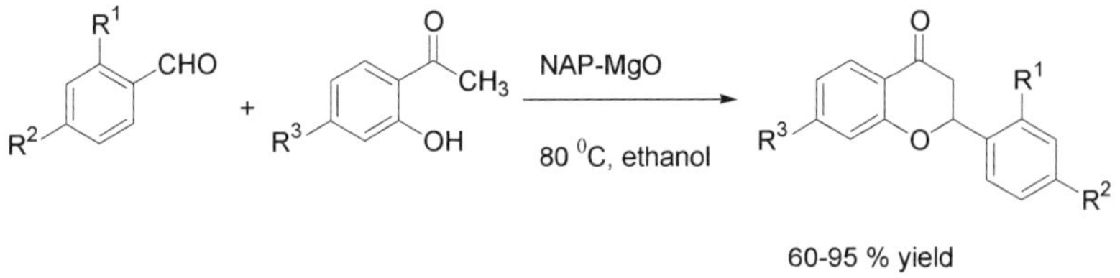

Nanoscale Oxides in Catalysis

R_1 = H, OH, NO_2 , R_2 = H, NO_2, OMe R_3 = H, OH, NO_2

Scheme 6.4   Synthesis of flavanones by nanocrystalline MgO

Interestingly, NAP-MgO appears to be the best catalyst among the other solid base catalysts known to catalyze this reaction.

The Baylis–Hillman (BH) reaction, which involves the coupling of activated alkenes with carbon electrophiles in the presence of Lewis base catalysts is particularly attractive as the resulting adducts serve as key synthetic intermediates for several biologically active molecules.[122] Both Lewis acids and Lewis bases are known to catalyze this reaction under varying reaction conditions. Kantam *et al.* recently reported[83] the use of NAP-MgO for the Baylis–Hillman reaction of cyclic enones with aryl aldehydes and *N*-benzylidine-4-methyl benzenesulfonamides affording the BH adducts in moderate to good yields with high selectivity under mild conditions (Scheme 6.6). Various magnesium oxide crystals CM-MgO, NA-MgO and NAP-MgO were initially screened for the BH reaction. NAP-MgO was found to be more active when compared to the other crystalline forms of MgO.

The Strecker reaction, nucleophilic addition of cyanide ion to the imines, is of great importance to modern organic chemistry as it offers one of the most direct and viable methods for the synthesis of α-amino nitriles.[123,124] α-Amino nitriles are versatile precursors for the synthesis of α-amino acids, various nitrogen and sulfur containing heterocycles such as imidazoles, thiadiazoles and pharmaceuticals. The Strecker reaction has been studied extensively by using various catalyst systems such as Lewis acid and Lewis base catalysts, metal complexes and metal–salen complexes. However, there are only few reports of catalysis of the Strecker reaction using heterogeneous catalysts. Recently, *Kantam et al.* reported[78] the use of nanocrystalline metal oxides as

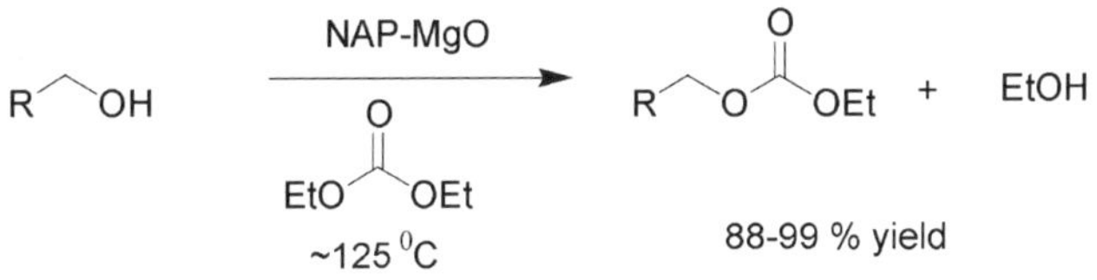

R = Aromatic, Aliphatic, Alicyclic

Scheme 6.5   Synthesis of organic carbonates using NAP-MgO as catalyst at ∼125 °C under solvent free conditions.

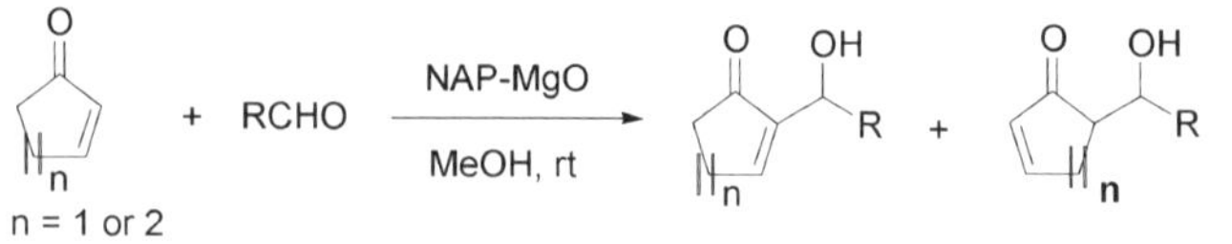

Scheme 6.6 Baylis–Hillman reaction of aryl aldehydes with cyclic enones using NAP-MgO.

catalysts for the Strecker reaction of aldimines and ketoimines to afford α-amino nitriles and α,α-disubstituted α-amino nitriles, respectively, in high yields by using NAP-MgO catalyst (Scheme 6.7).

α-Diazo carbonyl compounds are valuable intermediates for the synthesis of amino alcohols and amino acids.[125] Moreover, the carbene species generated from α-diazo carbonyl compounds are widely used in molecular insertion reactions forming C–C and/ or C–heteroatom bonds.[126] Despite their tendency to interconvert into the corresponding β-keto carbonyl compounds, they are useful synthetic intermediates for natural products.[127] These versatile α-diazo carbonyl compounds are generally prepared by the azido transfer reaction of carbonyl compounds. This can usually be achieved by reaction with a strong base, such as butyllithium, lithium diisopropylamide (LDA), sodium hydride or potassium hydroxide, 1,8-diazabicyclo[5.4.0]undec-7-ene (DBU) or quaternary ammonium hydroxide under controlled conditions.[128]

In 2007, Kantam and co-workers reported[86] the use of NAP-MgO for the synthesis of α-diazo-β-hydroxy esters in good to excellent yields with high selectivity under mild conditions (Scheme 6.8). Various magnesium oxide crystals, CM-MgO, NA-MgO and NAP-MgO were initially screened in the reaction of 4-chlorobenzaldehyde and ethyl diazoacetate at room temperature. NAP-MgO was found to be more active when compared to the other crystalline forms of MgO.

The cyanation reaction of carbonyl compounds is one of the most important methods to obtain poly-functional molecules in organic synthesis.[129] Cyanohydrins or cyanohydrin trimethylsilyl ethers are highly versatile synthetic intermediates, which can be easily converted into α-hydroxy carbonyl derivatives, β-amino alcohols and α-amino acids[130] for their application in pharmaceuticals and agrochemicals.[131] A variety of catalysts have been used to promote the cyanosilylation reactions, such as Lewis acids and Lewis bases. In

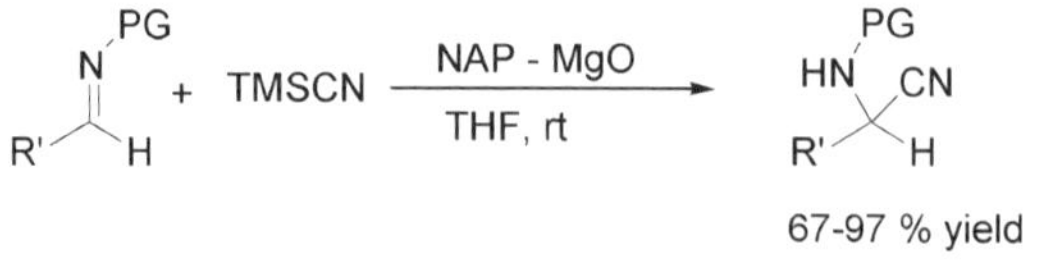

R^1 = 4-MeOPh, 4- NO_2PH, 2-Furyl, 2-Naphthyl, 4-ClPh, *i*-Bu, Ph

PG = Ts, Bzs, Ms, BOC, CBz, Bus, TMS, PMP

Scheme 6.7 NAP-MgO catalyzed Strecker reaction between various aldimines and TMSCN.

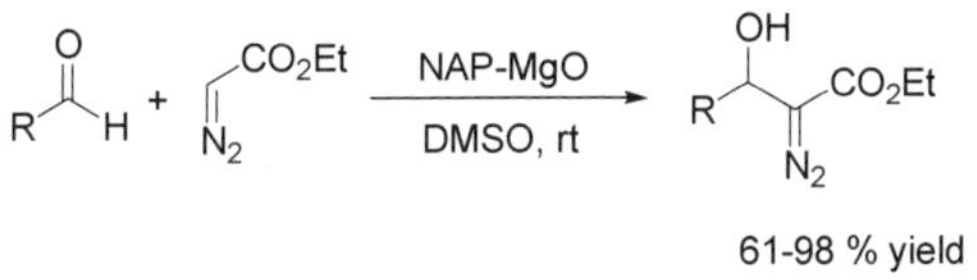

Nanoscale Oxides in Catalysis

$$R\text{-CHO} + \text{N}_2\text{=CH-CO}_2\text{Et} \xrightarrow[\text{DMSO, rt}]{\text{NAP-MgO}} R\text{-CH(OH)-C(N}_2\text{)-CO}_2\text{Et}$$

61-98 % yield

R = 4-CN-C_6H_4, 4-F-C_6H_4, 4-NO_2-C_6H_4, 4-OMe-C_6H_4, Cyclohexyl, 3-Pyridyl, 2-Furfuyl, 2-Thiophenyl

Scheme 6.8 Direct aldol type reaction of various aldehydes and ethyl diazoacetate using NAP-MgO.

2008, Kantam et al.[82] used the NAP-MgO as catalyst for the synthesis of cyanohydrins from carbonyl compounds at room temperature under recyclable conditions (Scheme 6.9).

β-Amino acids and their derivatives, although far less abundant in nature than their α-analogues, are pharmacologically important compounds. They can be used as precursors for medicinally important β-lactam antibiotics, antifungal cyclic β-amino acids and as constituents of biologically active unnatural peptides. The Mannich reaction is one of the direct approaches for the preparation of this useful class of compounds, and has been actively investigated for many years and generally catalyzed by Lewis acids or Lewis bases.[132–135] Homogeneous procedures used for the synthesis of β-amino acid derivatives often involve the use of undesirable toxic and hazardous metal salts. The development of suitable eco-friendly reagents is of great significance and the use of relatively benign heterogeneous catalysts has been reported recently.[136,137]

Kantam et al. reported[138] the use of NAP-MgO as catalyst for the synthesis of α-sulfanyl-β-amino acid derivatives at room temperature (Scheme 6.10). Various forms of magnesium oxide crystals CM-MgO, NA-MgO, NAP-MgO and silylated MgO, Sil-NAP-MgO, were tested in the reaction between N-tosyl benzaldimine and (ethoxycarbonylmethyl)dimethylsulfonium bromide. It was found that all forms of MgO catalyze the reaction in high yields; however, the high surface area NAP-MgO was found to be slightly better than other MgO analogues.

Synthesis of highly substituted pyridine derivatives, such as 2-amino-4-aryl-3,5-dicyano-6-sulfanylpyridines, have received much attention in recent years due their diverse medicinal utility. Kantam et al. developed[139] an eco-friendly and reusable catalytic system of three-component condensation of aldehydes, malononitrile and thiols to afford 2-amino-4-aryl-3,5-dicyano-6-sulfanylpyridines, P, and the corresponding 1,4-dihydropyridines, DP, in moderate to high yields by using nanocrystalline magnesium oxide (NAP-MgO) catalyst.

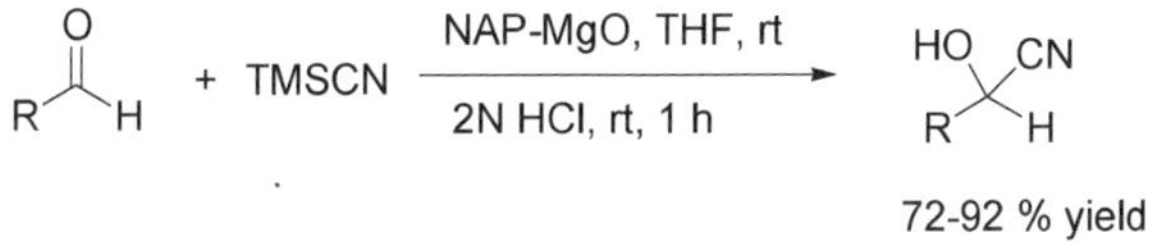

$$R\text{-CHO} + \text{TMSCN} \xrightarrow[\text{2N HCl, rt, 1 h}]{\text{NAP-MgO, THF, rt}} R\text{-CH(OH)-CN}$$

72-92 % yield

Scheme 6.9 Synthesis of various cyanohydrins by using NAP-MgO.

In 2003, Klabunde *et al.* reported[77] the benzylation of toluene using various MgO catalysts (Scheme 6.12). The activity profile of benzylation of toluene with polycrystalline CM-MgO, microcrystalline CP-MgO, and nanocrystalline AP-MgO is in the order CP-MgO > CM-MgO > AP-MgO-1 > AP-MgO-2. They observed less activity with the higher surface area samples (AP-MgO). In general, both CP-MgO and AP-MgO showed higher rates of benzylation of *o*-xylene over toluene, whereas in the case of benzene both were inert.

6.4.2 Copper Oxide (CuO).

Klabunde and co-workers reported the synthesis of nanocrystalline copper oxide (NC CuO).[140] The reactions involved in the preparation are shown below.

The preparation consists of two main steps:

$$CuCl_2 + 2NaOH \rightarrow Cu(OH)_2 + 2NaCl$$

$$Cu(OH)_2 \rightarrow CuO + H_2O$$

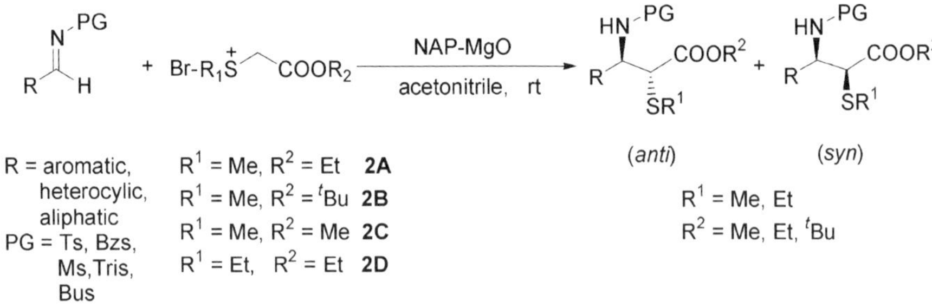
PG
N
R
H
+ Br-R1S+
COOR2
NAP-MgO
acetonitrile, rt
HN
PG
COOR2
R
SR1
+
(anti)
(syn)
R = aromatic, heterocylic, aliphatic
PG = Ts, Bzs, Ms,Tris, Bus
R1 = Me, R2 = Et 2A
R1 = Me, R2 = tBu 2B
R1 = Me, R2 = Me 2C
R1 = Et, R2 = Et 2D
R1 = Me, Et
R2 = Me, Et, tBu

Scheme 6.10 NAP-MgO catalyzed Mannich-type reaction between *N*-sulfonyl aldimines and ester sulfonium salts.

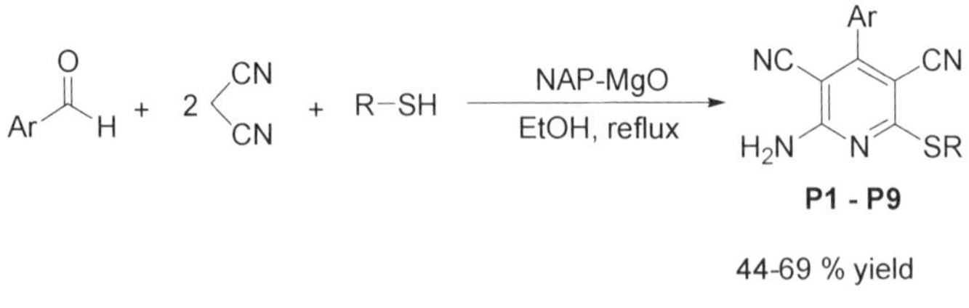
O
Ar
H
+ 2
CN
CN
+ R−SH
NAP-MgO
EtOH, reflux
Ar
NC
CN
H2N
N
SR
P1 - P9
44-69 % yield

Ar = C_6H_5, 4-MeO-C_6H_4, 4-NO_2-C_6H_4, 2-Furyl, 4-HO-C_6H_4, 4-HOOC-C_6H_4, 4-Cl-C_6H_4

R = C_6H_5, 4-Me-C_6H_4, Benzyl, Cyclohexyl

Scheme 6.11 One-pot synthesis of substituted pyridines catalyzed by NAP-MgO

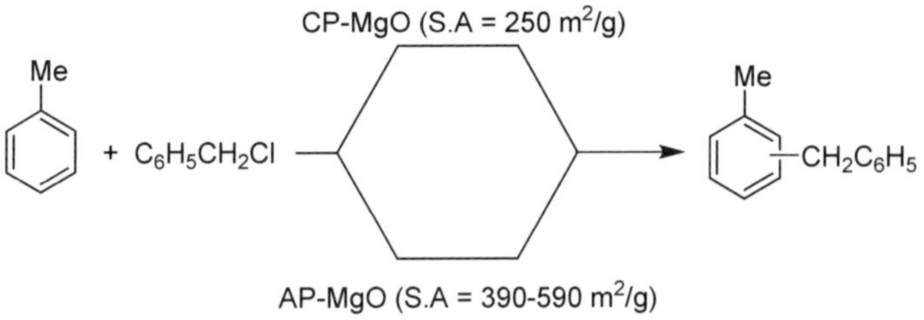

Nanoscale Oxides in Catalysis

$$\text{Me-C}_6\text{H}_5 + \text{C}_6\text{H}_5\text{CH}_2\text{Cl} \xrightarrow[\text{AP-MgO (S.A = 390-590 m}^2/\text{g)}]{\text{CP-MgO (S.A = 250 m}^2/\text{g)}} \text{Me-C}_6\text{H}_4\text{-CH}_2\text{C}_6\text{H}_5$$

Scheme 6.12 Benzylation of aromatic compounds.

6.4.2.1 Synthesis of the Copper Hydroxide Powder

Under argon, 1.5 g (0.0112 mol) of copper(II) chloride (Aldrich) was added to a 250 mL round bottom flask. This was dissolved with 70 mL of absolute ethanol (McCormick) to form a clear green solution. Then 0.0224 mol of sodium hydroxide (Fisher) was dissolved in absolute ethanol (McCormick) and was added dropwise to form the copper hydroxide gel. The reaction mixture was then stirred at room temperature for 2 h. During this time, the reaction mixture forms a blue-green gel. After the reaction was complete, the solution was filtered and washed with water to remove the sodium chloride. The copper hydroxide was then air-dried on the frit, to give a 90% yield.

6.4.2.2 Conversion of Copper Hydroxide to Copper Oxide

Data from thermal gravimetric analysis (TGA) confirmed the copper hydroxide to copper oxide conversion occurs between 190 and 220 °C. The dry copper hydroxide powder was then placed into a Schlenk tube, connected to a flow of argon and surrounded by a furnace, and was heated at 250 °C for 15 min; after heat treatment the copper oxide powder obtained is black. The resulting oxide is in the form of a powder, with the CuO having crystallites in the size range of 7–9 nm. These crystallites aggregate together to form larger spherical particles, which have been studied by transmission electron microscopy and Brunauer–Emmet–Teller methods, and were found to contain many pores and tunnels. The surface area of CuO is about 135 $m^2\ g^{-1}$.

Copper oxide nanoparticles have been of considerable interest not only due to their role in catalysis, but also in metallurgy and in high-temperature superconductors.[141] CuO nanoparticles were found to be effective catalysts for CO and NO oxidation as well as oxidation of volatile organic chemicals such as methanol.[142] Hyeon et al. reported Cu_2O coated Cu nanoparticles as catalysts for the coupling reaction using activated chloroarenes in the presence of Cs_2CO_3.[143]

N-Arylation of heterocycles with chloro- and fluoroarenes using Cu(II) fluoroapatite was also reported.[144] Arylheterocycles are important compounds as they find wide application in medicinal,[145] biological[146] and N-heterocyclic carbene chemistry.[147] There is thus considerable interest in developing efficient synthetic protocols for *N*-arylation of heterocyclic compounds. We have reported the *N*-arylation of heterocycles with activated chloro- and fluoroarenes in excellent yields using nanocrystalline copper oxide (nano-CuO) (Scheme 6.13).[148]

Asymmetric reduction of ketones represents one of the most important methodologies for obtaining enantiomerically enriched alcohols, which are key intermediates for numerous biologically active molecules.[149] In this attempt, asymmetric hydrosilylation of aryl alkyl ketones to afford chiral secondary alcohols with good yields and excellent enantioselectivity is realized by using nanocrystalline copper(II) oxide and BINAP in the presence of organosilanes as the stoichiometric reducing agents (Scheme 6.14).[150]

Further, one-pot multicomponent coupling reactions (MCR) where several organic moieties are coupled in one step is an attractive synthetic strategy.[151,152] Three-component coupling of aldehydes, amines and alkynes (A^3 coupling) is one of the best examples of MCR, and has received much attention in recent times. The resultant propargylamines obtained by A^3 coupling reactions are versatile synthetic intermediates for biologically active compounds such as β-lactams, conformationally restricted peptides, isosteres, natural products and therapeutic drug molecules.[153–156] An efficient three-component coupling of aldehydes, amines and alkynes to prepare propargylamines, in nearly quantitative yields using nanocrystalline CuO as catalyst any co-catalyst was achieved (Scheme 6.15).[157] Structurally divergent aldehydes and amines were successfully converted to the corresponding propargylamines.

The aldol reaction is widely used in synthetic organic chemistry to generate intermediates for anti-hypertensive drugs and calcium antagonists.[158] Chiral β-hydroxy carbonyl compounds can be readily converted to 1,3-*syn*- and *anti*-diols and amino alcohols, which are the building blocks for antibiotics, pheromones and many biologically active compounds.[159] The direct aldol

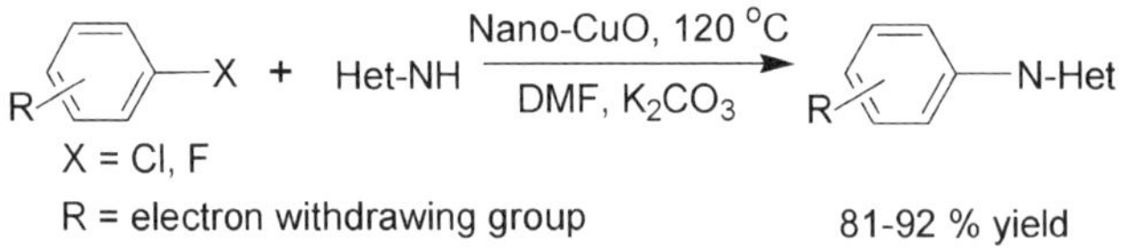

Scheme 6.13 *N*-Arylation of heterocycles with activated chloro- and fluoroarenes.

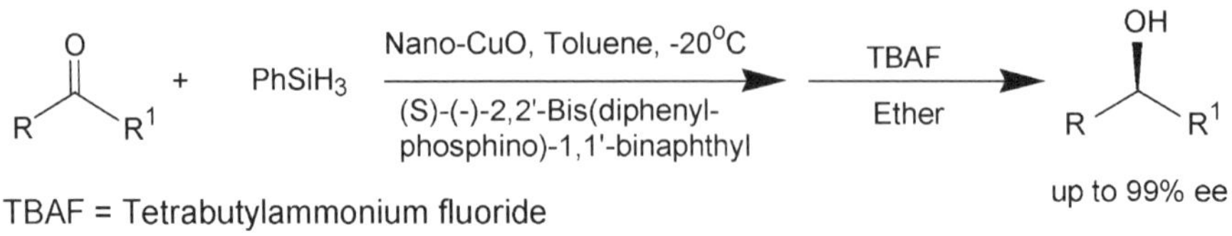

Scheme 6.14 Asymmetric hydrosilylation reaction of ketones using nano CuO in the presence of a chiral ligand.

reaction, starting from an aldehyde and an unmodified ketone, is highly atom efficient[160] compared with the process in which the preconversion of a ketone into a more reactive species such as an enol silyl ether, enol methyl ether or ketone silyl acetal as the aldol donor (Mukaiyama aldol reaction) is required.[161,162] Since the control of stereochemistry during aldol additions is a crucial problem, the metal catalyzed direct asymmetric aldol reaction of aldehydes with unmodified ketones still remains a challenge for synthetic chemists.[163] The direct asymmetric aldol reactions of aromatic and heteroaromatic aldehydes with acetone to afford chiral β-hydroxy carbonyl compounds in good yields and good to moderate enantioselectivities have recently been realized using nanocrystalline copper(II) oxide in the presence of (1S,2S)-(−)-1,2-diphenylethylenediamine at −30 °C (Scheme 6.16).[164]

6.4.3 Titania (TiO$_2$)

Among the various nano-metal oxides, the physics and chemistry of titania (titanium dioxide, TiO$_2$) has been the subject of a great deal of scientific and technological attention. Ancient paintings suggest titanium oxide was used as an artist's pigment because of its suitable refractive index, hiding characteristics and non-toxicity. Presently, TiO$_2$ finds usage in a wide range of applications, such as coatings, cosmetics, ceramics, textiles, food-contact material, leather, pharmaceuticals, photoelectrochromics, photoconductors, sensors, polymer electrolytes, as well as in several biological and biomedical areas. Titania is envisaged to be an effective photocatalyst and plays a major role in addressing several environmental and pollution challenges. The

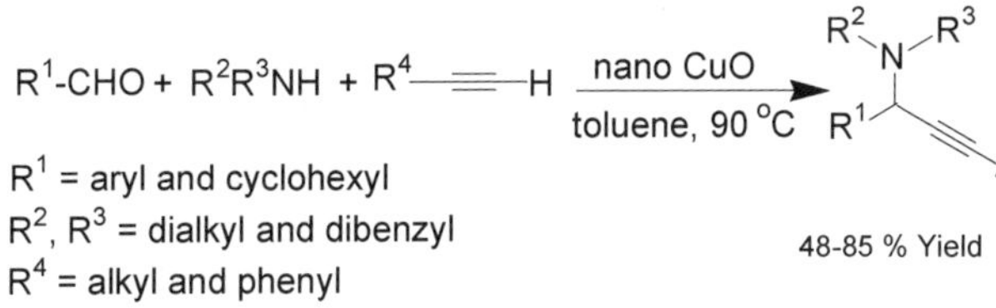

Scheme 6.15 Three-component coupling of aldehydes, amines and alkynes catalyzed by nano CuO.

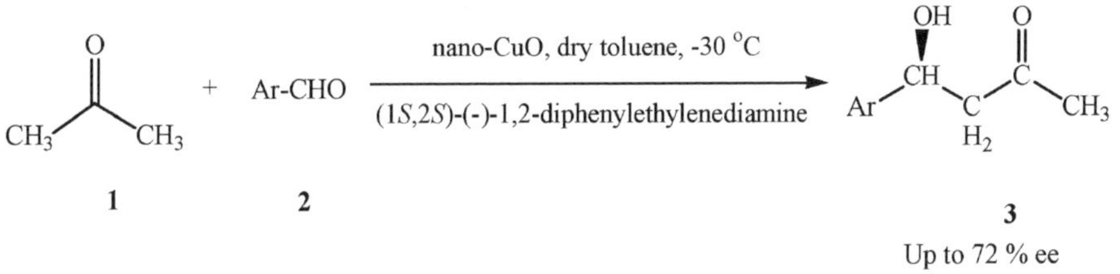

Scheme 6.16 Direct asymmetric aldol reactions of aromatic aldehydes with acetone using nanocrystalline copper(II) oxide

prospect of cost-effective harvesting of solar energy based on photovoltaic and water-splitting devices is also promising. The development of advanced hybrids of TiO_2 with metals, organics and polymers enriches the list of potentially new applications.[165–170]

Each of the three main crystallographic forms of TiO_2 (anatase, rutile, brookite), exhibit different physical properties, such as refractive index, chemical reactivity and photochemical activity,[171] thus offering a considerable degree of freedom to tailor the appropriate material for specific applications.[169,170] Rutile TiO_2 is thermodynamically the most stable of the three phases, possessing a smaller optical band gap (3.0 eV) compared to anatase (3.2 eV)[172] and offers higher chemical stability, relative permittivity, refractive index, dielectric constant and UV absorption. Owing to these qualities, the rutile form finds applications in high quality paints, cosmetics, photocatalysis, and optical and electronic devices.[173,174]

Preparation of TiO_2 involves the following steps:[175] hydrolysis of titanium alkoxides in presence of acid or base followed by condensation.

$$Ti-OR + H_2O \rightarrow Ti-OH + ROH$$

$$Ti-OR + HO-Ti \rightarrow Ti-O-Ti + ROH$$

A TiO_2 sol can be obtained through rapid addition of a solution containing 5.9 mL $Ti(OC_3H_7)_4$ in 75 mL 2-propanol to a mixture of 0.6 mL H_2O and 0.25 mL HNO_3 in 75 mL 2-propanol. The final molar ratios of starting compounds were: $[H_2O]/[Ti(OC_3H_7)_4] = 2$ and $[HNO_3]/[Ti(OC_3H_7)_4] = 0.2$. The TiO_2 powder was first dried at room temperature in air, then at 70 °C, and finally annealed at 200 °C in air for 4 h. Calcination of TiO_2 particles in the powder form at progressively higher temperatures caused morphological and structural changes, which were followed by XRD, UV-Vis spectroscopy, AFM, and SEM microscopy. In particular, it was found that annealing at 500 °C leads to the formation of the anatase phase TiO_2 in the powder form (self-supported). However, after annealing at 800 °C the anatase phase of self-supported TiO_2 particles was easily converted into the rutile form. Annealing TiO_2 powders at 500 and 800 °C led to bigger crystallites with their eventual sintering to very big particles of micrometre size. The TiO_2 powder structure can be inferred from BET surface measurement studies. According to that, the surface area of these powders were 340, 390, 1.1, and 0.7 $m^2\,g^{-1}$ for the samples annealed at 70, 200, 500, and 800 °C, respectively. Assuming for simplicity a spherical shape of TiO_2 particles involved, the sizes derived from the surface area studies are around 3, 2.6, 900, and 1400 nm.[175]

3-Alkylindole derivatives have significant biological and pharmacological importance[176] and can be prepared by Friedel–Crafts alkylation of indoles using epoxides. Epoxide ring opening with different nucleophiles[177] is generally carried out with acid or base catalysts. In a recent report, the use of recyclable nanocrystalline titanium(IV) oxide (nano TiO_2) catalyst for the Friedel–Crafts alkylation of indoles with epoxides affording 3-alkyl indole derivatives at room temperature with moderate to good yields and high regioselectivity was demonstrated (Scheme 6.17).[178]

Michael addition of indoles to α,β-unsaturated ketones is an important C–C bond forming reaction, as the resultant β-indolyl ketones are highly interesting building blocks for the synthesis of biologically active compounds and natural products.[179] They are usually carried out in the presence of catalytic or stoichiometric amounts of Lewis acids. Since the nanometal oxides are composed of Lewis acidic sites, use of nanometal oxides, in particular nanocrystalline titanium (IV) oxide, as catalysts for the Michael addition of indoles to α,β-unsaturated ketones to form β-indolyl ketones with high to good yields has been reported (Scheme 6.18).[180]

The physical and chemical properties of titania nanocrystals are greatly influenced by the methods of synthesis and processing. A variety of synthetic procedures with strict control over phase formation and morphology of TiO_2 nanoparticles have been documented. Simple hydrolysis of titanium salts results in TiO_2 nanoparticles.[181–184] Usually, these synthetic processes lead to a mixture of anatase and rutile polymorphs,[185–187] and the synthesis of monomineralic, well-shaped rutile nanocrystals via solution hydrolysis is relatively difficult in comparison to the synthesis of anatase nanocrystals.[188] Rutile TiO_2 powders are conventionally obtained by phase transformation of anatase to rutile above 450 °C[189] or by flame oxidation of $TiCl_4$.[190] However, these two methods usually result in larger titania particles with lower surface area. The only alternative to control the phase and size of rutile titania particles is by use of a precursor along with other mineralizers and inorganic

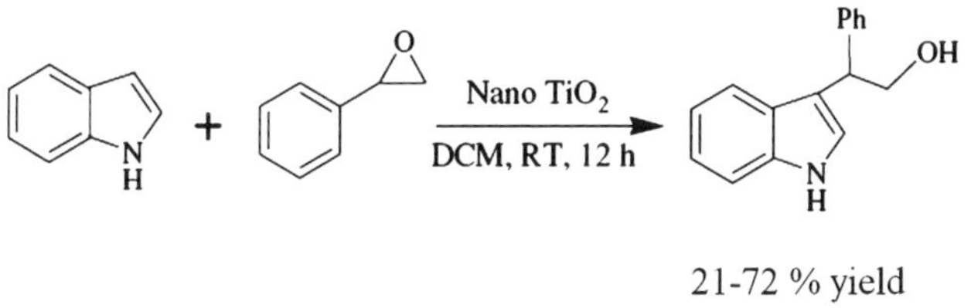

21-72 % yield

Scheme 6.17 Nano TiO_2 catalyzed Friedel–Crafts alkylation reaction of indole with styrene oxide.

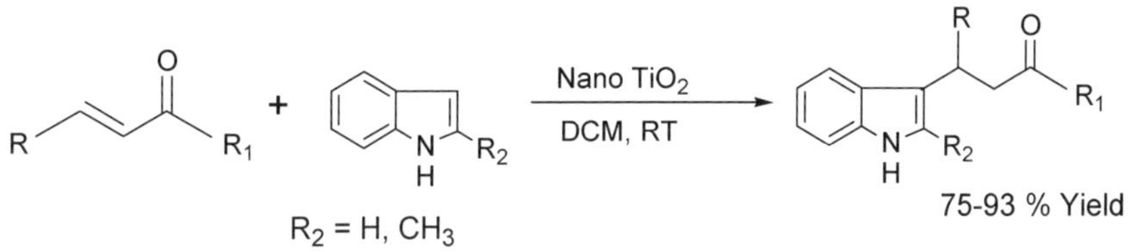

75-93 % Yield

R_2 = H, CH_3

Scheme 6.18 Nano TiO_2 catalyzed Michael addition of α,β-unsaturated ketones with indole and 2-methyl-indole.

additives, like $SnCl_4$, NH_4Cl, $NaCl$ or SnO_2[191] that are difficult to remove and have an undesirable influence in several applications.

The controlled synthesis of crystalline rutile nano-TiO_2 without the use of inorganic additives has also been recently reported.[192,193] All these processes invariably require high temperatures to get crystalline titania which results in an increase in crystallite size. A few studies have reported preparation of nano rutile TiO_2 at low temperature under acidic conditions with large specific surface areas.[194–196] Aruna et al.[197] and Zhang et al.[187,198] prepared nanosize rutile titania by a hydrothermal method using titanium(IV) isopropoxide and the thermal hydrolysis of titanium(IV) chloride, respectively. Pedraza et al.[199] have reported the preparation of large surface area rutile TiO_2 using $TiCl_3$ in O_2 at room temperature. All these processes lead to the formation of spherical or irregularly shaped rutile titania nanoparticles. Amorphous precipitates of rutile titania nanocrystals have been obtained by the hydrothermal treatment or peptization of titanium alkoxide at 433–573 K.[200,201] An improvement in the anatase-to-rutile phase transformation was reported by Bacsa and co-workers.[202] Wang et al.[203] prepared pure rutile titania by peptizing in HNO_3. Broadly speaking, all the synthetic processes described above are spontaneous, unless templates are employed, little or no control over the morphological properties of the products is achieved, and generally give a mixture of spherical, broomlike, or rod-shaped rutile titania.[204–206] Technically attempts to separate nanocrystals of any particular morphology from the others are very difficult.

Developing a method for one-step preparation of nanocrystalline phase-pure rutile TiO_2 powders at low temperature with proper control over morphology is a challenge. Recently, we have reported a simple, efficient low-temperature hydrothermal method to prepare ultrafine phase-pure rutile titania nanocrystallites with controlled morphologies from a readily available, low-cost starting material ($TiCl_4$).[207–209] This has been achieved by a simple variation of the hydrothermal reaction conditions with respect to temperature and time without using additives. Systematic variation of all the reaction parameters coupled with a detailed transmission electron microscopy study, lead us to demonstrate unambiguously the controlled transformation of rutile titania nanocrystals from nano-rods to bunched nano-spindles and finally to spherical nanoparticles, depending on the reaction conditions (Figure 6.4).

6.4.4 Zinc oxide

Zinc oxide is an important semiconductor and has been investigated widely for its catalytic, electrical, optical and photochemical properties.[210] Recently, it has been used as an efficient catalyst for the synthesis of cyclic ureas from diamines and for Friedel–Crafts acylation as well as Beckmann rearrangement reactions.[211] Klabunde reported earlier the synthesis, isolation and the unique chemical reactivity studies of nanocrystalline zinc oxide.[73,212]

The reactions involved in the preparation of Alkoxide Based Nanocrystalline (NC) ZnO are shown. The preparation consists of three main steps:

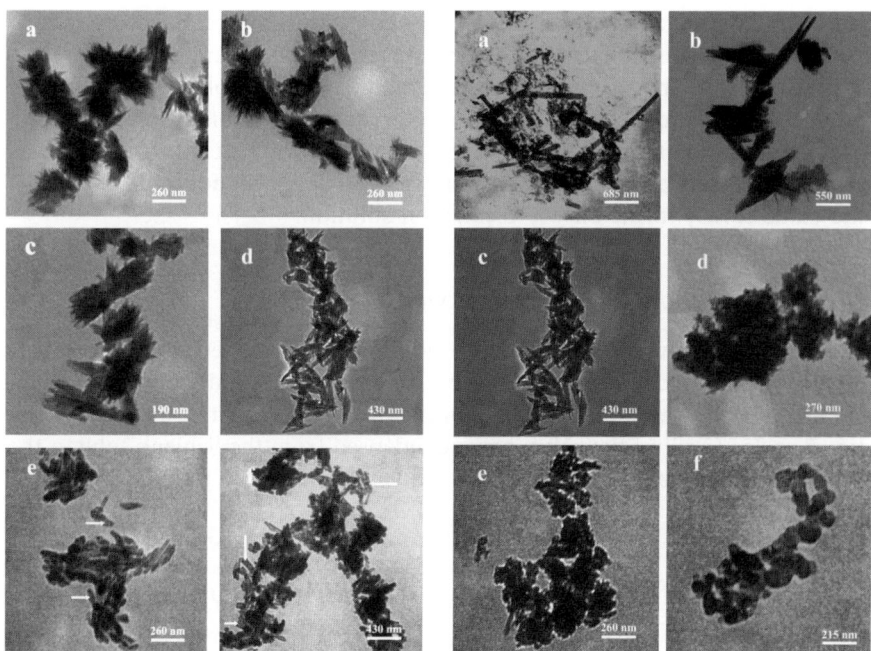

Figure 6.4 (Left) Transmission electron micrographs of hydrothermally synthesized phase-pure rutile TiO_2 nanocrystals obtained for a constant reaction time of 16 h at different reaction temperatures: (a) 40 °C, (b) 60 °C, (c) 80 °C, (d) 100 °C, (e) 120 °C and (f) 150 °C. (Right) Transmission electron micrographs of hydrothermally synthesized phase-pure rutile TiO_2 nanocrystals obtained at a constant reaction temperature of 100 °C for different reaction times: (a) 4 h, (b) 8 h, (c) 16 h, (d) 24 h, (e) 32 h and (f) 48 h. [Manaswita *et al.*, *Mater. Res. Bull.*, 2007, **42**, 1691–1704; reprinted with permission from Elsevier Publications.]

$$Zn(CH_2CH_3)_2 + 2(CH_3)_3COH \rightarrow Zn[OC(CH_3)_3]_2 + 2CH_3-CH_3$$

$$Zn[OC(CH_3)_3]_2 + 2H_2O \rightarrow Zn(OH)_2 + 2(CH_3)_3COH$$

$$Zn(OH)_2 \rightarrow ZnO + H_2O$$

6.4.4.1 Synthesis of the Zinc Oxide Powder

In a glove box, 40 mL (0.040 mol) of 1.0 M diethylzinc in hexane (Aldrich) was added to a 500 mL round-bottom flask. The RB flask was removed from the glove box and placed on a stir plate and cooled to 0 °C with an ice bath under argon. A solution of 5.8 g (0.080 mol) of *tert*-butyl alcohol (Fisher) in 60 mL of hexane (Fisher) was added to the cooled diethylzinc solution *via* a syringe. This solution was added over a time span so that the release of ethane was not extremely violent. Once the solution was completely added, the ice bath was removed and the reaction was allowed to come to room temperature (about 25 °C). The reaction mixture was then stirred at room temperature for 2 h. During this time the reaction mixture remained a clear colorless solution. A solution of 1.44 mL (0.080 mol) of water in 140 mL of absolute ethanol (McCormick) was then added to the reaction mixture. Once the solution was completely added, the reaction was allowed to stir for an additional 2 h. The zinc oxide slowly formed a white colloidal solution.

6.4.4.2 Isolation of the Zinc Oxide Powder

After the reaction was complete, the argon line was removed and the reaction mixture was poured into a Schlenk tube. This was then connected to a vacuum line attached to a second liquid nitrogen trap. While the reaction was being stirred at 25 °C, all of the solvent was condensed from the Schlenk tube to the second trap. This left a dry, white zinc oxide powder with a yield of 2.9 g (90% yield).

6.4.4.3 Activation Heat Treatment of the Zinc Oxide Powder

The dry powder was then placed in another Schlenk tube. This was connected to a flow of argon and surrounded by a furnace. The furnace was connected to a temperature controller, and the temperature was slowly raised from room temperature to 90 °C where it was heated for 15 min. Next the temperature was slowly raised from 90 to 250 °C where it was heated for an additional 15 min. After the heat treatment was complete, the furnace was turned off and allowed to cool to room temperature. The zinc oxide powder remained white. Heat treatment was conducted under dynamic vacuum and compared to the argon flow method, and both methods were found comparable. The resulting ZnO is in the form of a powder, made up of zincite crystallites in the size range of 3–5 nm. These spherical particles have been studied by transmission electron microscopy (TEM) and Brunauer–Emmet–Teller (BET) methods and were found to contain many pores and tunnels. It is because of this that an uncharacteristically high surface area is found, averaging about 120 $m^2\,g^{-1}$. As seen with other metal oxides, once they are made as nanoparticles, their reactivity is greatly enhanced. This is thought to be due to morphological differences, whereas larger crystallites have only a small percentage of reactive sites on the surface, smaller crystallites will possess much higher surface concentration of such sites.

Tetrazoles have a wide range of applications in pharmaceuticals as lipophilic spacers and carboxylic acid surrogates, in materials as specialty explosives and information recording systems, in coordination chemistry as ligands and also as precursors to a variety of nitrogen containing heterocycles.[213] Nanocrystalline ZnO with a specific surface area (BET) 120 $m^2\,g^{-1}$ was purchased from NanoScale Corporation, U.S.A. It was oven dried at 80 °C prior to use. The BET surface area obtained was 60 $m^2\,g^{-1}$. Nanocrystalline

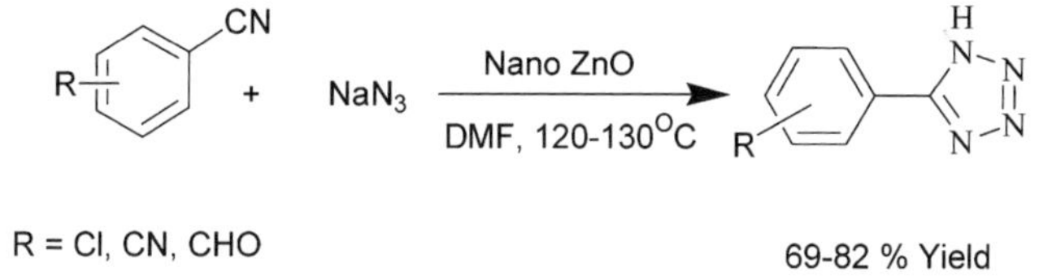

Scheme 6.19 Nanocrystalline ZnO is an efficient heterogeneous catalyst for the synthesis of 5-substituted 1H-tetrazoles.

ZnO is an effective heterogeneous catalyst for the (2+3) cycloaddition of sodium azide with nitriles to afford 5-substituted 1H-tetrazoles in good yields (Scheme 6.19).[214]

6.4.5 Iron Oxides (Fe_2O_3, Fe_3O_4, Mixed Ferrites)

Ferrites form an important class of ceramic-like ferromagnetic materials. Mainly composed of ferric oxide, α-Fe_2O_3, these materials have been considered as highly useful electronic materials for more than a century. Hilpert, in 1909, successfully demonstrated the preparation of spinel ferrites, containing manganese, copper, cobalt, magnesium and zinc. Since then, transition metal ferrites have been a family of oxides that have played an important role in a wide variety of fields. The range of applications is related to the variety of transition metal cations that can be incorporated into the lattice of the parent magnetite ($Fe^{2+}Fe_2^{3+}O_4$) structure. Magnetite, Fe_3O_4, a natural mineral, is the purest form of ferrites. Owing to their increasing importance, researchers in chemistry, materials science, ceramics and metallurgy are actively engaged in the development of new ferrites with enhanced characteristics, and improvements in their manufacturing processes.

Apart from their technological importance as magnetic and electric materials,[215] ferrites have also been well studied for their catalytic behavior,[216,217] especially in some industrially important reactions. Further, complex metal oxides are also used as good sensors for reducing gases such as H_2, H_2S, and LPG.[218]

Conventional semiconducting oxides can be prepared by a variety of methods, and different precautionary measures can be adopted to decrease the solid–solid interactions between particles during the calcination process.[219–222] Pure spinel ferrites require calcination at 800–1000 °C in a solid-state or sol–gel preparation process, probably because of the loose contact between metal and iron atoms and the possible low diffusion rate.[223] The preparation of different spinel ferrites (AFe_2O_4), where A is Zn, Co, Ni, Cu or Cd, at relatively low temperatures with high surface area have been reported, which show gas sensing behavior for reducing gases like LPG with high sensitivity and good selectivity. All these processes are directed towards obtaining very fine particles. In hydrothermal synthesis, because of the relatively high electrostatic potential difference between the solutions with the surface charge, it can result in hetero-coagulation of single-phase products.[221,224] Hydrothermal powder synthesis has been known as a powerful method for the preparation of fine, high-purity and homogeneous powders of various single-component and multi-component oxide powders. In addition, the dilution, pH, temperature and thereby the pressure and the reaction time can be controlled to obtain particles of required size. Therefore, the hydrothermal synthesis route to obtain nanocrystalline ferrites suitable for gas sensing applications at very low temperatures is sometimes preferable.

The development of the synthesis of ceramics at low temperatures is believed to be important in relation to energy conversion because most ceramic materials have been prepared at high temperatures. Patil *et al.*[225] reported an entirely different approach for the synthesis of simple and complex oxide materials, in which a new class of precursors containing a carboxylate anion, hydrazide, hydrazine or hydrazinium groups were found to ignite at low temperatures and decompose auto-catalytically to yield fine particles of oxides with large surface area. This is because hydrazine monohydrate $(NH_2)_2.H_2O$ is a strong reducing and basic agent and produces Fe^{2+} by reduction of Fe^{3+} from $Fe(NO_3)_3$ solution due to its strong reducing action. Hydrazine prevents the formation of ferric oxide in the synthesis of magnetite (Fe_3O_4),[226] and it complexes with metal nitrates and is also used as a precursor for the zinc–nickel mixed ferrite.[227] Moreover, as a reducing agent, hydrazine has been widely used for non-electrolytic plating (chemical plating) and for precipitation of metals from solution, for example, fixing solution of photo films. Ueda *et al.*[228] reported low temperature synthesis of zinc ferrite using hydrazine monohydrate with chlorides as the starting materials. Gopal Reddy *et al.* prepared γ-Fe_2O_3[229] and nanocrystalline $Fe_2O_{3(0.9)}$–$SnO_{2(0.1)}$[230] powders by the hydrazine method and used them as alcohol sensor materials. Hence, hydrazine monohydrate has been used to obtain nanostructured mixed nickel ferrite at a relatively low temperature.

Recently ferrites and mixed ferrites are generating lot of interest and investigation into their use as catalytic supports for heterogeneous reactions. The primary formidable task in using such nanoparticles for heterogeneous catalysis is the problem associated with the separation of the catalyst after the reaction. It is here that nanoparticles, especially superparamagnetic particles amenable for magnetic separation, come as a redeemer.[231–233] Magnetic separation is an attractive alternative to filtration or centrifugation as it prevents loss of the catalyst and the reusability increases. This makes the catalyst cost-effective and promising for industrial applications. Manorama and co-workers[232,233] have demonstrated such an application for hydrogenation reactions, Suzuki and Heck coupling reactions using anchored palladium on Ferrite nanoparticles (Figure 6.5).

6.5 Conclusions

The chapter is an attempt to provide the readers with a comprehensive overview on some of the new strategies practiced in the area of heterogeneous

Nanoscale Oxides in Catalysis 155

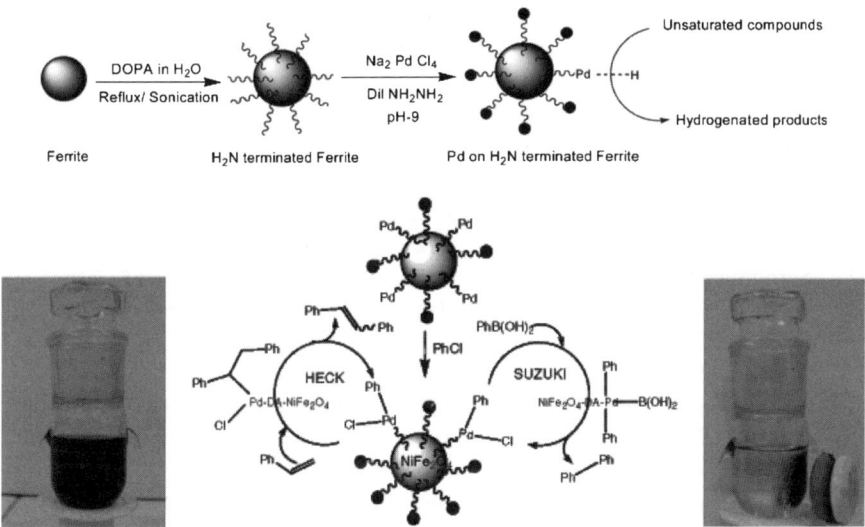

Figure 6.5 Schematic representation illustrating ease of the magnetically separable heterogeneous catalysis.

catalysis with special emphasis on nanostructured metal-oxides that have demonstrated prospects for future. Some of the model metal oxide systems which are of immense interest to the researchers and few of the industrially important organic transformations are discussed in this context. Though our sincere effort have been to provide all the details and appropriate citations relevant to our discussions, with the volume and rapid pace of research in this area, some gaps might remain. Despite phenomenal progress over the years, the challenge to control and manipulate the morphology at nanoscale, absolute control on the properties of these nanomaterials and oriented assemblies of preferred planes with selective catalytic activity would keep the academics, scientists and technologists motivated for years to come.

References

1. G. A. Somorjai and Y. Li, *Introduction to Surface Chemistry and Catalysis*, 2nd edn, John Wiley, Hoboken, 2010.
2. G. Ertl, H. Knozinger and J. Weitkamp, *Handbook of Heterogeneous Catalysis*, Wiley-VCH, Weinheim, 2008.
3. C. H. Bartholomew and R. J. Farrauto, *Fundamentals of Industrial Catalytic Processes*, 2nd edn, John Wiley, New York, 2005.
4. V. Smil, *Nature* 1999, **400**, 415.
5. L.-S. Zhong, J.-S. Hu, A.-M. Cao, Q. Liu, W.-G. Song and L.-J. Wan, *3D Chem. Mater.*, 2007, **19**, 1648.
6. A. T. Bell. *Science*, 2003, **299**, 1688.
7. D. R. Rolison. *Science*, 2003, **299**, 1698.

8. I. Chorkendorf and J. W. Niemantsverdriet, *Concepts of Modern Catalysis and Kinetics*. 2nd edn, Wiley-VCH, 2007.
9. A. T. Bell, B. C. Gates and D. Ray, *Basic Research Needs: Catalysis for Energy*, U. S. DOE Office of Science, 2007.
10. J. M. White and J. Bercaw, *Opportunities for Catalysis in the 21st Century*, U. S. DOE Office of Science, 2002.
11. R. Feynman, *Eng. Sci.*, 1960, **23**, 22.
12. S. Chikazumi, S. Taketomi, M. Ukita, M. Mizukami, H. Miyajima, M. Setogawa and Y. Kurihara, *J. Magn. Magn. Mater.*, 1987, **65**, 245.
13. A-H. Lu, W. Schmidt, N. Matoussevitch, H. B. Pnnermann, B. Spliethoff, B. Tesche, E. Bill, W. Kiefer and F. Schüth, *Angew. Chem., Int. Ed.*, 2004, **43**, 4303.
14. S. C. Tsang, V. Caps, I. Paraskevas, D. Chadwick and D. Thompsett, *Angew. Chem., Int. Ed.*, 2004, **43**, 5645.
15. A. K. Gupta and M. Gupta, *Biomaterials*, 2005, **26**, 3995.
16. S. Mornet, S. Vasseur, F. Grasset, P. Verveka, G. Goglio, A. Demourgues, J. Portier, E. Pollert and E.Duguet, *Prog. Solid State Chem.*, 2006, **34**, 237.
17. Z. Li, L. Wei, M. Y. Gao and H. Lei, *Adv. Mater.*, 2005, **17**, 1001.
18. T. Hyeon, *Chem. Commun.*, 2003, 927.
19. D. W. Elliott and W-X. Zhang, *Environ. Sci. Technol.*, 2001, **35**, 4922.
20. M. Takafuji, S. Ide, H. Ihara and Z. Xu, *Chem. Mater.*, 2004, **16**, 1977.
21. P. Tartaj, M.P. Morales, S. Veintemillas-Verdaguer, T. Gonzalez-Carreno and C. J. Serna, *Handbook of Magnetic Materials*; Elsevier, Amsterdam, 2006.
22. H. P. Beck, W. Eiser and R. Haberkorn, *J. Eur. Ceram. Soc.*, 2001, **21**, 687.
23. Z. X. Tang, C. M. Sorensen, K. J. Klabunde and G. C. Hadjipanayis, *J. Colloid. Interface Sci.*, 1991, **146**, 38.
24. R. Massart, *IEEE Trans. Magn.*, 1981, **17**, 1247.
25. Q. Chen, A. J. Rondinone, B. C. Chakoumakos and J. Z. Zhang, *J. Magn. Magn. Mater.*, 1999, **194**, 1.
26. S. Sun, H. Zeng, D. B. Robinson, S. Raoux, P. M. Rice, S. X. Wang and L. Guanxiong, *J. Am. Chem. Soc.*, 2004, **126**, 273.
27. J. Z. Zhang, Z. L. Wang, B. C. Chakoumakos and J. S. Yin, *J. Am. Chem. Soc.*, 1998, **120**, 1800.
28. V. Pillai, P. Kumar, M. J. Hou, P. Ayyub, D. O. Shah, *Adv. Colloid Interface Sci,.*, 1995, **55**, 241.
29. V. Pillai and D. O. Shah, *J. Magn. Magn. Mater.*, 1996, **163**, 243.
30. L. L. Hench and J. K. West, *Chem. Rev.*, 1990, **90**, 33.
31. C. Cannas, D. Gatteschi, A. Musinu, G. Piccaluga, and C. Sangregorio, *J. Phys. Chem.*, 1998, **102**, 7721.
32. G. Ennas, A. Musinu, G. Piccaluga, D. Zedda, D. Gatteschi, C. Sangregorio, J. L. Stanger, G. Concas and G. Spano, *Chem. Mater.*, 1998, **10**, 495.
33. N. N. Ghosh and P. Pramanik, *Mat. Sci. and Eng.*, 2001, **16**, 113.

34. Y. Bessekhouad, D. Robert and J. V. Weber, *J. Photochem. Photobiol.*, 2003, **157**, 47.
35. K. M. Reddy, C. V. G. Reddy and S. V. Manorama, *J. Solid State Chem.*, 2001, **158**, 180.
36. H. Zhang and J. F. Banfield, *Chem. Mater.*, 2005, **17**, 3421.
37. J. Tang, F. Redl, Y. Zhu, T. Siegrist, L. E. Brus and M. L. Steigerwald, *Nano Lett.*, 2005, **5**, 543.
38. J. T. Davis and E. K. Rideal, *Interfacial Phenomena*, New York Academic Press, New York, 1963.
39. R. K. Iler, *Silica Chemistry*, American Chemical Society, Washington D. C., 1982.
40. R. K. Iler, *Science of Ceramic Chemical Processing*, New York: Wiley, 1986.
41. D. Makovec, M. Drofenik and A. Znidarsic, *J. Am. Ceram. Soc.*, 1999, **82**, 1113.
42. A-H. Lu, E. L. Salabas and F. Schüth, *Angew. Chem., Int. Ed.*, 2007, **46**, 1222.
43. K. Sridhar, M. C. D'Arrigo, C. Leonelli, G. C. Pellacan and H. Katsuki, *J. Am. Ceram. Soc.*, 1998, **81**, 3041.
44. M. Rozman and M. Drofenik, *J. Am. Ceram. Soc.*, 1995, **78**, 2449.
45. X. Wang, J. Zhuang, Q. Peng and Y. Li, *Nature*, 2005, **437**, 121.
46. J. Xu, J. P. Ge and Y. D. Li, *J. Phys. Chem.*, 2006, **110**, 2497.
47. B. Wen, C. Liu and Y. Liu, *J. Phys. Chem.*, 2005, **109**, 12372.
48. S. W. Yang and L. Gao, *Mater. Chem. Phys.*, 2006, **99**, 437.
49. Q. Song and Z. J. Zhang, *J. Am. Chem. Soc.*, 2004, **126**, 6164.
50. K. S. Suslick, *Science*, 1990, **247**, 1439.
51. K. S. Suslick, *Ultrasound: Its chemical, physical, and biological effects*, Wiley-VCH, New York, 1998.
52. H. Xia and Q. Wang, *Chem. Mater.*, 2002, **14**, 2158.
53. J. C. Yu, L. Zhang, Q. Li, K. W. Kwong, A. W. Xu and J. Lin, *Langmuir*, 2003, **19**, 7673.
54. T. Prozorov, R. Prozorov, Y. Koltypin, I. Felner and A. Gendanken, *J. Phys. Chem.*, 1998, **102**, 10165.
55. Y. Lei, L. D. Zhang and J. C. Fan, *Chem. Phys. Lett.*, 2001, **338**, 231.
56. S. Liu and K. Huang, *Sol. Energy Mater. Sol. Cells*, 2004, **85**, 125.
57. A. B. Corradi, F. Bondioli, B. Focher, A. M. Ferrari, C. Grippo, E. Mariani and C. Villa, *J. Am. Ceram. Soc.*, 2005, **88**, 2639.
58. X. Wu, Q. Z. Jiang, Z. F. Ma, M. Fu and W. F. Shangguan, *Solid State Commun.*, 2005, **36**, 513.
59. T. Yamamoto, Y. Wada, H. Yin, T. Sakata, H. Mori and S. Yanagida, *Chem. Lett.*, 2002, **10**, 964.
60. T. Douglas and M. Young, *Nature*, 1998, **393**, 152.
61. T. Douglas and V.T. Stark, *Inorg. Chem.*, 2000, **39**, 1828.
62. M. Allen, D. Willits, M. Young, and T. Douglas, *Inorg. Chem.*, 2003, **42**, 6300.

63. G. Rice, L. Tang, K. Stedman, F. Roberto, J. Sphuler, E. Gillitzer, J. E. Johnson, T. Douglas and M.Young, *Proc. Natl. Acad. Sci. U. S. A.*, 2004, **101**, 7716.
64. M. T. Klem, D. Willits, M. Young and T. Douglas, *J. Am. Chem. Soc.*, 2003, **125**, 10806.
65. (a) E. Lucas, S. Decker, A. Khaleel, A. Seitz, S. Fultz, A. Ponce, W. Li, C. Carnes and K. J. Klabunde, *Chem.–Eur. J.*, 2001, **7**, 2505; (b) P. Jeevanandam and K. J. Klabunde, *Langmuir*, 2002, **18**, 5309.
66. R. Richards, W. Li, S. Decker, C. Davidson, O. Koper, V. Zaikovski, A. Voldin, T. Reiker and K. J. Klabunde, *J. Am. Chem. Soc.*, 2000, **122**, 4921.
67. M. S. Mel'gunov, V. B. Fenelonov, E. A. Mel'gunova, A. F. Bedilo and K. J. Klabunde, *J. Phy. Chem. B*, 2003, **107**, 2427.
68. M. Utiyama, H. Hattori and K. Tanabe, *J. Catal.*, 1978, **53**, 237.
69. A. Pelmenschikov, G. Morosi, A. Gamba and S. Coluccia, *J. Phys. Chem.*, 1995, **99**, 15018.
70. I. V. Mishakov, A. F. Bedilo, R. M. Richards, V. V. Chesnokov, A. M. Volodin, V. I. Zaikovskii, R. A. Buyanov and K. J. Klabunde, *J. Catal.*, 2002, **206**, 40.
71. K. J. Klabunde, J. V. Stark, O. Koper, C. Mohs, D. G. Park, S. Decker, Y. Jiang, I. Lagadic and D. Zhang, *J. Phys. Chem.*, 1996, **100**, 12142.
72. S. Utamapanya, K. J. Klabunde, J. R. Schlup, *Chem. Mater.*, **1991**, *3*, 175.
73. C. L. Carnes and K. J. Klabunde, *Langmuir*, 2000, **16**, 3764.
74. I. V. Mishakov, D. S.Heroux, V. V. Chesnokov, S. G. Koscheev, M. S. Mel'gunov, A. F. Bedilo, R. A. Buyanov and K. J. Klabunde, *J. Catal.*, 2005, **229**, 344.
75. V. V. Chesnokov, A. F. Bedilo, D. S. Heroux, I. V. Mishakov and K. J. Klabunde, *J. Catal.*, 2003, **218**, 438.
76. P. P. Gupta, K. L. Hohn, L. E. Erickson, K. J. Klabunde and A. F. Bedilo, *AIChE J.*, 2004, **50**, 3195.
77. B. M. Choudary, R. S. Mulukutla and K. J. Klabunde, *J. Am. Chem. Soc.*, 2003, **125**, 2020.
78. M. Lakshmi Kantam, K. Mahendar, B. Sreedhar and B. M. Choudary, *Tetrahedron*, 2008, **64**, 3351.
79. C. Pak, A. T. Bell and T. D. Tilley, *J. Catal.*, 2002, **23**, 51.
80. R. Kakkar, P. N. Kapoor and K. J. Klabunde, *J. Phys. Chem. B*, 2006, **110**, 25941.
81. B. M. Choudary, M. Lakshmi Kantam, K. V. S. Ranganath, K. Mahendar and B. Sreedhar, *J. Am. Chem. Soc.*, 2004, **126**, 3396.
82. M. Lakshmi Kantam, K. Mahendar, B. Sreedhar, K. V. Kumar and B. M. Choudary, *Synth. Commun.*, 2008, **38**, 3919.
83. M. Lakshmi Kantam, L. Chakrapani and B. M. Choudary, *Synlett.*, 2008, 1946.
84. B. M. Choudary, K. Mahendar, M. Lakshmi Kantam and K. V. S. Ranganath, *Adv. Syn. Catal.*, 2006, **348**, 1977.

85. M. Lakshmi Kantam, U. Pal, B. Sreedhar and B. M. Choudary, *Adv. Synth. Catal.*, 2007, **349**, 1671.
86. M. Lakshmi Kantam, L. Chakrapani and T. Ramani, *Tetrahedron Lett.*, 2007, **48**, 6121.
87. B. M. Choudary, K. V. S. Ranganath, J. Yadav and M. Lakshmi Kantam, *Tetrahedron Lett.*, 2005, **46**, 1369.
88. M. Lakshmi Kantam, R. Chakravarti, V. R. Chintareddy, B. Sreedhar and S. Bhargava, *Adv. Synth. Catal.*, 2008, **350**, 2544.
89. M. Lakshmi Kantam, J. Yadav, S. Laha, B. Sreedhar and S. Bhargava, *Adv. Synth. Catal.*, 2008, **350**, 2575.
90. M. Lakshmi Kantam, R. S. Reddy, U. Pal, B. Sreedhar and S. Bhargava, *Adv. Synth. Catal.*, 2008, **350**, 2231.
91. M. Lakshmi Kantam, U. Pal, B. Sreedhar, S. Bhargava, Y. Iwasawa, M. Tada and B. M. Choudary, *Adv. Synth. Catal.*, 2008, **350**, 1225.
92. M. Lakshmi Kantam, R. Chakravarti, U. Pal, B. Sreedhar and S. Bhargava, *Adv. Synth. Catal.*, 2008, **350**, 822.
93. M. Lakshmi Kantam, S. Roy, M. Roy, M. S. Subhas, P. R. Likhar, B. Sreedhar and B. M. Choudary, *Synlett.*, 2006, 2747.
94. B. M. Choudary, K. Jyothi, M. Lakshmi Kantam and B. Sreedhar, *Adv. Synth. Catal.*, 2004, **346**, 45.
95. J. M. Montero, P. Gai, K. Wilson and A. F. Lee, *Green Chem.*, 2009, **11**, 265.
96. L. Wang and J. Yang, *Fuel*, 2007, **86**, 328.
97. D. Kumar, V. B. Reddy, B. G. Mishra, R. K, Rana, M. N. Nadagouda and R. S. Varma, *Tetrahedron*, 2007, **63**, 3093.
98. G. Duan, X. Yang, J. Chen, G. Huang, L. Lu and X. Wang, *Powder Technol.*, 2007, **172**, 27.
99. K. Zhu, J. Hu, C. Kubel and R. Richards, *Angew. Chem., Int. Ed.*, 2006, **45**, 7277.
100. B. Nagappa and G. T. Chandrappa, *Mesopor. Micropor. Mater.*, 2007, **106**, 212.
101. W. S. Wadsworth and W. D. Emmons, *J. Am. Chem. Soc.*, 1961, **83**, 1733.
102. W. S. Wadsworth, *Org. React.*, 1977, **25**, 73.
103. B. Iorga, F. Eymery, V. Mouries and P. Savignac, *Tetrahedron*, 1998, **54**, 14637.
104. J. Boutagy and R. Thomas, *Chem. Rev.*, 1974, **74**, 87.
105. B. E. Maryanoff and A. B. Reitz, *Chem. Rev.*, 1989, **89**, 863.
106. K. C. Nicolaou, M. W. Harter, J. L. Gunzner and A. Nadin, *Liebigs Ann Recueil*, 1997, 1283.
107. B. M. Choudary, M. Lakshmi Kantam, Ch. Venkat Reddy and B. Bharathi, *J. Catal.*, 2003, **218**, 191.
108. M. Lakshmi Kantam, H. Kochkar, J. M. Clacens, B. Veldurthy, A. Garcia Ruiz and F. Figueras, *Appl. Catal., B*, 2005, **55**, 177.
109. B. M. Choudary, K. Mahendar and K. V. S. Ranganath, *J. Mol. Catal. A Chem.*, 2005, **234**, 25.

110. J. Boutagy and R. Thomas, *Chem. Rev.*, 1974, **74**, 87.
111. B. E. Maryanoff and A. B. Reitz, *Chem. Rev.*, 1989, **89**, 863.
112. R. W. Hoffmann, *Angew. Chem.*, 2001, 113, 1457 and references therein.
113. R. S. Al-Awar, J. E. Ray, R. M. Schultz, S. L Andis, J. H. Kennedy, R. E. Moore, J. Liang, T. Golakoti, G. V. Subbaraju and T. H .Corbett, *J. Med. Chem.*, 2003, **46**, 2985.
114. G. R. Pettit, C. R. Anderson, D. L. Herald, M. K. Jung, D. J. Lee, E. Hamel and R. K. Pettit, *J. Med. Chem.*, 2003, **46**, 525.
115. D. J. Burton, Z. Y. Yang and W. Qiu, *Chem. Rev.*, 1996, 96, 1641 and references therein.
116. C Pouget, C. Fagnere, J. P. Basly, G. Habrioux and A. J. Chulia, *Bioorg. Med. Chem. Lett.*, 2002, **12**, 1059.
117. H. K. Hsieh, T. H. Lee, J. P. Wang, J. J. Wang and C. N. Lin, *Pharm. Res.*, 1998, **15**, 39.
118. A. F. Hegarty, *Comprehensive Organic Chemistry*, I. O. Pergamon, London, vol. 2, 1979.
119. Y. Ono, *Appl. Catal., A*, 1997, **155**, 133.
120. J. P. Parrish, R. N. Salvatore and K. W. Jung, *Tetrahedron*, 2000, **56**, 8207.
121. S. Gryglewicz, F. A. Oko and G. Gryglewicz, *Ind. Eng. Chem. Res.*, 2003, **42**, 5007.
122. D. Basavaiah, P. D. Rao and R. S. Hyma, *Tetrahedron*, 1996, **52**, 8001.
123. H. Groger, *Chem. Rev.*, 2003, **103**, 2795 and references cited therein.
124. L. Yet, *Angew. Chem., Int. Ed.*, 2001, **40**, 875 and references cited therein.
125. M. P. Doyle and M. A. McKervey, *Modern Catalytic Methods for Organic Synthesis with Diazo Compounds*, Wiley-Interscience: New York, 1998.
126. D. J. Miller and C. J. Moody, *Tetrahedron*, 1995, **51**, 10811.
127. R. Pellicciari, B. Natalini, S. Ceccheti and R. Fringnelli, *Steroids*, 1987, **49**, 433.
128. V. Ravi, E. Ramu, N. Sreelatha and A. S. Rao, *Tetrahedron Lett.*, 2006, **47**, 877 and cited therein.
129. R. J. H. Geregory, *Chem. Rev.*, 1999, **99**, 3649.
130. T. Ziegler, B. Horsch and F. Effenberger, *Synthesis*, 1990, 575.
131. C. G. Kruse, *Chirality in Industry*, Wiley, Chichester, 1992.
132. D. C. Cole, *Tetrahedron*, 1994, **50**, 9517.
133. A. Cordova, *Acc. Chem. Res.*, 2004, **37**, 102.
134. E. Uaristi, D. Quintana and J. Escalante, *Aldrichim Acta*, 1994, **27**, 3.
135. M. Arend, B. Westermann and N. Risch, *Angew. Chem., Int. Ed.*, 1998, **37**, 1044.
136. N. Azizi, L. Torkiyan and M. R. Saidi, *Org. Lett.*, 2006, **8**, 2079.
137. M. Kidwai, N. K. Mishra, V. Bansal, A. Kumar and S. Mozumdar, *Tetrahedron Lett.*, 2009, **50**, 1355.
138. M. Lakshmi Kantam, K. Mahendar, B. Sreedhar, B. M. Choudary, S. K. Bhargava and S. H. Priver, *Tetrahedron*, 2010, 66, 5042.

139. M. Lakshmi Kantam, K. Mahendar and S. Bhargava, *J. Chem. Sci.*, 2010, **122**, 63.
140. C. L. Carnes, J. Stipp and K. J. Klabunde, *Langmuir*, 2002, **18**, 1352.
141. (a) P. Larsson and A. Andersson, *J. Catal.*, 1998, **179**, 72; (b) V. Chikan, A. Molnar and K. Balazsik, *J. Catal.*, 1999, **184**, 134; (c) B. Raveau, C. Michel, M. Herview and D. Groult, *Crystal Chemistry of High-T_c Superconducting Copper Oxides*; Springer-Verlag, Berlin, 1991; (d) C. P. Poole, T. Datta, H. A. Farach, M. M. Rigney and C. R. Sanders, *Copper Oxide Superconductors*, John Wiley & Sons, NewYork, 1988.
142. (a) Y. Li u, Q. Fu and M. F. Stephanopoulos, *Catal. Today*, 2004, **93–95**, 241; (b) A. Martinez-Arias, A. B. Hungria, M. Fernandez-Garcia, J. C. Conesa and G. Munuera, *J. Phys. Chem. B*, 2004, **108**, 17983; (c) Z. Gu and K. L. Hohn, *Ind. Eng. Chem. Res.*, 2004, **43**, 30.
143. S. U. Son, I. K. Park, J. Park and T. Hyeon, *Chem. Commun.*, 2004, 778.
144. B. M. Choudary, Ch. Sridhar, M. L. Kantam, G. T. Venkanna and B. Sreedhar, *J. Am. Chem. Soc.*, 2005, **127**, 9948.
145. (a) P. Cozzi, G. Carganico, D. Fusar, M. Grossoni, M. Menichincheri, V. Pinciroli, R. Tonani, F. Vaghi and P. Salvati, *J. Med. Chem.*, 1993, **36**, 2964; (b) C. Almansa, J. Bartroli, J. Belloc, F. L. Cavalcanti, R. Ferrando, L. A. Gomez, I. Ramis, E. Carceller, M. Merlos and J. Garcia-Rafanell, *J. Med. Chem.*, 2004, **47**, 5579.
146. T. Güngör, A. Fouquet, J. M. Teulon, D. Provost, M. Cazes and A. Cloarec, *J. Med. Chem.*, 1992, **35**, 4455.
147. W. A. Herrmann, *Angew. Chem., Int. Ed.*, 2002, **41**, 1290.
148. M. Lakshmi Kantam, J. Yadav, S. Laha, B. Sreedhar and S. Jha, *Adv. Synth. Catal.*, 2007, **349**, 1938.
149. (a) M. Breuer, K. Ditrich and T. Zelinski, *Angew. Chem., Int. Ed.*, 2004, **43**, 788; (b) B. A. Astelford and L. O. Weigel, *Chirality in Industry II*, Wiley, New York, 1997.
150. M. Lakshmi Kantam, S. Laha, J. Yadav, P. R. Likhar, B. Sreedhar and B. M. Choudary, *Adv. Synth. Catal.*, 2007, **349**, 1797.
151. R. W. Armstrong, A. P. Combs, P. A. Tempst, S. D. Brown and T. A. Keating, *Acc. Chem. Res.*, 1996, **29**, 123.
152. S. Kamijo and Y. Yamamoto, *J. Am. Chem. Soc.*, 2002, **124**, 11940.
153. B. Ringdahl, *The Muscarinic Receptors*, Humana Press, Clifton, NJ, 1989.
154. M. Miura, M. Enna, K. Okuro and M. Nomura, *J. Org. Chem.*, 1995, **60**, 4999.
155. A. Jenmalm, W. Berts, Y. L. Li, K. Luthman, I. Csoregh and U. Hacksell, *J. Org. Chem.*, 1994, **59**, 1139.
156. G. Dyker, *Angew. Chem., Int. Ed.*, 1999, **38**, 1698.
157. M. Lakshmi Kantam, S. Laha, J. Yadav and S. Bhargava, *Tetrahedron Lett.*, 2008, **49**, 3083.
158. G. Marciniak, A. Delgado, J. Velly, G. Leclerc, N. Decker and J. Schwartz, *J. Med. Chem.*, 1989, **32**, 1402.

159. D. Enders, S. Muller and A. S. Demir, *Tetrahedron Lett.*, 1988, **29**, 6437.
160. B. M. Trost, *Science*, 1991, **254**, 1471.
161. C. Palomo, M. Oiarbide and J. Garcia, *Chem. Soc. Rev.*, 2004, **33**, 65.
162. T. D. Machajewski and C.-H. Wong, *Angew. Chem., Int. Ed.*, 2000, **39**, 1352.
163. M. Shibasaki, *Modern Aldol Reactions,*, Wiley-VCH, Weinheim, vol. 11, 2004.
164. M. Lakshmi Kantam, T. Ramani, L. Chakrapani and K. Vijay Kumar; *Tetrahedron Lett*, 2008, **49**, 1498.
165. B. O'Regan and M. Grätzel, *Nature*, 1991, **353**, 737.
166. N. Fujishima, T. Rao and D. A.Tryk, *J. Photochem. Photobiol., C*, 2000, **1**, 1.
167. T. Rajh, Z. Saponjic, J. Liu, N. M. Dimitrijevic, N. F. Scherer, M. Vega-Arroyo, P. Zapol, L. A. Curtiss and M. C. Thurnauer, *Nano Lett.*, 2004, **4**, 1017.
168. H. Tokuhisa and P. T. Hammond, *Adv. Funct. Mater.*, 2003, **13**, 831.
169. T. Gerfin, M. Grätzel and L. Walder, *Molecular and Supermolecular Surface Modification of Nanocrystalline TiO$_2$ Films: Charge Separating and Charge Injecting Devices*, in, ed. K. D. Karlin, John Wiley and Sons Inc., New York, 1997, p.345.
170. P. V. Kamat, *Chem. Rev.*, 1993, **93**, 267.
171. F. Wells, *Structural Inorganic Chemistry*, Clarendon Press, Oxford, 1975.
172. C. Kormann, D. W. Bahnemann and M. R. Hoffmann, *J. Phys. Chem.*, 1988, **92**, 5196.
173. K. Prasad, A. R. Bally, P. E. Schmid, F. Levy, J. Benoit, C. Barthou and P. J. Benalloul, *J. Appl. Phys.*, 1997, **36**, 5696.
174. S. Kim, S. Park and Y. H. Jeong, *J. Am. Ceram. Soc.*, 1999, **82**, 927.
175. I. N. Martyanov and K. J. Klabunde, *J. Catalysis*, 2004, **225**, 408.
176. (a) M. E. Jung and F. Slowinski, *Tetrahedron Lett.*, 2001, **42**, 6835; (b) T. Walsh, R. B. Toupence, F. Ujjainwalla, J. R. Young and M. T. Goulet, *Tetrahedron*, 2001, **57**, 5233; (c) H.-C. Zhang, H. Ye, A. F. Moretto, K. K. Brumfield and B. E. Maryanoff, *Org. Lett.*, 2000, **2**, 89.
177. (a) A. S. Rao, S. K. Paknikar and J. G. Kirtane, *Tetrahedron*, 1983, **39**, 2323; (b) J. G. Smith, *Synthesis*, 1984, 629; (c) T. Hirose, T. Sunazuka, T. Zhi-ming, M. Handa, R. Vchida, K. Shiomi, Y. Hrigaya and S. Omura, *Heterocycles*, 2000, **53**, 777.
178. M. Lakshmi Kantam, S. Laha, J. Yadav and B. Sreedhar, *Tetrahedron Lett.*, 2006, **47**, 6213.
179. (a) J. Szmuszkovicz, *J. Am. Chem. Soc.* 1957, **79**, 2819; (b) W. E. Noland, G. M. Christensen, G. L. Sauer and G. G. S. Dutton, *J. Am. Chem. Soc.*, 1955, **77**, 456; (c) Z. Iqbal, A. H. Jackson and K. R. N. Rao, *Tetrahedron Lett.*, 1988, **29**, 2577.
180. M. Lakshmi Kantam, S. Laha, J. Yadav, B. M. Choudary and B. Sreedhar, *Adv. Synth. Catal.*, 2006, **348**, 867.
181. W. Wang, B. Gu, L. Liang, W. A. Hamilton and D. J. Wesolowski, *J. Phys. Chem.*, 2004, **108**, 14789.
182. M. Visca and E. .J Matijević, *J. Colloid Interface Sci.*, 1979, **68**, 308.

183. H. K. Park, Y. T. Moon, D. K. Kim and C. H. Kim, *J. Am. Ceram. Soc.*, 1996, **79(10)**, 2727.
184. E. A. Barringer and H. K. Bowen, *Langmuir*, 198, **1**, 414.
185. H. Zhang and J. F. Banfield, *J. Phys. Chem.*, 2000, **104**, 3481.
186. M. M. Wu, G. Lin, D. H. Chen, G. G Wang, D. He, S. H. Feng and R. R. Xu, *Chem. Mater.*, 2002, **14**(5), 1974.
187. Q.-H. Zhang, L. Gao and J. -K. Guo, *Nanostruct. Mater.*, 1999, **11**, 1293.
188. F. Cavani, E. Foresti, F. Parrinello and F. Trifiro, *Appl. Catal.*, 1988, **38**, 311.
189. K. –N. P. Kumar, K. Keizer and A. Burggraaf, *Nature*, 1992, **358**, 48.
190. O. Manuel, G. R. Josev and S. Carlos, *J. Am. Ceram. Soc.*, 1992, **75**, 2010.
191. K. –N. P. Kumar, K. Keizer and A. J. Burggraaf, *J. Mater. Sci. Lett.*, 1994, **13**, 59.
192. H. Luo, C. Wang and Y. Yan, *Chem. Mater.*, 2003, **15**, 3841.
193. W. Liu, A. Chen, J. Lin, Z. Dai, W. Qiu, W. Liu, M. Zhu and S. Usuday, *Chem. Lett.*, 2004, **33**, 390.
194. S. Yin, R. Li, Q. He and T. Sato, *Mater. Chem. Phys.*, 2002, **75**, 76.
195. R. Chu, J. Yan, S. Lian, Y. Wang, F. Yan and D. Chen, *Solid State Commun.*, 2004, **130**, 789.
196. S. Yin, H. Hasegawa, D. Maeda, M. Ishitsukan and T. Sato, *J. Photochem. Photobiol., A*, 2004, **163**, 1.
197. S. T. Aruna, S. Tirosh and A. J. Zaban, *J. Mater. Chem.*, 2000, **10**, 2388.
198. Q. Zhang and L. Gao, *Langmuir*, 2003, **19**, 967.
199. F. Pedraza and A.Vazquez, *J. Phys. Chem. Solids.*, 1999, **60**, 445.
200. J. Yang, S. Mei and J. M. F. Ferreira, *J. Am. Ceram. Soc.*, 2000, **83**, 1361.
201. H. Cheng, J. Ma, Z. Zhou and L. Qi, *Chem. Mater.*, 1995, **7**, 663.
202. R. R. Bacsa and M. Grätzel, *J. Am. Ceram. Soc.*, 1996, **79**, 2185.
203. C. -C. Wang and J. Y. Ying, *Chem. Mater.*, 1999, **11**, 3113.
204. H. Cheng, J. Ma, Z. Zhou and L. Qi, *Chem. Mater.*, 1995, **7**, 663.
205. T. Sasamoto, S. Enomoto, Z. Shimoda and Y. Saeki, *J. Ceram. Soc.*, 1993, **101**, 230.
206. N. -G. Park, G. Schlichthörl, J. van de Lagemaat, H. M. Cheong, A. Mascarenhas and J. Frank, *J. Phys. Chem.*, 1999, **103**, 3308.
207. M. Nag, P. Basak and S. V. Manorama, *Mater. Res. Bull.*, 2007, **42**(9), 1691.
208. M. Nag, D. Guin, P. Basak and S. V. Manorama, *Mater. Res. Bull.*, 2008, **43**(12), 3270.
209. M. Nag, S. Ghosh, R. K. Rana and S. V. Manorama, *J. Phys. Chem. Lett.*, 2010, **1**, 2881.
210. (a) L. Vayssieres, K. Keis, A. Hagfeldt and S. E. Lindquist, *Chem. Mater.*, 2001, **13**, 4395; (b) D. S. King and R. M. Nix, *J. Catal.*, 1996, **160**, 76; (c) Z. W. Pan, Z. R. Dai and Z. L. Wang, *Science*, 2001, **291**,1947; (d) Q. Li and C. R. Wang, *Chem. Phys. Lett.*, 2003, **375**, 525; (e) X. Kong, X. Sun, X. Li, Y. Li, *Mater. Chem. Phys.*, 2003, **82**, 997; (f) C. X. Xu, X. W. Sun, B. J. Chen, P. Shum, S. Li and X. Hu, *J. Appl. Phys.*, 2004, **95**, 661; (g) K. Hara, T. Horiguchi, T. Kinoshita, K.

Sayama, H. Sugihara and H. Arakawa, *Sol. Energy Mater. Sol. Cells*, 2000, **64**, 115.
211. (a) Y. J. Kim and R. S. Varma, *Tetrahedron Lett.*, 2004, **45**, 7205; (b) M. H. Sarvari and H. Sharghi, *J. Org. Chem.*, 2004, **69**, 6953; (c) H. Sharghi and M. H. Sarvari, *Synthesis*, 2002, **8**, 1057.
212. C. L. Carnes and K. J. Klabunde, *Chem. Mater.*, 2002, **14**, 1806.
213. (a) R. N. Butler, *Comprehensive Heterocyclic Chemistry*, Pergamon, Oxford, U.K., 1996; (b) H. Singh, A. S. Chala, V. K. Kapoor, D. Paul and R. K. Malhotra, *Prog. Med. Chem.*, 1980, **17**, 151; (c) V. A. Ostrovskii, M. S. Pevzner, T. P. Kofmna, M. B. Sheherbinin and I. V. Tselinskii, *Targets Heterocycl. Syst.*, 1999, **3**, 467; (d) M. Hiskey, D. E. Chavez, D. L. Naud, S. F. Son, H. L. Berghout and C. A. Bome, *Proc. Int. Pyrotech. Semin.*, 2000, **27**, 3.
214. M. L. Kantam, K. B. Shiva Kumar and C. Sridhar, *Adv. Synth. Catal.*, 2005, **347**, 1212.
215. G. R. Dube and V. S. Darshana, *J. Mol. Catal.*, 1993, **79**, 285.
216. R. J. Rennard and W. L. Kehl, *J. Catal.*, 1971, **21**, 282.
217. Z. Yuan and L. Zhang, *Mat. Res. Bull.*, 1998, **33**, 1587.
218. T. Zhang, Y. Shen and R. Zhang, *Material Mater. Lett.*, 1995, **23**, 69.
219. C. C. Wang and J. Y. Ying, *Chem. Mater.*, 1999, **11**, 3113.
220. C W. Turner, *Am. Ceram. Soc. Bull.*, 1991, **70**, 1487.
221. A. Ataie, M. R. Piramoon, I. R. Harris and C. B. Ponton, *J. Mater. Sci.*, 1995, **30**, 5600.
222. G. Y. Adachi and N. Imanaka, *Chem. Rev.*, 1998, **98**, 1479.
223. S. Tao, F. Gao, X. Liu and O. T. Sorensen, *Mat. Sci. Eng. B*, 2000, **77**, 172.
224. S. Somiya and B. Roy, *Bull. Mater. Sci.*, 2000, **23**, 453.
225. P. Ravindranathan, K. C. Patil, *Am. Ceram. Soc. Bull.*, 1987, **66**, 688.
226. E. Regazzoni, G. A. Urutia, M. A. Blesa and A. J. G. Maroto, *J. Inorg. Nucl. Chem.*, 1981, **43**, 1489.
227. J. Chen, K. Brinder, S. R. Winzer and V. Paivernerker, *J. Appl. Phy.*, 1988, **63**, 3786.
228. M. Ueda, S. Shimada and M. Inagaki, *J. Eur. Ceram. Soc.*, 1995, **15**, 265.
229. C. V. G. Reddy, K. Seela and S. V. Manorama, *Int. J. Inorg. Mater.*, 2000, **12**, 301.
230. C. V. G. Reddy, W. Cao, O. K. Tan and W. Zhu, *Sensor Actuators B: Chem.*, 2002, **81**, 170.
231. (a) H. Bönnemann, W. Brijoux, R. Brinkmann, E. Dinjus, T. Jouâen and B. Korall, *Angew. Chem., Int. Ed. Engl.*, 1991, **30**, 1312. (b) P. Lu and N. Toshima, *Bull. Chem. Soc. Jpn.*, 2000, **73**, 751. (c) S. C. Tsang, V. Caps, I. Paraskevas, D. Chadwick and D. Thompsett, *Angew. Chem., Int. Ed.*, 2004, **43**, 5645. (d) P. D. Stevens, J. Fan, H. M. R. Gardimalla, M. Yen and Y. Gao, *Org. Lett.*, 2005, **7**, 2085. (e) A.-H. Lu, W. C. Li, A. Kiefer, W. Schmidt, E. Bill, G. Fink and F. Schüth, *J. Am. Chem. Soc.*, 2004, **126**, 8616.
232. D. Guin, B. Baruwati and S. V. Manorama, *Org. Lett.*, 2007, **9**, 1419.
233. B. Baruwati, D. Guin and S. V. Manorama, *Org. Lett.*, 2007, **9**, 5377.

CHAPTER 7
Nanoarchitechtonics of Photocatalytic Materials

JINHUA YE*[a,b,c] and Hua Tong[a,b,c]

[a] International Center for Materials Nanoarchitectonics (MANA), National Institute for Materials Science (NIMS), 1-2-1 Sengen, Tsukuba, Ibaraki 305-0047, Japan; [b] Environmental Remediation Materials Unit, National Institute for Materials Science (NIMS), 1-2-1 Sengen, Tsukuba, Ibaraki 305-0047, Japan; [c] TU-NIMS Joint Research Center, School of Materials Science and Engineering, Tianjin University, 92 Weijin Road, Nankai District, Tianjin, P.R. China
*E-mail: jinhua.ye@nims.go.jp

7.1 Introduction

In 1972 Fujishima and Honda reported the water photolysis on a TiO_2 electrode.[1] This discovery has been recognized as the landmark event that stimulated extensive and intense research into the utilization of the energy of light by photocatalytic methods. Currently, semiconductor photocatalysis has emerged as one of the most promising technologies to the solution of the worldwide energy shortage and for counteracting environmental degradation.[2–5] Its most anticipated potential involves use in hydrogen fuel production from water photolysis, photo-degradation of inorganic/organic pollutants, artificial photosynthesis, photo-induced self-cleaning, and photoelectrochemical conversion, *etc.*

From the point of view of photochemistry, the role of semiconductor photocatalysis is to initiate or accelerate specific reduction and oxidation (redox) reactions in the presence of irradiated semiconductors, where the excited electrons and holes can act as reductants and oxidants, respectively. As is illustrated in Figure 7.1A, a typical photocatalytic process comprises three steps: first, photoexcitation of electron–hole pairs; second, migration of carriers to the surface; third, redox reactions at the surface. This illustration reflects the main aspects that are associated with the photocatalytic activity of semiconductors. First of all, the energy band configuration of semiconductors determines the photoexcitation of electron–hole pairs, the migration of carriers and the redox capabilities of excited-state electrons and holes (see Figure 7.1B).[6] Next, the nature of the semiconductor surface plays crucial roles in governing the selectivity, rate and overpotential of redox reactions.[7,8] Another key issue influencing the photocatalytic capability is the size of the photocatalysts. In general, smaller particles can result in a larger amount of reactive sites and a shorter distance for the carriers migrating to the surface.

The above aspects point out the significant possibilities of promoting the photocatalytic activity of semiconductors, such as by energy band engineering, morphology control, hierarchical composite/assembly, *etc*. It is worth noting that nanoarchitectonics of photocatalytic materials can provide adjustable energy band configuration, large surface areas, abundant surface states, diverse morphologies and easy device modeling, all of which are properties beneficial to photocatalysis. As a typical example, nanostructured materials with large percentages of highly reactive facets arouse increasing interest. But attempts to deliberately fabricate such materials must break the barrier of the thermodynamic growth mechanisms of the crystals. It was found that the selective absorption of polymers or ions on high-energy facets could suppress the growth rate along their axes. In this way, it is possible to create specific nanostructured materials comprised of nearly 100% high-energy facets on the surface, such as ultra-thin sheets and highly symmetric polyhedral particles.

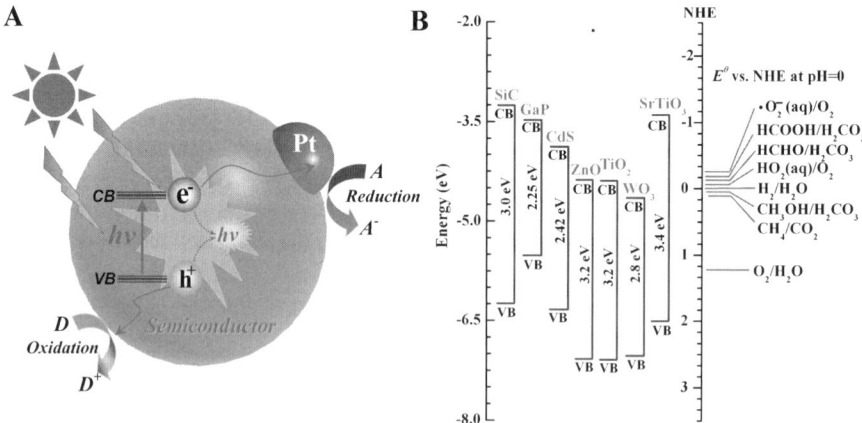

Figure 7.1 (A) Schematic illustration of basic mechanism of a semiconductor photocatalytic process. (B) Band-edge positions of semiconductor photocatalysts relative to the energy levels of various redox couples in water. Copyright 2011, Wiley-VCH, adapted with permission.

Further significant opportunities offered by nanoarchitectonics of photocatalytic materials lie in hierarchical composite/assembly, which can not only facilitate the separation of photo-excited electron–hole pairs but also present novel collective properties. For example, recently a new band gap narrowing mechanism has been found in the electronic coupling assembly of semiconductor nanoparticles; as a result, the utilization efficiency of incident photons can be improved greatly in such photocatalytic materials.[9,10] Moreover, better light sensitization to photocatalytic materials may be fulfilled by hierarchical composite/assembly methods, such as the inclusion of quantum dots,[11] plasmon-exciton coupling between anchored noble metal nanoparticle co-catalysts and the host semiconductor[12] and photon coupling in semiconductor photonic crystals.[13]

In the past decade, a great deal of progress in the photocatalysis field has been made alongside the rapid development of nanotechnology.[14] Nanoarchitectonics has emerged as the most important and pioneering means of searching for novel highly efficient photocatalytic materials. This chapter provides a review of the main recent achievements in nanostructured photocatalytic materials, and highlights the relevance of the approaches and principles involved in nanoarchitectonics with respect to the development of environmentally acceptable visible-light and solar applications.

7.2 Morphology Control

It is obvious that reducing the size of photocatalysts is generally beneficial for photocatalytic reactions because of increased reactive sites as well as a shortened distance to favour carriers migrating to the surface. This migration still requires a suitable concentration gradient or potential gradient from the inner to the surface, which is closely correlated with the morphology, structure and surface properties of photocatalysts. Over the past years, intense study has been carried out on the shape-controlled synthesis of photocatalytic materials and the relationship between the morphological or structural characteristics and photocatalytic properties.

7.2.1 One-dimensional Nanostructures

Conventional photocatalysts are typically in the form of irregular particles. As nanostructured materials are employed as photocatalysts, in addition to the size, surface chemistry and crystal structure, the morphology often plays a crucial role in determining their photocatalytic properties. In this regard, one-dimensional (1D) nanostructures such as wires, belts and tubes have attracted considerable attention for photocatalytic applications due to their distinct electronic, optical and chemical properties, which differ from their bulk counterparts.[15–18] These nanostructures have an extremely large aspect ratio (the ratio of length to diameter or surface to volume), allow the lateral confinement of electrons and guide the movement of electrons in the axial direction, and show significantly enhanced light sensitization. Furthermore, single-crystal nanowires exhibit more efficient ballistic charge transport along their axis compared with irregular particles. Accordingly, significant enhancement in photocatalytic activity can be expected from 1D nanostructured materials.

The properties of 1D nanostructured materials are sensitively variable with the size and morphology, leading to diverse strategies to optimize the photocatalytic reactivity. For example, increasing the aspect ratio can result in an enhancement of activity. Such successful examples include 1D nanostructures of CdS, ZnO and GaN.[19–21] In other cases, well-designed surfaces on 1D nanostructures, such as the {101} facet on TiO_2 nanobelts or the {010} facet on Zn_2GeO_4 nanobelts,[22,23] have been shown to yield superior photocatalytic activity. Wang *et al.* made remarkable progress in the ordered alignment of 1D GaN nanowire arrays for the construction of water photolysis devices.[24] This work showed that the water splitting reaction primarily occurred on the Ga-face nonpolar lateral surfaces, allowing control over the photocatalyst crystal plane to promote either H_2 or O_2 evolution.

1D hetero-nanostructures arouse considerable interest because of their synergetic effects on photocatalytic performance.[16,25,26] Based on junction modes, there are three types of 1D hetero-nanostructures: axial hetero-nanostructures, radial hetero-nanostructures, and hierarchical hetero-nanostructures. Studies have shown that the photocatalytic performance of 1D coaxial hetero-nanostructures depends on the interactions that occur between different chemical species in the core and shell. Therefore, appropriately modulating the compositions and interfaces may lead to significant improvement in photocatalytic performance. Bi and Ye presented such an instance of Ag/AgCl core–shell nanowires, which yielded the best photocatalytic activity in the decomposition of methyl orange (MO) (Figure 7.2) when the core–shell ratio was 8/92.[16]

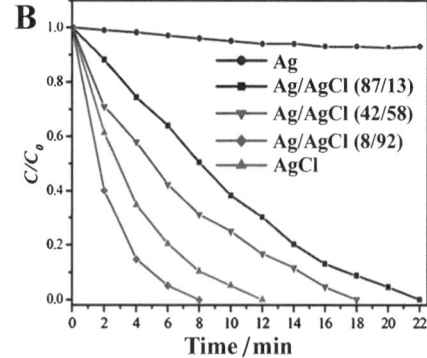

Figure 7.2 (A) SEM images of Ag/AgCl core–shell nanowires. (B) Photocatalytic activities of Ag/AgCl core–shell nanowires in the decomposition of MO dye under visible light irradiation ($\lambda > 400$ nm). Adapted with permission. Copyright 2010, Royal Society of Chemistry.

7.2.2 Facet-controlled Nanostructures

Considering that photocatalytic reactions take place on the surfaces of photocatalysts, the nature of their surface/interface chemistry, such as their surface energy and chemisorption properties, plays a crucial role in determining the photocatalytic reactivity and efficiency. Usually, a higher surface energy is desired, because it is most likely to accompany higher catalytic activity. The stronger the sorption of reagents and desorption of products on the photocatalyst surface, the easier and faster the photocatalytic process. In this sense, nanoarchitectonic surface engineering has ample scope in its abilities to aid in the exploration and development of photocatalysts with advanced performance.

The synthesis of single crystals with exposed highly reactive facets is a challenging and promising method for the further improvement of photocatalytic performance. According to the thermodynamic mechanism of the crystals, however, highly reactive facets tend to disappear during the growth process due to their higher surface energy. To address this issue, the employment of both organic polymers and inorganic ions has been tried to selectively control the growth rates of planes of interest. For example, using hydrofluoric acid (HF) as a capping agent under hydrothermal conditions, Yang and co-workers prepared anatase TiO_2 single crystals for which 47% of the surface was comprised of highly reactive {001} facets.[27] Later they made use of 2-propanol as a synergistic capping agent and reaction medium together with HF to synthesize high-quality anatase single-crystal nanosheets with surfaces comprised of 64% {001} facets.[28] Liu *et al.* developed a modified fluoride-mediated self-transformation strategy for the synthesis of hollow TiO_2 microspheres composed of polyhedral anatase crystallites with ~20% {001} facets.[29] The photocatalytic selectivity of these microspheres toward the decomposition of azo dyes could be tuned by varying the percentage of {001} facets as well as by varying the concentration of adsorbed species on the surfaces of the anatase polyhedra.[29]

In the above cases, the capping agents adsorbed on the surfaces of semiconductor crystals during synthesis must be removed before they can be used in photocatalytic applications. This often requires complex treatments and the highly reactive facets can then lose many of their active sites as a result of surface reconstruction. Therefore, the capping-agent-free fabrication of photocatalysts with highly reactive facets exposed can ideally favour experimental exploration of the reactivity of different facets, although this is still a great challenge. Recently, well-defined m-$BiVO_4$ nanoplates (Figure 7.3A, B) with exposed {001} facets have been synthesized by a straightforward hydrothermal route in which no template or organic surfactant was used.[30] These m-$BiVO_4$ nanoplates emerged as energetic photocatalysts for the degradation of rhodamine B (RhB) (Figure 7.3C) and for the evolution of oxygen from an aqueous $AgNO_3$ solution (Figure 7.3D).

Over the past years, the chemical exfoliation of layered compounds has become a well explored process to allow the synthesis of various nanosheet

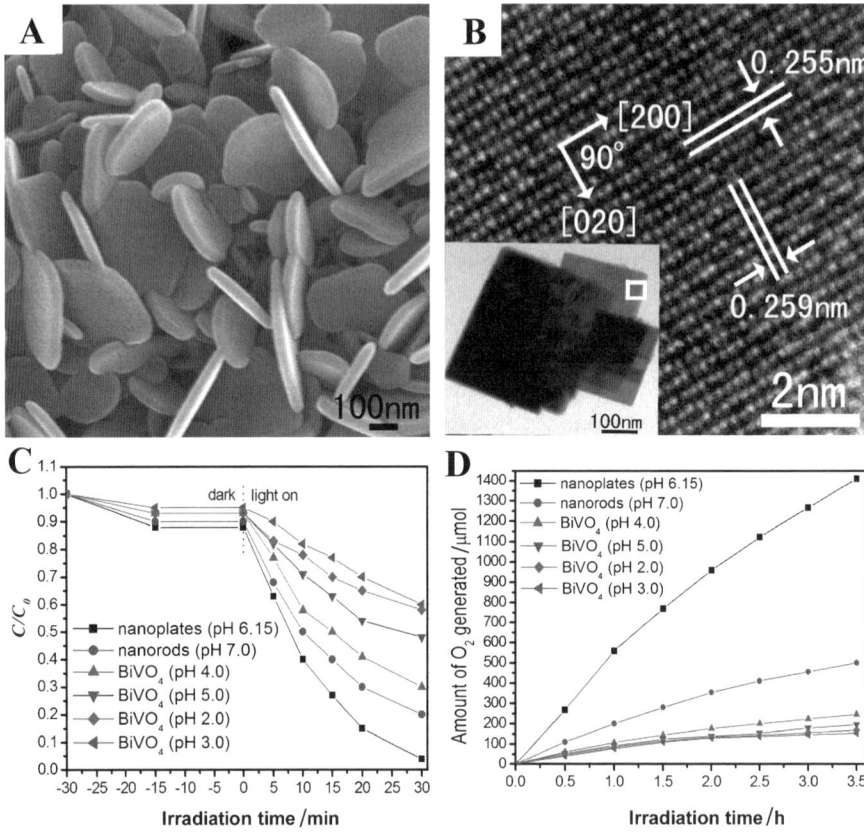

Figure 7.3 (A) SEM image and (B) high-magnification HRTEM image of m-BiVO$_4$ nanoplates. (C) Photocatalytic degradation of RhB over various BiVO$_4$ photocatalysts under visible light ($\lambda > 400$ nm) irradiation. (D) Photocatalytic evolution of O$_2$ from an aqueous AgNO$_3$ solution (0.05 M, 270 mL) over various BiVO$_4$ photocatalysts under visible-light irradiation ($\lambda > 400$ nm). Adapted with permission. Copyright 2010, Royal Society of Chemistry.

materials.[31] Such materials usually appear as sheet-shaped crystallites of molecular thickness and extremely high two-dimensional anisotropy, so they can provide ratios of reactive sites on the surface that approach 100% and show intriguing charge-bearing properties due to large-area nonstoichiometry on the surface.[32] So far a variety of promising photocatalysts, such as KCa$_2$Nb$_3$O$_{10}$,[33,34] Ba$_5$Ta$_4$O$_{15}$,[35] HCa$_2$Nb$_3$O$_{10}$,[36] TaO$_3^-$,[37] and TiNbO$_5^-$,[38] have been found to belong to this category.

Highly symmetric polyhedral particles provide an important model for investigating the effects of the shapes and facets on the photocatalytic properties. A good illustration is found for single-crystals of Ag$_3$PO$_4$ in two forms: rhombic dodecahedrons with exposed {110} facets, and cubes bounded by {100} facets.[39] The Ag_3PO_4 dodecahedrons are formed by 12 well-defined {110} planes with cubic crystal symmetry (Figure 7.4A), whereas the Ag_3PO_4 cubes display sharp corners, edges, and smooth surfaces (Figure 7.4B). When used in the photocatalytic degradation of MO and RhB dyes under visible-light irradiation (Figure 7.4C, D), the dodecahedrons exhibited higher activity than the cubes, revealing that the {110} facet is more reactive than the {100} facet.

In order to clarify the underlying mechanism with regard to the higher reactivity of the {110} facet, further theoretical study has been carried out on the surface energies of Ag_3PO_4 {110} and {100} planes using DFT calculations.[39] The surface models were based on slabs comprised of 192 atoms with a vacuum region of the same thickness for both {100} and {110} facets. The surface energy γ was computed using the following formula:

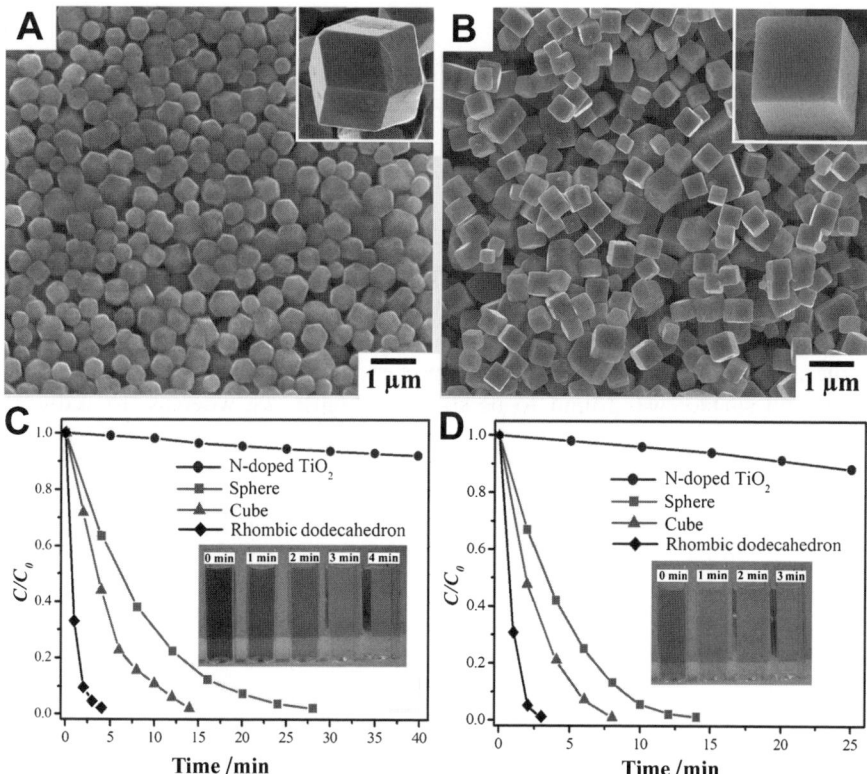

Figure 7.4 SEM images of Ag_3PO_4 sub-microcrystals with different morphologies: (A) rhombic dodecahedrons and (B) cubes. The photocatalytic activities of Ag_3PO_4 rhombic dodecahedrons, cubes, spheres, and N-doped TiO_2 are shown for the degradation of (C) MO and (D) RhB under visible light irradiation ($\lambda > 400$ nm). Adapted with permission. Copyright 2011, American Chemical Society.

$$\gamma = (E_{\text{slab}} - nE_{\text{bulk}})/2A \qquad (7.1)$$

where E_{slab} is the total energy of the slab, E_{bulk} is the total energy of the bulk per unit cell, n is the number of bulk unit cells contained in the slab, and A is the surface area of each side of the slab. The computational results showed that the surface energy of the {110} facet (1.31 J m^{-2}) is higher than that of the {100} facet (1.12 J m^{-2}).

The abundance of oxygen vacancies of the {110} plane is another possible origin of the high reactivity of Ag$_3$PO$_4$ rhombic dodecahedral crystals according to the computational results on the energies of reduced surfaces. This study was based on a model structure created by removing an oxygen atom from each side of the {100} and {110} slabs and relaxing the geometries of the reduced surfaces with the mid-layer fixed in the same way as the ideal surfaces. The energy γ of the reduced surface was computed using the following formulas:

$$\gamma = (E_{\text{surf}} + \mu_{\text{O}})/A \qquad (7.2)$$

$$E_{\text{surf}} = (E_{\text{slab}} - nE_{\text{bulk}})/2 \qquad (7.3)$$

where E_{slab} is the total energy of the reduced slab and μ_{O} is the chemical potential of oxygen, which is directly related to the O$_2$ partial pressure in the atmosphere. The surface energy is shown in Figure 7.5 as a function of μ_{O}, which was referenced with respect to the total energy of a single oxygen molecule. The surface structure with lower energy is more likely to be realized. The ideal surface was found to be stable at higher μ_{O}, whereas the reduced

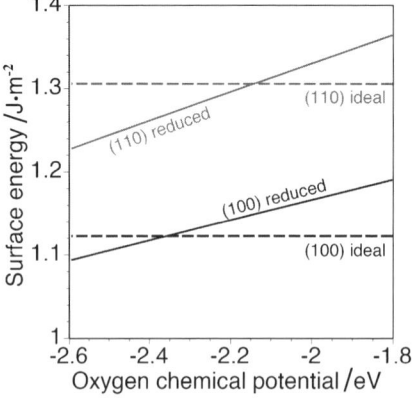

Figure 7.5 Energies of reduced and ideal {110} and {100} surfaces of Ag$_3$PO$_4$ as a function of oxygen chemical potential. Adapted with permission. Copyright 2011, American Chemical Society.

surface predominated at lower μ_O. The intersection of the two lines corresponding to the ideal and reduced surfaces gives the transition value of μ_O, which is higher for the {110} plane than for {100}. This indicates that oxygen vacancies start to form on the {110} surface even under a relatively high O_2 partial pressure.

7.2.3 Hierarchical Composite Nanostructures

In a practical photocatalytic system, a series of sub-tasks such as optoelectronic conversion, surface/interface catalysis and the post-treatment or recycling of the photocatalyst should be dealt with. No single composition or structure is currently able to fulfill all such tasks satisfactorily by itself. Thus, all the relevant functions have to be optimized and integrated into composite materials. This viewpoint leads to the concept of designing and fabricating advanced photocatalytic materials based on hierarchical composite nanostructures.

In order to facilitate charge rectification and carrier separation in photocatalysts, many attempts have been made to organize hierarchical nanostructures of semiconductor–metal or semiconductor–semiconductor composites.[40–42] It is a universal approach to bind photocatalysts with co-catalysts of noble metal or oxide nanoparticles, such as Pt, Pd, Au, Ir NiO, and RuO_2, etc. The hetero-junctions formed between photocatalysts and co-catalysts produce an internal electric field, which can facilitate separation of the electron–hole pairs and induce faster carrier migration. The nanoparticle co-catalysts usually serve as a reservoir for photogenerated electrons. Moreover, they often exhibit better conductivity, lower overpotential and higher catalytic activity than the host photocatalysts, thus can act as ideal active sites for photocatalytic reactions to proceed.

Another important kind of hierarchical composite nanostructure is composed of two types of semiconductors, both of which can undertake the task of harvesting incident photons. These nanostructures often exhibit synergetic effects on both carrier separation and photocatalytic efficiency, and are able to perform stoichiometric water splitting under UV light irradiation; examples include Zn-doped Lu_2O_3/Ga_2O_3 and Cr-doped $Ba_2In_2O_5/In_2O_3$ composites.[43,44] The so-called Z-scheme systems may be considered as special hierarchical composite nanostructures, in which the carrier transfer between two semiconductors proceeds via a pair of redox electrolyte mediators such as IO_3^-/I^+. There have been a few successful cases of overall water photolysis through Z-scheme systems such $TaON/WO_3$ and $SrTiO_3/BiVO_4$.[45–48] Here the two semiconductors take on different functions: one for the evolution of H_2 and the other for O_2. In other cases, a narrow band gap semiconductor, such as CdS, CdSe, PbS, or WO_3, serves as a visible light sensitizer to transfer the excited-state electrons to a large band gap semiconductor, such as TiO_2, SnO_2, or ZnO.[49–51] Such systems can exhibit enhanced photocatalytic activity under visible light or sunlight irradiation.

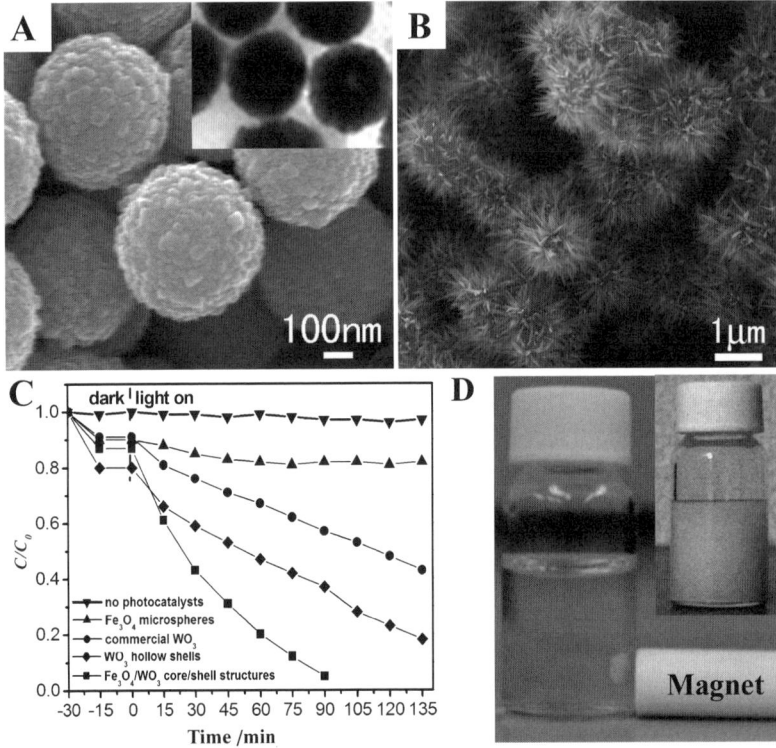

Figure 7.6 SEM images of (A) Fe_3O_4 microspheres and (B) Fe_3O_4/WO_3 core–shell microspheres. (C) Photocatalytic degradation of RhB over Fe_3O_4/WO_3 core–shell microspheres, WO_3 hollow microspheres, Fe_3O_4 microspheres, and commercial WO_3 powder. (D) Digital photographs showing that the Fe_3O_4/WO_3 core–shell microspheres were being recycled by applying a magnetic field. Adapted with permission. Copyright 2011, Wiley-VCH.

Recyclability of photocatalysts is an important property for practical photocatalytic applications in aqueous solution, such as water photolysis or water purification. But it is extremely difficult to recycle nanostructured photocatalysts due to their small size. Developing hierarchical nanostructures of magnetic composites is one of the feasible solutions. Recently, Xi *et al.* presented a successful example of chestnut-like core–shell structures, in which WO_3 nanoplates were grown on the surfaces of Fe_3O_4 spheres (Figure 7.6A, B).[41] This material not only showed considerably high photocatalytic activity for the degradation of RhB under visible light irradiation (Figure 7.6C), but also could be recycled under an applied magnetic field after the reaction was complete (Figure 7.6D).

Hierarchical composite nanostructures also hold a potential advantage in promoting the stability of photocatalysts. For example, Ag_3PO_4 has relatively poor stability in aqueous solution due to its degree of solubility and relatively positive Ag^+/Ag reduction potential (E^θ_{Ag/Ag_3PO_4} = 0.45 eV *versus* NHE). An effective solution is to epitaxially grow AgX (X = Cl, Br, I) nanoshells on the surfaces of Ag_3PO_4 crystals.[42] In this way, the AgX nanoshells can protect the core Ag_3PO_4 crystals from dissolution due to their much lower solubility. In addition, the conduction band and valence band potentials of AgX semiconductors (except AgCl) are more negative than that of Ag_3PO_4, which can promote the transfer and separation of photoexcited electron–hole pairs through the hetero-junctions.

7.3 Nano-assembly

One of the most significant opportunities offered by nanoarchitectonics with respect to practical applications lies in the collective properties of nano-assembly, which also represents a feasible and low-cost approach for the fabrication of hierarchical nanostructures with wide-ranging architectures, dimensions, morphologies and patterns. Going beyond systems comprised of separate individuals, nano-assembly may exhibit new collective properties resulting from the inter-particle coupling of surface electrons (excitons), plasmons or magnetic moments.[52–55] In this field, remarkable progress has been made in the construction of photocatalytic functional materials in which the interactions between nanocrystalline building blocks become key elements of the electronic and optical properties. To date, at least three kinds of interactions, including electronic coupling, plasmon–exciton coupling and optical coupling, have been explored by designing the configuration of a nano-assembly and the method of inter-particle electron and energy transfer. All these attempts can come to a focused aim to enhance the optoelectronic conversion efficiency of photocatalytic materials.

7.3.1 Electronic Coupling Assembly

In a chemical sense, semiconductor nanoparticles appear to be very big and complex molecules assembled from several hundreds to ten of thousands of atoms, thus they exhibit discrete electronic structures and are often referred to as artificial atoms. Electronic coupling assembly of semiconductor nanoparticles has been used to induce a substantial alteration of the electronic

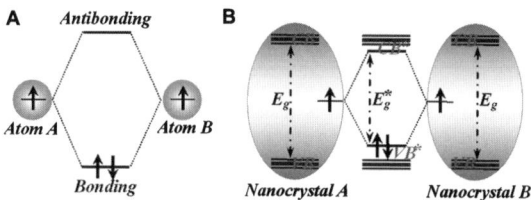

Figure 7.7 Illustration of an electronic bond formed between (A) two atoms and (B) two nanocrystals. Adapted with permission. Copyright 2011, Wiley-VCH.

structures of the nanoparticle ensemble.[9,10] Figure 7.7 illustrates that bonding and anti-bonding levels are formed between two electronically coupled nanoparticles, yielding a new electronic structure.

To make this concept clear, a recent achievement in electronic structure calculations is presented here based on surface contact between two zinc blende CdS crystals, as shown in Figure 7.8. Model I (Figure 7.8A) represents an interface modeled by the contact of two mirror symmetric (110) surfaces, where the interface distance is 20% greater than the layer-to-layer distance in the bulk. In model II (Figure 7.8B), half of the sulfur atoms in Model I have been removed from each surface. The relaxed geometry of Model II (Figure 7.8B) shows that the introduction of sulfur vacancies contributes to the formation of interfacial Cd–Cd bonds. According to the bond counting concept, the removal of one sulfur atom leaves two electrons behind in dangling bonds consisting of cadmium sp^2 hybrid orbital. The surfaces of

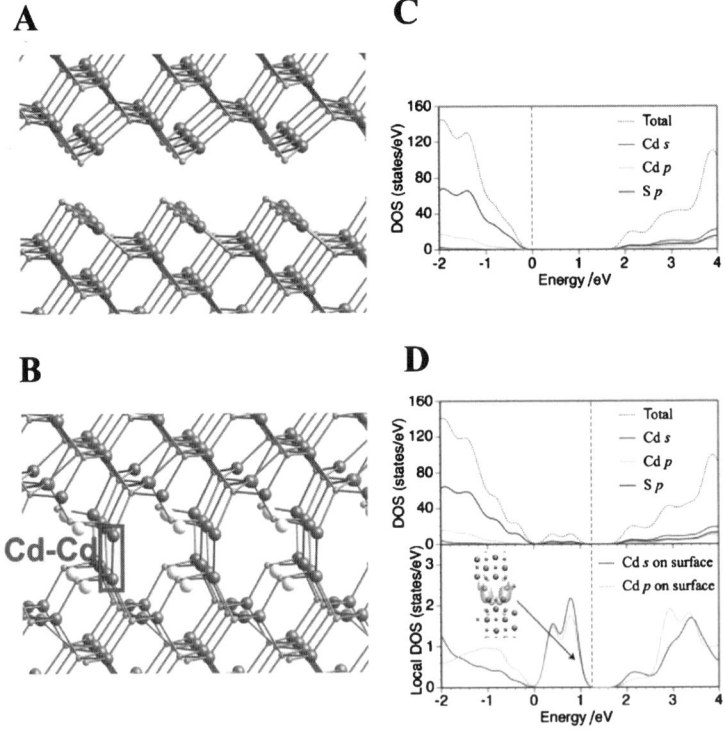

Figure 7.8 Relaxed model structures used for electronic structure calculations. (A) Two (110) surfaces in contact with mirror symmetry, separated by 120% of the layer-to-layer distance in the bulk (Model I). (B) Half of the sulfur atoms were removed from the interface (Model II). (C) and (D) Calculated total and local DOS. The DOS projected on the cadmium atoms at the surface or interface is also shown in (D). The vertical dashed line denotes the highest occupied state in each system. Adapted with permission. Copyright 2011, Royal Society of Chemistry.

covalent materials are usually stabilized by the termination of dangling bonds, and thus the two electrons will naturally tend to form a Cd–Cd bond. The interfacial Cd–Cd bond length (2.7 Å) is somewhat shorter than that calculated for hexagonal bulk cadmium (3.0 Å). Figures 7.8C and D show the total density of states (DOS) for models I and II; the local DOS projected on all of the cadmium atoms is also shown for model II. The band gap is dramatically decreased in model B due to the formation of a new band in the gap of model I, and hence an effective upward shift of the valence band maximum. The inset in Figure 7.8D shows the electron charge density corresponding to an energy of 1 eV in the DOS plot of Figure 7.8D. It is clear that the eigenstate mainly consists of Cd–Cd bonding states, although it also possesses some sulfur p-state character. As it is based on simple bonding chemistry, this calculation result is surely applicable to the nanoparticle system, irrespective of the orientations of the bonded surfaces.

Electronic coupling interaction is strongly dependent on the inter-particle spacing and surface characteristics of assembled nanoparticles. As nanoparticles are embedded in a passivating medium (*e.g.*, surfactant ligands, polymers and dielectric substances), the degree of electronic coupling rapidly decreases with the increased spacing of nanoparticles imposed by the passivating medium. In the case of CdTe, overlap of the wave functions of paired nanoparticles was observed as the nanoparticles were separated by less than 2 nm, leading to a red-shift of the absorption and photoluminescence emission bands.[54] When the inter-particle separation of CdTe nanoparticles was increased to a range of ~2–10 nm, long-range dipole–dipole interactions were found to induce resonance energy transfer.[52]

The nanoparticles without passivating medium attachment may be closely aggregated with negligible spacing between them. For example, covalent Se–Se bonds could be formed between assembled CdSe nanoparticles in the presence of unpassivated Se^{2-} ions on the surface.[56] In this case, short-range bonding interactions between surface defect states have been found to strongly influence the properties of the nanoparticle assembly. Figure 7.9 shows an exemplification in electronic coupling assembly of TiO_2 nanoparticles with an average size of 5 nm.[9] Due to a special surfactant-free chemical synthesis, the TiO_2 nanoparticle surface contained abundant active titanium ions that were combined with ethanol molecules, thus they were able to disperse in absolute ethanol colloid, exhibiting a white colour (Figure 7.9B). In contrast, the TiO_2 nanoparticle assemblies showed a bright yellow colour (Figure 7.9C), implying the change of optical band gap. It can be clearly observed in the UV-vis absorption spectra that the optical absorption edge of the TiO_2 nanoparticle assemblies red-shifted to 450 nm with a tail extending to 500 nm (Figure 7.9D). According to the wavelength-to-energy relationship, the band gap was narrowed by about 0.6 eV from TiO_2 nanoparticle colloid (3.4 eV) to assemblies (2.8 eV). Such band gap narrowing has been attributed to energy band reconstruction as proposed in Figure 7.7, and has shown great potential to improve the solar photocatalysis of TiO_2 materials (Figure 7.9E). This

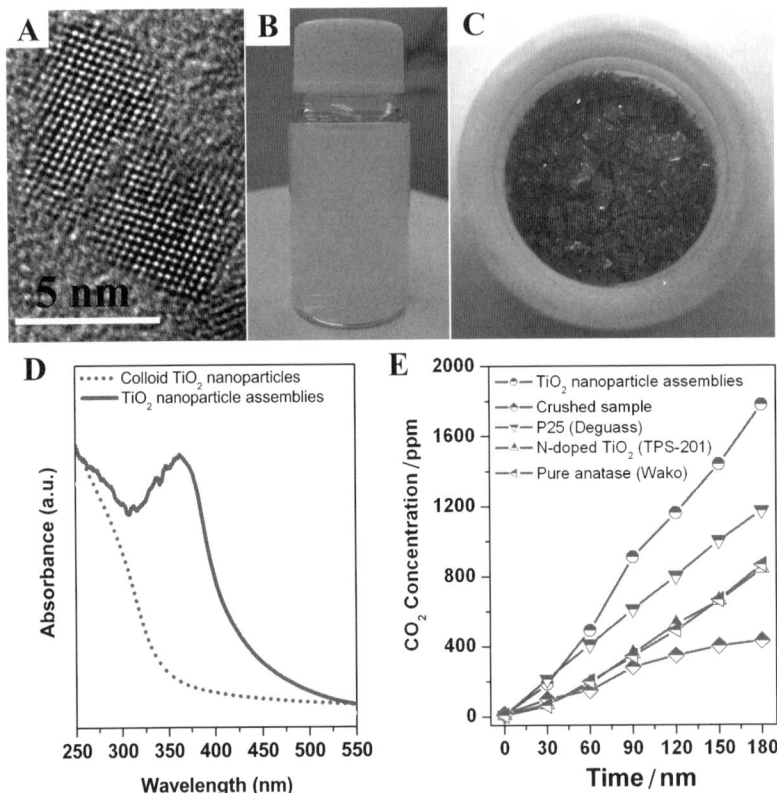

Figure 7.9 (A) HRTEM image of TiO_2 nanoparticles. Photographs of TiO_2 nanoparticle colloid (B) and assemblies (C). (D) UV-visible absorption spectra of the the two samples. (E) Evolution of CO_2 during the decomposition of 2-proponol (initially ~650 ppm) over TiO_2 photocatalysts (0.2 g, irradiated area 8 cm^2) in a 500 mL (total volume) glass vessel under simulated sunlight irradiation (AM 1.5 G, 100 $mW \cdot cm^{-2}$).

finding presents a novel doping-free band gap narrowing strategy, which is applicable to not only oxides, *e.g.* TiO_2, ZnO, Nb_2O_5, but also chalcogenides, *e.g.* CdS, CdSe, ZnS. In addition to photocatalysis, it also can benefit other applications that prefer semiconductor materials with a narrower band gap for harvesting more photons in sunlight spectrum.

In principle, electronic coupling in assemblies of nanoparticles should be found for all inorganic semiconductors with any configurations. To fabricate hierarchical nanostructures with controllable configurations, a number of nanoparticle assembly strategies have been proposed, including self-assembly in presence of external inducements, template-assisted assembly, Langmuir–Blodgett assembly, and so on.[57–62] Electronic coupling assembly of CdS nanoparticles has been recently realized *in situ* on sacrificial $KCdCl_3$ nanowire (Figure 7.10A) and $Cd(OH)_2$ nanosheet templates (Figure 7.10B), forming

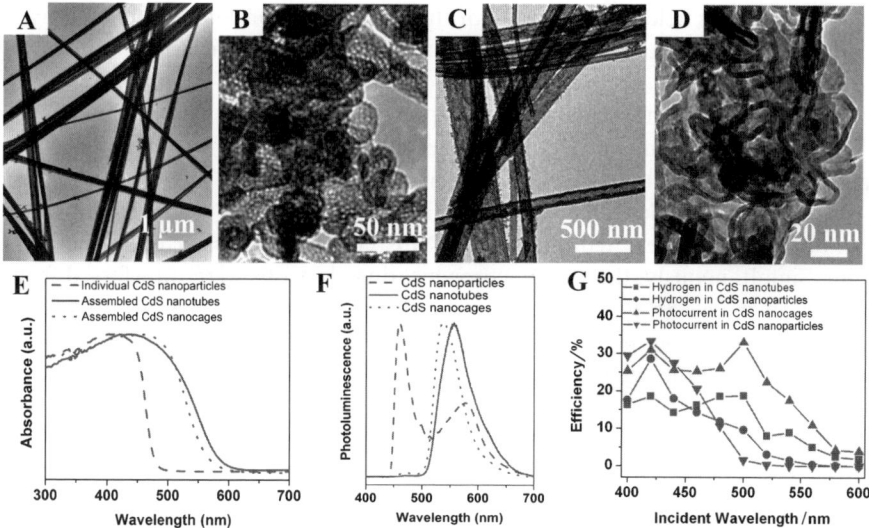

Figure 7.10 TEM image of KCdCl$_3$ (A) nanowires, (B) Cd(OH)$_2$ nanosheets, (C) CdS nanotubes and (D) CdS nanocages. (E) UV-visible absorption spectra. (F) Photoluminescence spectra. (G) Wavelength-dependent efficiency of sunlight utilization by CdS nano-assemblies for the production of hydrogen by photocatalytic water splitting and for photovoltaic current generation. Adapted with permission. Copyright 2011, Royal Society of Chemistry.

nanotubes (Figure 7.10C) and nanocages (Figure 7.10D), respectively.[10] These CdS nanoparticle assemblies exhibited much narrower band gaps and red-shifted optical absorption bands compared to individual CdS nanoparticles and bulk CdS (Figure 7.10E). Consistently, the intrinsic photoluminescence bands were red-shifted (Figure 7.10F). The narrower band gaps of these CdS nanoparticle assemblies allow the system to more efficiently take advantage of longer wavelength light in the solar spectrum for applications (Figure 7.10G) such as hydrogen production from photocatalytic water splitting and photovoltaic current generation.

7.3.2 Plasmon–Exciton Coupling Assembly

Plasmonic metal nanoparticles show their optical absorption bands in the visible to infrared region of the spectrum due to coherent electron oscillation; this is known as localized surface plasmon resonance (LSPR). Accordingly, plasmonic metal nanoparticles can act as photo-sensitizers to enhance optical absorption in metal–semiconductor hybrid assemblies.[63] In addition, plasmon–exciton coupling interactions often occur in such nanostructures under light irradiation. The main characteristics of plasmon–exciton coupling interactions are as follows: the plasmon resonance of the metal nanoparticles induces an electromagnetic field in the vicinity of the semiconductor, and exciton energy is transferred from the semiconductor to the metal nanoparticles.[64]

Plasmon–exciton coupling interactions may result in many interesting effects, such as the enhancement, quenching or wavelength shifting of fluorescent emission, and nonlinear resonance.[65–67] Tian et al. studied the plasmon-induced photo-electrochemical performance of TiO_2 films loaded with Au nanoparticles.[68] Under visible light irradiation, the electrons in the Au nanoparticles were excited due to the LSPR effect and transferred to the TiO_2 film. Simultaneously, the charge-compensating electrons were transferred from donors in the electrolyte solution to the Au nanoparticles. These results show that metal–semiconductor hybrid assemblies can possess not only enhanced optical absorption characteristics but in addition, the electronic heterojunction facilitates the separation of photo-generated electrons and holes. Thus, such nanoarchitectonics has great potential to improve the optoelectronic conversion efficiency.

For example, Chen et al. studied a photoelectrode consisting of Au nanoprisms on a nanocrystalline WO_3 film.[12] The Au nanoprisms were extremely uniform in their size and shape distribution and formed a periodic hexagonal pattern on the WO_3 film (Figure 7.11A). The Au-WO_3 hybrid film

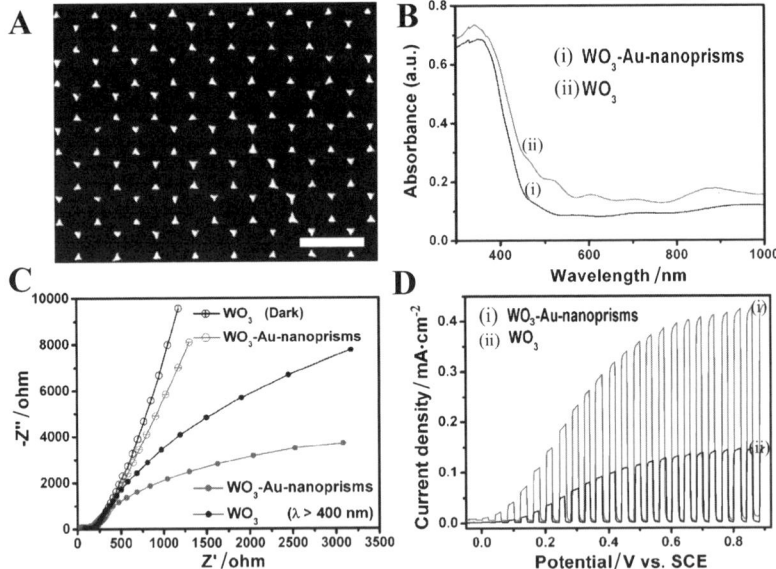

Figure 7.11 (A) SEM image of a hexagonal array of Au nanoprisms deposited on a WO_3 film. The scale bars represent 1 μm. (B) Optical absorption spectra of a bare WO_3 film (black curve) and a WO_3 film loaded with a Au nanoprism array (grey curve). (C) Electrochemical impedance spectroscopy plots of WO_3 photoelectrodes under illumination at $\lambda > 400$ nm. (F) Photocurrent response of the WO_3 photoelectrodes under illumination at $\lambda > 400$ nm. Adapted with permission. Copyright 2011, IOP Science.

can not only harvest photons in the visible to infrared region of the spectrum but also promote the intrinsic absorption of WO_3 (Figure 7.11B) due to the plasmon–exciton coupling interactions. It was revealed by electrochemical impedance spectroscopy (Figure 7.11C) that the Schottky junctions formed between Au and WO_3 can facilitate the separation of photo-generated carriers as well as the interfacial transfer of carriers. Consequently, this photoelectrode exhibited substantially enhanced optoelectronic conversion efficiency (Figure 7.11D).

Recent studies have also indicated that the practical performance of plasmon–exciton coupling assemblies strongly depends on the mutual arrangement, sizes and shapes of the metal nanoparticles and semiconductor, as well as on the ratio of metal nanoparticles to semiconductor and the distance between them.[12,54,64,69] In order to achieve harmonious behaviour, metal nanoparticles are frequently required to have a homogeneous size distribution, highly symmetric morphology and specific exposed facets.

7.3.3 Optical Coupling Assembly

The utilization of solar energy *via* photonic crystals has aroused a great deal of interest in recent years, because fine control over the characteristics of photonic crystals can dramatically modulate light–matter interactions.[70,71] Photonic crystals may be viewed as specially assembled nanostructures that are composed of regularly architectures with a high refractive index contrast. Usually, the periodicity of photonic crystals is of a length scale compatible with the wavelength of the electromagnetic waves of interest (*i.e.*, ~350–700 nm). The propagation of electromagnetic waves in photonic crystals is affected in the same way as electron motion is affected by the periodic potential in semiconductor crystals.[72] Analogous to the electronic band gap in a semiconductor crystal, a photonic band gap can be defined in a photonic crystal; here electromagnetic waves with a certain frequency range are forbidden due to Bragg scattering from the dielectric interfaces.[72] For inverse opal photonic crystals, the photonic band gap is given by the following equation:[73]

$$\lambda = 2\sqrt{\frac{2}{3}}D\sqrt{n_s^2 f + n_{air}^2(1-f) - \sin^2\theta} \quad (7.4)$$

where λ is the wavelength of forbidden light, D is the pore size of the inverse opal structure, n_s and n_{air} are the refractive indices of the solid matter and the air voids, respectively, f is the volume percentage of the solid matter, and θ is the incident angle of light. Based on eqn (7.4), appropriate selection of the structural parameters of the crystal may offer an optical coupling effect where the harmonic resonance occurs at a particular frequency around the electronic band gap. For a given geometry and composition, theoretical calculations are already able to predict how the optical properties of photonic crystals can be tuned by minor variations of the lattice parameters, filling fraction, topology and refractive indices.[71] In this way, it is possible to tune the spectral range in which the absorption enhancement is expected. Therefore, optical coupling effect occurring in photonic crystals can be employed to greatly enhance the solar energy conversion efficiency of materials and devices.

The light in a photonic crystal undergoes strong coherent multiple scattering and travels with a very low group velocity near the photonic band edges.[74] This is known as the slow-light effect and increases the effective optical path length within the photonic crystal, leading to the delay and storage of light. A number of new photo-functional materials have been designed using this principle. For example, Chen *et al.* made use of colloidal templates of polystyrene spheres with diameters of 200, 260 and 360 nm to fabricate WO_3 inverse opals with pore sizes of 140, 170 and 230 nm (Figures 7.12A–C).[13] The $\sim 30\%$ reduction in pore size relative to the polystyrene spheres was due to the annealing process used for removal of the polystyrene template and formation of the WO_3 phase. Under white light, the inverse opals reflected brilliant monochromatic colors (insets to Figures 7.12A–C), a phenomenon caused by pore-size-dependent Bragg diffraction of visible light. An absorption band characteristic of photonic crystals was observed in the inverse opal samples, but not in the disordered porous or non-porous WO_3 samples (Figure 7.12D). The photonic band edges of the three inverse opal samples, in order of increasing pore size,

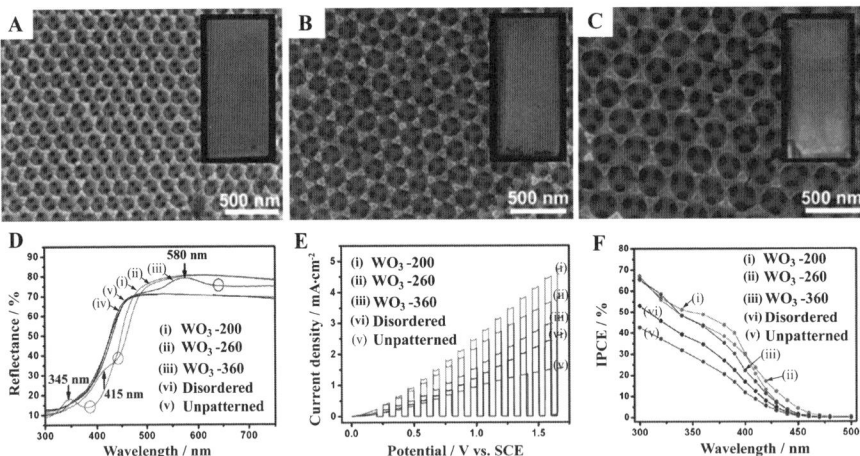

Figure 7.12 (A–C) SEM images of WO_3 inverse opals fabricated using colloidal crystal templates comprised of polystyrene spheres with diameters of 200, 260, and 360 nm, respectively. The insets show photographs of the structures under white light. (D) Light reflectance spectra of the WO_3 inverse opals. The black arrows indicate the photonic band edges and the open circles show the locations of slow light. (E) Photocurrent–potential curves of WO_3 electrodes under light irradiation at $\lambda > 300$ nm. (F) Incident photon-to-electron conversion efficiency. Adapted with permission. Copyright 2011, American Chemical Society.

were located on the blue side of the electronic band edge, at the electronic band edge, and on the red side. Measurement of the photocurrent indicated that the WO_3 inverse opals with coinciding photonic and electronic band edges exhibited the largest current output (Figure 7.12E) and optoelectronic conversion efficiency (Figure 7.12F).

Photonic crystals have led to many other achievements in the field of solar cells. Mallouk *et al.* introduced TiO_2 inverse opal photonic crystals into a dye-sensitized solar cell and obtained a 26% increase in the short-circuit current.[75] This enhancement was attributed to optical coupling between the photonic crystal layer and the disordered nanoparticle layer, resulting in standing waves that improved the absorption of light by the dye molecules. Progress has also been made in silicon solar cells, where a photonic crystal was used as a high refractive mirror instead of a common metallic mirror.[76]

7.4 Conclusion

This chapter has selectively discussed and summarized the main recent achievements in the field of nano-photocatalytic materials, and demonstrated the diverse and flexible ways of promoting photocatalytic or photoelectro-chemical conversion efficiency. With the significant fundamental scientific and technological knowledge already accumulated by past research, nanoarchitec-tonics has emerged as the most important and pioneering way of designing and fabricating advanced photocatalytic materials, and accounts for most of the current research in the field.

Recent research interest is mainly focused on the shape-controlled synthesis of photocatalytic materials as well as on the investigations of the relationship between the morphological or structural characteristics and the photocatalytic properties. Typically, nanostructured materials with large percentages of highly reactive facets have aroused increasing interest. In view of synergetic effects and multi-functions, hierarchical composite nanostructures are most likely to have ample scope of their abilities in a practical photocatalytic system. To date, strong emphasis has been placed on special hierarchical nanos-tructures made by nano-assembly, which show novel collective properties resulting from the interactions between nanocrystalline building blocks or harmonic light–matter interactions due to regularly periodic structures.

Despite important developments being made in nano-photocatalytic materials, many challenges remain in the areas of materials science and engineering from the point of view of practical applications. First of all, the solar energy conversion efficiency of available photocatalysts have not sufficiently satisfied industrial or commercial demands. To reach this goal requires a good understanding of the mechanisms involved in photocatalysis at the atomic level. Both *in situ* observations by advanced microscopy and theoretical calculations and simulations are strongly encouraged to take part in this project to provide guidance toward the practical improvement of photocatalytic materials and their applications.

Acknowledgements

This work was supported by the World Premier International Research Center Initiative on Materials Nanoarchitectonics (MANA), MEXT, Japan.

References

1. A. Fujishima and K. Honda, *Nature*, 1972, **238**, 37–38.
2. M. R. Hoffmann, S. T. Martin, W. Y. Choi and D. W. Bahnemann, *Chem. Rev.*, 1995, **95**, 69–96.
3. A. Hagfeldt and M. Gratzel, *Chem. Rev.*, 1995, **95**, 49–68.
4. M. A. Fox and M. T. Dulay, *Chem. Rev.*, 1993, **93**, 341–357.
5. Z. G. Zou, J. H. Ye, K. Sayama and H. Arakawa, *Nature*, 2001, **414**, 625–627.
6. D. E. Scaife, *Solar Energy*, 1980, **25**, 41–54.
7. T. L. Thompson and J. T. Yates, *Chem. Rev.*, 2006, **106**, 4428–4453.
8. R. Osgood, *Chem. Rev.*, 2006, **106**, 4379–4401.
9. H. Tong, N. Umezawa and J. H. Ye, *Chem. Commun.*, 2011, **47**, 4219–4221.
10. H. Tong, N. Umezawa, J. H. Ye and T. Ohno, *Energy Environ. Sci.*, 2011, 1684–1689.
11. Q. Y. Li, T. Kako and J. H. Ye, *J. Mater. Chem.*, 2010, **20**, 10187–10192.
12. X. Q. Chen, P. Li, H. Tong, T. Kako and J. H. Ye, *Sci. Technol. Adv. Mater.*, 2011, **12**, 044604.
13. X. Q. Chen, J. H. Ye, S. X. Ouyang, T. Kako, Z. S. Li and Z. G. Zou, *ACS Nano*, 2011, 4310–4318.
14. H. Tong, S. Ouyang, Y. Bi, N. Umezawa, M. Oshikiri and J. Ye, *Adv. Mater.*, 2011, **24**, 229–251.
15. Y. P. Bi and J. H. Ye, *Chem.–Eur. J.*, 2010, **16**, 10327–10331.
16. Y. P. Bi and J. H. Ye, *Chem. Commun.*, 2009, 6551–6553.
17. I. S. Cho, S. Lee, J. H. Noh, D. W. Kim, D. K. Lee, H. S. Jung, D. W. Kim and K. S. Hong, *J. Mater. Chem.*, 2010, **20**, 3979–3983.
18. S. Chatterjee, K. Bhattacharyya, P. Ayyub and A. K. Tyagi, *J. Phys. Chem. C*, 2010, **114**, 9424–9430.
19. H. S. Jung, Y. J. Hong, Y. Li, J. Cho, Y. J. Kim and G. C. Yi, *ACS Nano*, 2008, **2**, 637–642.
20. J. S. Jang, U. A. Joshi and J. S. Lee, *J. Phys. Chem. C*, 2007, **111**, 13280–13287.
21. G. Kenanakis and N. Katsarakis, *Appl. Catal., A*, 2010, **378**, 227–233.
22. N. Q. Wu, J. Wang, D. Tafen, H. Wang, J. G. Zheng, J. P. Lewis, X. G. Liu, S. S. Leonard and A. Manivannan, *J. Am. Chem. Soc.*, 2010, **132**, 6679–6685.
23. Q. Liu, Y. Zhou, J. H. Kou, X. Y. Chen, Z. P. Tian, J. Gao, S. C. Yan and Z. G. Zou, *J. Am. Chem. Soc.*, 2010, **132**, 14385–14387.

24. D. F. Wang, A. Pierre, M. G. Kibria, K. Cui, X. G. Han, K. H. Bevan, H. Guo, S. Paradis, A. R. Hakima and Z. T. Mi, *Nano Lett.*, 2011, **11**, 2353–2357.
25. L. Wang, H. W. Wei, Y. J. Fan, X. Z. Liu and J. H. Zhan, *Nanoscale Res. Lett.*, 2009, **4**, 558–564.
26. M. Shahid, I. Shakir, S. J. Yang and D. J. Kang, *Mater. Chem. Phys.*, 2010, **124**, 619–622.
27. H. G. Yang, C. H. Sun, S. Z. Qiao, J. Zou, G. Liu, S. C. Smith, H. M. Cheng and G. Q. Lu, *Nature*, 2008, **453**, 638–U634.
28. G. Liu, C. H. Sun, H. G. Yang, S. C. Smith, L. Z. Wang, G. Q. Lu and H. M. Cheng, *Chem. Commun.*, 2010, **46**, 755–757.
29. S. W. Liu, J. G. Yu and M. Jaroniec, *J. Am. Chem. Soc.*, 2010, **132**, 11914–11916.
30. G. C. Xi and J. H. Ye, *Chem. Commun.*, 2010, **46**, 1893–1895.
31. R. Z. Ma and T. Sasaki, *Adv. Mater.*, 2010, **22**, 5082–5104.
32. T. Sasaki, M. Watanabe, H. Hashizume, H. Yamada and H. Nakazawa, *J. Am. Chem. Soc.*, 1996, **118**, 8329–8335.
33. Y. Ebina, T. Sasaki, M. Harada and M. Watanabe, *Chem. Mater.*, 2002, **14**, 4390–4395.
34. Y. Ebina, N. Sakai and T. Sasaki, *J. Phys. Chem. B*, 2005, **109**, 17212–17216.
35. T. G. Xu, C. Zhang, X. Shao, K. Wu and Y. F. Zhu, *Adv. Funct. Mater.*, 2006, **16**, 1599–1607.
36. K. Maeda, M. Eguchi, S. H. A. Lee, W. J. Youngblood, H. Hata and T. E. Mallouk, *J. Phys. Chem. C*, 2009, **113**, 7962–7969.
37. J. H. Huang, R. Ma, Y. Ebina, K. Fukuda, K. Takada and T. Sasaki, *Chem. Mater.*, 2010, **22**, 2582–2587.
38. T. Shibata, G. Takanashi, T. Nakamura, K. Fukuda, Y. Ebina and T. Sasaki, *Energy Environ. Sci.*, 2011, **4**, 535–542.
39. Y. P. Bi, S. X. Ouyang, N. Umezawa, J. Y. Cao and J. H. Ye, *J. Am. Chem. Soc.*, 2011, **133**, 6490–6492.
40. P. V. Kamat, *J. Phys. Chem. C*, 2007, **111**, 2834–2860.
41. G. C. Xi, B. Yue, J. Y. Cao and J. H. Ye, *Chem.–Eur. J.*, 2011, **17**, 5144–5153.
42. Y. P. Bi, S. X. Ouyang, J. Y. Cao and J. H. Ye, *Phys. Chem. Chem. Phys.*, 2011, **13**, 10071–10075.
43. D. F. Wang, Z. G. Zou and J. H. Ye, *Chem. Phys. Lett.*, 2004, **384**, 139–143.
44. D. F. Wang, Z. G. Zou and J. H. Ye, *Chem. Mater.*, 2005, **17**, 3255–3261.
45. R. Abe, T. Takata, H. Sugihara and K. Domen, *Chem. Commun.*, 2005, 3829–3831.
46. K. Maeda, M. Higashi, D. L. Lu, R. Abe and K. Domen, *J. Am. Chem. Soc.*, 2010, **132**, 5858–5868.
47. Y. Sasaki, A. Iwase, H. Kato and A. Kudo, *J. Catal.*, 2008, **259**, 133–137.

48. Y. Sasaki, H. Nemoto, K. Saito and A. Kudo, *J. Phys. Chem. C*, 2009, **113**, 17536–17542.
49. X. Zong, H. J. Yan, G. P. Wu, G. J. Ma, F. Y. Wen, L. Wang and C. Li, *J. Am. Chem. Soc.*, 2008, **130**, 7176–7177.
50. J. Hensel, G. M. Wang, Y. Li and J. Z. Zhang, *Nano Lett.*, 2010, **10**, 478–483.
51. G. M. Wang, X. Y. Yang, F. Qian, J. Z. Zhang and Y. Li, *Nano Lett.*, 2010, **10**, 1088–1092.
52. S. A. Crooker, J. A. Hollingsworth, S. Tretiak and V. I. Klimov, *Phys. Rev. Lett.*, 2002, **89**, 186802.
53. H. Zeng, J. Li, J. P. Liu, Z. L. Wang and S. H. Sun, *Nature*, 2002, **420**, 395–398.
54. R. Koole, P. Liljeroth, C. D. Donega, D. Vanmaekelbergh and A. Meijerink, *J. Am. Chem. Soc.*, 2006, **128**, 10436–10441.
55. J. A. Fan, C. H. Wu, K. Bao, J. M. Bao, R. Bardhan, N. J. Halas, V. N. Manoharan, P. Nordlander, G. Shvets and F. Capasso, *Science*, 2010, **328**, 1135–1138.
56. I. R. Pala, I. U. Arachchige, D. G. Georgiev and S. L. Brock, *Angew. Chem., Int. Ed.*, 2010, **49**, 3661–3665.
57. E. Rabani, D. R. Reichman, P. L. Geissler and L. E. Brus, *Nature*, 2003, **426**, 271–274.
58. S. Srivastava, A. Santos, K. Critchley, K. S. Kim, P. Podsiadlo, K. Sun, J. Lee, C. L. Xu, G. D. Lilly, S. C. Glotzer and N. A. Kotov, *Science*, 2010, **327**, 1355–1359.
59. B. D. Korth, P. Keng, I. Shim, S. E. Bowles, C. Tang, T. Kowalewski, K. W. Nebesny and J. Pyun, *J. Am. Chem. Soc.*, 2006, **128**, 6562–6563.
60. F. A. Aldaye, A. L. Palmer and H. F. Sleiman, *Science*, 2008, **321**, 1795–1799.
61. J. H. Weaver and G. D. Waddill, *Science*, 1991, **251**, 1444–1451.
62. A. R. Tao, J. X. Huang and P. D. Yang, *Acc. Chem. Res.*, 2008, **41**, 1662–1673.
63. Z. W. Liu, W. B. Hou, P. Pavaskar, M. Aykol and S. B. Cronin, *Nano Lett.*, 2010, **11**, 1111–1116.
64. A. O. Govorov, G. W. Bryant, W. Zhang, T. Skeini, J. Lee, N. A. Kotov, J. M. Slocik and R. R. Naik, *Nano Lett.*, 2006, **6**, 984–994.
65. O. Kulakovich, N. Strekal, A. Yaroshevich, S. Maskevich, S. Gaponenko, I. Nabiev, U. Woggon and M. Artemyev, *Nano Lett.*, 2002, **2**, 1449–1452.
66. W. Zhang, A. O. Govorov and G. W. Bryant, *Phys. Rev. Lett.*, 2006, **97**, 146804.
67. J. Lee, P. Hernandez, J. Lee, A. O. Govorov and N. A. Kotov, *Nat. Mater.*, 2007, **6**, 291–295.
68. Y. Tian and T. Tatsuma, *J. Am. Chem. Soc.*, 2005, **127**, 7632–7637.
69. J. Lee, T. Javed, T. Skeini, A. O. Govorov, G. W. Bryant and N. A. Kotov, *Angew. Chem., Int. Ed.*, 2006, **45**, 4819–4823.
70. M. Honda, T. Seki and Y. Takeoka, *Adv. Mater.*, 2009, **21**, 1801–1804.

71. K. A. Arpin, A. Mihi, H. T. Johnson, A. J. Baca, J. A. Rogers, J. A. Lewis and P. V. Braun, *Adv. Mater.*, 2010, **22**, 1084–1101.
72. J. D. Joannopoulos, P. R. Villeneuve and S. H. Fan, *Nature*, 1997, **386**, 143–149.
73. M. M. Ren, R. Ravikrishna and K. T. Valsaraj, *Environ. Sci. Technol.*, 2006, **40**, 7029–7033.
74. T. Baba, *Nat. Photon.*, 2008, **2**, 465–473.
75. S. Nishimura, N. Abrams, B. A. Lewis, L. I. Halaoui, T. E. Mallouk, K. D. Benkstein, J. van de Lagemaat and A. J. Frank, *J. Am. Chem. Soc.*, 2003, **125**, 6306–6310.
76. T. Suezaki, P. G. O'Brien, J. I. L. Chen, E. Loso, N. P. Kherani and G. A. Ozin, *Adv. Mater.*, 2009, **21**, 559–563.

Materials Nanoarchitechtonics for Bio-Conjugates and Bio-Applications

CHAPTER 8

Design, Synthesis and Application of Bio-conjugate Nanostructures

J. B. FEI AND J. B. LI*

Beijing National Laboratory for Molecular Sciences (BNLMS), Institute of Chemistry, Chinese Academy of Sciences, Beijing, 100190, China
*E-mail: jbli@iccas.ac.cn

8.1 Introduction

Saccharides, phospholipids, proteins and nucleic acids have been well investigated and some special features are well known in biomedical applications.[1-4] The combination of these typical biomoelcules with organic or inorganic compounds enables the formation of various advanced bioconjugates. With the development of nanoscience and nanotechnology, bioconjugate nanostructures have been widely investigated. Here we will summarise the recent achievements of in the development of functional nanostructures conjugated with biomolecules through covalent or non-covalent bonds.

8.2 Saccharides

8.2.1 Monosaccharides

Monosaccharides serve as important fuel molecules, selective recognition units and essential building blocks for nucleic acids.[1] As is well known, many bacteria use saccharides on the surface of mammalian cells as anchors for

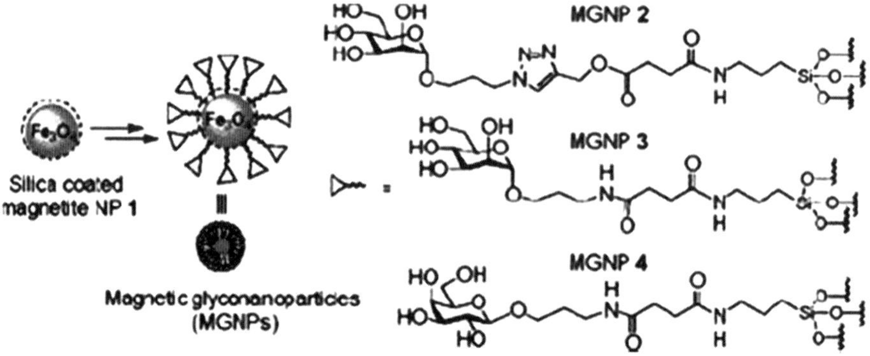

Figure 8.1 Functionalization of silica-coated magnetite NP (NP 1) with D-mannose (Man) through either a triazole linker (MGNP 2) formed by the [2+3] Huisgen reaction or an amide linkage (MGNP 3 and 4). Reprinted with permission from ref. 7. Copyright 2007, American Chemical Society.

attachments, subsequently resulting in serious infection.[5,6] Using novel click chemistry and a traditional amide reaction, El-Boubbou *et al.* constructed highly efficient magnetic glyco-nanoparticle (MGNP) systems, which could detect *Escherichia coli* (*E. coli*) within 5 min and remove up to 88% of the target bacteria from the medium (Figure 8.1).[7]

8.2.2 Polysaccharides

Compared with monosaccharides, polysaccharides are longer carbohydrate molecules of repeated monomer units linked through glycosidic bonds.[1] In general, hydroxyl polysaccharides can be modified with carboxyl terminal compounds. In particular, many polysaccharides can be selelctively oxidized by $NaIO_4$ to form related derivatives with dialdyhade and then functionalized by proteins or nanoparticles containing amine. Typical polysaccharides such as chitosan (CHI), hyaluronan (HA), dextran (DEX) have been used widely as building blocks for constructing functional bioconjugate nanostructures, due to their excellent biocompatibility, biodegradability and special biological features.

8.2.2.1 CHI

About ten years ago, Illum *et al.* reviewed CHI-based delivery systems for vaccines.[8] Recently, Ma and co-workers have conducted research on CHI and their derivatives.[9–12] For example, they fabricated a thermo- and pH-sensitive hydrogel system through electrostatic interaction between a cationic CHI derivative (HTCC) and glycerophosphate (GP) for smart drug delivery.[9] Furthermore, they developed another thermo-responsive hydrogel by simply mixing HTCC and poly(ethylene glycol) (PEG) with a small amount of GP, which could be used to improve the absorption of hydrophilic macromolecular drugs.[10] They also introduced the synthesis of smooth and solid monodisperse CHI microspheres cross-linked with glutaraldehyde (GL) by Shirasu porous-glass (SPG) membrane emulsification.[11] These microspheres exhibited autofluorescence, which could be attributed to the n–p* transitions of C=N bonds in Schiff's bases. Subsequently, they prepared hollow quaternized CHI microspheres by a similar approach.[12] Further evaluation *in vivo* showed an optimal reduction in blood glucose level, indicating excellent therapeutic effects, after treatment with the designed microspheres encapsulating insulin.

In our group, by templating with porous polycarbonate, polysaccharide-based nanotubes with good biodegradability and low cytotoxicity were fabricated by alternately adsorbing CHI and alginate (ALG) *via* a powerful layer-by-layer (LbL) technique (Figure 8.2).[13] In another study, by using melamine formaldehyde (MF) microspheres as a template, biodegradable microcapsules were assembled through electrostatic interaction between CHI and ALG (Figure 8.3).[14] Remarkably, the microcapsules unloading photosensitizer hypocrellin B (HB) showed good biocompatibility under darkness and high cytotoxicity after irradiation. We also prepared covalently cross-linked periodate-oxidized heparin (O-HEP) nanotubes through Schiff's base

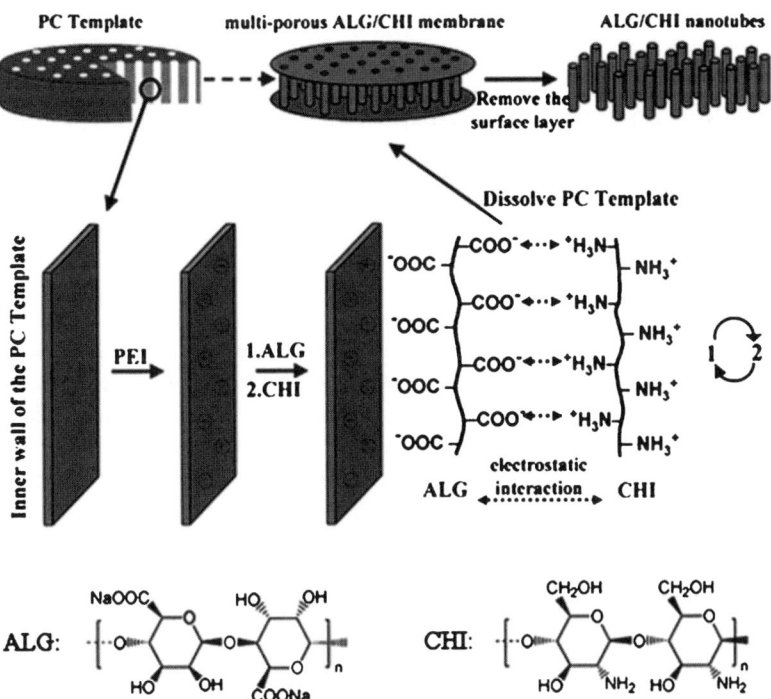

Figure 8.2 The schematic illustration for the fabrication of ALG/CHI nanotubes and multi-porous ALG/CHI membrane. Reprinted with permission from ref. 13. Copyright 2007, Elsevier.

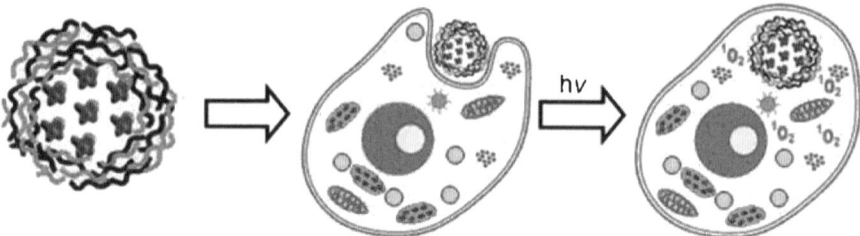

Figure 8.3 Proposed mechanism for photodynamic therapy of HB-loaded CHI/ALG microcapsules. Reprinted with permission from ref. 14. Copyright 2007, Royal Society of Chemistry.

by combining the template method with a LbL technique.[15] The obtained nanotubes conserved anticoagulation activity and exhibited autofluorescence.

Incorporating inorganic nanocrystallines into CHI-based bioconjugation nanostructures could create multifunctional integrated systems. For instance, supermagnetic Fe_3O_4 nanoparticles were decorated with carboxymethylated CHI for the covalent conjugation of papain.[16] Compared with free papain, the conjugated papain exhibited enhanced enzyme activity, better tolerance and improved storage stability, as well as good reusability. In addition, Lee et al. reported that oleyl-chitosan could self-assemble into core–shell structures, providing the effective core compartment for loading iron nanoparticles (ION).[17] After intravenous injection of ION-loaded Cy5.5-conjugated oleyl-CHI (ION-Cy5.5-oleyl-CHI) nanoparticles in tumor-bearing mice, both near-infrared (NIR) and magnetic resonance imaging (MRI) showed the detectable signal intensity and enhancement in tumor tissues *via* an enhanced permeability and retention (EPR) effect.

8.2.2.2 HA

HA is an anionic glycosaminoglycan distributed widely throughout connective, epithelial and neural tissues.[18] Many HA-based bioconjugates have been developed for the enhancement and modulation of its therapeutic action. It should be noted that very recently, Schanté et al. reviewed the various chemical modification methods and synthetic routes to obtain HA derivatives and their bioconjugate nanostructures.[19] Here we will focus on some recent excellent cases.

Bodnár et al. prepared novel biodegradable HA-based nanoparticles, using 2,2′(ethylenedioxy) bis(ethylamine) as a cross-linker.[20] The synthesized nanosystems were stable, transparent or mildly opalescent systems depending on the cross-linking ratio. Choi et al. synthesized amphiphilic HA nanoparticles by grafting hydrophobic 5β-cholanic acid to the backbone of HA.[21] Preliminarily, it could be found that HA nanoparticles were efficiently taken up by SCC7 cancer cells which over-expressed CD44 (the receptor for HA). Furthermore, when HA nanoparticles were systemically administrated into the tail vein of tumor-bearing mice, most of them were selectively concentrated in tumor and liver sites. High tumor targeting ability of HA nanoparticles may result from their prolonged circulation in blood and high affinity to tumor cells.

Lee *et al.* fabricated novel HA nanogels encapsulating small interfering RNA (siRNA) by an inverse water-in-oil emulsion method.[22] In detail, thiol-conjugated HA dissolved in aqueous emulsion droplets was cross-linked under ultrasonication *via* the formation of disulfide linkages to form HA nanogels. The HA nanogels entrapping green fluorescence protein (GFP) siRNA were readily taken up by HCT-116 cells having HA-specific CD44 receptors on the surface. When they were co-transfected with GFP plasmid/lipofectamine to HCT-116 cells, a significant extent of GFP gene silencing could be observed in both serum and serum-free conditions.

Additionally, utilizing the insolubility of paclitaxel (PTX), Rivkin *et al.* constructed nanoparticle-like clusters by simply mixing it with lipids.[23] After coating with HA and a glycosaminoglycan (GAG), these nanoparticles (PTX-GAGs) could deliver PTX selectively into tumor cells over-expressing CD44. Moreover, injected systemically to mice bearing solid tumors, these nanoparticles exhibited a high safety profile and tumor accumulation. Remarkably, PTX-GAGs induced tumor arrest and were as potent as a 4-fold higher PTX dose.

8.2.2.3 DEX

DEX is composed of branched glucan with various chains. It has been widely employed to reduce blood viscosity. Nanostructured DEX and relative derivates have also been exploited for novel biomedical applications. For instance, a lauric acid modified dextran–agmatine nanobioconjugate (DEX-L-AGM) was prepared through 1,10-carbonyldiimidazole (CDI) activation and the nucleophilic reaction between the tosyl of tosylated dextran and primary amine of agmatine.[24] DEX-L-AGM as-obtained could condense DNA into nanocomplexes through electrostatic interaction. Importantly, cells co-cultured with the degraded products of DEX-L-AGM retained more than 80% viability, suggesting their great potential application for gene therapy.

8.3 Phospholipids

Phospholipids are a major component of all cell membranes. Most phospholipids, except sphingomyelin, contain a diglyceride, a phosphate group, as well as a simple organic molecule.[2] Up to now, lipids and their derivates have been used to construct biocompatible interfaces and encapsulate drugs or genes to form nanosystems for treatment of serious diseases.

8.3.1 Lipid

Supported lipid bilayers are significant well-defined model systems used to mimic a cell membrane for developing advanced drug delivery systems, biosensor and biodevices. Recently, we reported that supported lipid bilayers could spontaneously wrap around the LbL-assembled luminescent nanotubes with a hydrophilic polymer-cushioned outer surface (Figure 8.4).[25] The results showed that one-dimensional lipid membranes were laterally mobile and had a high mobile fraction. Furthermore, a simply diffusion model was also presented to estimate the mobility of lipid molecules. In another of our studies, by coating with a lipid layer, the efficiency of mesoporous silica nanoparticles (MSNs) with photosensitizers taken up by cells could be improved (Figure 8.5).[26]

Additionally, we also demonstrated that human serum albumin (HSA) patterns constructed in a solid substrate by a micro-contact printing (μCP) technique supported the deposition of a phospholipid bilayer containing the glycolipid 10-tetradecyloxymethy-3,6,9,12-tetraoxahexacosyl 2- acetamido-2-deoxy-b-D-glucopyranoside (PB1124).[27] The results revealed that the obtained glycolipid patterns were well-defined and could be utilized to recognize and immobilize *E. coli*. We believed that this strategy would be promising for bacterial detection through solid surface recognition.

Liposome-based bioconjugate nanostructures are very promising vehicles for theranostics of Alzheimer's disease (AD). For instance, Gobbi *et al.* developed a liposome composed of phosphatidic acid and cardiolipin to target $A\beta_{1-42}$ which is released as monomer and then undergoes different aggregation to form oligomers, fibrils and plaques in diseased brains.[28] Relative surface plasmon resonance (SPR) studies, evaluating the binding ability of flowing liposomes to immobilized $A\beta_{1-42}$, revealed that the bioconjugate displayed high affinity for $A\beta_{1-42}$ fibrils (22–60 nM). Outstandingly, Antimisiaris group have also constructed liposome-based nanoassemblies with very high affinity for $A\beta_{1-42}$.[29,30] They used click chemistry to assemble nanoliposomes decorated with a curcumin derivative which was designed to maintain the planar structure required for interaction with $A\beta_{1-42}$.[29] They demonstrated high integrity during incubation in the presence of plasma proteins. SPR experiments indicated that the liposomes exposing the curcumin derivative had extremely high affinity for $A\beta_{1-42}$ fibrils (1–5 nM). By another approach,

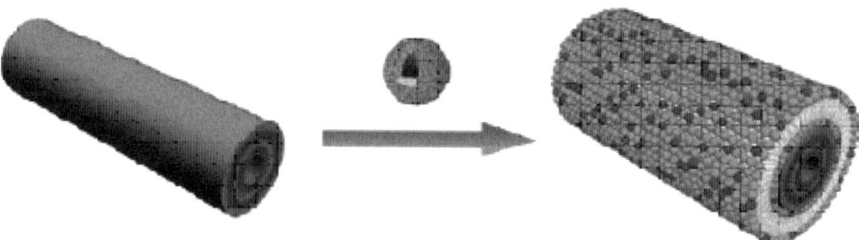

Figure 8.4 Schematic of lipid bilayer coating on LbL assembled luminescent nanotube by liposome fusion. Reprinted with permission from ref. 25. Copyright 2009, Royal Society of Chemistry.

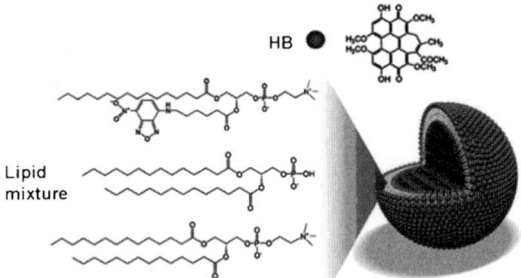

Figure 8.5 Chemical structures of lipids and HB; the model of lipid coated MSN-HB structure. Reprinted with permission from ref. 26. Copyright 2010, Royal Society of Chemistry.

they further fabricated highly efficient liposomes decorated with an anti-Aβ monoclonal antibody.[30]

8.3.2 PEG–Lipid

PEG modified or doped liposomes have been constructured as highly efficient nanosystems for *in vivo* anticancer treatment and gene therapy. For example, the Liang group constructed a novel self-assembled nanocarrier to delivery doxorubicin (Dox) for cancer therapy.[31] In detail, Dox was incorporated into PEG phosphatidylethanolamine (PEG-PE) block copolymer micelles to form nanoassemblies of Dox and PEG-PE (M-Dox). Excitingly, M-Dox could increased Dox internalization by A549 cells into lysosomes and enhance cytotoxicity. Furthermore, M-Dox was more effective in inhibiting tumor growth in the subcutaneous LLC tumor model than free doxorubicin. In addition, M-Dox treatment prolonged survival in both mouse models and reduced metastases in the pulmonary model.

Very recently, PEG doped nanoporphysomes made through self-assembly of porphyrin bilayers were developed, which could generate tunable extinction coefficients, and exhibit structure-dependent fluorescence self-quenching and unique photothermal and photoacoustic properties (Figures 8.6).[32] Interestingly, porphysomes enabled the sensitive visualization of lymphatic systems using photoacoustic tomography. Importantly, porphysomes were enzymatically biodegradable and induced minimal acute toxicity in mice with intravenous doses of 1000 mg kg^{-1}. Following systemic administration, porphysomes accumulated in tumours of xenograft-bearing mice and laser irradiation induced photothermal tumour ablation (Figure 8.7).

Noticeably, Harashima developed many PEGylated liposomes (PEG-LPs) to deliver genes with high efficiency.[33–38] It was worthy of note that they constructed a dual-ligand based PEGylated liposome delivery system that had target specificity and enhanced cellular uptake.[33] In detail, PEG-LPs were assembled by adding distearoyl phosphoethanolaminepolyethylene-glycol-2000 conjugate (DSPE-PEG2000) to a lipid mixture. Then, the cyclic RGD

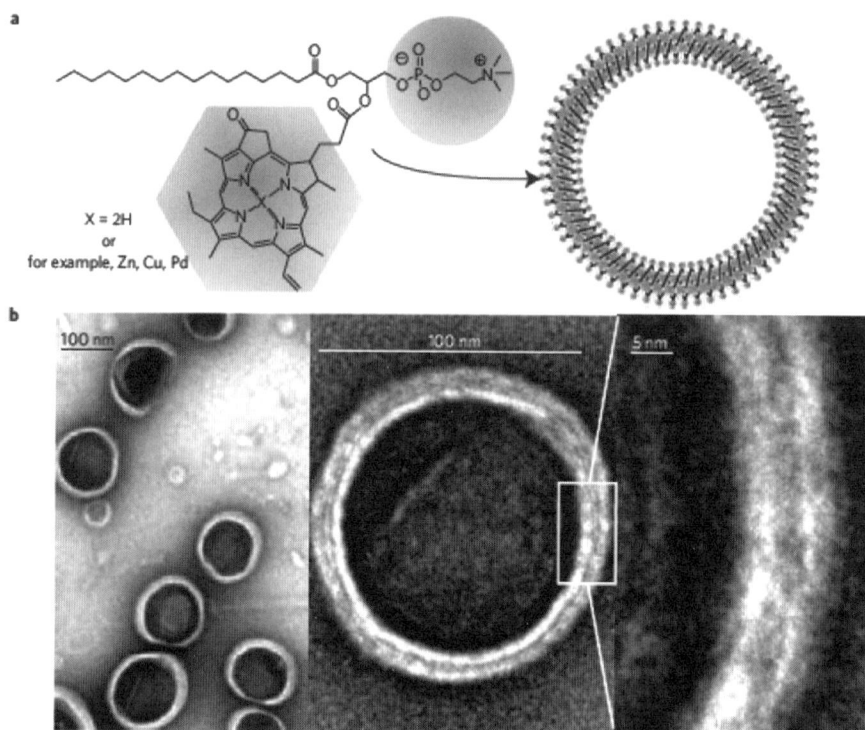

Figure 8.6 Porphysomes are optically active nanovesicles formed from porphyrin bilayers. (a) Schematic representation of a pyropheophorbide–lipid porphysome. (b) Electron micrographs of negatively stained porphysomes (5% PEG–lipid, 95% pyropheophorbide–lipid). Reprinted with permission from ref. 32. Copyright 2011, Nature Publishing Group.

(Arg–Gly–Asp) peptide, a specific ligand with high affinity for Integrin $\alpha v\beta 3$ was grafted to the distal end of the PEG on the PEG-LP to synthesize RGD-PEG-LP. Stearylated octaarginine (STR-R8) was incorporated on the surface of the RGD-PEG-LP to form R8/RGD-PEG-LP. Compared with RGD-PEG-LP, R8/RGD-PEG-LP showed an enhanced cellular uptake as well as higher transfection efficiency in Integrin $\alpha v\beta 3$ expressing cells.

8.3.3 Calcium Phosphate–Lipid

Biocompatibile and biodegradable inorganic nanoparticles have been used to support lipid layers for gene therapy. For instance, Li *et al.* developed a lipid-coated calcium phosphate (LCP) nanoparticle to deliver siRNA to a xenograft tumor model by intravenous administration.[39] Anisamide functionalized LCP nanoparticles uploading luciferase siRNA silenced about 70% and 50% of luciferase activity for the tumor cells in culture and those grown in a xenograft

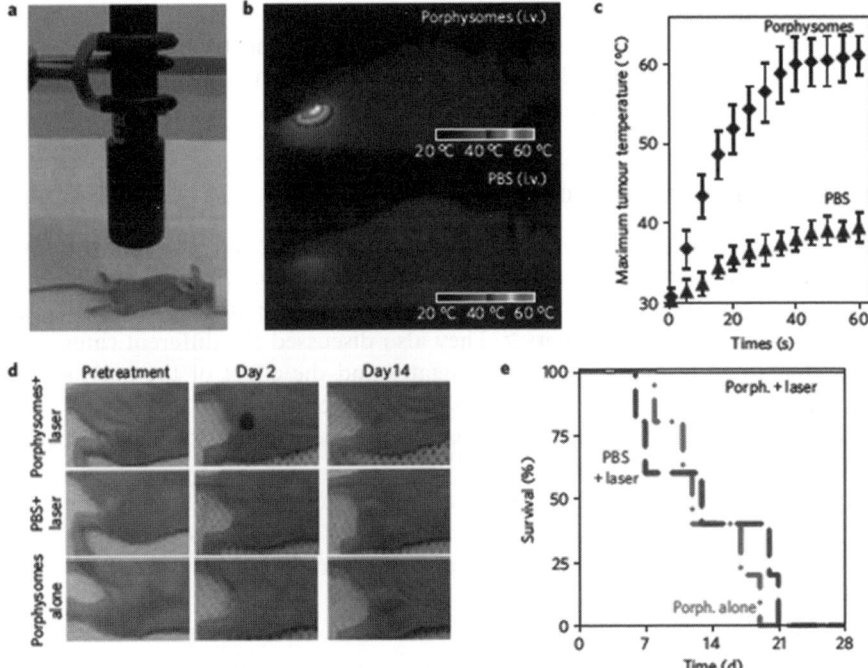

Figure 8.7 Porphysomes as photothermal therapy agents. (a) Photothermal therapy set-up showing laser and tumor-bearing mouse. (b) Representative thermal response in KB tumor-bearing mice injected intravenously 24 h before with 42 mg kg^{-1} porphysomes or PBS. Thermal image was obtained after 60 s of laser irradiation (1.9 W cm^{-2}). (c) Maximum tumor temperature during 60 s laser irradiation (mean ± s.d. for five mice per group). (d) Photographs showing therapeutic response to photothermal therapy using porphysomes. (e) Survival plot of tumor-bearing mice treated with the indicated conditions. Mice were euthanized when tumors reached 10 mm in size ($n = 5$ for each group). Reprinted with permission from ref. 32. Copyright 2011, Nature Publishing Group.

model, respectively. Similarly, Zhou *et al.* reported a highly stable gene transfer vector composed of lipid and calcium phosphate nanoparticles.[40] Importantly, the bioconjugates transfected pDNA 24 times greater than the naked pDNA and 10-fold greater relative to the standard calcium-phosphate precipitation preparations.

8.4 Proteins

Proteins composed of one or more polypeptides are typically folded into a globular or fibrous form to facilitate relative catalytic, structural and mechanical functions. Some proteins are of significance in cell adhesion, cell differentiation and cell proliferation.[3] Bioconjugation techniques have been developed to produce many novel protein-based nanosystems by assembling functional nanomaterials (*e.g.* polymers, gold nanostructures, magnetic nanoparticles, quantum dots and carbon nanomaterials) with various physicochemical and biological properties.

8.4.1 Polymer–Protein

About ten years ago, Hoffman produced a review on bioconjugates of intelligent polymers and recognition proteins for diagnostics and separations.[41] In 2006, Thordarson *et al.* highlighted the synthetic pathways to conjugate proteins to intelligent polymers.[42] They also discussed the different categories of well-defined protein–polymer conjugates and the effect of the polymer on protein function. Very recently, Salmaso *et al.* reviewed the few examples of grafting to and growing from PEGylation for the preparation of therapeutically effective protein bioconjugates.[43]

The development of protein-based nanosystems relies heavily on new chemical reactions and related novel reagents. In recent years, click chemistry has been widely used to construct polymer–protein nanosystems.[44–48] For instance, Li *et al.* synthesized thermo-responsive polymer–protein conjugates by combining reversible addition–fragmentation chain transfer (RAFT) polymerization and a grafting-to approach by a highly efficient click chemistry strategy.[47] First of all, bovine serum albumin (BSA), was modified with an alkyne moiety. Secondly, azido-terminated poly(*N*-isopropylacrylamide) (PNIPAM-N_3) was prepared *via* RAFT. Finally, polymer–protein bioconjugates could be obtained by copper-catalyzed azide–alkyne cycloaddition. When heated above the PNIPAM lower critical solution temperature (LCST), the PNIPAM-BSA bioconjugates formed stable nanoparticles composed of dehydrated polymer and hydrophilic protein. In another study, using click chemistry, Droumaguet *et al.* designed and synthesized many tri-block protein–polymer giant amphiphiles, which showed interesting aggregation patterns.[48]

8.4.2 Gold–Protein

Immobilized functional proteins have attracted much attention in many fields, such as biomedicine, food conservation and environmental protection. Gold nanostructures are a superior support because of their simplicity of preparation and modification, low toxicity and easy detection.[49,50] In our recent study, a glucose oxidase (GOD) monolayer was covalently immobilized on the surface of gold nanoparticles (GNPs) through a classical EDC/NHS coupling reaction (Figure 8.8).[51] Interestingly, the enzyme activity assays of the prepared bioconjugates displayed an enhanced thermostability, compared with that of free enzyme.

Similar to the use of GNPs, Ma and Ding reported that bioconjugate nanostructures consisting of an anionic GOD and positively charged gold

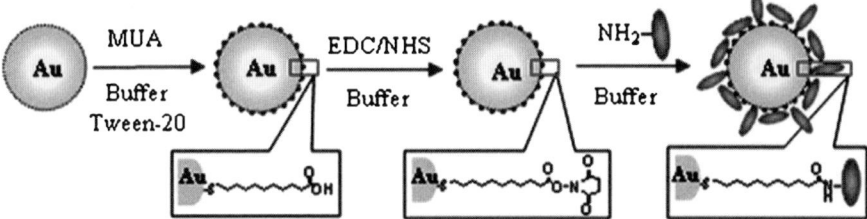

Figure 8.8 Schematic illustration of the fabrication of the GOD/GNPs bioconjugates. Reprinted with permission from ref. 51. Copyright 2007, Elsevier.

nanorods (GNRs) that were assembled through electrostatic interactions *via* LbL technique.[52] Surprisingly, the immobilized GOD on GNRs still retained about 39.3% activity even at 90 °C, while free GOD in solution only retained about 22% of its relative activity after the same treatment.

8.4.3 Quantum Dot–Protein

Since the early reports on the biological use of quantum dots (QDs) about ten years ago, QDs and their bioconjugate nanostructures have been successfully applied in different imaging applications, such as fluorescence detection, imaging of live cell dynamics, *in situ* tissue profiling, as well as *in vivo* animal imaging.[53–55] In a recent review, Rao *et al.* covered the optical properties of various QDs, the relative biofunctionization strategies, *in vitro* diagnostic and *in vivo* imaging applications.[56]

Water-solubility and buffer-stability are required for QDs and their bioconjugates in bioapplications. However, it is still a great challenge because present syntheses are difficult and take a long time. To address this problem, Wolcott *et al.* developed a simple aqueous fabrication of silica-capped CdTe QDs with high fluorescence.[57] The silica shell could prevent the leakage of toxic Cd(II) and provide a modifiable surface for ease of conjugation to bioactive molecules at the same time.

New approaches are desirable to detect the interaction of nanoparticles and cellular proteins at the molecular level. Recently, Liu and Vu applied a modification of PAGE co-immunoprecipitation (QD-based PAGE electrophoresis blotting) to identify QD bioconjugate–cellular protein association.[58] The authors demonstrated that this method could provide the opportunity to isolate and evaluate the action of QD bioconjugate–protein complexes in intact cells and to correlate these identified interactions with their location in cells.

Clapp *et al.* used hybrid QD-peptide substrates for monitoring enzymatic proteolysis and protein–protease interactions.[59] Detection of proteolysis was based on changes in the rate of fluorescence resonance energy transfer between the QDs and the proximal dye-labeled proteins following protein digestion by added enzyme. The authors claimed that this approach could be extensive for drug discovery assays and *in vivo* cellular monitoring of enzymatic activity.

Noticeably, BSA was covalently conjugated to fluorescent YVO_4:Eu nanocrystals through primary amine groups by EDC/NHS chemistry.[60] Immunoblots verified that BSA coupled to nanocrystals still remained immunoreactive.

8.4.4 Iron Oxide–Protein

Integrating multiple functionalities into individual nanoscale complexes is of tremendous significance in biomedicine. Bora and Deb developed a novel bioconjugate of stearic acid capped maghemite nanoparticles (γ-Fe_2O_3) with BSA through a traditional amide reaction.[61] Magnetic measurement showed a retention of the magnetic property by significant values of saturation magnetization and other hysteretic parameters. Bardhan *et al.* combined the ability to enhance two different imaging technologies simultaneously—fluorescence optical imaging and MRI with antibody targeting and photothermal therapeutic actuation all within the same nanoshell-based bioconjugates.[62] The bioconjugate nanostructures were designed and synthesized by coating a gold nanoshell with a silica epilayer doped with Fe_3O_4 and the fluorophore ICG, which led to a high T_2 relaxivity (390 mM^{-1} s^{-1}) and 45× fluorescence enhancement of ICG. The experimental results also showed that bioconjugate nanostructures could target HER2+ cells and cause photothermal cell death upon NIR radiation.

Very recently, by conjugation with fetoprotein antibodies, multifunctional manganese carbonate microspheres with superparamagnetic and fluorescent properties were fabricated.[63] The related electrochemical immunosensor showed high sensitivity and selectivity with a detection limit of 0.3 pg mL^{-1} for fetoprotein. Johnson *et al.* demonstrated the feasibility of specifically attaching a haloalkane dehalogenase enzyme to silica-coated or uncoated iron oxide superparamagnetic nanoparticles using high-affinity peptides.[64] In detail, the enzyme was cloned from Xanthobacter autotrophicus strain GJ10 into *E. coli* to generate fusion proteins containing dehalogenase sequences with C-terminal polypeptide repeats that could specially bind the surface of either silica or iron oxide. As a result, the degree of fusion protein adsorption to nanoparticle surfaces was found to exceed that of enzymes not activated with an affinity sequence.

8.4.5 Carbon–Protein

Carbon nanotubes (CNTs) are low-dimensional sp^2 carbon nanomaterials with many unique physical and chemical properties that have attracted much research interest in a wide range of areas, especially in bioconjugated nanomedicine. For instance, Cui *et al.* developed an electrochemical immunosensor for detecting human IgG (HIgG) by using a GNPs/CNTs bioconjugate platform with a horseradish peroxidase (HRP)-functionalized GNP label.[65] The GNPs/CNTs bioconjugates were coated on the glass carbon electrode to construct an effective antibody immobilization matrix, which allowed the immobilized biomolecules to retain high stability and bioactivity. Subsequent results revealed that using bioconjugates featuring HRP labels and secondary antibodies (Ab_2) linked to GNPs could enhance sensitivity. The final results showed that the approach could provide a linear response range between 0.125 and 80 ng mL^{-1} with a detection limit of 40 pg mL^{-1} for HIgG. Furthermore, the immunosensor exhibited good precision, acceptable stability and reproducibility and could be employed for the detection of HIgG in real samples, indicating a potential tool for detecting the protein in clinical laboratories.

Very recently, the Liu group comprehensively highlighted recent research on CNTs for applications in drug delivery and cancer therapy, and introduced the relative opportunities and challenges.[66] Meanwhile, they have also done excellent work on bioconjugates composed of CNTs and proteins.[67,68] For instance, they constructed multi-enzyme LbL assembled SWCNT composite labels for amplified ultrasensitive electrochemical detection of a cancer biomarker, carcinoembryonic antigen (CEA).[67] In detail, CEA was sandwiched between an electrode surface-confined capture anti-CEA antibody and the secondary signal anti-CEA/enzyme-LbL/SWCNT bioconjugate. The biocatalytic signal amplification for monitoring CEA was realized through both the enzymes loaded on the CNTs and redox-recycling of the enzymatic products in the presence of the secondary enzyme and the corresponding substrate. Their novel and remarkable signal amplification strategy with a detection limit of 0.04 pg mL^{-1} showed about 2–4 orders of magnitude improvement in sensitivity for CEA detection, compared with other common signal amplified assays.

8.5 Nucleic Acids

Deoxyribonucleic acid (DNA) and ribonucleic acid (RNA) are two essential macromolecules for life.[4] Compared with RNA, much more attention has been paid to DNA in nanoscience and nanotechnology. There have been many excellent reviews on DNA-based nanotechnology. In particular, combination of DNA with novel functional nanomaterials has attracted remarkable interest and has led to a wide variety of biomedical applications.

8.5.1 DNA

Pioneered by the Seeman group, structural DNA nanotechnology has become a booming field of research.[69–80] For instance, they have constructed nanosized walking devices, and biological replication of DNA nanostructures with simple topologies has also been realized.[75,76] They also extended self-assembled DNA crystalline systems from 2D to 3D and accomplished 2D algorithmic assembly.[78–80] These perfect nanosystems can be expected to have a very bright future in many areas.

Notably, Pei *et al.* designed a DNA tetrahedron structure with pendant probe DNA at one vertex and three thiol groups at the other three vertices.[81] In detail, this tetrahedron was assembled from three thiolated DNA fragments of 55 nucleotides (55-nt) and one probe-containing DNA fragment of 80-nt, which were mixed in stoichiometric equivalents in buffer, heated, and then rapidly cooled to 4 °C. Remarkably, the tetrahedron assembly process was extremely fast (within 2 min) with a high yield of over 85%.

In addition, aptamers, which are *in vitro* selected functional oligonucleotides, have been employed to design advanced biosensors (*i.e.* aptasensors) due to their inherent selectivity, affinity and multifarious advantages over traditional recognition elements. Ren and co-workers developed many DNA-based nanosystems to bind and release drugs and detect enzymatic/ oxidative cleavage of single-stranded DNA.[82–86]

8.5.2 Metal–DNA

DNA conjugated GNPs as biosensors with high sensitivity and selectivity have been developed. For instance, Fan and co-workers prepared a multi-component nanoprobe by co-assembling thiolated oligonucleotide, HRP and BSA at the surface of 15 nm GNPs.[87] The results showed that only in the presence of target DNA, HRP confined at the surface of GNPs could catalytically oxidize the substrate and generate optical signals that reflected the quantity of target DNA. Subsequently, they reported a novel sandwich type assay for optically detecting DNA using cross-linked GNP aggregates.[88] In a typical sensing process, the GNP aggregates probe was brought into the proximity of magnetic particles through DNA hybridization. Once a magnetic field was added, these sandwich complexes were magnetically separated. As a result, HRP that was confined at the surface of GNP aggregates could catalyze the enzyme substrate and generate an optical signal. Furthermore, this assay was utilized to detect breast cancer-associated BRCA-1 gene with the detection limit of 1 fmol.

Additionally, using DNA as building blocks, noble metal nanoclusters (NCs) exhibiting strong, robust and size-dependent fluorescence emission have been developed as many kinds of fluorophore probes.[89–91] For example, Guo *et al.* developed hybrid DNA duplexes as capping scaffolds for the fabrication of fluorescent Ag NCs.[89] They found that the formation of fluorescent Ag NCs in DNA duplex scaffolds was highly dependent on the sequence and could identify a typical single-nucleotide mutation. Moreover, the identification of single-nucleotide differences using this strategy has also been extended to more general types of single-nucleotide mismatches. Yang *et al.* in the same group synthesized a new kind of silver micro-dendrites, without surfactant protection, for separating and detecting DNA merely by gravity (Figure 8.9).[90] The authors demonstrated that through this approach, the DNA of human T-lymphotropic virus type I (HTLV-I) could be detected down to 10 pM. Furthermore, Han *et al.* reported a simple and sensitive method to detect biothiols (cysteine, homocysteine and glutathione) by using fluorescent silver nanoclusters (Ag

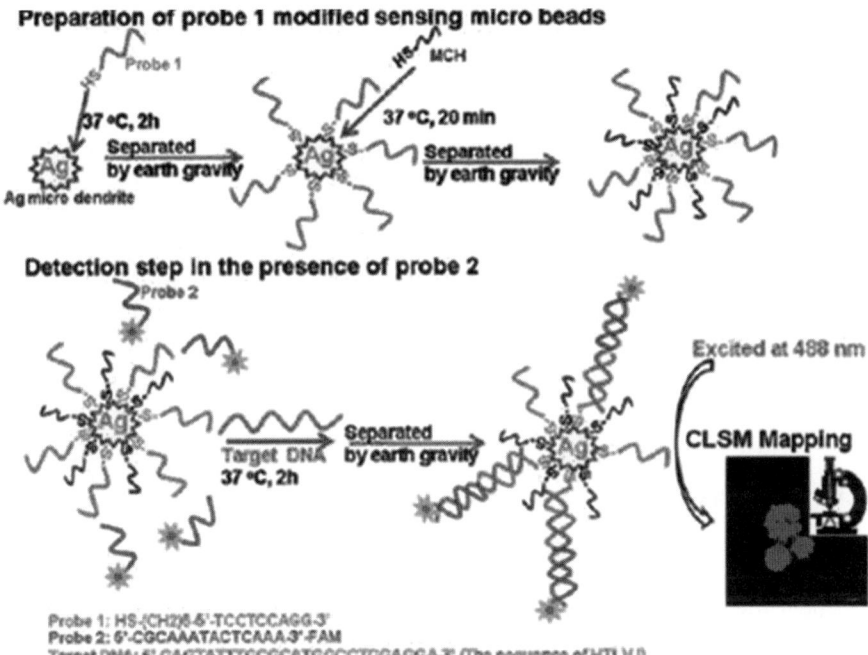

Figure 8.9 The overall separation and detection strategy for separating and detecting DNA by gravity. Reprinted with permission from ref. 90. Copyright 2010, Royal Society of Chemistry.

NCs) stabilized by single-stranded DNA (DNA-Ag NCs) as probes.[91] The photoluminescence intensity of DNA-Ag NCs was found to be quenched effectively with an increase of biothiols concentration due to the formed non-fluorescent coordination complex between DNA-Ag NCs and biothiols, resulting in the shift-to-red of emission wavelength. Satisfactory detection limits and linear relationships of three biothiols above could be obtained.

Outstandingly, the Tan group have produced many excellent reviews and reports on aptamers for biomolecular sensor, cell detection and capture, and drug delivery.[92–107] Typically, they also constructed multifunctional aptamer-conjugated nanoparticles for the collection and detection of multiple cancer cells.[102] The authors demonstrated the wide applicability of this methodology for medical diagnostics, cell enrichment and separation. Furthermore, they prepared a therapeutic aptamer-conjugated liposome drug delivery system, which could deliver the loaded drug to target cells with high specificity and excellent efficiency.[103]

8.5.3 Graphene–DNA

Graphene, a two-dimensional single-layer sheet of sp^2 hybridized carbon atoms, has attracted much attention and research interest in a short time.

Huang et al. reported a regenerating and reusable electrochemical sensor that could realize label-free cancer cell detection by using the first clinical trial II used aptamer, AS1411, and functionalized graphene.[108] Du et al. demonstrated the use of graphene–mesoporous silica–GNP hybrids (GSGHs) as an enhanced element of an integrated sensing platform for the ultrasensitive and selective detection of DNA by using strand-displacement DNA polymerization and parallel-motif DNA triplex system as dual amplifications.[109] The results showed that the present new sensing strategy based on GSGHs was capable of detecting target DNA with a fairly high sensitivity of 10 fM through the hybridization of duplex DNA to the acceptor DNA and even had good capability to investigate single-nucleotide polymorphisms. The authors stated that the detection limit for target DNA was the lowest, compared with those by other methods.

8.6 Extension

In real biosystems, well-defined integration of special biomolecules can result in a more complex functional network. The Mann group have provided excellent reviews on artificial cells.[110,111] Our group have tried to mimic these systems in order to exploit the mechanisms, examine how they work and further develop novel advanced bionanodevices and smarter drug carriers.[112–116] Recently, we fabricated a new biomimetic energy converter via the assembly of CFoF1-ATPase on lipid-coated hemoglobin capsules.[113] As a whole, ATP could be continuously synthesized in the protein–lipid microcapsules by utilizing proton gradients, which were generated from the oxidation and catalytic hydrolysis of glucose by GOD. In a subsequent study, we directly used GOD as a shell component to construct lipid–protein microcapsules for creating another continuous proton gradient (Figure 8.10).[115] In our system, once glucose solution was added into the microcapsule suspension, the catalytic hydrolysis of glucoses by GOD generated a proton gradient between the outside and the inside of the capsule walls, and then CFoF1-ATPase was activated and thus the synthesis of ATP was started. Interestingly, ATP could be partly synthesized in the interior solution. This might be attributed to unidirectional arrangement of CFoF1-ATPases. In the future, we will improve these biomimetic nanosystems to address the problem of incorporating CFoF1-ATPases directionally.

8.7 Conclusions

This review reports the recent progress in the design, synthesis and preparation of biomolecule-based nanostructured conjugates, including saccharides, phospholipids, proteins and nucleic acids. We also highlight untraditional chemical reactions (e.g. click chemistry) or self-assembly strategies to construct novel bionanocomplexes. Modulation and control of these chemical, physical and biological properties of biomolecule-based building blocks is of significance in optimizing and integrating the biofunctions of nanostructured

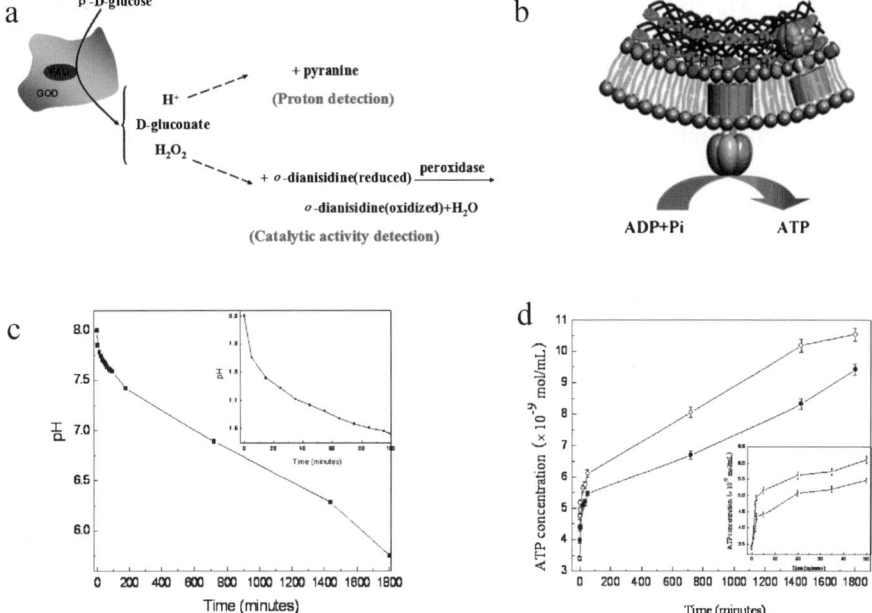

Figure 8.10 (a) The biocatalytic oxidation process of β-D-glucose by glucose oxidase; (b) schematic representation of the arrangement of CFoF1-ATPase in lipid-coated GOD capsule; (c) pH change with time in the lipid-modified GOD microcapsule interior (the inset image shows the pH change within 100 min); (d) ATP biosynthesis as a function of reaction time in CFoF1/lipid-modified GOD microcapsule solutions: exterior solutions (●), interior and exterior solution (○) after addition of 0.1% Triton X-100. The ATP synthesis within 50 min is displayed in the inset image. Reprinted with permission from ref. 115. Copyright 2009, American Chemical Society.

conjugates in future applications such as diagnostic imaging and therapy, biosensors, and biomimetic production.

Acknowledgements

We gratefully acknowledge financial support from the National Nature Science Foundation of China (Project No. 91027045 and 21007075) and National Basic Research Program of China (973 Program 2009CB930101).

References

1. http://en.wikipedia.org/wiki/Saccharide, date accessed 16/06/2012.
2. http://en.wikipedia.org/wiki/Phospholipids, date accessed 16/06/2012.
3. http://en.wikipedia.org/wiki/Protein, date accessed 16/06/2012.
4. http://en.wikipedia.org/wiki/DNA, date accessed 16/06/2012.

5. Z. F. Ma, J. R. Li, M. H. Liu, J. Cao, Z. Y. Zou, J. Tu and L. Jiang, *J. Am. Chem. Soc.*, 1998, **120**, 12678.
6. Z. F. Ma, J. R. Li, L. Jiang, J. Cao and P. Boullanger, *Langmuir*, 2000, **16**, 7801.
7. K. El-Boubbou, C. Gruden and X. F. Huang, *J. Am. Chem. Soc.* 2007, **129**, 13392.
8. L. Illum, I. Jabbal-Gill, M. Hinchcliffe, A. N. Fisher and S. S. Davis, *Adv. Drug Delivery Rev.*, 2001, **51**, 81.
9. J. Wu, Z. G. Su and G. H. Ma, *Int. J. Pharm.*, 2006, **315**, 1.
10. J. Wu, W. Wei, L. Y. Wang, Z. G. Su and G. H. Ma, *Biomaterials*, 2007, **28**, 2220.
11. W. Wei, L. Y. Wang, L. Yuan, Q. Wei, X. D. Yang, Z. G. Su and G. H. Ma, *Adv. Funct. Mater.*, 2007, **17**, 3153.
12. W. Wei, L. Yuan, G. Hu, L. Y. Wang, J. Wu, X. Hu, Z. G. Su and G. H. Ma, *Adv. Mater.*, 2008, **20**, 2292.
13. Y. Yang, Q. He, L. Duan, Y. Cui and J. B. Li, *Biomaterials*, 2007, **28**, 3083.
14. K. W. Wang, Q. He, X. H. Yan, Y. Cui, W. Qi, L. Duan and J. B. Li, *J. Mater. Chem.*, 2007, **17**, 4018.
15. W. Cui, Y. Cui, P. L. Zhu, J. Zhao, Y. Su, Y. Yang and J. B. Li, *Chem–Asian J.*, 2011, **7**, 127.
16. Y. Y. Liang and L. M. Zhang, *Biomacromolecules*, 2007, **8**, 1480.
17. C. Lee, D. Jang, J. Kim, S. Cheong, E. Kim, M. Jeong, S. Kim, D. Kim, S. Lim, M. Sohn, Y. Jeong and H. Jeong, *Bioconjugate Chem.*, 2011, **22**, 186.
18. http://en.wikipedia.org/wiki/Hyaluronic_acid
19. C. E. Schanté, G. Zubera, C. Herlinb and T. F. Vandammea, *Carbohydr. Polym.*, 2011, **85**, 469.
20. M. Bodnár, L. Daróczi, G. Batta, J. Bakó, J. Hartmann and J. Borbély, *Colloid Polym. Sci.*, 2009, **287**, 991.
21. K. Choi, K. Min, J. Na, K. Choi, K. Kim and J. Park, *J. Mater. Chem.*, 2009, **19**, 4102.
22. H. Lee, H. Mok, S. Lee, Y. Oh and T. Park, *J. Controlled Release*, 2007, **119**, 245.
23. I. Rivkin, K. Cohen, J. Koffler, D. Melikhov, D. Peer and R. Margalit, *Biomaterials*, 2010, **31**, 7106.
24. J. H. Yang, Y. Liu, H. B. Wang, L. Liu, W. Wang, C. D. Wang, Q. Wang and W. G. Liu, *Biomaterials*, 2012, **33**, 604.
25. Q. He, Y. Tian, H. Möhwald and J. B. Li, *Soft Matter*, 2009, **5**, 300.
26. Y. Yang, W. X. Song, A. H. Wang, P. L. Zhu, J. B. Fei and J. B. Li, *Phys. Chem. Chem. Phys.*, 2010, **12**, 4418.
27. X. M. Zhang, Q. He, X. H. Yan, P. Boullanger and J. B. Li, *Biochem. Biophys. Res. Commun.*, 2007, **358**, 424.
28. M. Gobbi, F. Re, M. Canovi, M. Beeg, M. Gregori, S. Sesana, S. Sonnino, D. Brogioli, C. Musicanti, P. Gasco, M. Salmona and M. E. Masserini, *Biomaterials*, 2010, **31**, 6519.

29. S. Mourtas, M. Canovi, C. Zona, D. Aurilia, A. Niarakis, B. L. Ferla, M. Salmona, F. Nicotra, M. Gobbi and S. G. Antimisiaris, *Biomaterials*, 2011, **32**, 1635.
30. M. Canovi, E. Markoutsa, A. N. Lazar, G. Pampalakis, C. Clemente, F. Re, S. Sesana, M. Masserini, M. Salmona, C. Duyckaerts, O. Flores, M. Gobbi and S. G. Antimisiaris, *Biomaterials*, 2011, **32**, 5489.
31. N. Tang, G. J. Du, N. Wang, C. C. Liu, H. Y. Hang and W. Liang, *J. Natl. Cancer Inst.*, 2007, **99**, 1004.
32. J. F. Lovell, C. S. Jin, E. Huynh, H. L. Jin, C. Kim, J. L. Rubinstein, W. C. W. Chan, W. G. Cao, L. V. Wang and G. Zheng, *Nat. Mater.*, 2011, **10**, 324.
33. G. Kibria, H. Hatakeyama, N. Ohga, K. Hida and H. Harashima, *J. Controlled Release*, 2011, **153**, 141.
34. Y. Sakurai, H. Hatakeyama, Y. Sato, H. Akita, K. Takayama, S. Kobayashi, S. Futaki and H. Harashima, *Biomaterials*, 2011, **32**, 5733.
35. T. Masuda, H. Akita, K. Niikura, T. Nishio, M. Ukawa, K. Enoto, R. Danev, K. Nagayama, K. Ijiro and H. Harashima, *Biomaterials*, 2009, **30**, 4806.
36. K. Kogure, H. Akita, Y. Yamada and H. Harashima, *Adv. Drug Deliv. Rev.*, 2008, **60**, 559.
37. H. Akita, A. Kudo, A. Minoura, M. Yamaguti, I. A. Khalil, R. Moriguchi, T. Masuda, R. Danev, K. Nagayama, K. Kogure and H. Harashima, *Biomaterials*, 2009, **30**, 2940.
38. A. Homhuan, K. Kogure, H. Akaza, S. Futaki, T. Naka, Y. Fujita, I. Yano and H. Harashima, *J. Controlled Release*, 2007, **120**, 60.
39. J. Li, Y. C. Chen, Y. C. Tseng, S. Mozumdar and L. Huang, *J. Controlled Release*, 2010, **142**, 416.
40. C. G. Zhou, B. Yu, X. J. Yang, T. Y. Huo, L. J. Lee, R. F. Barth and R. J. Lee, *Inte. J. Pharm.*, 2010, **392**, 201.
41. A. S. Hoffman, *Clinical Chem.*, 2000, **46**, 1478.
42. P. Thordarson, B. L. Droumaguet and K. Velonia, *Appl. Microbiol. Biotechnol.*, 2006, **73**, 243.
43. S. Salmaso, S. Bersani, A. Scomparin, F. Mastrotto and P. Caliceti, *Isr. J. Chem.*, 2010, **50**, 160.
44. M. E. B. Smith, F. F. Schumacher, C. P. Ryan, L. M. Tedaldi, D. Papaioannou, G. Waksman, S. Caddick and J. R. Baker, *J. Am. Chem. Soc.*, 2010, **132**, 1960.
45. C. P. Ryan, M. E. B. Smith, F. F. Schumacher, D. Grohmann, D. Papaioannou, G. Waksman, F. Werner, J. R. Baker and S. Caddick, *Chem. Commun.*, 2011, **47**, 5452.
46. F. F. Schumacher, M. Nobles, C. P. Ryan, M. E. B. Smith, A. Tinker, S. Caddick and J. R. Baker, *Bioconjugate Chem.*, 2011, **22**, 132.
47. M. Li, P. De, S. R. Gondi and B. S. Sumerlin, *Macromol. Rapid Commun.*, 2008, **29**, 1172.

48. B. L. Droumaguet, G. Mantovani, D. M. Haddleton and K. Velonia, *J. Mater. Chem.*, 2007, **17**, 1916.
49. V. Mani, B. V. Chikkaveeraiah, V. Patel, J. S. Gutkind and J. F. Rusling, *ACS Nano*, 2009, **3**, 585.
50. J. D. Keighron and C. D. Keating, *Langmuir*, 2010, **26**, 18992.
51. D. X. Li, Q. He, Y. Cui, L. Duan and J. B. Li, *Biochem. Biophys. Res. Commun.*, 2007, **355**, 488.
52. Z. F. Ma and T. Ding, *Nanoscale. Res. Lett.*, 2009, **4**, 1236.
53. W. C. W. Chan and S. M. Nie, *Science*, 1998, **281**, 2016.
54. L. Medintz, J. H. Konnert, A. R. Clapp, I. Stanish, M. E. Twigg, H. Mattoussi, J. M. Mauro and J. R. Deschamps, *Proc. Natl. Acad. Sci. U. S. A.*, 2004, **101**, 9612.
55. A. R. Clapp, I. L. Medintz, J. M. Mauro, B. R. Fisher, M. G. Bawendi and H. Mattoussi, *J. Am. Chem. Soc.*, 2004, **126**, 301.
56. Y. Xing and J. H. Rao, *Cancer Biomarkers*, 2008, **4**, 307.
57. A. Wolcott, D. Gerion, M. Visconte, J. Sun, A. Schwartzberg, S. Chen and J. Z. Zhang, *J. Phys. Chem. B*, 2006, **110**, 5779.
58. H. Y. Liu and T. Q. Vu, *Nano Lett.*, 2007, **7**, 1044.
59. A. R. Clapp, E. R. Goldman, H. T. Uyeda, E. L. Chang, J. L. Whitley and I. L. Medintz, *J. Sens.*, 2008, 797436.
60. J. Kang, X. Y. Zhang, L. D. Sun and X. X. Zhang, *Talanta*, 2007, **71**, 1186.
61. D. K. Bora and P. Deb, *Nanoscale Res. Lett.*, 2009, **4**, 138.
62. R. Bardhan, W.X. Chen, C. Perez-Torres, M. Bartels, R. M. Huschka, L. L. Zhao, E. Morosan, R. G. Pautler, A. Joshi and N. J. Halas, *Adv. Funct. Mater.*, 2009, **19**, 3901.
63. J. Peng, L. N. Feng, K. Zhang, J. J. Li, L. P. Jiang and J. J. Zhu, *Chem–Eur J.*, 2011, **17**, 10916.
64. A. K. Johnson, A. M. Z. Lee, A. Deobald, R. L. Crawford and A. J. Paszczynski, *J. Nanopart. Res.*, 2008, **10**, 1009.
65. R. J. Cui, H. P. Huang, Z. Z. Yin, D. Gao and J. J. Zhu, *Biosens. Bioelectron.*, 2008, **23**, 1666.
66. Z. Liu, J. T. Robinson, S. M. Tabakman, K. Yang and H. J. Dai, *Mater. Today*, 2011, **14**, 316.
67. L. Z. Feng and Z. Liu, *Nanomedicine*, 2011, **6**, 317.
68. Z. Liu, S. Tabakman, S. Sherlock, X. L. Li, Z. Chen, K. L. Jiang, S. S. Fan and H. J. Dai, *Nano Res.*, 2010, **3**, 222.
69. E. Winfree, F. R. Liu, L. A. Wenzler and N. C. Seeman, *Nature*, 1998, **394**, 539.
70. N. C. Seeman, *Mol. Biotechnol.*, 2007, **37**, 246.
71. T. Wang, R. J. Sha, R. Dreyfus, M. E. Leunissen, C. Maass, D. J. Pine, P. M. Chaikin and N. C. Seeman, *Nature*, 2011, **478**, 225.
72. B. Chakraborty, R. J. Sha and N. C. Seeman, *Proc. Natl. Acad. Sci. U. S. A.*, 2008, **105**, 17245.

73. H. Z. Gu, J. Chao, S. J. Xiao and N. C. Seeman, *Nat. Nanotechnol.*, 2009, **4**, 245.
74. N. C. Seeman, *Annu. Rev. Biochem.*, 2010, **79**, 65.
75. H. Z. Gu, J. Chao, S. J. Xiao and N. C. Seeman, *Nature*, 2010, **465**, 202.
76. C. Lin, X. Wang, Y. Liu, N. C. Seeman and H. Yan, *J. Am. Chem. Soc.*, 2007, **129**, 14475.
77. W. Y. Liu, H. Zhong, R. S. Wang and N. C. Seeman, *Angew. Chem., Int. Ed.*, 2011, **50**, 264.
78. B. Q. Ding and N. C. Seeman, *Science*, 2006, **314**, 1583.
79. J. P. Zheng, J. J. Birktoft, Y. Chen, T. Wang, R. J. Sha, P. E. Constantinou, S. L. Ginell, C. D. Mao and N. C. Seeman, *Nature*, 2009, **461**, 74.
80. N. C. Seeman, *Nano Lett.*, 2010, **10**, 1971.
81. H. Pei, N. Lu, Y. L. Wen, S. P. Song, Y. Liu, H. Yan and C. H. Fan, *Adv. Mater.*, 2010, **22**, 4754.
82. C. Xu, C. Q. Zhao, J. S. Ren and X. G. Qu, *Chem. Commun.*, 2011, **47**, 8043.
83. X. J. Yang, F. Pu, J. S. Ren and X. G. Qu, *Chem. Commun.*, 2011, **47**, 8133.
84. K. G. Qu, J. S. Ren and X. G. Qu. *Mol. Biosyst.*, 2011, **7**, 2681.
85. L. Y. Feng, Y. Chen, J. S. Ren and X. G. Qu, *Biomaterials*, 2011, **32**, 2930.
86. C. Chen, C. Q. Zhao, X. J. Yang, J. S. Ren and X. G. Qu, *Adv. Mater.*, 2010, **22**, 389.
87. J. Li, S. P. Song, X. F. Liu, L.H. Wang, D. Pan, Q. Huang, Y. Zhao and C. H. Fan, *Adv. Mater.*, 2008, **20**, 497.
88. J. Li, S. P. Song, D. Li, Y. Su, Q. Huang, Y. Zhao and C. H. Fan, *Biosens. Bioelectron.*, 2009, **24**, 3311.
89. W. W. Guo, J. P. Yuan, Q. Z. Dong and E. K. Wang, *J. Am. Chem. Soc.*, 2010, **132**, 932.
90. X. Yang, X. P. Sun, Z. Z. Lv, W. W. Guo, Y. Du and E. K. Wang, *Chem. Commun.*, 2010, **46**, 8818.
91. B. Y. Han and E. K. Wang, *Biosens. Bioelectron.*, 2011, **26**, 2585.
92. X. H. Fang and W. H. Tan, *Acc. Chem. Res.*, 2010, **43**, 48.
93. T. Chen, M. I. Shukoor, Y. Chen, Q. Yuan, Z. Zhu, Z. L. Zhao, B. Gulbakana and W. H. Tan, *Nanoscale*, 2011, **3**, 546.
94. L. Yang, X. B. Zhang, M. Ye, J. H. Jiang, R. H. Yang, T. Fu, Y. Chen, K. M. Wang, C. Liu and W. H. Tan, *Adv. Drug Deliv. Rev.*, 2011, **63**, 1361.
95. Z. W. Tang, D. H. Shangguan, K. M. Wang, H. Shi, K. Sefah, P. Mallikratchy, H. W. Chen, Y. Li and W. H. Tan, *Anal. Chem.*, 2007, **79**, 4900.
96. S. Bamrungsap, M. I. Shukoor, T. Chen, K. Sefah and W. H. Tan, *Anal. Chem.*, 2011, **83**, 7795.

97. Y. Pu, Z. Zhu, H. X. Liu, J. N. Zhang, J. Liu and W. H. Tan, *Anal. Bioanal. Chem.*, 2010, **397**, 3225.
98. Y. F. Huang, H. T. Chang and W. H. Tan, *Anal. Chem.*, 2008, **80**, 567.
99. X. L. Chen, M. C. Estevez, Z. Zhu, Y. F. Huang, Y. Chen, L. Wang and W. H. Tan, *Anal. Chem.*, 2009, **81**, 7009.
100. J. A. Martin, J. A. Phillips, P. Parekh, K. Sefah and W. H. Tan, *Mol. Biosyst.*, 2011, **7**, 1720.
101. J. E. Smith, C. D. Medley, Z. W. Tang, D. H. Shangguan, C. Lofton and W. H. Tan, *Anal. Chem.*, 2007, **79**, 3075.
102. C. D. Medley, S. Bamrungsap, W. H. Tan and J. E. Smith, *Anal. Chem.*, 2011, **83**, 727.
103. H. Z. Kang, M. B. O'Donoghue, H. P Liu and W. H. Tan, *Chem. Commun.*, 2010, **46**, 249.
104. L. Yang, L. Meng, X. B. Zhang, Y. Chen, G. Z. Zhu, H. P. Liu, X. L. Xiong, K. Sefah and W. H. Tan, *J. Am. Chem. Soc.*, 2011, **133**, 13380.
105. Y. R. Wu, K. Sefah, H. P. Liu, R. W. Wang and W. H. Tan, *Proc. Natl. Acad. Sci. U. S. A.*, 2010, **107**, 5.
106. T. Chen, M. I. Shukoor, R. W. Wang, Z. L. Zhao, Q. Yuan, S. Bamrungsap, X. L. Xiong and W. H. Tan, *ACS Nano*, 2011, **5**, 7866.
107. Z. Wu, L. J. Tang, X. B. Zhang, J. H. Jiang and W. H. Tan, *ACS Nano*, 2011, **5**, 7696.
108. Z. Z. Huang, F. Pu, Y, H. Lin, J. S. Ren and X. G. Qu, *Chem. Commun.*, 2011, **47**, 3487.
109. Y. Du, S. J. Guo, S. J. Dong and E. K. Wang, *Biomaterials*, 2011, **32**, 8584.
110. R. K. Kumar, X. X. Yu, A. J. Patil, M. Li and S. Mann, *Angew. Chem., Int. Ed.*, 2011, **50**, 9343.
111. A. J. Dzieciol and S. Mann, *Chem. Soc. Rev.*, 2012, **41**, 79.
112. L. Duan, Q. He, K. W. Wang, X. H. Yan, Y. Cui, H. Möhwald and J. B. Li, *Angew. Chem., Int. Ed.*, 2007, **46**, 6996.
113. W. Qi, L. Duan, K. W.i Wang, X. H. Yan, Y. Cui, Q. He and J. B. Li, *Adv. Mater.*, 2008, **20**, 601.
114. Q. He, L. Duan, W.Qi, K. W. Wang, Y. Cui, X. H. Yan and J. B. Li, *Adv. Mater.*, 2008, **20**, 2933.
115. L. Duan, W. Qi, X. H. Yan, Q. He, Y. Cui, K. Wang, D. X. Li and J. B. Li, *J. Phys. Chem. B*, 2009, **113**, 395.
116. Q. He, Y. Cui and J. B. Li, *Chem. Soc. Rev.*, 2009, **38**, 2292.

CHAPTER 9
Architectonics of Active Sites: Life Processes at Nanodimensions

S. DUTTA BANIK AND N. NANDI*

Department of Chemistry, University of Kalyani, Kalyani, Nadia, West Bengal, 741235, India
*E-mail: nilashisnandi@yahoo.com

9.1 Enzymes, Active Sites and Vital Biological Reactions

Enzymes have an amazing capacity to perform critical life processes. They control a number of important chemical reactions in every moment of the life-cycle of living organisms. Although the tasks performed by enzymes are of bewildering variety, they can be classified in six broad categories (EC1–EC6) depending on the reactions they perform.[1] The tremendous importance of enzymes is that they determine the transformations of chemicals and energies in numerous vital biological reactions. These macromolecules are functional molecular machines with surprising catalytic efficiency and specificity. An enzyme usually catalyzes a single chemical reaction or a set of closely related reaction and is highly specific to the substrate on which it works.[2]

The molecular study of enzyme action has been a subject of intense experimental and theoretical research for a long time due to their vital importance.[2–13] What has particularly fascinated scientists about enzymes for decades is their capacity to speed up a reaction a million times compared to the uncatalyzed one.[2] The catalytic action of a specific enzyme is dependent on the conformational fluctuations in its structure. Specific regions of the protein are dynamically coupled, and a network of correlated motions extended throughout the protein structure facilitates the reaction.[14] A given residue may be dynamically coupled with a residue far away in the primary sequence, which is a consequence of the structural hierarchy of the protein. The coupled motions are related to equilibrium, thermally averaged conformational changes leading to configurations favorable for the reaction. While the conformational fluctuations are essential for catalytic activities, the architectonics of the enzyme is the key for enzyme function, which is an outcome of the structure–function relationship in biological structure. The interaction between the substrate and the residues of the enzyme in close proximity to the substrate are favorable for lowering the activation barrier of the system and this is governed by the principle of the architectonics of the enzymes designed by evolution.

Since all enzymes are designed and developed by evolution to carry out either a specific reaction or a class of reaction efficiently, it is worth looking into the region of enzyme which is the molecular vicinity of the chemical reaction to understand the principles behind their remarkable efficiency and selectivity. Despite the macromolecular structure of enzymes, the enzymatic reactions occur within active sites. An active site is a region in the enzyme, relatively small compared to the overall structure, and is a nanosized cavity or cleft. The active site residues not only form a perfectly complementary binding pocket for the substrates but also carry the reactants to the corresponding product state *via* a transition state. The preorganized electrostatic environment of the active site is the origin of the catalytic efficiency of the enzyme compared to the same reaction in the bulk solvent (Figures 9.1a and 9.1b).[2,13] It is instructive to look into the organizational pattern of the active site residues since these residues are in close proximity of the reactants, although such a consideration does not imply that the overall structure of the enzyme is irrelevant in understanding the enzyme function. The organization of the active site residues has a dominant influence on lowering the activation barrier of the system. Consequently, understanding of the principle of architectonics of enzymatic active site structure has fundamental and biotechnological importance.

Due to the remoteness of the positions of distant residues, they are unable to directly participate in the reaction mechanism. An importance of the active site is that it modifies the speed and efficiency of the reaction compared to the same reaction in bulk. Substrates bind to the active site of the enzyme to form the enzyme–substrate complex, mostly through nonbonded interactions such as electrostatic (including hydrogen bond), hydrophobic interactions as well as van der Waals interactions. The active site structure and the chemical properties of the residues therein are responsible for the recognition and substrate specificity. The residues bind the reactants to a specific location at a suitable orientation to carry out the reaction efficiently. According to the lock-and-key model, the active site is a perfect fit for a specific substrate, and once the substrate binds to the enzyme no further modification is necessary. On the other hand, the induced fit model assumes that an active site is flexible and

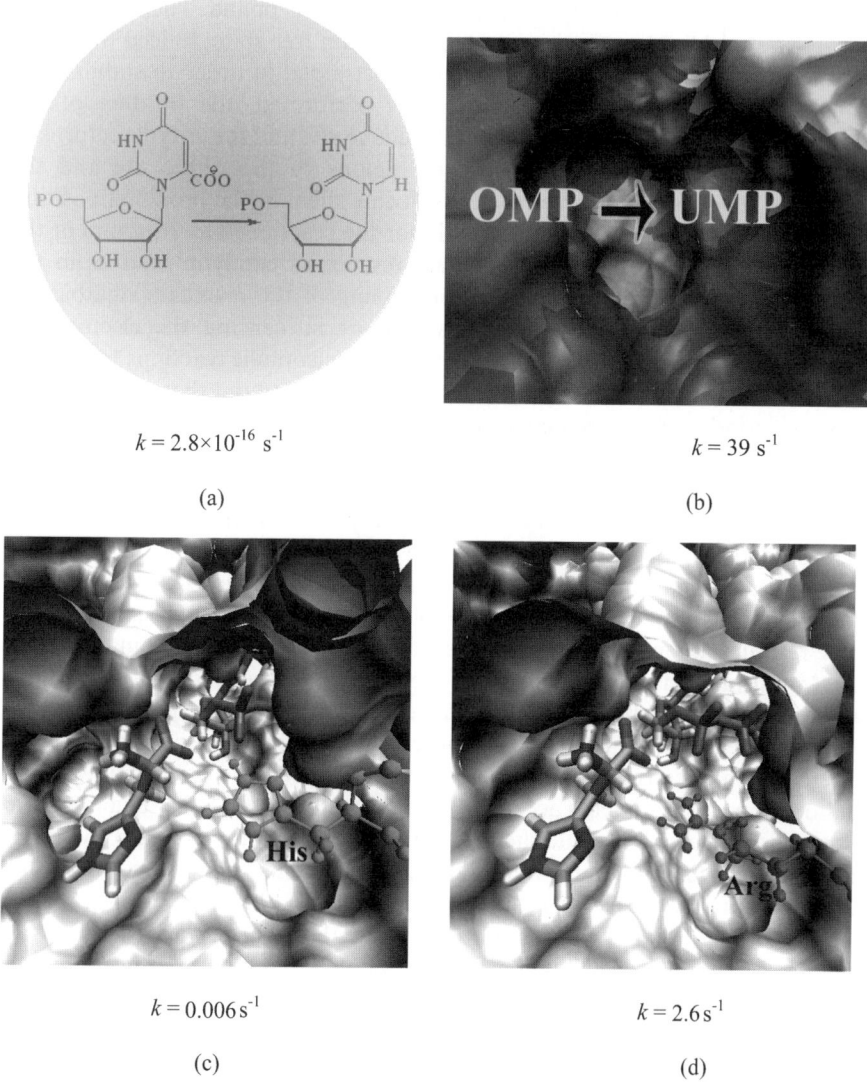

Figure 9.1 Examples of the influences of few active sites of enzymes on the corresponding reaction rates. The rate constants show the enhancement by the active site of enzyme compared to that in water for decarboxylation of orotidine monophosphate (OMP) to uridine monophosphate (UMP) (a) in aqueous medium and (b) in the active site of orotidine 5'-phosphate decarboxylase (PDB code: 1DBT.PDB),[112] respectively. The decrease in rate constant by mutation of the catalytic active site residue is shown for aminoacylation reaction in (c) the active site of mutated (R259H) HisRS (mutated His is shown in green) and in (d) active site of wild type HisRS (catalytic residue Arg is shown in pink). The rate constant (k) of the decarboxylation of OMP[2] and the rate constant for aminoacylation[26] are also indicated below the respective figures.

conformational changes may occur as the substrate is bound. The flexibility of the active site should not be interpreted as a strain free state. Indeed the active site is a preorganized state and is highly constrained in order to stabilize the transition state.[15–17] Once the reaction is performed, the product of the reaction is subsequently released from the active site and the enzyme returns to its initial unbound state. The different regions of the active site and the respective residues enclose the reactants, locate them in the proximal position and orient them in a way that is suitable for the reaction. The structure of the active site is important in terms of the presence of catalytic residues in the immediate vicinity of the reactants, positioning of the reactants suitable for reaction by molecular recognition, as well as influencing the electrostatic potential between the substrates in the case of nucleophilic reactions to make it favorable for the reaction. It has been rightfully pointed out that the "catalytic effect is entirely due to the active site environment".[15]

The correlation between the architectonics of the active site and the enzymatic function is far from understood. It is useful to note the distinction between the reaction chemistry and the catalytic mechanism.[18] The reaction chemistry principally includes the overall strategy of changing the substrate into product, and the nature of the intermediates involved, whereas the catalytic mechanism describes the roles played by specific residues. Commonly these residues are in close proximity to the substrates and located in the active site. Mutations of these residues affect the progress of the reaction significantly (Figures 9.1c and 9.1d). A related and equally important aspect of enzyme action is their specificity. The enzyme can be remarkably specific to subtle differences in the structure of the substrate. The specificity of enzymes is such that only one enantiomer is accepted as a substrate and the other enantiomeric form will be rejected. Despite the fact that the remarkable stereoselectivities of enzymes has been known for a long time, the molecular understanding of this stringent chiral specificity is sparse.[13]

In the present review, we attempt to correlate the architectonics of active site and the reaction performed therein. In section 9.2, we give a brief introduction to active site structures as reaction centers and discuss a few examples of the dependence of the rate of enzymatic reactions on active site residues. In section 9.3, we discuss the influence of the confinement in a nanodimension on the enzymatic reactions and chiral discrimination therein. This is followed by a specific example of the influence of active site architectonics on the process of a biological reaction, namely the aminoacylation reaction in section 9.4. Finally we mention a few future prospects of the study of the architectonics of active sites in enzymes.

9.2 Influence of the Architectonics of Active Sites on Enzymatic Reactions

Enzymes have diversification in their macromolecular structures and it is not straightforward to relate the architecture with function. Proteinous enzymes belong to diverse superfamilies and are classified into six groups based on their reaction chemistry. Enzyme superfamilies exhibit remarkable variations concerning conservation and variation of substrate specificity, reaction chemistry and catalytic residues.[18] Different classes or even subclasses of enzymes show diversity and conservation in their sequence and the organization of structural elements in the structural hierarchy. Variation of the composition of the active site structure is noted in enzymes from different species carrying a particular type of reaction.

Recent years have witnessed a great deal of understanding of various active site structures of enzymes from crystallographic analysis, which is complemented by computational analysis.[3-10,13-16,19-25] These studies indicate that despite the apparent structural diversity of the enzymes,[26] the active site structure can have conserved features in different species. Conservation of key residues and residues of similar types in the active site are noted in various aminoacyl tRNA synthetases (aaRSs), which have importance in carrying out the related reaction efficiently.[13] The aaRSs belong to the ligase class of enzymes[26] and catalyzes the adenylation and tRNA charging of at least 20 natural amino acids. Although all the aaRSs catalyze the same basic reaction and share common substrates, their diversities in sequence and three-dimensional structures are significant.[13,22,27] However, there are a number of common features that can be noted in the architectonics of active sites of different aaRSs, which will be discussed in details in section 9.4.1. A set of residues can be conserved in enzymes of similar function. A typical example is the Ser–His–Asp catalytic triads found in a variety of hydrolases.[28] Similar conservations in the other enzyme classes can be noted. However, there is a lacuna in understanding of the correlation between the conservation of residues and its influence on the enzymatic reaction.

Conclusive proof that the specific active site residues strongly influence the substrate binding, reaction rate and specificity comes from the mutation experiments. Mutation experiments and computational analysis are ideal for the purpose, where one can modify a given residue and can observe the concomitant rate enhancement or decrement. It is pointed out that distant mutations are more probable than closer mutation due to significantly higher number of residues present beyond the first shell compared to those present in the first shell.[97] Consequently, the mutations in the regions beyond the active site may influence the catalytic activity, albeit in a different way from those in the active site residues. A few examples of the influence of the active site residues on the rate, studied by mutation experiments, are mentioned below.

A steady state kinetics experiment on aaRS shows that the replacement of Arg259 by His in the active site of histidyl tRNA synthetase (HisRS) from *E. Coli* (*EC*) reduces the rate of exchange and transfer by a factor of 1000 (rate from 130 s^{-1} to 0.103 s^{-1}) and 500 (rate from 2.6 s^{-1} to 0.006 s^{-1}).[22] The aminoacylation rate in the active site of HisRS reduces by a factor of 3000-fold when the active site residue Arg259 is replaced by His. The rate is reduced by a factor of 4500-fold when another active site residue Glu83 is replaced by Ala.[27]

However, the effect becomes less pronounced if Glu83 is replaced by Gln. The substitution of active site Gln127 by Ala decreases the pyrophosphate exchange by 54-fold and aminoacylation by 70-fold.[27] In the case of the first step of the aminoacylation reaction in TyrRS, mutation of active site Lys230, Lys233 and Thr234 by Ala destabilizes the transition state complex by 2.8–3.1 kcal mol^{-1}.[29] Binding of the substrate is also affected by mutation in this case. The forward rate constant for the formation of enzyme bound tyrosyl–adenylate decreases by two to three orders of magnitude.[29] Mutation of active site residues also shows changes in the catalytic activity of ProCysRS, which is a special synthetase[30] with a capacity to attach two different amino acids (Pro and Cys) to their cognate tRNA. The mutation of Gln103 by Ala leads to complete loss of the prolylation activity and reduce the cysteinylation activity. While the mutation of the active site residue Pro100 by Ala decreases the cysteinylation activity by 10-fold, it does not alter the prolylation activity. The mutation of an active site Gly by Arg in GlyRS blocks the binding site of the substrate making the synthetase inactive towards the first step of the aminoacylation reaction.[31] Similarly, mutation (Tyr to Val) in an erythromycin polyketide synthetase derived enoyl reductase (belonging to the oxidoreductase class) caused a switch in the methyl branch configuration in the product from S to R.[32] The stereospecificity of the hydride transfer reaction of the substrate by alcohol dehydrogenase (belonging to the oxidoreductase class) depends on the active sites of the respective species. The active site of *S. cerevisiae* contains a Zn^{2+} ion which is absent in *L. brevis*. The *S. cerevisiae* shows more stereospecificity compared to the later.[33]

Computational studies show that mutation of two conserved active site Tyr residues of human soluble epoxide hydrolase (belonging to the hydrolase class) increases the activation barrier for the first step (alkylation step) of the epoxide hydrolysis reaction by 17 kcal mol^{-1}.[34] The site-directed mutagenesis in hydroxynitrile lyase (belonging to the hydrolase class) shows that the replacement of Ser or Asp or His leads to the complete loss of enzyme activity. The replacement of Cys (by serine) also leads to loss of more than 95% of catalytic activity.[35] Ribosomal peptide synthesis occurs in the active site (peptidyl transferase center, PTC), which is mostly composed of nucleobases. A high level of suppression in the presence of D-Met (23 per cent) and D-Phe (12 per cent) is observed in modified ribosomes with mutations in regions 2447–2450 (belonging to the PTC region) and 2457–2462 (belonging to the helix 89 region) of 23S rRNA of *EC* and cell-free protein synthesizing systems prepared from mutant ribosomes. This indicates that in ribosome the active site structure is, at least partly, responsible for the discriminating mechanism between D-aminoacyl-tRNA$_{CUA}$'s in the ribosomal A site. Putative alterations may lead to enhanced incorporation of non-cognate D-amino acid.[13] Computational studies show that the removal of the active site residue in PTC can reduce the chiral preference. It is also observed that the mutual rotation of A- and P-terminal amino acids are significantly dependent on the active site residues. Active site bases such as U2620, A2486, G2618 and C2487 have a favorable influence on rotatory path while the residues such as U2541, C2104, C2105, A2485, C2542, C2608, U2619 and A2637 have lesser influence.[36] The results can be understood from the fact that once an active site residue is mutated, the interaction pattern of the substrate with the residues in the mutated enzyme and that in the wild type differ, leading to the variation in enzymatic activity. The perturbed network of interaction is not only inefficient for binding of the cognate substrate, but also raises the activation barrier of the reaction due to loss of favorable interactions.

The foregoing studies strongly indicate that the architectonics of the active site is important for enzymatic reactions. While the overall dynamics of the enzyme is important for fluctuations conducive to the reaction, it may not be vastly different for two individual closely similar residues, for example two enantiomeric species. However, such a difference is sufficient to greatly influence the progress of the reaction or even bring it to a halt. Consequently, the chemical composition of the active site structure is important for enzymatic reactions. Recent studies have confirmed that the specific three-dimensional chemical structures of the active site residues (amino acids or nucleobases) and their spatial organization are important for the reaction. The confinement of the substrate by the active site nanospace is also an important factor for the discrimination specificity. This is discussed in the following section.

9.3 Nanodimension of Active site, Confinement and Chiral Discrimination

It is now well established that the orientation-dependent minima in the intermolecular energy profile of a pair of chiral molecules are different for enantiomeric and racemic pairs when such molecules are at nanoscale separation.[37] This difference is responsible for chiral discrimination. When the intermolecular separation increases significantly, the dissymmetric nature of the orientation-dependent interaction is diminished and the discrimination vanishes. In such large separations, the preferential mutual orientation is observed over the mesoscopic length scale and the manifestation of chiral discrimination vanishes. This is observed in biomimetic systems such as monolayers and bilayers. The reacting molecular segments approach within nanoscale proximity during the course of a biological reaction in the active site. This proximity is important for the chiral discrimination.

Studies on chiral discrimination in biomimetic systems have indicated that the specific orientation and nanoscale separation of the chiral amphiphilic molecules is important for the manifestation of discrimination. It has been mentioned that the discrimination in biological molecules is stringent. The reactions which lead to the synthesis of biomolecules or involve them might have specific position and orientation dependence on the chirality of the active site as a whole or of its components. This suggests that the chirality of the active site residues might be correlated with the chiral specificity of reactions. Enzymes are molecular machines, where stringent chiral discrimination occurs within the active site. Analysis of the molecular organization of the active site structure and the related interactions are useful to understand the chiral specificity of enzymatic reactions.

Active sites are nanoflasks where efficient chiral discriminations take place.[13] Most commonly, they are composed of chiral moieties such as amino acids and sugars. Active sites can have diverse types of chiral specificity depending on the different mutual spatial arrangements of the residues. Chiral discrimination can be significant when the molecules are confined (by other molecules, for example) within the nanometre range. It vanishes if the chiral molecules are separated over a longer scale (tens of nanometres or larger). Since the reacting moieties are confined within the active site, it is possible that the confinement within a nanosized cavity might have an important influence on discrimination. Consequently, the organization of the active site residues (close to the reacting moieties) have a more important role in the chiral discrimination in reaction than the residues far from the active site.[13]

The active site residues not only enclose the reactants in a close separation but also confine them, which is important for the specificity. The restriction of degrees of freedom is an important factor in steric recognition. It has long been known that diffusion in reduced dimensions can speed up biological interaction over the limits normally set by three-dimensional diffusion processes.[111] The confined space of the active site is thus more effective in driving the reaction than the bulk process. This is a state of reduced degrees of freedom for the reactants compared to the free state of the same moieties in the bulk. The reduction in rotational and translational degrees of freedom enhances the stereoselectivity. Due to the reduction in rotational and translational degrees of freedom, reactants with specific mutual orientations are more probable than the random mutual arrangements. The orientation dependent intermolecular interaction is more effective in the state of confinement in active site and this augments the chiral discrimination in biological reactions. It has long been recognized that enzymes, which are inherently chiral, are needed to orient both substrates and catalytic residues to preferentially stabilize one stereoisomeric transition state over another in the enzymatic reactions. The effectiveness of the catalytic activity of an enzyme in the rate enhancement significantly depends on the realization of tight binding, which is coupled with a high degree of steric recognition.[38,39]

Notably, the mechanism of discrimination of one substrate from the other is the key to the ability of the enzymes to act as efficient biocatalysts, a property which is hard to match by synthetic design but provides impetus for many studies. Although different enzymes and even enzymes in a given class can catalyze reactions of bewildering variety, studies so far are greatly promising in understanding the role of chirality in the biological reactions and to follow the fundamental correlation between chirality and life. In the following, we mention few examples of the influence of the active site in chiral discrimination. This indicates that diverse stereospecific reactions can be effectively carried out within such sites composed of chiral biological subunits, like amino acids and sugars. Rearrangement of a few chiral moieties is sufficient to obtain diverse interaction profiles, which can create specificity for different substrates. This principle may be a novel way, as utilized by nature, to perform myriads of biological reactions using a limited set of amino acids and nucleotides.

The human cytochrome P450 binds only S-warfarin as a substrate (Figure 9.2).[40] Crystallographic study indicates that the loss of hydrogen bonding interactions with the oxo group of warfarin and Ala103, or unfavorable interaction with a second ligand bound in the available space (as suggested in the study), could be responsible for the chiral discrimination. Two separate enzymes of bacterial alcohol dehydrogenase (namely, 2-[R/S-2-hydroxypropylthio] ethanesulfonate dehydrogenases) are available for the reversible oxidation of two enantiomers of 2-[2-hydroxypropylthio]-ethanesulfonate (HPC) to 2-(2-ketopropylthio) ethanesulfonate (2-KPC). Although the two enantiomers of HPC can bind in the active site with same affinity, only the cognate enantiomer orients properly for catalysis, which is an important factor for the reaction.[41(a)] The chiral recognition at the active site of nitric oxide synthases during catalysis is exhibited in the presence of the active effector, L-Arg.[41(b)]

The R-enantiomer of styrene oxide is completely hydrolyzed by epicholorohydrin epoxide hydrolase before the S-enantiomer, and the latter is subsequently converted at a much higher rate. This feature is expected to arise from the capacity of the active site to discriminate the interactions with the R- and S-enantiomers.[42] Chiral discrimination is noted in the inhibition of cutinase from *Fusarium solani pisi* and *Staphylococcus hyicus* lipase by 1,2-dioctylcarbamoylglycero-3-O-p-nitrophenyl methyl/octyl-phosphonates.[43] Both lipases exhibit high chiral selectivity of these compounds in glycerol and phosphorus. For the methylphosphonates, the difference in the reactivity of the faster and slower reacting isomers is about 250-fold. For octylphosphonates the difference is 60-fold. This high level of discrimination indicates the higher selectivity of the enzyme for the phosphorus chirality. It is indicated that these phosphonates can be regarded as true active site-directed inhibitors.

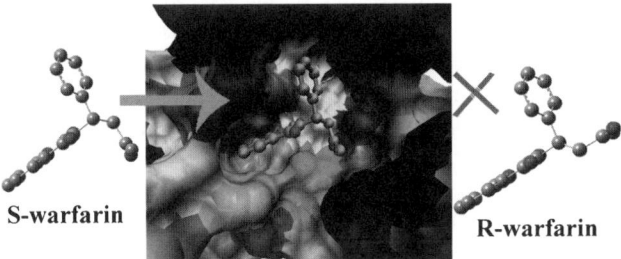

Figure 9.2 The active site structure of CYP450 2C9 complexed with S-warfarin.[40] The corresponding R-enantiomer of warfarin is inappropriate for the active site. The image is generated using VMD.[113]

Stringent chiral discrimination is observed in enzymes involved in various steps of protein biosynthesis. Although the racemic compound (crystalline) is more stable than the enantiomeric form, racemization does not happen in the natural biosynthetic pathway. The mechanism of retention of enantiopurity is not straightforward to understand, since D-amino acids are involved in non-ribosomal peptide synthesis, post-translational modifications and are present in bacterial cell walls and the human brain, for example. The D-enantiomers are esterified to alanyl-, leucyl- and valyl-tRNA at $<10^{-5}$ the rate of L-enantiomers in the first step of protein synthesis (amino acid activation step). Barely detectable amount of D-phenyl alanyl tRNA are noted to be formed using phenylalanyl tRNA synthetases, and this was only at $\sim 6 \times 10^{-5}$ times the rate found for L-phenylalanyl tRNA.[44] The preference for acylation of L-Asp over D-Asp by Aspartyl-tRNA synthetase (AspRS) is studied using detailed classical molecular dynamics simulations.[45] The binding energy difference between L-Asp and D-Asp is ~ 5.9 kcal mol^{-1} favoring L-Asp.[46] Kinetic experiments also show that the formation of D-aspartyl tRNA occurs at a rate 4000 times lower than L-aspartyl-tRNA. Electronic structure based analysis shows that the L-His is located in a network of interactions within the nanospace enclosed by the active site residues of HisRS, which is perturbed when the D-amino acid is incorporated. Consequently the approach of D-His becomes more unfavorable than that of L-His within the active site of HisRS during the course of the reaction.[24,25] The peptidyl transferase reaction is one of the most intensively studied partial reactions of protein biosynthesis. Peptidyl transferase is an aminoacyl transferase enzyme in ribosomes, and is essentially a ribozyme. The peptide bond formation occurs in the ribosome, and the active site is termed as PTC. Despite the reaction being remarkably fast, the chance of incorporation of the wrong (D-enantiomer, for example) amino acid is only 10^{-4} per amino acid for a 1000-residue protein in the ribosomal pathway. Modification of the active site structure (mutation) is necessary to enhance D-amino acid incorporation during elongation.[37]

The origin of chiral discriminations in the active site of PTC in ribosome has been studied in detail.[36,47,48] Several factors are noted to be responsible for the discrimination and to explain the high level of stereospecificity of the process. The chiralities of the amino acids at the A- and P-terminals are most important. The rotatory path for the approach of D-amino acids towards the P-terminal is unfavorable due to steric hindrance between the amino acids themselves. Their nanoscale separation is important for discrimination, and at larger separation no discrimination is noted. The second factor is the restricted nature of the range of mutual orientations of the terminals during the rotatory path for the approach to form the peptide bond. This factor makes the resultant interaction profiles for L–L and D–L pair different and is the cause of chiral discrimination. The natural chirality (D-form) of the sugar ring has a favorable influence on the long length scale organization of the tRNA structure and is another factor for influencing discrimination. Alteration of the chirality of the sugar ring is unfavorable as it requires large structural rearrangements of tRNA. Favorable influence of the D-sugar ring is noted on the rotatory path of the process of approach of substrate to form the peptide bond. The removal of the sugar ring makes the rotatory path for approach to form the peptide bond unfavorable, which indicates that the interaction of the D-sugar with the amino acids are more favorable than other homo- or heteropair combinations of the sugar amino acid pair. The stereochemistry of the 2' center of the D-sugar has vital influence as it catalyzes the peptide bond formation by ensuring proper placement of the OH group, which is involved in the catalysis. The analysis of the transition state structure revealed that the alteration and removal of chirality of the 2' center destabilizes the transition state and makes the formation of the peptide bond unfavorable. Finally, the nanoscale proximity of some of the surrounding bases present in PTC with the A- and P-terminals and their restricted orientation have influence on the discrimination. Thus, multiple factors control the discrimination in the peptide synthesis in PTC and allow accurate retention of the biological homochirality in the reaction.

In the following, we continue to limit our discussions about the architectonics of active sites to the class of enzyme that carries out the first couple of steps of one of the most important life processes, namely protein biosynthesis. We present the analysis of the molecular organization of active site of various aaRSs where aminoacylation reaction takes place. We focus on the recent developments of molecular understanding of the structure of active sites of aminoacyl synthetase and the relationship with catalytic activity and specificity. In recent years, the combination of detailed crystallographic structures of the enzymes combined with computational analysis have revealed the influences of the active site structure in guiding the specificity and efficiency of enzymatic reactions.

9.4 Example of a Life Process in a Nanospace: Aminoacylation Reaction in the Active Site of Aminoacyl tRNA Synthetase (aaRS)

Protein biosynthesis takes place in successive stages: activation of the amino acid, or aminoacylation reaction, followed by initiation, elongation, termination, release, folding, and posttranslational processing.[2] The aminoacylation reaction is an important biological reaction as it is the necessary prerequisite for peptide bond formation. This step correlates the realm of the protein with the RNA world. The reaction involves adenosine triphosphate (ATP) and Mg^{2+} ion dependent aaRS to form aminoacyl adenylate followed by attachment of the amino acid to the tRNA. The twenty aaRSs are specific for twenty cognate natural amino acids. Although all the aaRSs catalyze the same basic reaction and share common substrates, the amino acid sequence, topology and most importantly their active site structure have similarities as well as differences. Since the invariant parts of substrates, such as amino group, carboxylic acid group and ATP itself, are common molecular fragments or molecules in all aaRS, it is expected that common features might be observed in the active site structure. Although each of the aaRS carries out the same reaction, twenty synthetases (classified based on few conserved features into two separate classes, respectively) lack overall homology. Even for a given synthetase, the residues constituting the active site differ from one species to the other.[2] Several other differences, such as number of monomer units, mode of binding of acceptor arm of tRNA, the attachment of the amino acid with the 2′ or 3′ –OH group of the terminal ribose moiety of tRNA, structural resemblance of catalytic and non-catalytic domain and their compositions, are noted among the aaRSs. Consequently, it is important to look into the microscopic organization of enzyme structure with a focus to address the forgoing puzzles.

9.4.1 Architectonics of the Active Site of aaRS

It is not straightforward to identify an active site within an enzyme which contains several cavities or clefts. Efforts have been made to identify the active site residues among several other residues in enzymes.[49–64] Once identified, it is natural to ask which special features of the region make it capable of carrying out the reaction while other cavities fail to do so. Most naturally, one would look into the construction of the active site. The walls of the active site cavity are principally composed of catalytic and other residues (amino acids or nucleotides). There are many complications in assigning the function of a catalytic residue, due to the multistep nature of the chemical reactions. One residue can play more than one role and can be involved in different steps of the reaction. A feature of the catalytic residues present in the active site in the vicinity of the substrates is that they are highly conserved. It is plausible that the conservation is a consequence of evolution, which was targeted to developing and retaining the efficiency of the enzyme. By doing such multitasking, conserved active site residues can discriminate non-cognate substrates, as noted in different classes of enzymes.[13] In other words, a network of interactions exist between the substrate and the active site residues which are favorable for the cognate substrate. Several residues of the active site might take part in formation of such a network. In this section, the architectonics of the active site structures of aaRSs are discussed.

The active site cavity of aaRSs can be subdivided into three regions: the ATP binding pocket, the intervening region which is the region closest to the reaction center (region close to the ester linkage) and the amino acid binding pocket (Figure 9.3).[65] The residues in the three pockets have close interactions with the substrates during the course of the reaction. The molecules involved in the aminoacylation reaction are ATP complexed with Mg^{2+} ion, an amino acid and tRNA. The molecular fragments of ATP and amino acids are (i) the adenosine group, (ii) the ribose sugar, (iii) the triphosphate group bearing a negative charge (belonging to ATP), (iv) the positively charged amino group of the substrate amino acids, (v) the negatively charged carboxylic acid group in the zwitterionic form of the amino acid and (vi) the side chain (except glycine) which differs in polarity or charge or hydrophobicity for 20 amino acids and may contain a functional group (positively or negatively charged or polar). In corroboration with the chemical structure of the foregoing groups of the substrates, the active site residues of the corresponding aaRS have a complementary chemical structure so that favorable interactions can bind the substrates effectively and fidelity. This complementary nature of the interaction between substrate and active site is also responsible for carrying out the reactants to form the product *via* the transition state through the reaction pathway. The pattern of the interaction is dependent on the chiral structure of both the active site residue and substrate. Hence, it is not surprising that the chiral discrimination is significantly observable in this reaction. Below we refer the chemical structure of the substrates and various active site residues of the corresponding aaRS that have complementary favorable interactions. We collectively mention those residues which have favorable interactions with ATP and substrate amino acids in different aaRSs as observed from the available crystallographic data. The various species for which the information

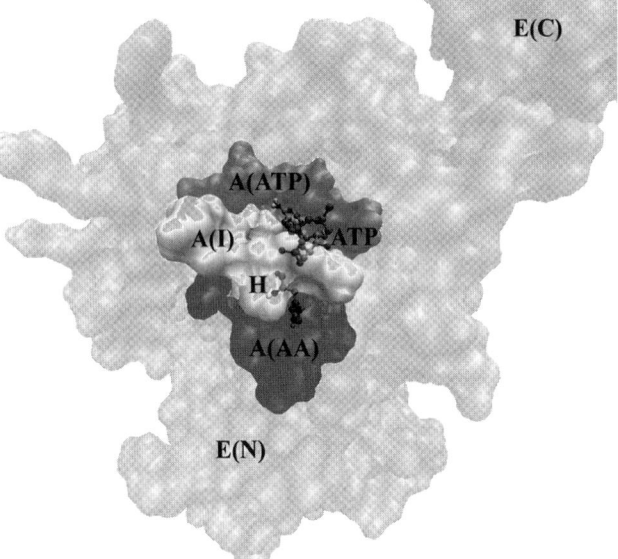

Figure 9.3 The schematic representation of the active site HisRS from *EC* (1KMN.PDB) located within the N-terminal domain (catalytic domain). The N-terminal domain and C-terminal domain are indicated by E(N) and E(C), respectively. The ATP binding pocket, intervening region and the amino acid binding pocket are indicated by A(ATP), A(I) and A(AA), respectively. The reactants (histidinol, an inhibitor and ATP) are shown by ball and stick representations and the surrounding active site cavity is shown by surface representation. The images are prepared using VMD.[113]

of the active site of different aaRS available are *Bacillus stearothermophilus* (*BS*), *EC*, *Thermus thermophilus* (*TT*), *Staphylococcus aureus*; (*SA*) and *Saccharomyces cerevisiae* (yeast).

In the ATP binding region, the π-electron cloud present in the adenosine group of ATP can have favorable π-stacking interactions with the electron cloud of active site amino acids, such as Phe. The presence of such residues are noted in different aaRSs. Interaction of the π-electron cloud of active site Phe group with adenosine group is noted in AsnRS,[66] SerRS,[67,68] HisRS[22] as well as in GlyRS.[69] Further hydrogen bonding between the nitrogen atoms of the adenosine base and active site residues such as His, Pro, Asp, Asn, Met and Gln are observed. Such hydrogen bonding are present in IleRS, with active site His,[70] and with Met in ValRS.[71] The α-phosphorus atom of ATP directly participates in the reaction and is considered to be located in the intervening region. We discuss the interaction between the active site residues and the α-phosphorus atom of ATP in the following.

In the intervening region, the presence of negative charge density over the triphosphate group of ATP as well as the carboxylic acid group of amino acid could give rise to an unfavorable electrostatic potential prohibitive for the reaction. The reduction of the unfavorable electrostatic potential between the α-phosphorus atom of ATP and the carboxylic acid group of the substrate amino acid is a prerequisite to carry out the nucleophilic reaction. The positively charged amino acids (such as His, Lys, Arg), polar residues such as Thr, or one/more divalent Mg^{2+} cations carry out the task of reduction of the electrostatic potential. The presence of one or more such residue or ion is observed in the ATP binding site of different aaRSs. Interactions with His residues and triphosphate group of ATP are observed in ArgRS[72,73] and CysRS.[74] Similarly, interactions with the triphosphate group of ATP and different positively charged amino acid residues and cations are noted in GlnRS with His and Lys,[65,75] in GluRS with Arg,[76] in IleRS with His and Lys,[70] in TyrRS with Lys,[77–79] and in AsnRS with Arg and His residues as well as three Mg^{2+} ions.[69] The triphosphate group of ATP in ProRS interacts with active site Arg, His and Thr.[80] Interactions between Arg residues and Mg^{2+} ions and the triphosphate group of ATP are noted in class II aaRSs, such as GlyRS, HisRS, LysRS and SerRS.[22,67–69,81,82] The triphosphate groups are also involved in hydrogen bonding with polar active site groups in some aaRS. In a few cases, the triphosphate group is involved in hydrogen bonding with the amide hydrogen of the main chain of a peptide linkage located nearby in the active site.

The α-carboxylic acid group of the substrate amino acid in the intervening region can have favorable interactions with the positively charged active site amino acid residues, such as Arg/Lys/His or active site amino acid containing polar side chain such as Asn or Gln. Various active site residues bearing a positive charge or with a polar nature, present in the proximity of the α-carboxylic acid group of the substrate amino acid of different aaRS active sites, are noted. The examples are residues such as Asn, Gln and His in

ArgRS,[72,73] Arg and Lys in LysRS,[81,82] Arg in HisRS,[22] Gln in IleRS,[70] and His in LeuRS.[83] Other than the polar groups (as mentioned before), the amino group of a peptide linkage can form a hydrogen bond with the carboxylic acid group. The positively charged α-amino group of the substrate amino acid, also considered to be located in the intervening region (based on its proximity with the reaction center) can have favorable interactions with negatively charged active side residues such as Asp or Glu. The negatively charged active site Asp residue interacts with positively charged α-amino group of the substrate amino acid in GlnRS[65,75] and in LeuRS.[83] Similarly, interaction between negatively charged active site Glu and the α-amino group of the substrate amino acid is noted in GluRS,[75] LysRS,[81,82] SerRS,[67,68] and HisRS.[22] Active site residues with polar side chains such as Asn, Ser, Gln, Tyr and Thr are also observed in the vicinity of the α-amino group of the substrate amino acid, and can form a stabilizing hydrogen bond interaction. Presence of such hydrogen bonds is noted between the α-amino group of the substrate amino acid and Asn as well as Ser in ArgRS[72,73] and with Thr in CysRS.[74]

Unlike the invariant carboxylic acid and amino groups, the side chains are variable in the structure of amino acids (except glycine that has no side chain). The related active site residues are considered to be in the amino acid binding pocket of the active site. For example, the side chains could be positively charged, negatively charged, polar, non-polar or may have a sulfur containing group. When the side chain of the substrate amino acid is positively charged, like those in Arg or Lys, the vicinity of the active site close to the side chain is composed of negatively charged amino acids like Glu, Asp or polar amino acids such as Tyr. Presence of active site Glu, Asp and Tyr is noted in the proximity of the positively charged side chain of the substrate Arg.[72,73] Active site Tyr and Glu are present in LysRS.[81,82] Similar interactions with Tyr and Glu are noted in HisRS where the side chain of the substrate amino acid can be in protonated state (ionic) or non-protonated (polar) state.[22] The binding pocket for the substrate amino acid bearing a negatively charged side chain (such as Glu, Asp) may contain positively charged active site amino acids, such as Arg or Lys, or amino acids which can form hydrogen bonds (such as Gln, Tyr, Ser or Asn) with the carboxylic acid group of the substrate amino acid side chain. The residues such as Arg, Tyr, Asn and Arg interact with the side chain in GluRS.[76] In AspRS, interaction with the active site Arg, Lys, Gln, Asp and Ser are noted in the vicinity of the negatively charged side chain of substrate amino acid.[84]

When the side chain of the substrate amino acid is polar (for example, side chains of Gln, Asn, Ser, Cys), hydrogen bonding with polar active site residues such as Tyr, Thr or water molecules, or interaction with charged active site residues such as Glu, Arg, is observed. Such interactions with Tyr and water molecules are noted in GlnRS,[65,75] interaction with Arg and Glu are noted in AsnRS,[66] interaction with Thr and Glu are noted in SerRS.[67,68] In the case of the substrate amino acid with a non-polar side chain, the corresponding active site pocket is composed of non-polar amino acids such as Pro, Trp, Phe and

Ile, which might have favorable van der Waals and hydrophobic interaction with the side chain of the substrate. Further, presence of the hydrophobic part of the side chain of amino acids such as Met, His and Tyr are observed near the non-polar side chain of the substrate. Examples of the non-polar active site residue and the hydrophobic part of the side chain of charged or polar amino acids near the non-polar side chain of the substrate amino acid are as follows: Pro and Trp are present in active site of IleRS,[69] Pro, Ile and Trp are present in ValRS,[71] Trp and Gly are present in ProRS,[80] Met, Phe, Tyr, His are present in LeuRS[83] near the non-polar side chain of the substrate amino acid.

In the case of a few amino acids such as Cys or Gly, customized interactions by the active site residues or ions to form favorable interactions with these amino acids are noted. The side chain binding pocket of CysRS contains a Zn^{2+} ion which is involved in a coordination interaction. The Zn^{2+} ligates the Cys substrate *via* the side chain thiolate, while the remaining four coordination sites of Zn^{2+} ion are satisfied by the active site amino acids such as multiple Cys, His and Glu.[74] Gly is the smallest amino acid and is achiral. Due to the absence of a side chain, the specificity of GlyRS depends exclusively on the carboxylic acid and amino acid group. The negatively charged carboxylic acid group of two Glu residues interacts directly with the α-NH^{3+} group. Another Glu fills up the space in the binding pocket which would have been occupied by the side chain of substrate amino acids other than Gly, and helps in retaining the fidelity exhibited by GlyRS. The pro-L α-hydrogen atom interacts with the carboxyl oxygen atom of a second Glu residue. The oxygen atom of the side chain of serine is involved in the hydrogen bonding with the α-ammonium group. This interaction specifies the orientation of the substrate glycine moiety and prevents binding of amino acids with similar chemical structures, such as alanine, by creating a steric block for the methyl group of alanine.[69]

Since the electrostatic nature of the groups of the substrate amino acid and ATP are limited to polar, non-polar, positively or negatively charged or groups with a π-electron cloud, the pattern of interactions between these moieties and active site residues bear commonalities, as mentioned above. Various aaRSs use different residues or set of residues, which are analogous in terms of their interaction and recognition of the substrate, positioning and carrying out of the related function (for example, reduction of the unfavorable potential or by catalytic action). In other words, although the active site walls of different aaRSs have diversification in composition, one or clusters of such residues have common roles in different aaRSs. For example, the reduction of unfavorable electrostatic potentials between the amino acid and ATP is carried out by two Mg^{2+} ions and two Arg in HisRS; one Arg residues, one Lys and two Mg^{2+} in ThrRS; one Arg, one His and two Mg^{2+} ions in ProRS; one Arg and three Mg^{2+} ions in LysRS and PheRS; one His, one Lys and one Mg^{2+} ion in GlnRS *etc*. Such commonality can be noted in the case of other types of interactions played by different active site residues, as mentioned before, using crystallographic studies and recent computational analysis.[13] The study indicates that the types or nature of the interactions (electrostatic, hydrogen bonding, π-stacking and hydrophobic) between the active site residues and substrates are common in all aaRSs, although the particular residue/ion (or clusters of residues) participating in these interactions may differ. Presumably, the residues forming such common interaction patterns for an amino acid are conserved through evolution in different species.

Of course, the generic interactions are not limited to the above-mentioned types. Long range influences of the distant residues (located in the second shell around the active site or beyond) present in the enzyme must be there due to the long range nature of electrostatic interactions. It is, however, expected that the recognition and reaction will be significantly influenced by interactions with the nearest neighbors. The structural information described above reveals that the substrates are located in a network of interactions (principally electrostatic such as ionic, dipolar, hydrogen bonding or van der Waals and hydrophobic interactions). These interactions between the active site residues of a given aaRS and its cognate amino acid are responsible for recognition, lowering of unfavorable electrostatic potential and catalytic action. While the network of interaction is favorable for the cognate substrate amino acid, it is unfavorable for a non-cognate amino acid. In the latter case, such interactions are either absent or unfavorable for the non-cognate substrate.

Recently, the primary and secondary structures of a particular aaRS (HisRS) from different organisms (three prokaryotic species, *EC*, *TT* and *SA*) have been compared with a focus on their respective active site structures.[85] Similarities and dissimilarities in the respective primary sequences are compared. More than 30% of the residues within the catalytic domain are conserved in the three organisms. Despite this sequence similarity in the catalytic domain, diversities are observed in the domains and loops that are relatively far from the reaction center. The common features of the secondary structures (the arrangements of the anti-parallel β-sheet and α-helices) of the active sites for the HisRS of *EC*, *TT* and *SA* are observed. In contrast, comparison of the molecular organization of the first shell active site residues of HisRS in *EC*, *TT* and *SA* indicates that the active site wall of HisRS in the vicinity of the reactants is composed of both conserved and non-conserved residues in all three cases (Figure 9.4).

Among the conserved residues in *EC*, *TT* and *SA* near the reaction center, Arg259 (numbering scheme as HisRS of *EC*) has a catalytic role and Arg113 reduces the electrostatic potential of phosphate group of ATP, which is essential for the progress of the reaction. The Glu83 is responsible for the placement of His *via* favorable interactions with the amino group of His and is conserved. Pro82 and Gly84 make the conformation of the peptide chain turn in such a way that the side chain of the intervening Glu83 is projected towards the reaction center. The proline residue (Pro82 in HisRS) is known to be responsible for creating a turn.[86] The Gly (Gly84, in HisRS) is a highly conformationally flexible residue and is also observed in a turn. These two residues are also conserved. Gln127 forms a hydrogen bond with the ribose sugar ring of ATP as well as the carboxylic group of His and is conserved. The

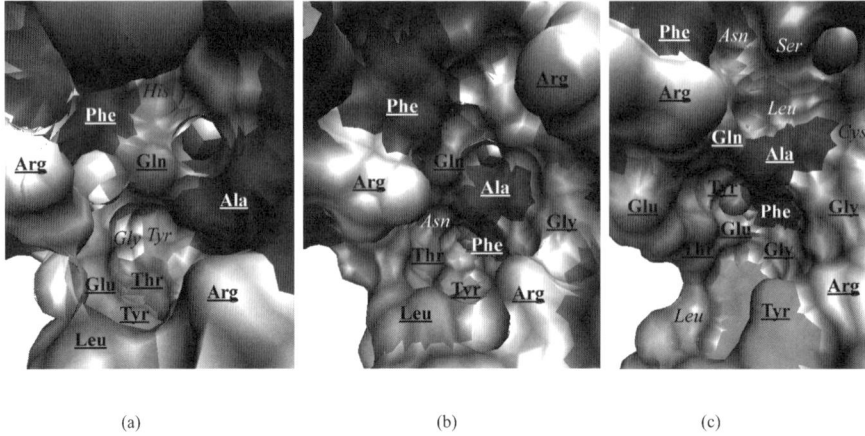

Figure 9.4 Comparison of the active site residues (both conserved as well as non-conserved) in HisRS of three prokaryotic species *EC*, *TT* and *SA*. The residues that are non-conserved in three prokaryotic HisRS (a) *EC* (b) *TT* and (c) *SA* are shown in *italics* and the conserved residues are underlined. The images are prepared using VMD.[113]

Ala306, Gly84, Gly260 and Leu261 have van der Waals interaction with the non-polar part of His and are conserved in the three species. These residues are parts of a hydrophobic patch close to the non-polar region of His and are conserved. The Ala284 is close to the chiral center and is non-conserved (a Gly is present in *TT* and *SA* instead of Ala). The polar side chain of Thr60 is at the remote side of the wall near the ATP and is conserved. The Val62 (having a non-polar side chain) is non-conserved and is replace by Ile (in *TT*) which is also has a nonpolar side chain. In many cases, the influence of non-conservation on the reaction mechanism and catalysis is expected to be low as the peptide linkage corresponding to the non-conserved residue is relatively close to the reaction center. The linkage has an invariant structure and is not dependent on whatever variation occurs in the nature of the side chain. In such cases, because the side chain of the non-conserved residues is relatively far from the reaction center, their influence is expected to be less compared to the conserved ones.

It is noted above that the active site residues in the three regions, namely the ATP binding pocket, intervening region and amino acid binding pocket are complementary to the substrate located in their vicinity. Below we mention how the related favorable interactions influence the catalytic efficiency and chiral specificity of the aminoacylation reaction. It is also observed that the networks of interactions formed by the conserved residues in the active sites with the substrates in the three species have significant commonality. This is responsible for the basic reaction mechanism and catalytic action being same in the three species despite the sequence and structural diversity.

9.4.2 Influence of Active Site of aaRS on Aminoacylation Reaction

The crystallographic analysis of the network of interaction indicates that the progress of the reaction is dictated by the architectonics of the active site nanospace of synthetase. While crystallographic studies are indispensable for the essential structural information, they can provide only average structural information. In principle, catalytic mechanisms can only be quantitatively modeled by quantum mechanical methods, by studying the changes in the electronic structure involved during the course of the reaction. We discuss here the recent electronic structure based analysis of the origin of the discrimination capacity by the active site of aaRS. The specificity of the aaRS to incorporate a specific amino acid and its ability to discriminate between several competing substrates is well known.[87]

A number of experimental studies unambiguously established that the process of aminoacylation is stringent about incorporation of the L-enantiomer.[37] Experimental studies indicate that the specificity is present at a level sufficient to discriminate the enantiomeric species.[88–94] Neither the acylation site nor the editing site would be able to discriminate the enantiomers if the size factor is essentially controlling the proofreading process. The discrimination by aaRS can be better understood in terms of the difference in interactions (principally electrostatic) between the aaRS and the enantiomers of the amino acid. This view is supported by recent computational studies on aminoacylation of AspRS[45] and HisRS.[24,25] The preference for acylation of L-Asp over D-Asp and L-Asn by Aspartyl-tRNA synthetase (AspRS) has been studied using molecular dynamics simulation.[45,95] The studies point out that the influence of the network of electrostatic interaction present in AspRS protects against most binding errors.

A recent combined quantum mechanical/semiempirical study is carried out using a model based on the crystal structure of the oligomeric complex of histidyl-tRNA synthetase (HisRS) from EC.[24,25] The variation in the energy during the mutual approach of the His and ATP to form adenylate shows that the surrounding nanospace of synthetase confines the reactants (L-His and ATP) and proximally places them in a geometry suitable for in-line nucleophilic attack. The energy surface of the model containing D-His is significantly higher than that of L-His, as shown in Figure 9.5. This is due to unfavorable interactions of D-His with ATP and the surrounding residues. This indicates that the network of interaction (principally electrostatic) is highly unfavorable when the D-amino acid is incorporated. The reorganization of the surrounding nanospace can lower the unfavorable nature of the intermolecular energy surface of D-His and surrounding residues. However, such a rearrangement requires large-scale structural reorganization of the synthetase structure and might be unfavorable. The variations in the interaction energy as a function of the intermolecular distances between His, ATP and the respective surrounding residues, such as Glu83, Arg113, Arg259 and Tyr264, and the concomitant effect on discrimination have been studied.

The variations in energy reveal the relative stability of the cognate amino acid (L-His) compared to the D-His as a function of the separation between the surrounding residues and His. The results emphasize the role of the active site residues in controlling the fidelity of the reaction (Figure 9.6).

The variation in the bond angles and distances during the progress of the aminoacylation reaction are calculated from the optimized geometries of the reactant, product and transition state.[24,25] The study demonstrates the attack of the caroboxylic oxygen of the substrate amino acid at the α-phsophorous of ATP and the concomitant inversion of oxygen atoms around the α-phosphorus of ATP, corroborating proposals from the crystallographic studies. The transition state structure containing D-His lacks the hydrogen bond network present in the L-His case and is highly unfavorable. Thus, the origin of specificity about the L-enantiomer is at least two-fold. First, the interaction of the non-cognate amino acid with the residues constructing the active site nanospace is unfavorable. Secondly, the transition state geometry with the cognate amino acid (L-His) is lower in energy than the transition state geometry with the non-cognate one (D-His).

We have yet to address the relation between the conservation of active site residues and the molecular mechanism of aminoacylation reaction. Recently the influences of the conserved active site residues on the reaction mechanism, as well as the electrostatic potential near the reaction center, have been analyzed for HisRS from *EC*, *TT* and SA.[85] In order to understand this issue, two different models are considered for electronic structure based computation of the progress of reaction. The relatively large model includes Thr60, Val62, Pro82, Glu83, Gly84, Arg113, Gln127, Arg259, Gly260, Leu261, Ala284 and

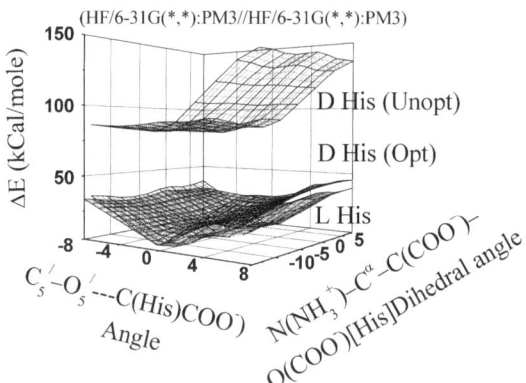

Figure 9.5 Comparison of the variation of the interaction energy as a function of the orientation of the His moiety (expressed in degrees) and the orientation of the carboxylic acid group of His relative to the ATP (expressed in degrees) for $Model^{Opt}_{Reactant\ (L)}$, $Model^{Opt}_{Reactant\ (D)}$ and $Model^{Unopt}_{Reactant\ (D)}$.[25] The variation in the energy is computed using two level ONIOM (HF/6-31G**:PM3//HF/6-31G**:PM3). All calculations are performed with the Gaussian 03W suite of programs.[114] (Reprinted with permission from the American Chemical Society. © American Chemical Society.)

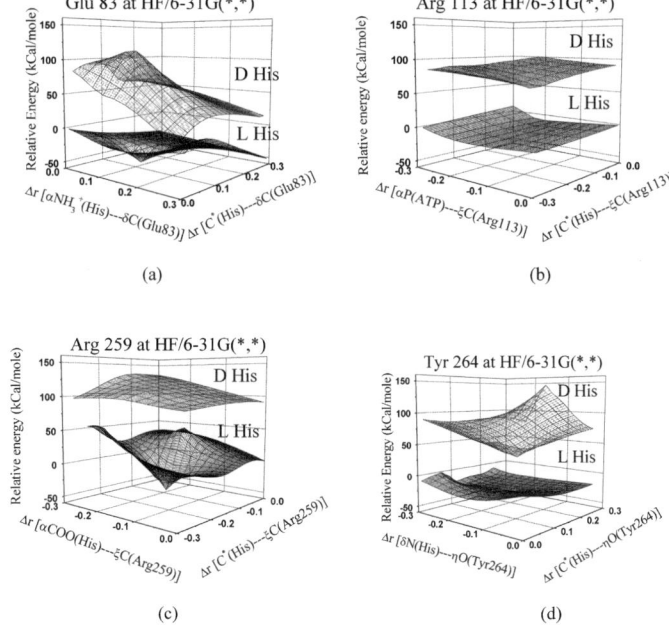

Figure 9.6 Variation in interaction energy as a function of relative distance between nitrogen atom of α-amino group of His and δC of Glu83 (Å) as well as the relative distance between the chiral center of His and δC of Glu83 (Å) for Model$_{Reactant\ (L)}$ and Model$_{Reactant\ (D)}$ with rigid geometry. Starting from the mutual arrangement of His and Glu83 as in the crystal structure, mutual separation is varied as present in the product state Model$_{Product\ (L)}$.[25] (b) Variation in interaction energy as a function of relative distance between α phosphorus atom of ATP and ξC of Arg113 (Å), as well as the relative distance between the chiral center of His and ξC of Arg113 (Å) for Model$_{Reactant\ (L)}$ and Model$_{Reactant\ (D)}$ with rigid geometry. Starting from the mutual arrangement of His or ATP and Arg113 as in the crystal structure mutual separation is varied as present in the product state Model$_{Product\ (L)}$.[25] (c) Variation in interaction energy as a function of relative distance between the carbon atom of the α-carboxylic acid group of His and ξC of Arg259 (Å), as well as the relative distance between the chiral center of His and ξC of Arg259 (Å) for Model$_{Reactant\ (L)}$ and Model$_{Reactant\ (D)}$ with rigid geometry. Starting from the mutual arrangement of His and Arg259 as in the crystal structure, mutual separation is varied as present in the product state Model$_{Product\ (L)}$.[25] (d) Variation in interaction energy as a function of relative distance between δN of His and ηO of Tyr264 (Å), as well as the relative distance between the chiral center of His and ηO of Tyr264 (Å) for Model$_{Reactant\ (L)}$ and Model$_{Reactant\ (D)}$ with rigid geometry. Starting from the mutual arrangement of His and Tyr264 as in the crystal structure mutual separation is varied as present in the product state Model$_{Product\ (L)}$.[25] Computations are performed using two level ONIOM method (HF/6-31G**:PM3) with respective active site residues (Glu83, Arg113, Arg259 and Tyr264) at the HF/6-31G** level. All calculations are performed with the Gaussian 03W suite of programs.[114] (Reprinted with permission from the American Chemical Society. © American Chemical Society.)

Ala306, two Mg^{2+} ions, two water molecules (present in the crystal structure) and substrates. The second one is a smaller model of the active site which includes Glu83, Arg113, Gln127 and Arg259 active site residues, two Mg^{2+} ions, two water molecules as well as the substrate amino acid (His) and ATP. ONIOM (HF/6-31G*: PM3) calculation is performed using both models. Validation of the smaller model is carried out by comparing the energy surfaces of large and small models as a function of reaction coordinates. The reaction coordinate q is given by, $q = r_{P-O(PPi)} - (r_{P-O(His)} + r')$. Here, the $r_{P-O(PPi)}$ is the relative separation between oxygen atom joining the α- and β-phosphorus atom of ATP and the α-phosphorus atom of ATP; $r_{P-O(His)}$ is the relative separation between the carboxylic oxygen atom of His and the α-phosphorus atom of ATP. The variable r' is the projection of the $αP-O_5'$ bond of ATP on the bond which is forming between the carboxylic oxygen atom of His and the α-phosphorus atom of ATP. The magnitude of the $r_{P-O(PPi)}$ gradually increases and the magnitude of the r' decreases during the progress of the reaction. As a result q increases as the reaction progresses towards the product state. The variable measuring the influence of Arg259 on the reaction is given by Ω_{Arg}, the orientation of the catalytic residue Arg259 relative to the substrate His.

The result indicates that the nature of the energy surfaces are not significantly affected as the distant residues such as Thr60, Val62, Pro82, Gly84, Gly260, Leu261, Ala284 and Ala306 are removed from the large model, and the progress of the reaction is not significantly dependent on these residues. Interestingly, many of these residues are non-conserved while the conserved residues are close to the substrate. This confirms that the non-conservation has little effect on the course of the reaction. Further, the electrostatic potential near the reaction center for the large and small model are comparable, which validates the small model. The results also show that the lowering of the electrostatic potential in the intervening region between His and ATP in *EC, TT* and *SA* is a necessary precondition for the reaction. The lowering of the potential is controlled by the Mg^{2+} and the conserved positively charged active site residues.

The progress of the reaction and the transition state structures of the activation step of the aminoacylation reaction for *EC, TT* and *SA* are calculated using combined *ab-initio*/semi-empirical calculations. The variation of the energies of the reactant states of *EC, TT* and *SA* as a function of reaction coordinate q and Ω_{Arg} is shown in Figure 9.7. The energy profiles as a function of the relevant reaction coordinate and the orientation of the catalytic residue (Arg259) are quite similar for three species. The similarity indicates that the reaction mechanisms are the same and they are principally guided by the strikingly similar structural pattern formed by the conserved residues for three species. The energy surfaces have close resemblance in three species and present a clear perspective of how the reaction proceeds with the aid of different conserved residues, as shown in Figure 9.7.

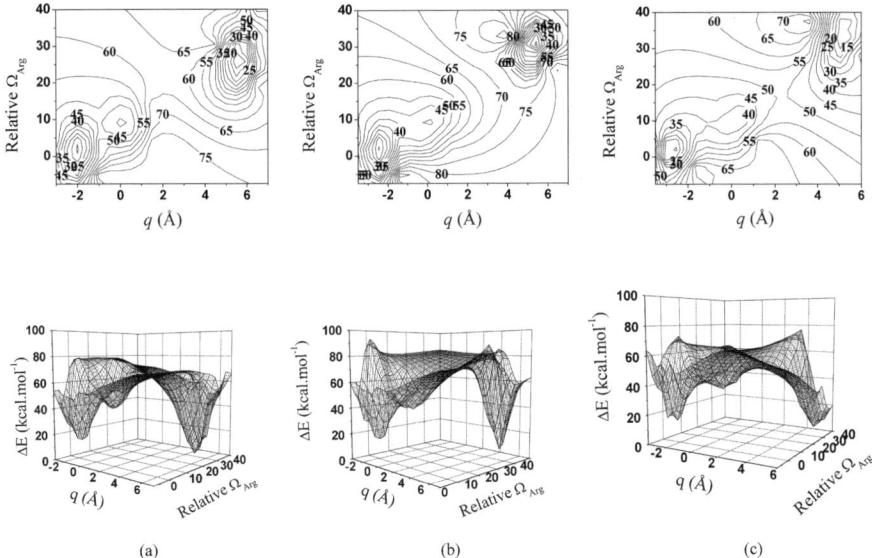

Figure 9.7 Schematic representation of the contour (top) and three dimensional schematic representation of the energy surface (bottom) of the variation of energy in the reactant state, product state and the transition state for each of (a) *EC* (b) *TT* and (c) *SA*.[85] The variation of interaction energy is shown as a function of reaction coordinate q and relative orientation of the Arg259. Details are described in the text. Calculation is carried out at ONIOM (HF/6-31G*:PM3//HF/6-31G*:PM3) level of theory. The relative orientation of the Arg259 with respect to the His and ATP as present in the reactant state is taken as the initial orientation (referred as zero degrees). The ΔE represents the relative energy in kcal mol^{-1} with respect to the optimized product of the respective models. All calculations are performed using the Gaussian 03W suite of programs.[114] (Reprinted with permission from the Elsevier, © Elsevier.)

In the present review we have discussed the issue of whether enzymes working on similar substrates (such as aaRSs acting on various amino acids) use common networks of interaction or not? And how the structures of the active sites compare and which residues are conserved in different species? A related question is the conservation of residues in the active site space and their role in the molecular mechanism of a particular enzymatic reaction. The analysis presented here shows that architectonics of the active site of aaRS is such that the network of interaction between the active site residues and their complementary parts of the substrates are very similar. Various aaRSs utilize different residues (or sets of residues) to carry out a given task during the course of reaction. These tasks are as follows. The electrostatic interaction between the phosphate group of ATP and the neighboring active site residues reduces the unfavorable charge over the phosphate group of ATP. The hydrogen bonding interaction between the adenosine base of ATP and its adjacent active site residues recognize the adenosine base and discriminate it against possible errors, such as incorporation of GTP. Hydrogen bonding and/or electrostatic interaction with the α-phosphate group of ATP as well as the carboxylic acid group of amino acid and the nearby active site residues anchors the substrate moieties in the proper position as well as orientation. Electrostatic interaction between the α-amino group of substrate amino acid and the active site residues in its proximity is vital for chiral discrimination as well as proper placement of the substrate moieties. Despite the bewildering diversity and non-conservation of residues in the enzyme as a whole and in the active site region, the analysis of the interaction provides a clue to the key factors of architectonics of the aaRSs which play decisive role in the reaction.

9.5 Future Prospects

The lessons learned from the active site structural features can be utilized in effective enzyme engineering. Knowledge of the structure and improved understanding of the properties of enzyme active sites, as well as their catalytic mechanisms, are essential for novel protein design and predicting protein function from structure. Once the network of interaction in an active site is understood, then rational modification of the active site residues can alter the interaction between the active site and reactants in a desired manner. Such an understanding is expected to develop control over the reaction in a more efficient manner. A rational approach towards improved enzymatic activity can significantly enhance the enantioselectivity by focusing on mutations close to the active site.[96] Compared to the random trial and error approach, closer mutations are more effective than distant ones in many cases (but not in all cases).[97] Preparation of the desired enzyme can be achieved either by random trial and error (a stochastic method in which the random change of amino acids are made without caring about their position or function) or through rational design.[98] Part of the rational design is the combination of crystal structure analysis and modeling. If the principles of stereoselctivity and fidelity of the enzyme catalysis is understood then it will be possible to develop new enzymes using either minimal modification of natural enzymatic structure or developing *de novo* scaffolds. If it is possible to correlate the network of interaction in the active site structure with the possible reaction, clues for constructing structural motifs as effective mimics of the archetypal biological active site can be obtained.

The development of the minimal model of the template for targeted reactions (with the aim of efficient miniaturization over the large biological macro-molecular machines) could be another interesting arena of future development. It is possible to create a molecular environment capable to support the higher level structural arrangement of the functional biological molecular fragments which are the hallmark of the protein. This is possible by mutation of a protein or by synthesizing short oligopeptides.[99] This procedure may reproduce the function of much larger proteins, by transferring the active sites to small and stable natural scaffolds. Further impetus in research in this direction is provided by the success of computational analysis of the interaction in the active site combined with knowledge of crystal structure, leading to the design of enzymes with rate enhancement and multiple turnovers.[100–102]

Enzymes are most efficient biocatalysts that carry out natural product metabolism reactions of diverse types. Enzyme engineering strategies have allowed the exploration of metabolic engineering of biosynthetic pathways to create new products which have biotechnological and medicinal implications. The analysis of the interaction pattern in the active site can be useful to develop new biocatalysts. The development of the microscopic understanding of the active site has biomedical prospects: the confined space could be a potential target of drugs. Improved understandings of drug and active site interaction have potential in developing novel drugs.[13] For example, large proportions of clinically useful antibiotics exert their effects by blocking protein synthesis at PTC in ribosome.[103]

The principles learned from active site architectonics can be utilized in tunable synthesis of useful non-biological materials. Novel functional nanomaterials can be developed once the molecular understanding of the correlation between the enzyme action and the structural organization is known using the related understanding. Numerous nanomaterials have been designed through the process of self-assembly and molecular recognition, including chiral recognition.[104–108] Excellent control of enantioselectivity in molecular recognition can be achieved,[109] and controlled fabrication of nanometre-scale objects can gain momentum once the principles behind the structural organization of nanoscale cavities are understood.[110]

We have discussed some of the topics concerning the principles behind nature's technique in building functional biological architectures. Combining these principles with the technological advancement already made might lead to the possible development of new synthetic nanovessels with desired functionality. The understanding of the architectonics of biological nanospaces has vast potential in building up newer functional structures with desired physico-chemical characteristics and biological function.

Acknowledgements

Authors thank D.S.T., Govt. of India for financial support in their research. SDB thanks University of Kalyani for a senior research fellowship.

References

1. G. P. Moss, 'Recommendations of the Nomenclature Committee', International Union of Biochemistry and Molecular Biology on the Nomenclature and Classification of Enzymes by the Reactions they Catalyze. <http://www.chem.qmul.ac.uk/iubmb/enzyme/>. Retrieved 14 March 2006.
2. J. M. Berg, J. L. Tymoczko, S. Stryer, *Biochemistry*, W. H. Freeman & Co., New York, 2002.
3. A. Warshel, *Computer Modeling of Chemical Reactions in Enzymes and Solutions*, Wiley Interscience, New York, 1997.

4. A. Fersht, *Structure and Mechanism in Protein Science: A Guide to Enzyme Catalysis and Protein Folding*, W. H. Freeman, San Francisco, 1999.
5. A. Warshel, *Proc. Natl. Acad. Sci. U. S. A.*, 1978, **75**, 5250.
6. A. Warshel, R. Weiss and D. Greenberg, Electrostatic Interactions in Enzyme Catalysis, in *Molecular Structure and Dynamics*, ed. M. Balaban, Balaban International Science Services, Philadelphia, 1980, **297**.
7. M. Garcia-Viloca, D. G. Truhlar and J. L. Gao, *Biochemistry*, 2003, **42**, 13558.
8. J. Z. Pu, S. H. Ma, J. L. Gao and D. G. Truhlar, *J. Phys. Chem. B*, 2005, **109**, 8551.
9. C. Alhambra, J. C. Corchado, M. L. Sanchez, J. L. Gao and D. G. Truhlar, *J. Am. Chem. Soc.*, 2000, **122**, 8197.
10. C. Alhambra, J. Corchado, M. L. Sanchez, M. Garcia-Viloca, J. L. Gao and D. G. Truhlar, *J. Phys. Chem. B*, 2001, **105**, 11326.
11. D. E. Koshland Jr and K. E. Neet, *Annu. Rev. Biochem*, 1968, **37**, 359.
12. J.J. Falke and D. E. Koshland Jr, *Science*, 1987, **237**, 1596.
13. N. Nandi, *Chirality in Biological Nanospaces: Reaction in Active sites*, CRC press, Boca Raton, FL, 2011.
14. S. Hammes-Schiffer and S. J. Benkovic, *Annu. Rev. Biochem.* 2006. **75**, 519.
15. A. Warshel, P. K. Sharma, M. Kato, Y. Xiang, H. Liu and M.H.M. Olsson, *Chem. Rev.*, 2006, **106**, 3210.
16. A. Warshel, *Proc. Natl. Acad. Sci. U. S. A.*, 1978, **75**, 5250.
17. S. Marti, J. Andres, V. Moliner, E. Silla, I. Tunon and J. Bertran, *Chem. Soc. Rev.*, 2008, **37**, 2634.
18. A. E. Todd, C. A. Orengo, and J. M. Thornton, *J. Mol. Biol.*, 2001, **307**, 1113.
19. A. Yonath and A. Bashan, *Annu. Rev. Microbiol.*, 2004, **58**, 233.
20. A. Gindulyte, A. Bashan, I. Agmon, L. Massa, A. Yonath and J. Karle, *Proc. Natl. Acad. Sci. U. S. A.*, 2006, **103**, 13327.
21. A. Yonath, *ChemBioChem*, 2003, **4**, 1008.
22. J. G. Arnez, J.G. Augstine, D. Moras and C. S. Francklyn, *Proc. Natl. Acad. Sci. U.S.A.*, 1997, **94**, 7144.
23. D. Benedicte, D. Moras and J. Cavarelli, *EMBO J.*, 2000, **19**, 5599.
24. S. Dutta Banik and N. Nandi, *Coll. Surf. B: Biointerfaces*, 2009, **74**, 468.
25. S. Dutta Banik and N. Nandi, *J. Phys. Chem. B.*, 2010, **114**, 2301.
26. J. W. Torrance, G. J. Bartlett, G. J. Porter and J. M. Thornton, *J. Mol. Biol.*, 2005, **347**, 565.
27. E. Guth, S. H. Connolly, M. Bovee and C. S. Francklyn, *Biochemistry*, 2005, **44**, 3785.
28. M. Cygler, P. Grochulski, R. J. Kazlauskas, J. D. Schrag, F. Bouthillier, B. Rubin, A. N. Serreqi and A. K. Gupta, *J. Am. Chem. Soc.*, 1994, **116**, 3180.
29. Y. Xin, W. Li and E. A. First, *Biochemistry*, 2000, **39**, 340.

30. M. Merle, V. Trezeguet, J. C. Gandar and B. Labouesse, *Biochemistry*, 1988, **27**, 2244.
31. W. Xie, L. A. Nangle, W. Zhang, P. Schimmel and X. L. Yang, *Proc. Natl. Acad. Sci. U. S. A.*, 2007, **104**, 9976.
32. D. H. Kwan, Y. Sun, F. Schulz, H. Hong, B. Popovic, C. Joalice, C. Sim-Stark, S. F. Haydock and P. F. Leadlay, *Chem. Biol.*, 2008, **15**, 1231.
33. R. A. Kwiecien, F. Ayadi, Y. Nemmaoui, V. Silvestre, B. L. Zhang and R. J. Robins, *Arch. Biochem. Biophys.*, 2009, **482**, 42.
34. K. H. Hopmann and F. Himo, *J. Phys. Chem. B.*, 2006, **110**, 21299.
35. H. Wajant and K. Pfizenmaier, *J. Biol. Chem.*, 1996, **271**, 25830.
36. K. Thirumoorthy and N. Nandi, *J. Phys. Chem. B*, 2007, **111**, 9999.
37. N. Nandi, *Int. Rev. Phys. Chem.*, 2009, **28**, 111.
38. T. C. Bruice, *Annu. Rev. Biochem.*, 1976, **45**, 331.
39. J. F. Kirsch, G. Eichele, G. C. Ford, M. G. Vincent, J. N. Jansonius, H. Gehring and P. Christen, *J. Mol. Biol.*, 1984, **174**, 497.
40. P. A. Williams, J. Cosme, A. Ward, H. C. Angove, D. M. Vinkovic and H. Jhoti, *Nature*, 2003, **424**, 464.
41. (a) D. D. Clark, J. M. Boyd and S. A. Ensign, *Biochemistry*, 2004, **43**, 6763. (b) K. Nakano, I. Sagami, S. Daff and T. Shimizu, *Biochem. Biophys. Res. Commun.*, 1998, **248**, 767.
42. J. H. Lutje Spelberg, R. Rink, R. M. Kellogg and D. B. Janssen, *Tetrahedron: Asymmetry*, 1998, **9**, 459.
43. M. L. M. Mannesse, J.-W. P. Boots, R. Dijkman, A. J. Slotboom, H. T. W. M. van der Hijden, M. R. Egmond, H. M. Verheij and G. H. de Haas, *Biochim. Biophys. Acta*, 1995, **1259**, 56.
44. R. Calendar and P. Berg, *J. Mol. Biol.*, 1967, **26**, 39.
45. D. Thompson, C. Lazennec, P. Plateau and T. Simonson, *J. Biol. Chem.*, 2007, **282**, 30856.
46. J. Soutourina, P. Plateau and S. Blanquet, *J. Biol. Chem.*, 2000, **275**, 32535.
47. K. Thirumoorthy and N. Nandi, *J. Phys. Chem. B*, 2006, **110**, 8840.
48. K. Thirumoorthy and N. Nandi, *J. Phys. Chem. B*, 2008, **112**, 9187.
49. W. R. Taylor, *J. Theor. Biol.*, 1986, **119**, 205.
50. O. Lichtarge, H.R. Bourne and F.E. Cohen, *J. Mol. Biol.*, 1996, **257**, 342.
51. R. A. Laskowski, N. M. Luscombe, M. B. Swindells and J. M. Thornton, *Protein Sci.*, 1996, **5**, 2438.
52. T.D. Schneider, *J. Theor. Biol.*, 1997, **189**, 427.
53. M. J. Ondrechen, J. G. Clifton and D. Ringe, *Proc. Natl. Acad. Sci. U. S. A.*, 2001, **98**, 12473.
54. A. Armon, D. Graur and N. Ben-Tal, *J. Mol. Biol.*, 2001, **307**, 447.
55. A.H. Elcock, *J. Mol. Biol.*, 2001, **312**, 885.
56. R. Landgraf, I. Xenarios and D. Eisenberg, *J. Mol. Biol.*, 2001, **307**, 1487.
57. S. Madabushi, H. Yao, M. Marsh, D. M. Kristensen, A. Philippi, M. E. Sowa and O. Lichtarge, *J. Mol. Biol.*, 2002, **316**, 139.

58. T. Kortemme and D. Baker, *Proc. Natl. Acad. Sci. U. S. A.*, 2002, **99**, 14116.
59. J. Pei, N. V. Dokholyan, E.I. Shakhnovich and N. V. Grishin, *Proc. Natl. Acad. Sci. U. S. A.*, 2003, **100**, 11361.
60. A. Gutteridge, G.J. Bartlett and J.M. Thornton, *J. Mol. Biol.*, 2003, **330**, 719.
61. S. Jones and J.M. Thornton, *Curr. Opin. Chem. Biol.*, 2004, **8**, 3.
62. V. Chelliah, L. Chen, T. L. Blundell and S.C. Lovell, *J. Mol. Biol.*, 2004, **342**, 1487.
63. K. Wang and R. Samudrala, *Bioinformatics*, 2005, **21**, 2969.
64. G. Cheng, B. Qian, R. Samudrala and D. Baker, *Nucleic Acids Res.*, 2005, **33**, 5861.
65. V. L. Rath, L. F. Silvian, B. Beijer, B. S. Sproat and T. A. Steitz, *Structure*, 1998, **6**, 439.
66. C. Colominus, L. Seignovert, M. Härtlein, M. Grotli, S. Cusack and R. Leberman, *EMBO J.*, 1998, **17**, 2947.
67. V. Biou, A. Yaremchuk, M. Tukalo and S. Cusack, *Science*, 1994, **263**, 1404.
68. H. Belrhali, A. Yaremchuk, M. Tukalo, C. Berthet-Colominas, B. Rasmussen, P. Bosecke, O. Diat and S. Cusack, *Structure*, 1995, **3**, 341.
69. J. G. Arnez, A. Dock-Bregeon and D. Moras, *J. Mol. Biol.*, 1999, **286**, 1449.
70. T. Nakama, O. Nureki and S. Yokoyama, *J Biol. Chem.*, 2001, **276**, 47387.
71. S. Fukai, O. Nureki, S. Sekine, A. Shimada, J. Tao, D. G. Vassylyev and S. Yokoyama, *Cell*, 2000, **103**, 793.
72. D. Benedicte, D. Moras and J. Cavarelli, *EMBO J.*, 2000, **19**, 5599.
73. J. Cavarelli, B. Delagoutte, G. Eriani, J. Gangloff and D. Moras, *EMBO J.*, 1998, **17**, 5438.
74. K. J. Newberry, Y. M. Hou and J. J. Perona, *EMBO J.*, 2002, **21**, 2778.
75. J. J. Perona, M. A. Rould and T. A. Steitz, *Biochemistry*, 1993, **32**, 8758.
76. S. Sekine, O. Nureki, D. Y. Dubois, S. Bernier, R. Chenevert, J. Lapointe, D. G. Vassylyev and S. Yokoyama, *EMBO J.*, 2003, **22**, 676.
77. A. Yaremchuk, I. Kriklivyi, M. Tukalo and S. Cusack, *EMBO J.*, 2002, **21**, 3829.
78. A. R. Fresht, J. W. Knill-Jones, H. Bedouelle and G. Winter, *Biochemistry*, 1988, **27**, 1581.
79. Y. Xin, W. Li and E. A. First, *J. Mol. Biol.*, 2000, **303**, 299.
80. A. Yaremchuk, M. Tukalo, M. Grøtli and S. Cusack, *J. Mol. Biol.*, 2001, **309**, 989.
81. G. Desogus, F. Todone, P. Brick and S. Onesti, *Biochemistry*, 2000, **39**, 8418.
82. S. Onesti, G. Desogus, A. Brevet, J. Chen, P. Plateau, S. Blanquet and P. Brick, *Biochemistry*, 2000, **39**, 12853.
83. S. Cusack, A. Yaremchuk and M. Tukalo, *EMBO J.*, 2000, **19**, 2351.
84. S. Eiler, A. C. Dock-Bregeon, L. Moulinier, J.C. Thierry and D. Moras, *EMBO J.*, 1999, **18**, 6532.
85. S. Dutta Banik and N. Nandi, *Biophys. Chem.*, 2011, **158**, 61.
86. D. L. Nelson and M. M. Cox, *Lehninger Principles of Biochemistry*, W.H. Freeman & Co., New York, 4th edn, 2002.

87. M. Ibba and D. Söll, *Annu. Rev. Biochem.*, 2000, **69**, 617.
88. E. W. Davie, V. V. Konigsberger and F. Lipman, *Arch. Biochem. Biophys.*, 1956, **65**, 21.
89. Calendar. R, Berg. P, *Biochemistry.* **1966**, 5, 1690.
90. J. Soutourina, P. Plateau and S. Blanquet,. *J. Biol. Chem.*, 2000, **275**, 32535.
91. F. Bergmann, P. Berg and M. Dieckmann, *J. Biol. Chem.*, 1961, **236**, 1735.
92. S. Norton, J. Ravel, C. Lee and W. Shive, *J. Biol Chem.*, 1963, **238**, 269.
93. T. Yamane, L. Miller and J. J. Hopfield, *Proc. Natl. Acad. Sci. U. S. A.*, 1974, **71**, 4135.
94. K. Tamura and P. Schimmel, *Science*, 2004, **305**, 1253.
95. G. Archontis, T. Simonson, D. Moras and M. Karplus, *J. Mol. Biol.*, 1998, **275**, 823.
96. G. P. Horsman, A. M. F. Liu, E. Henke, U. T. Bornscheuer and R. J. Kazlauskas, *Chem. Eur. J.*, 2003, **9**,1933.
97. K. L. Morley and R. J. Kazlauskas, *Trends Biotechnol.*, 2005, **23**, 231.
98. D. Rõthlisbeger, O. Khersonsky, A. M. Wollacott, L. Jiang, J. DeChancie, J. Betker, J.L. Gallaher, E. A. Althoff, A. Zanghellini, O. Dym, S. Albeck, K. A. Houk, D. S. Tawfik and D. Baker, *Nature*, 2008, **453**, 190.
99. V.A. Tatsis, A. Stavrakoudis and I.N. Demetropoulos, *Biophys. Chem.*, 2008, **133**, 36.
100. L. Jiang, E. A. Althoff, F. R. Clemente, L. Doyle, D. Rõthlisbeger, A. Zanghellini, J.L. Gallaher, J. Betker, F. Tanaka, C. F. Barbas, H. Hilvert, K. A. Houk, B. L. Stoddard and D. Baker, *De Novo Computational Design of Retro-aldol Enzymes*, 2008, **319**, 1387.
101. K. W. Hahn, W. A. Klis and J. M. Stewart, *Science*, 1990, **248**, 1544.
102. W.-D. Woggon, *Acc. Chem. Res.*, 2005, **38**, 127.
103. T. Hermann, *Curr. Opin. Struct. Biol.*, 2005, **15**, 355.
104. K. Ariga, T. Nakanishi and J. P. Hill, *Soft Matter*, 2006, **2**, 465.
105. K. Ariga and T. Kunitake, *Acc. Chem. Res.*, 1998, **31**, 371.
106. T. Kunitake, *Thin Solid Films*, 1996, **284–285**, 9.
107. M. Sakurai, H. Tamagawa, Y. Inous, K. Ariga and T. Kunitake, *J. Phys. Chem. B*, 1997, **101**, 4810.
108. M. Sakurai, H. Tamagawa, Y. Inous, K. Ariga and T. Kunitake, *J. Phys. Chem. B*, 1997, **101**, 4817.
109. T. Michinobu, S. Shinoda, T. Nakanishi, J. P. Hill, K. Fuji, T. N. Player, H. Tsukube and K. Ariga, *J. Am. Chem. Soc.*, 2006, **128**, 14478.
110. K. Ariga, J. P. Hill, M. V. Lee, A. Vinu, R. Charvet and S. Acharya, *Sci. Technol.*, 2009, Adv. Mater., **9**, 014109.
111. P.H. von Hippel and O.G. Berg, *J. Biol. Chem.*, 1989, **264**, 675.
112. T. C. Appleby, C. Kinsland, T. P. Begley and S. E. Ealick, *Proc. Natl. Acad. Sci. U. S. A*, 2000, **97**, 2000.
113. W. Humphrey, A. Dalke and K. Schulten, *J. Molec. Graphics*, 1996, **14**, 33.
114. M. J. Frisch *et al.*, *Gaussian 03, Revision C.02*, Gaussian, Inc., Wallingford, CT, 2004.

CHAPTER 10
Nanotechnology in Drug Delivery Systems

KOHSAKU KAWAKAMI* AND MITSUHIRO EBARA

National Institute for Materials Science, International Center for Materials Nanoarchitectonics, Biomaterials Unit, 1-1 Namiki, Tsukuba, Ibaraki 305-0044, Japan
*E-mail: kawakami.kohsaku@nims.go.jp

10.1 Introduction

The general strategy in the pharmaceutical industry is selection of compounds that are free from physicochemical problems and their development using simple dosage forms. Thus, research into favorable physicochemical properties for drug candidates has occurred throughout the last decade.[1–4] However, formulators still must deal with many challenging compounds that survive in discovery studies.[5] Nanotechnologies are playing important roles for overcoming physicochemical and pharmacokinetic problems of drug molecules.

Advances in nanoscale particle design and fabrication provide various options for drug delivery.[6] Nanoparticles are usually defined as particles with a diameter smaller than 100 nm. However, the definition for biomedical use is usually more liberal,[7–9] partially due to distinctive features of particles of a few hundred nanometres, including 'enhanced permeability and retention (EPR) effect' for injectable formulations.[7,10] Nanosuspensions for oral drug delivery typically contain particles with a diameter of a few hundred nanometres, which enhances dissolution rate.[11,12] Pulmonary drug delivery is a non-invasive method for systemic drug administration, for which particle size plays an important role.[13,14] For dry powder administration, particles larger than 5 μm are trapped in the upper portion of the respiratory system. The optimal particle size for achieving maximum deposition in the deep lung is thought to be a few micrometres,[15,16] and thus marketed inhalable formulations usually fall within this size range. However, both theoretical models and experimental case studies suggest that more effective deposition may be achieved with particles having a diameter smaller than 100 nm.[17–19]

Inorganic nanomaterials, including metallic nanoparticles,[20–24] magnetic nanoparticles,[25–29] porous materials,[30–33] and quantum dots,[28,34–37] may also be used for biomedical applications. Inorganic materials exhibit special characteristics that are different from those of bulk materials when their size is reduced down to the nanoscale (< 100 nm) due to the quantum size effect, producing various functions. Diagnosis is also an important application for the inorganic nanoparticles, although it is not the main scope of this chapter.

10.2 Nanocrystals

Use of nanoparticulate formulations is a powerful option for poorly soluble drugs. The dissolution rate can be enhanced through two mechanisms: an increase in surface area and an increase in solubility due to the increase in surface curvature. Surface area is inversely proportional to particle size. Thus, if a particle size is reduced from 1 μm to 100 nm, the surface area increases 10-fold, which should lead to a 10-fold enhancement of dissolution rate. Assuming that a particle is spherical, dependence of solubility on particle size can be described by the Ostwald–Freudlich equation:[11]

$$C(r) = C(\infty) \exp\left(\frac{2\gamma M}{r\rho RT}\right) \quad (10.1)$$

where $C(r)$ and $C(\infty)$ are the solubilities of a particle of radius r and of infinite size, respectively, and γ, M, and ρ are the interfacial tension at the particle surface, molecular weight of the solute, and density of the particle, respectively. According to this equation, solubility increases with a decrease in particle size, i.e., an increase in surface curvature. However, model calculations show that an increase in solubility is almost negligible above 100 nm.[38] Thus, an increase in surface area can be regarded as a dominant factor for increasing dissolution rate in nanosizing technology, and the "solubility advantage" is only marginal.

An enhanced dissolution rate may not be the only factor for nanoparticles to achieve better oral absorption. Peyer's patches may uptake nanoparticles.[39] In addition, control of surface properties may be very important for improving oral absorption of nanoparticles. Lai et al. directly investigated the diffusion of nanoparticles in a mucous layer[40] and found that 500 nm particles diffused very rapidly when coated with a poly(ethylene glycol) layer, although the mesh size of the mucus had been believed to be much smaller than that: 10–200 nm. In contrast, uncoated 100 nm nanoparticles diffused very slowly, indicating greater importance of the surface properties and a larger mesh size of the mucous layer than previously believed.

Solid nanoparticles can be produced *via* either a top-down or bottom-up procedure. The industrial top-down procedure includes media-milling and high-pressure homogenizer technologies,[41,42] which have produced the nanoparticulate solid formulations currently on the market. The bottom-up procedure requires control of the crystallization process,[12,43] which may be followed by a homogenizing process similar to the top-down production. Nanoparticles can be prepared by spray-drying or supercritical fluid technology as well.[44,45] Electrospray deposition is also a bottom-up method for producing nanoparticulate formulations,[46,47] which allows independent design of the surface and inner composition.[48]

Although nanosizing itself is not difficult using commercial equipment, physical stability during storage may be poor. Since the nanosized particles have very high surface energy, aggregation occurs easily. Media should be removed for manufacturing tablets. However, the resultant formulation may not disintegrate into nanoparticles after oral administration. Thus, polymers or surfactants are usually added before size reduction to improve stability.[11,12] Charged polymers or surfactants can enhance stability by giving electrostatic charges to the particles, which results in electric repulsion.[11,12] Sugars may work as cryoprotectants.[49]

The biopharmaceutical advantages of oral nanoparticles include increased bioavailability and elimination of the food effect. Jinno *et al.* compared oral absorption of cilostazol using 220 nm, 2.4 µm, and 13 µm particles[50] to find that the AUC (area under concentration) in beagle dogs increased with decreasing particle size, and that the effect was more pronounced in the fasted state. As a result, the food effect was diminished by using 220 nm nanoparticles. Wu *et al.* investigated the effect of nanosizing MK-0869, which is an active agent of Emend®.[51] Again, oral absorption of the drug increased with decreasing particle size. The AUC ratio from 5.5µm and 120nm particles was 4.3. The food effect was 3.2fold for the 5.5µm particles; however, it decreased to 0.96fold by reducing particle size to 120 nm. Since Emend® is used to control chemotherapy-induced nausea and vomiting, patients cannot have meals before taking of the drug. Thus, an improvement in oral absorption in the fasted state is very beneficial for this drug.

10.3 Surfactant/Polymer Micelles

Micelles are representative nano-structured carriers[52–54] that can entrap hydrophobic drugs in their interior spaces. For surfactant micelles, drug molecules are entrapped by hydrophobic interactions. Very hydrophobic drug molecules are usually located in the central part of the micelles, while drugs with relatively high polarity may be partitioned to the palisade layer. Surfactant molecules may cause serious side effects, including anaphylactic shock. Thus, although surfactants are common excipients in formulation research, the amount used in the formulations is usually not enough to form micelles.

Polymer micelles possess some advantages over surfactant micelles as a drug carrier, and clinical studies are ongoing for various drugs including doxorubicin,[55] paclitaxel,[56] and cisplatin.[57] The capacity of block copolymer micelles to increase the solubility of hydrophobic molecules stems from their unique structural composition, which is characterized by a hydrophobic core sterically stabilized by a hydrophilic corona. The core serves as a reservoir for incorporation of the drug molecules through chemical, physical or electrostatic interactions, depending on their physicochemical properties. Beyond solubilizing hydrophobic drugs, their nanoscale sizes are also attractive because they can escape from the reticuloendotherial system (RES) and allow the accumulation of loaded drug preferentially in solid tumor tissues through the enhanced permeability and retention (EPR) effect.[7,10] Indeed, polymeric micelles are sufficiently large to avoid renal excretion, yet small enough to bypass filtration by interendothelial cell slits in the spleen.[58] In addition, polymeric micelles are generally more stable, with a remarkably lower critical micelle concentration (CMC) compared to surfactant micelles. They have a slower rate of dissociation, allowing retention of loaded drugs for a longer period of time, and can accumulate greater amounts of drug at the target site.[59] Critical insight into the self-assembly mechanism was provided by Kataoka group's seminal work on poly(ethylene glycol) (PEG)-based block copolymers.[60] Biocompatibility is assured by the dense PEG shell, which endows the micelle with a stealth character in the blood, and ensures a long circulation.

A challenge in the development of novel micellar carrier systems is to design targetable polymeric micelles that have pilot molecules on their surface to achieve specific-binding to target cells. Stayton *et al.* proposed an attractive targeting system for tumor-selective therapies using folate receptor-targeted diblock copolymers.[61] The diblock copolymers were polymerized using a folate-functionalized chain transfer agent (CTA) to introduce telechelic chain ends reflecting the chemistry of the CTA. This approach is highly versatile for functionalizing polymer chains because many polymeric therapeutics can be functionalized using this technique. There is increasing interest in designing block copolymer micelles of which the assembly state in aqueous solution can be controlled by external stimuli. One of the most studied stimuli-responsive polymers is poly(*N*-isopropylacrylamide) (PNIPAAm), which exhibits a lower critical solution temperature (LCST) in aqueous solution at 32 °C (Figure 10.1). When block copolymers are designed with PNIPAAm as the outer shell and hydrophobic polymers as the inner core, they form micelles below the LCST, whereas intermicellar aggregation occurs above the LCST of PNIPAAm. This type of thermo-responsive micelle has demonstrated successful drug release and anti-cancer effects upon heating both *in vitro* and *in vivo*.[62–64] When block copolymers are designed with PNIPAAm as the inner-core and hydrophilic polymers as the outer-shell, they form micelles in aqueous solution and permit hydrophobic drug loading above the LCST.[65,66]

Increased complexity has also been demonstrated with double-responsive polymers with two stimuli-responsive blocks (*i.e.*, block copolymers that acquire and lose amphiphilicity upon application of stimuli). PNIPAAm-b-poly(acrylic acid) (AAc), for example, forms micelles upon conversion of one of the blocks into a water-insoluble block at temperatures greater than 32 °C or pH values less than 4.[67,68] In addition, Aoyagi *et al.* developed a series of double thermo-responsive block copolymers with tunable LCSTs, which allows the construction of reversible nano-assembly-forming species.[69–71] The block copolymer can be transformed from a double-hydrophilic to an amphiphilic and finally to a double-hydrophobic block copolymer under bio-inert conditions. Therefore, nano-assembly and drug encapsulation occurred through simple mixing of polymer solutions with the drug below the LCSTs of both blocks, and upon heating at temperatures greater the second LCST, the nano-assembly aggregated and the drug was released from the micelles (Figure 10.1(C)). This process was completely reversible. The protocol is simple and quick, and enables customization of the function of the assembly by a simple mixing and heating procedure of the selected block copolymers. Furthermore, Aoyagi *et al.* also proposed a protocol to produce stimuli-responsive self-assemblies using two block copolymers, poly(*N*-isopropylacrylamide) (PNIPAAm)-β-P(NIPAAm-co-*N*-(hydroxymethyl)acrylamide (HMAAm)) and PNIPAAm-β-P(NIPAAm-co-sodium 2-acrylamido-2-methylpropane sulfonic acid (AMPS)). This proposed protocol enables the facile preparation of a multi-functional assembly and customization of their size by simple mixing of plural functional block copolymers that contain PNIPAAm as the common block.[72]

10.4 Emulsions and Microemulsions

Although emulsions and microemulsions have structural similarities, they are totally different molecular assemblies from a thermodynamic point of view. Emulsions are non-equilibrium molecular assemblies that are produced by applying external energy, such as shear force. In contrast, microemulsions are in the equilibrium state. Thus, no external energy is required for preparation. Microemulsions are defined as isotropic solutions composed of, at least, an organic phase, aqueous phase, and surfactants. Note that no size requirements are defined for particles in microemulsions. However, microemulsions typically have a particle size less than 100 nm.

Emulsions composed of phospholipids and triglycerides have been used as traditional injectable nutrition; the size of particles in these emulsions is usually *ca.* 200 nm. Many attempts have been made to utilize them as drug carriers,[73–77] since hydrophobic drugs are easily incorporated in the oil core to exert various effects, including passive targeting to the disease site, prolonged retention in the circulation, and reduction in side effects. Many drugs, including prostaglandin E_1, dexamethasone palmitate, and flurbioprofen axetil, have been marketed as injectable emulsion formulations. The mixture of drug, oil, and surfactant,

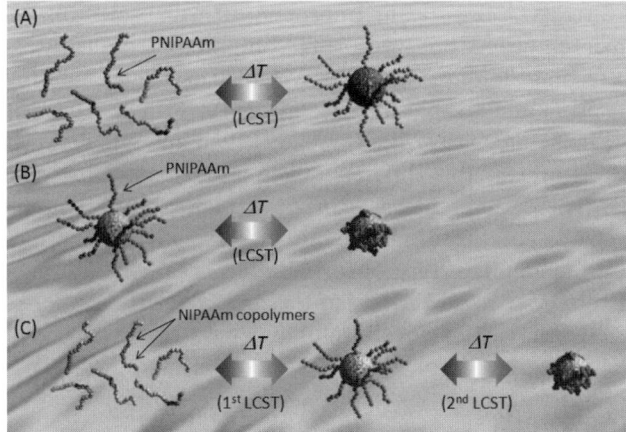

Figure 10.1 Thermo-responsive micellization behavior of the PNIPAAm block copolymers with PNIPAAm as the inner core (A), PNIPAAm as the outer shell (B), and double LCSTs (C) in aqueous solution.

designed to form an emulsion spontaneously in the stomach or small intestine after oral administration, is called a self-emulsifying drug delivery system. The oral absorption of poorly soluble drugs, and its reproducibility, can be improved by skipping the dissolution process and offering a wide oil–water interfacial area.

Well-designed combinations of oils and surfactants provide fine oil-in-water microemulsion droplets in the small intestine to increase the oil/water interfacial area significantly.[38,78–81] Figure 10.2 shows oral absorption of

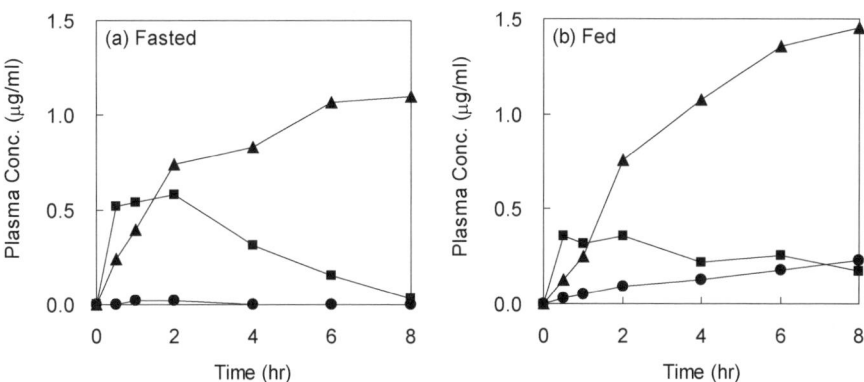

Figure 10.2 Oral administration study of nitrendipine self-microemulsifying formulations to rats under (a) fasted and (b) fed conditions (12 mg kg^{-1}). Formulation type: (●) Suspension in 0.5% methylcellulose solution, (■) Tween 80/propyleneglycol dicaprylic ester/glycerol monocaprylic ester self-microemulsifying formulation, (▲) HCO 60/propyleneglycol monocaprylic ester self-microemulsifying formulation.[81]

nitrendipine in rats.[81] Significant improvement in absorption was achieved using the microemulsion dosage forms compared to aqueous suspensions, regardless of feeding state. Addition of Tween 80 as a surfactant invoked more prompt absorption and a slight increase in the AUC. The AUC ratio between the fed and fasted state was 21 times greater when an aqueous suspension was administered. However, this difference was almost unity for microemulsion formulations. These observations can be explained by elimination of the dissolution process of the drug and an increase in the interfacial area by addition of surfactant. This type of formulation is called a self-microemulsifying drug delivery system, and is one of the most powerful options to improve oral absorption behavior of poorly soluble drugs.

10.5 Liposomes

Liposomes are spherical vesicles of colloidal dimension, usually composed of phospholipids and cholesterol. Liposomes have characteristics favorable for drug carriers, including biocompatibility, biodegradability, ease of surface modifications, and ability to entrap both hydrophilic and hydrophobic drugs, although an effort, such as creation of a pH-gradient between the inner and outer phases, must be made for entrapping hydrophilic drugs because spontaneous entrapment efficiency of such drugs is typically low. Some liposomal formulations have reached the market already and others are in clinical trials.[82]

The use of saturated phospholipids and cholesterol in the formulation of liposome delivery systems cannot completely eliminate binding with serum components, which consequently decreases mononuclear phagocyte system (MPS) uptake of the vesicles. Several different strategies have been developed to overcome these difficulties by coating the surface of the liposomes with inert molecules to form a spatial barrier. A basic concept is that a hydrophilic polymer, possessing a flexible chain that occupies the space immediately adjacent to the liposome surface, tends to exclude other macromolecules from this space. Consequently, access and binding of blood plasma opsonins to the liposome surface are hindered, and interactions of MPS macrophages with the liposomes are inhibited. Among the different polymers investigated in an attempt to improve the blood circulation time of liposomes, polyethyleneglycol (PEG) has been widely used as a polymeric steric stabilizer. It can be incorporated on the liposomal surface using different methods, but the most widely used method is to anchor the polymer in the liposomal membrane *via* a cross-linked lipid.[83,84] The most obvious characteristic of PEG-grafted liposomes is their circulation longevity, regardless of surface charge or inclusion of stabilizing agent such as cholesterol. The ability of PEG to increase the circulation lifetime of the vehicle depends on both the amount of grafted PEG and the length or molecular weight of the polymer.[84] In most cases, the longer-chain PEGs produce the greatest improvements in blood residence time.

Drug targeting using liposomes as carriers holds much promise, especially in reducing toxicity and targeting delivery to disease sites. By grafting different glycosides on the surface of liposomes, it is possible to direct the liposomes to different cell types in the rat liver.[85] Galactosylated liposomes are mainly taken up by liver hepatocytes, whereas mannosylated liposomes are mainly taken up nanoparenchymal cells. Specific ligands can be grafted to the liposome with target cells by endocytosis, thus releasing the material to be delivered. Targeted delivery to cancer cells could be achieved by coating monoclonal antibodies (MAbs) raised against tumor-cell specific antigens. Both *in vitro* and *in vivo* studies provided evidence that antibody-coated polyethyleneglycol (PEG) liposomes containing doxorubicin were more effective and less toxic than the free drug, drugs incorporated into antibody-free liposomes.[86]

Stimuli-responsive liposomes are a promising and extensively studied class of liposomes with tunable drug release properties. The pH-sensitive liposomes were designed to promote efficient release of entrapped agents in response to low pH for intracellular gene delivery.[87] Kikuchi *et al.* developed a photocontrolled release system using liposomes and caged antimicrobial peptides.[88] The caged peptide was activated by UV irradiation, resulting in the formation of pores on the liposome surface to release the fluorophores. Temperature-responsive liposomes are also extensively studied liposomes with tunable lipid bilayers permeable at elevated temperatures, resulting in rapid release of the encapsulated drug upon heating.[89] Ta *et al.* developed a novel temperature-responsive liposome containing a pH/temperature-sensitive copolymer of *N*-isopropylacrylamide (NIPAAm) and propylacrylic acid (PAA).[90] These copolymers were membrane-disruptive in a pH/temperature-dependent manner (Figure 10.3). The copolymer-modified liposomes demonstrated an enhanced release profile and significantly lower thermal dose threshold when compared to traditional thermoresponsive formulations and were stable in serum with minimal drug leakage over time. These liposomes can be used for temperature-triggered drug release with thermal dose requirements well under the accepted thresholds for the onset of tissue necrosis, and these carriers can remain stable in

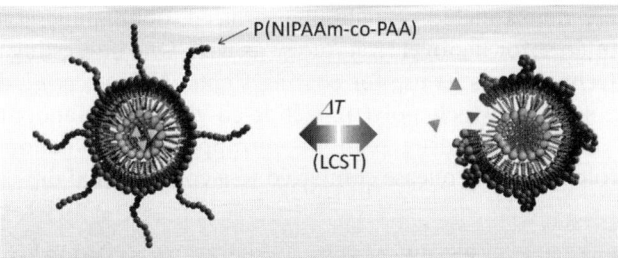

Figure 10.3 Illustration of drug release from temperature-responsive liposome decorated with a pH/temperature-sensitive copolymer of *N*-isopropylacrylamide (NIPAAm) and propylacrylic acid (PAA). Heating results in the collapse of polymer chains, which results in disruption of the bilayer and release of encapsulated drug.

circulation long enough to accumulate in the tumor, where their pH-sensitivity can then exploit the slightly acidic environment of the tumor interstitium.

10.6 Polymer Nanogels

Various polymer nanogels have been prepared using both natural and synthetic polymers.[91,92] Natural polymers included chitosan, alginate, gelatin, and albumin, while poly(lactic acid), poly(lactic acid)–poly(glycolic acid) copolymer, poly(methylmethacrylate), poly(methylcyanoacrylate), and poly(ε-caprolactone) were representative synthetic polymers for nanogel formation. The term "nanogel" is not clearly defined, and it sometimes includes polymer micelles that have a biphasic core–corona structure. However, the term should be applied to nanoparticles that have a random network structure, which is formed by either chemical or physical cross-linking. These particles are of interest in this section. The drug molecules are usually entrapped in the interior; however, they may be physically adsorbed or chemically linked on the surface.

Preparation methods of nanogels are well summarized in the review of Vauthier et al.[92] Briefly, polymer components are first emulsified, followed by condensation using various procedures, including precipitation, gelation, and polymerization. If the polymer molecules are condensed spontaneously, an emulsification procedure is not necessary. For example, cholesterol-bearing pullulan forms nanoparticles simply by applying sonication due to self-association of the cholesteryl group.[93] Alkyl chain-bearing dextran and β-cyclodextrin polymer form nanoparticles spontaneously due to host–guest interaction between the alkyl chain and the cyclodextrin cavity.[94] Some preparation methods are available for pilot-scale productions for conducting clinical studies.[92]

Although most of the nanogel formulations are supposed to be administered *via* injection, appealing *in vivo* results have been reported for oral absorption studies. Makhlof et al. developed chitosan–hydroxypropyl methylcellulose phthalate nanoparticles, which are formed spontaneously *via* ionic cross-linking, for oral insulin delivery.[95] The nanoparticles enhanced the hypoglycemic effect of insulin by more than 9.8-fold compared to an insulin solution. Other delivery routes including pulmonary, nasal, ophthalmic, and transdermal, are of interest for nanogel technology as well. One interesting application of nanogel technology is its use for coating stents. Nakano et al. developed a drug-eluting stent in which poly(DL-lactide-co-glycolide) nanoparticles were adsorbed.[96] Hydrophilic drugs can be entrapped in the nanoparticles for achieving prolonged drug release compared to a conventional dip-coated stent.

10.7 Molecular Conjugates

Problems of solubility and/or distribution of drug in the body may be overcome by forming physical conjugates with other molecules. One of the most promising molecular carriers is albumin,[97] which has already been utilized for formulation of paclitaxel (Abraxane®). Since aqueous solubility of

Table 10.1 Plasma clearance time of proteins and their conjugates [7]

Protein / Conjugate	Molecular weight (kDa)	$t_{1/2}$	$t_{1/10}$	Animals
Neocarzinostatin	12	1.8 min	15 min	Mouse
conjugate with poly (styrene-co-maleic acid)	16	19 min	5 h	
Ribonuclease	13.7	5 min	30 min	Mouse
cross-linked dimer	27	18 min	5 h	
Soybean trypsin inhibitor	20	< 2 min	3 min	Rabbit
conjugate with dextran	127	20 min	> 80 min	
Superoxide dismutase	30	4 min	30 min	Rat
conjugate with poly (styrene-co-maleic acid)	40	> 300 min	> 10 h	
Bilirubin oxidase	50	< 10 min	1.8 min	Rat
conjugate with poly (ethylene oxide)	70	5 min	48 h	

paclitaxel is very low, it requires solubilizing agents for injection. Although Cremophor EL and ethanol have been used for solubilization, the formulation often exhibited an allergic reaction, including anaphylactic shock, after injection. The albumin-conjugated formulation is free from this problem.

Drug molecules may also be chemically bound to carriers for overcoming these problems. The pioneering work in this field was done by Maeda et al., involving conjugation of neocarzinostatin with poly(styrene-co-maleic acid) to improve its retention in circulation.[7] Examples of improvement in plasma retention time using the molecular conjugation approach are summarized in Table 10.1. The plasma half-life of neocarzinostatin increased by more than a magnitude by forming the conjugate, and a similar effect was observed for other protein molecules. Since the rapid clearance of small proteins is basically due to renal clearance, it can be overcome by increasing molecular size.

10.8 Dendrimers

Dendrimers are highly branched globular macromolecules with a size on the order of nanometers, and can physically/chemically entrap drug molecules for improving solubility[98,99] and permeability.[100] Details on dendrimer–drug interactions are described in the review of D'Emanuele et al.[101] Application of dendrimers for drug delivery has been activated by an increase in the number of water-soluble and biocompatible backbones including polyamidoamine (PAMAM) and peptide-based dendrimers. Their physical properties, such as size and entrapment efficiency, can easily be controlled, while the structure is retained after dilution, unlike micelles. Introduction of poly(-ethylene oxide) chains can extend release rates of the incorporated drugs, increase drug-loading capacity, enhance retention in the circulation, and reduce hemolytic toxicity.[102–104] Due to their unique properties as drug carriers, various molecular designs of the dendrimers are ongoing.

10.9 Carbon Nanomaterials

Representative carbon nanomaterials being investigated as drug carriers include fullerenes, carbon nanotubes, and carbon nanohorns. In general, carbon materials are regarded as too stable to function as drug carriers. Besides, severe toxicity is a possibility for carbon nanotubes due to its morphological similarity to asbestos.[105] Carbon nanohorns have a unique cone-shaped structure and form spherical aggregates of 80–100 nm spontaneously,[106,107] which enables utilization of the EPR effect. Some anticancer drugs and the anti-inflammatory drug, dexamethasone, were successfully incorporated into the structure to exhibit slow-release behavior.[106,107] Attempts have also made to utilize nanodiamonds as drug carriers[108] on which poorly soluble drugs were adsorbed for effective dispersion.

10.10 Inorganic Nanomaterials

Although most of the nanomaterials used for drug delivery systems are composed of organic materials, inorganic nanomaterials including metallic nanoparticles,[20–24] magnetic nanoparticles,[25–29] porous materials,[30–33] and quantum dots,[28,34–37] may be used as well. Some metallic nanoparticles are considered potential drug carriers and their ability to penetrate into cells has actively been investigated to prove that their size, shape, charge, and surface properties are important parameters for determining their uptake efficiency and fate in the cells. Jiang *et al.* investigated the effects of the size of gold and silver nanoparticles on internalization *via* the cell membrane receptor and found that 40 and 50 nm diameter particles exhibited the greatest effect.[109] Verma *et al.* discovered that a slight difference in the arrangement of surface molecules produced a large variation in the uptake behavior by cells.[110] The drug molecules may be loaded onto the nanoparticles by either covalent conjugation or non-covalent interactions. However, these strategies may prevent precise release of drug molecules in the body, where the types of stimuli available to release drugs are limited. Kim *et al.* designed a functional monolayer on gold nanoparticles that had a hydrophobic alkane–thiol interior and a hydrophilic shell composed of a tetra(ethylene glycol) unit.[111] Hydrophobic drugs can be entrapped in the hydrophobic pocket, and entrapment and release of the drug molecules can be expected to be easier than direct binding on nanoparticles.

Iron oxides are the most representative magnetic nanoparticles for biomedical use.[25,26] As with metallic nanoparticles, magnetic particles are usually subjected to surface coating and functionalization. The biggest advantage of magnetic particles as drug carriers is the ability to control them using a magnetic field from outside the body. However, their application as contrast agents for magnetic resonance imaging has been the main focus of application rather than drug delivery.

Porous materials have very large surface areas, and so they can be used as carriers that adsorb a large amount of a drug. The drug molecules may be stabilized or transformed into an amorphous state by adsorption. If the entrances of the pores are capped with stimuli-responsive materials, drug molecules may be released upon the stimuli. Silica carriers are the most representative materials for this purpose. For example, if ibuprofen molecules with a size of approximately 1 nm are adsorbed into the MCM-41 pore, with a pore size of 2.5 or 1.8 nm, the release of ibuprofen continues for three days.[112] Lai *et al.* applied a CdS nanoparticle cap on the pore entrance that could be removed by disulfide bond-reducing molecules to produce a stimuli-responsive porous carrier.[113]

Quantum dots are usually regarded as imaging agents, and do not possess great advantages as drug carriers. However, they can be used as a part of a smart drug delivery system by utilizing their sensing and imaging properties.

10.11 Inhalable Particles

Inhalation therapy is a promising method for systemic drug delivery as well as treatment of pulmonary diseases such as asthma. Particle size is one of the most important factors in determining the deposition site of the formulated drug in the lung.[13,14] There are several deposition models available. The most widely known one is the ICRP (International Commission on Radio Protection) model,[17] which was developed by combining experimental observations with model calculations that considered complicated lung morphology. According to this model, particles smaller than 100 nm most effectively reach the alveoli region.

Unfortunately, it is very difficult to prepare organic nanoparticles composed of drug molecules that are free from aggregation problems. Nevertheless, nanoparticles have other advantages including avoidance of mucocilary clearance and uptake by macrophages. Thus, some attempts have been made to disperse nanoparticulate drugs in the very hydrophilic microparticles that are easily dissolved upon contact with the mucus layer in the lung.[114]

10.12 Summary

This chapter has summarized the applications of nanotechnology for drug delivery systems. Some technologies, including nanosizing of drug crystals and

Table 10.2 Advantages and requirements of nanotechnologies for drug delivery systems.

Advantages
Protect drugs from degradation
Control pharmacokinetic properties and tissue distribution
Improve intracellular penetration
Enable active/passive targeting
Requirements
Components must be biocompatible and well-defined
Accumulation in the body must be avoided
Must be functionalized easily
Must be manufactured on an industry scale

use of (micro)emulsions, liposomes, and molecular conjugates, are already being used in practice, while others are still in preliminary stages. Table 10.2 shows the advantages and requirements of nanotechnologies in this field. Although the advantages of the nanotechnologies are frequently discussed among academic researchers, a precise understanding of the requirements is sometimes neglected. Successful application of a new technology is always accompanied by a thorough understanding of the practical requirements. All of the nanotechnologies introduced in this chapter can make revolutionary contributions to human health; therefore, further development in this field is of great importance.

References

1. C. A. Lipinski, F. Lombardo, B. W. Dominy and P. J. Feeney, *Adv. Drug Delivery Rev.*, 1997, **23**, 3.
2. M. M. Hann and T. I. Oprea, *Curr. Opin. Chem. Biol.*, 2004, **8**, 255.
3. M. Q. Zhang and B. Wilkinson, *Curr. Opin. Biotechnol.*, 2007, **18**, 478.
4. G. Vistoli, A. Pedretti and B. Testaet, *Drug Discovery Today*, 2008, **13**, 285.
5. D. Brown, *Drug Discovery Today*, 2007, **12**, 1007.
6. G. A. Hughes, *Nanomedicine: Nanotechnol. Biol. Med.*, 2005, **1**, 22.
7. H. Maeda, *Adv. Drug Delivery Rev.*, 2001, **46**, 169.
8. C. P. Reis, R. J. Neufeld, A. J. Ribeiro and F. Veiga, *Nanomed.: Nanotechnol. Biol. Med.*, 2006, **2**, 8.
9. C. Vauthier and K. Bouchemal, *Pharm. Res.*, 2009, **26**, 1025.
10. V. P. Torchilin, *Adv. Drug Delivery Rev.*, 2006, **58**, 1532.
11. F. Kesisoglou, S. Panmai and Y. Wu, *Adv. Drug Delivery Rev.*, 2007, **59**, 631.
12. B. van Eerdenbrugh and G. van den Mooter, *Int. J. Pharm.*, 2008, **364**, 64.
13. J. S. Patton and P. R. Byron, *Nat. Rev. Drug Discovery*, 2007, **6**, 67.
14. J. C. Sung, B. L. Pulliam and D. A. Edwards, *Trends. Biotechnol.*, 2007, **25**, 563.
15. W. Yang, J. I. Peters and R. O. Williams III, *Int. J. Pharm.*, 2008, **356**, 239.
16. Y. Xie and J. Castracane, *IEEE Eng. Med. Biol. Magn.*, 2009, January/February, 23.
17. H. Smith, ICRP Publication 66: Human Respiratory Tract Model for Radiological Protection, Pergamon, New York, 1994.
18. P. A. Jaques and C. S. Kim, *Inhalation Toxicol.*, 2000, **12**, 715.
19. C. C. Daigle, D. C. Chalupa, F. R. Gibb and P. E. Morrow, *Inhalation Toxicol.*, 2003, **15**, 539.
20. R. Bhattacharya and P. Mukherjee, *Adv. Drug Delivery Rev.*, 2008, **60**, 1289.

21. P. Ghosh, G. Han, M. De, C. K. Kim and V. M. Rotello, *Adv. Drug Delivery Rev.*, 2008, **60**, 1307.
22. P. Cherukuri, E. S. Glazer and S. A. Curley, *Adv. Drug Delivery Rev.*, 2010, **62**, 339.
23. C. R. Patra, R. Bhattacharya, D. Mukhopadhyay and P. Mukherjee, *Adv. Drug Delivery Rev.*, 2010, **62**, 346.
24. B. S. Sekhon and S. R. Kamboj, *Nanomed.: Nanotechnol. Biol. Med.*, 2010, **6**, 612.
25. J. R. McCarthy and R. Weissleder, *Adv. Drug Delivery Rev.*, 2008, **60**, 1241.
26. C. Sun, J. S. H. Lee and M. Zhang, *Adv. Drug Delivery Rev.*, 2008, **60**, 1252.
27. O. Veiseh, J. W. Gunn and M. Zhang, *Adv. Drug Delivery Rev.*, 2010, **62**, 284.
28. B. S. Sekhon and S. R. Kamboj, *Nanomed.: Nanotechnol. Biol. Med.*, 2010, **6**, 516.
29. M. P. Marszall, *Pharm. Res.*, 2011, **28**, 480.
30. S. J. Son, X. Bai and S. B. Lee, *Drug Discovery Today*, 2007, **12**, 650.
31. S. J. Son, X. Bai and S. B. Lee, *Drug Discovery Today*, 2007, **12**, 657.
32. E. J. Anglin, L. Cheng, W. R. Freeman and M. J. Sailor, *Adv. Drug Delivery Rev.*, 2008, **60**, 1266.
33. I. I. Slowing, J. L. Vivero-Escoto, C. W. Wu and V. S. Y. Lin, *Adv. Drug Delivery Rev.*, 2008, **60**, 1278.
34. I. L. Medintz, H. T. Uyeda, E. R. Goldman and H. Mattoussi, *Nat. Mater.*, 2005, **4**, 435.
35. X. Michalet, F. F. Pinaud, L. A. Bentolila, J. M. Tsay, S. Doose, J. J. Li, G. Sundaresan, A. M. Wu, S. S. Gambhir and S. Weiss, *Science*, 2005, **307**, 538.
36. J. M. Klostranec and W. C. W. Chan, *Adv. Mater.*, 2006, **18**, 1953.
37. A. M. Smith, H. Duan, A. M. Mohs and S. Nie, *Adv. Drug Delivery Rev.*, 2008, **60**, 1226.
38. K. Kawakami, *Adv. Drug Delivery Rev.*, 2012, **64**, 480.
39. M. P. Desai, V. Labhasetwar, G. L. Amidon and R. J. Levy, *Pharm. Res.*, 1996, **13**, 1838.
40. S. K. Lai, D. E. O'Hanlon, S. Harrold, S. T. Man, Y. Y. Wang, R. Cone and J. Hanes, *Proc. Natl. Acad. Sci. U. S. A.*, 2007, **104**, 1482.
41. E. Merisko-Liversidge, G. G. Liversidge and E. R. Cooper, *Eur. J. Pharm. Sci.*, 2003, **18**, 113.
42. N. Rasenack, H. Hartenhauer and B. W. Müller, *Int. J. Pharm.*, 2003, **254**, 137.
43. H. De Waard, W. L. J. Hinrichs and H. W. Frijlink, *J. Controlled Release*, 2008, **128**, 179.
44. F. Qian, J. Tao, S. Desikan, M. Hussain and R. L. Smith, *Pharm. Res.*, 2007, **24**, 1551.
45. Y. Tozuka, Y. Miyazaki and H. Takeuchi, *Int. J. Pharm.*, 2010, **386**, 243.

46. S. Chakraborty, I. C. Liao, A. Adler and K. W. Leong, *Adv. Drug Delivery Rev.*, 2009, **61**, 1043.
47. S. Zhang and K. Kawakami, *Int. J. Pharm.*, 2010, **397**, 211.
48. S. Zhang, K. Kawakami, M. Yamamoto, Y. Masaoka, M. Kataoka, S. Yamashita and S. Sakuma, *Mol. Pharm.*, 2011, **8**, 807.
49. W. Abdelwahed, G. Degobert, S. Stainmesse and H. Fessi, *Adv. Drug Delivery Rev.*, 2006, **58**, 1688.
50. J. Jinno, N. Kamada, M. Miyake, K. Yamada, T. Mukai, M. Odomi, H. Toguchi, G. G. Liversidge, K. Higaki and T. Kimura, *J. Control. Release*, 2006, **111**, 56.
51. Y. Wu, A. Loper, E. Landis, L. Hettrick, L. Novak, K. Lynn, C. Chen, K. Thompson, R. Higgins, U. Batra, S. Shelukar, G. Kwei and D. Storey, *Int. J. Pharm.*, 2004, **285**, 135.
52. R. G. Strickley, *Pharm. Res.*, 2004, **21**, 201.
53. K. Kawakami, K. Miyoshi and Y. Ida, *J. Pharm. Sci.*, 2004, **93**, 1471.
54. K. Kawakami, N. Oda, K. Miyoshi, T. Funaki and Y. Ida, *Eur. J. Pharm. Sci.*, 2006, **28**, 7.
55. Y. Matsumura, T. Hamaguchi, T. Ura, K. Muro, Y. Yamada, Y. Shimada, K. Shirao, T. Okusaka, H. Ueno, M. Ikeda and N. Watanabe, *Br. J. Cancer*, 2004, **91**, 1775.
56. T. Hamaguchi, K. Kato, H. Yasui, C. Morizane, M. Ikeda, H. Ueno, K. Muro, Y. Yamada, T. Okusaka, K. Shirao, Y. Shimada, H. Nakahama and Y. Matsumura, *Br. J. Cancer*, 2007, **97**, 170.
57. R. Plummer, R. H. Wilson, H. Calvert, A. V. Boddy, M. Griffin, J. Sludden, M. J. Tilby, M. Eatock, D. G. Pearson, C. J. Ottley, Y. Matsumura, K. Kataoka and T. Nishiya, *Br. J. Cancer*, 2011, **104**, 593.
58. G. S. Kwon, *Crit. Rev. Ther. Drug Carr. Syst.*, 2003, **20**, 357.
59. K. Kataoka, G. S. Kwon, M. Yokoyama, T. Okano and Y. Sakurai, *J. Controlled Release*, 1993, **24**, 119.
60. H. Otsuka, Y. Nagasaki and K. Kataoka, *Adv. Drug Delivery Rev.*, 2003, **55**, 403.
61. D. S. W. Benoit, S. Srinivasan, A. D. Shubin and P. S. Stayton, *Biomacromolecules*, 2011, **12**, 2708.
62. C. Chang, H. Wei, C.-Y. Quan, Y.-Y. Li, J. Liu, Z.-C. Wang, S.-X. Cheng, X.-Z. Zhang and R.-X. Zhuo, *J. Polym. Sci. A: Polym. Chem.*, 2008, **46**, 3048.
63. G. Fundueanu, M. Constantin and P. Ascenzi, *Biomaterials*, 2008, **29**, 2767.
64. J. Zhao, G. Zhang and S. Pispas, *J. Polym. Sci. A: Polym. Chem.*, 2009, **47**, 4099.
65. C. Hong, Y. You and C. Pan, *J. Polym. Sci. A: Polym. Chem.*, 2004, **42**, 4873.
66. S. Qin, Y. Geng, D. E. Discher and S. Yang, *Adv. Mater.*, 2006, **18**, 2905.
67. G. Li, S. Song, L. Guo and S. Ma, *J. Polym. Sci. A: Polym. Chem.*, 2008, **46**, 5028.

68. C. M. Schilli, M. Zhang, E. Rizzardo, S. H. Thang, B. Y. K. Chong, K. Edwards, G. Karlsson and A. H. E. Muller, *Macromolecules*, 2004, **37**, 7861.
69. Y. Kotsuchibashi, Y. Kuboshima, K. Yamamoto and T. Aoyagi, *J. Polym. Sci. A: Polym. Chem.*, 2008, **46**, 6142.
70. Y. Kotsuchibashi, K. Yamamoto and T. Aoyagi, *J. Colloid Interface Sci.*, 2009, **336**, 67.
71. Y. Kotsuchibashi, M. Ebara, K. Yamamoto and T. Aoyagi, *J. Polym. Sci. A: Polym. Chem.*, 2010, **48**, 4393.
72. Y. Kotsuchibashi, M. Ebara, K. Yamamoto and T. Aoyagi, *Polym. Chem.*, 2011, **2**, 1362.
73. T. Yamaguchi, *Adv. Drug Delivery Rev.*, 1996, **20**, 117.
74. S. A. Wissing, O. Kayser and R. H. Müller, *Adv. Drug Delivery Rev.*, 2004, **56**, 1257.
75. P. Blasi, S. Giovagnoli, A. Schoubben, M. Ricci and C. Rossi, *Adv. Drug Delivery Rev.*, 2007, **59**. 454.
76. A. J. Almeida and E. Souto, *Adv. Drug Delivery Rev.*, 2007, **59**. 478.
77. H. L. Wong, R. Bendayan, A. M. Rauth, Y. Li and X. Y. Wu, *Adv. Drug Delivery Rev.*, 2007, **59**. 491.
78. J. M. Kovarik, E. A. Mueller, J. B. van Bree, W. Tetzloff and K. Kutz, *J. Pharm. Sci.*,1994, **83**, 444.
79. M. J. Lawrence and G. D. Rees, *Adv. Drug Delivery Rev.*, 2000, **45**, 89.
80. K. Kawakami, T. Yoshikawa, Y. Moroto, E. Kanaoka, K. Takahashi, Y. Nishihara and K. Masuda, *J. Control. Release*, 2002, **81**, 65.
81. K. Kawakami, T. Yoshikawa, T. Hayashi, Y. Nishihara and K. Masuda, *J. Control. Release*, 2002, **81**, 75.
82. M. L. Immordino, F. Dosio and L. Cattel, *Intl. J. Nanomed.*, 2006, **1**, 297.
83. C. Allen, S. N. Dos, R. Gallagher, G. N. C. Chiu, Y. Shu, W. M. Li, S. A. Johnstone, A. S. Janoff, L. D. Mayer, M. S. Webb and M. B. Bally, *Biosci. Rep.*, 2002, **22**, 225.
84. T. M. Allen, C. Hansen, F. Martin, C. Redemann and A. Yau-Young, *Biochim. Biophys. Acta*, 1991, **1066**, 29.
85. P. Ghosh, B. K. Bachhawat and A. Surolia, *Arch. Biochem. Biophys.*, 1981, **206**, 454.
86. I. Ahmad, M. Longenecker, J. Samuel and T. M. Allen, *Cancer Res.*, 1993, **53**, 1484.
87. G. Shi, W. Guo, S. M. Stephenson and R. J. Lee, *J. Control. Rel.*, 2002, **80**, 309.
88. S. Mizukami, M. Hosoda, T. Satake, S. Okada, Y. Hori, T. Furuta and K. Kikuchi, *J. Am. Chem. Soc.*, 2010, **132**, 9524.
89. D. Needham, G. Anyarambhatla, G. Kong and M. W. Dewhirst, *Cancer Res.*, 2000, **60**, 1197.
90. T. Ta, A. J. Convertine, C. R. Reyes, P. S. Stayton and T. M. Porter, *Biomacromolecules*, 2010, **11**, 1915.
91. C. P. Reis, R. J. Neufeld, A. J. Ribeiro and F. Veiga, *Nanomed. Nanotechnol. Biol. Med.*, 2006, **2**, 8.

92. C. Vauthier and K. Bouchemal, *Pharm. Res.*, 2009, **26**, 1025.
93. K. Akiyoshi, S. Deguchi, N. Moriguchi, S. Yamaguchi and J. Sunamoto, *Macromolecules*, 1993, **26**, 3062.
94. R. Gref, C. Amiel, K. Molinard, S. Daoud-Mahammed, B. Sebille, B. Gillet, J. C. Beloeil, C. Ringard, V. Rosilio, J. Poupaert and P. Couvreur, *J. Control. Release*, 2006, **111**, 316.
95. A. Makhlof, Y. Tozuka and H. Takeuchi, *Eur. J. Pharm. Sci.*, 2011, **42**, 445.
96. K. Nakano, K. Egashira, S. Masuda, K. Funakoshi, G. Zhao, S. Kimura, T. Matoba, K. Sueishi, Y. Endo, Y. Kawashima, K. Hara, H. Tsujimoto, R. Tominaga and K. Sunagawa, *J. Am. Coll. Cardiol. Intv.*, 2009, **2**, 277.
97. F. Kratz, *J. Control. Release*, 2008, **132**, 171.
98. B. Devarakonda, D. P. Otto, A. Jude feind, R. A. Hill and M. M. de Villiers, *Int. J. Pharm.*, 2007, **345**, 142.
99. M. Ma, Y. Cheng, Z. Xu, P. Xu, H. Qu, Y. Fang, T. Xu and L. Wen, *Eur. J. Med. Chem.*, 2007, **42**, 93.
100. M. Najlah, S. Freeman, D. Attwood and A. D'Emanuele, *Int. J. Pharm.*, 2007, **336**, 183.
101. A. D'Emanuele and D. Attwood, *Adv. Drug Delivery Rev.*, 2005, **57**, 2147.
102. D. Bhadra, S. Bhadra, S. Jain and N. K. Jain, *Int. J. Pharm.*, 2003, **257**, 111.
103. E. R. Gillies and J. M. J. Frechet, *Drug Discovery Today*, 2005, **10**, 35.
104. C. Kojima, C. Regino, Y. Umeda, H. Kobayashi and K. Kono, *Int. J. Pharm.*, 2010, **383**, 293.
105. C. A. Poland, R. Duffin, I. Kinloch, A. Maynard, W. A. H. Wallace, A. Seaton, V. Stone, S. Brown, W. MacNee and K. Donaldson, *Nat. Nanotechnol.*, 2008, **3**, 423.
106. T. Murakami, K. Ajima, J. Miyawaki, M. Yudasaka, S. Iijima and K. Shiba, *Mol. Pharm.*, 2004, **1**, 399.
107. K. Ajima, M. Yudasaka, Murakami, A. Maigne, K. Shiba and S. Iijima, *Mol. Pharm.*, 2005, **2**, 475.
108. M. Chen, E. D. Pierstorff, R. Lam, S. Y. Li, H. Huang, E. Osawa and D. Ho, *ACS Nano*, 2009, **3**, 2016.
109. W. Jiang, B. Y. S. Kim, J. T. Rutka and W. C. W. Chan, *Nat. Nanotechnol.*, 2008, **3**, 145.
110. A. Verma, O. Uzun, Y. Hu, H. S. Han, N. Watson, S. Chen, D. J. Irvine and F. Stellacci, *Nat. Mater.*, 2008, **7**, 588.
111. C. K. Kim, P. Ghosh, C. Pagliuca, Z. J. Zhu, S. Menichetti and V. M. Rotello, *J. Am. Chem. Soc.*, 209, **131**, 1360.
112. M. Vallet-Regi, A. Ramila, R. P. Del Real and J. Perez-Pariente, *Chem. Mater.*, 2001, **13**, 308.
113. C. Y. Lai, B. G. Trewyn, D. M. Jeftinija, K. Jeftnija, S. Xu, S. Jeftnija and V. S. Y. Lin, *J. Am. Chem. Soc.*, 2003, **125**, 4451.
114. K. Ohashi, T. Kabasawa, T. Ozeki and H. Okada, *J. Control. Release*, 2009, **135**, 19.

CHAPTER 11
Separation of Medically Useful Radioisotopes: Role of Nano-sorbents

RUBEL CHAKRAVARTY[a], RAKESH SHUKLA[b], AVESH KUMAR TYAGI*[b] AND ASHUTOSH DASH*[a]

[a] Radiopharmaceuticals Division, Bhabha Atomic Research Centre, Trombay, Mumbai – 400 085, India; [b] Chemistry Division, Bhabha Atomic Research Centre, Trombay, Mumbai – 400 085, India
*E-mail: adash@barc.gov.in; aktyagi@barc.gov.in

11.1 Radionuclides for Use in Nuclear Medicine

The discovery of radioactivity by Henri Becquerel in 1896 and the subsequent discoveries of radioactive elements by Marie Curie and Pierre Curie had a profound effect on understanding the biological effects of radiation. One of the consequences was the dawn of medical applications of radionuclides and radiation. Nuclear medicine is the medical specialty that involves the use of radioisotopes in the diagnosis and treatment of diseases.[1] It employs the nuclear properties of radioisotopes to evaluate metabolic, physiologic and pathologic conditions of the human body. Production of radioisotopes is an important aspect in the field of nuclear medicine and there has always been considerable interest in the standardization of easy and economically viable processes for obtaining promising radionuclides. The invention of the cyclotron in 1930s and the building of nuclear reactors in 1940s paved the way for the commercial production of wide variety of medically useful radioisotopes like ^{3}H, ^{14}C, ^{35}S, ^{32}P, ^{125}I, ^{131}I and ^{60}Co.[2] However, the widespread clinical applications of nuclear medicine started only after the development of the ^{99}Mo/^{99m}Tc generator by Walter Tucker and Margaret Greene at Brookhaven National Laboratory in 1950s.[3] Soon, ^{99m}Tc occupied a dominant position among the diagnostic tools in modern nuclear medicine and a variety of ^{99m}Tc-based radiopharmaceuticals became available commercially. This radioisotope is referred to as the 'work horse of nuclear medicine'[3] and is currently being used in ~80% of the diagnostic nuclear medicine procedures world-wide.[4] Subsequently, various other medically useful radionuclide generators were also developed.[5–7] Today, nuclear medicine is one of the fastest growing medical branches and it offers procedures that are immensely helpful to a broad span of medical specialties ranging from oncology and cardiology to psychiatry, bringing solace to millions of patients all over the world annually.

11.2 Classifications of Radionuclides Based on Their Application: Diagnostic and Therapeutic

The radionuclides in nuclear medicine can be broadly classified into two types, namely diagnostic radionuclides and therapeutic radionuclides, depending on their specific applications.[8]

11.2.1 Diagnostic Radionuclides

Diagnostic radionuclides are utilized for the preparation of radiolabeled molecules that are used for imaging organs in the human body in order to identify any abnormalities at the early stage of the disease, as well as to determine the functioning of organs. The essential criteria for choosing a radionuclide for diagnostic applications are: (a) suitable physical properties of the radionuclide, *i.e.* high detection efficiency for the radionuclide with minimum possible radiation dose to the patient (b) suitable chemical and biochemical properties, especially organ selectivity and compatibility with the biokinetics.

Based on the imaging modalities, the diagnostic radionuclides may be further categorized into two types; namely, the radionuclides used for single photon emission computed tomography (SPECT) imaging and the ones used for positron emission tomography (PET) imaging.[8]

(i) **Radionuclides used for SPECT imaging:** SPECT is a nuclear medicine tomographic imaging technique using γ-rays amd providing 3 dimensional images of the organs. A list of commonly used radionuclides used for SPECT imaging is given in Table 11.1 along with their physical decay characteristics.

(ii) **Radionuclides used for PET imaging:** PET imaging is carried out using radionuclides which decay by emission of positrons (β^+). It is based on the detection of two 511 keV γ-photons emitted in opposite directions after annihilation of a positron by an electron of the medium. The two photons are

Table 11.1 Physical characteristics of radionuclides used for SPECT imaging.

Radionuclide	Half-life ($t_{1/2}$)	Mode of decay[a]	Principal γ-component E/keV (% abundance)
^{99m}Tc	6.02 h	IT	140.47 (88.97)
^{123}I	13.27 h	EC	158.97 (82.80)
^{131}I	8.02 d	β^- & γ	364.48 (81.20)
^{67}Ga	3.26 d	EC	93.31 (38.30)
^{111}In	2.805 d	EC	245.39 (94.17)
^{201}Tl	72.91 h	EC	167.43 (10.00)

[a]EC: electron capture, IT: isomeric transition

Table 11.2 Physical characteristics of radionuclides used for PET imaging.

Radionuclide	Half-life ($t_{1/2}$)	Mode of decay[a]	Principal γ-component E/keV (% abundance)
^{18}F	109.77 min	β^+	511 (200.00)
^{68}Ga	68.3 min	β^+, EC	511 (176.00)
^{11}C	20.39 min	β^+	511 (199.52)
^{13}N	9.97 min	β^+	511 (199.64)
^{15}O	122.24 s	β^+	511 (199.77)
^{82}Rb	1.2 min	β^+, EC	511 (192)

[a]EC: electron capture

detected by two detectors in coincidence. PET has the ability to record images with superior sensitivity and high spatial resolution, compared to the SPECT technique. PET can be very helpful in detecting very small lesions, which may be missed in SPECT imaging. The physical characteristics of some commonly used PET radionuclides are shown in Table 11.2.

11.2.2 Therapeutic Radionuclides

Therapeutic radionuclides are utilized for the preparation of radiolabeled molecules, which are designed to deliver therapeutic dose of ionizing radiation to the disease site, with high specificity in the body. Just like diagnostic radionuclides, therapeutic radionuclides are also chosen on the basis of their suitable physical characteristics. However, the radionuclides used for therapy should be particulate emitters, *i.e.* they should emit α, β or Auger electrons. Apart from the particulate radiation, emission of low energy γ-rays will be advantageous to follow the pharmacokinetics and localization properties, as well as to determine the dosimetry in patients. A list of commonly used therapeutic radionuclides along with their physical characteristics is shown in Table 11.3.

Both therapeutic as well as diagnostic radionuclide should form a strong and irreversible bonding with the carrier molecule resulting in the formation of a highly stable complex.

Table 11.3 Physical characteristics of some therapeutic radionuclides.

Radionuclide	Half-life ($t_{1/2}$)	Mode of decay	Max. β^- energy ($E_{\beta max}$)/keV	Principal γ-component E/keV (% abundance)
^{32}P	14.26 d	β^-	1710.6	Nil
^{89}Sr	50.53 d	β^-	1496.6	Nil
^{90}Y	64.10 h	β^-	2282.0	Nil
^{131}I	8.020 d	β^-, γ	970.8	364.48 (81.20)
^{153}Sm	46.27 h	β^-, γ	808.4	103.18 (28.3)
^{166}Ho	26.83 h	β^-, γ	1854.5	80.57 (6.20)
^{177}Lu	6.734 d	β^-, γ	498.2	208.36 (11.00)
^{188}Re	16.98 h	β^-, γ	2120.4	155.04 (14.90)

11.3 Production of Radioisotopes for Nuclear Medicine Applications

The radioisotopes used for nuclear medicine are produced either in cyclotron or nuclear reactors or obtained directly at the hospital radiopharmacies from the radionuclide generators.[2]

11.3.1 Cyclotron-produced Radioisotopes

A wide variety of radionuclides can be produced in a cyclotron by varying the type of target and its position in a cyclotron.[9,10] Cyclotron produced radionuclides such as ^{67}Ga, ^{111}In, ^{123}I and ^{201}Tl which decay by electron capture, followed by emission of γ-photons are used for SPECT imaging while β^+ emitting radionuclides such as, ^{18}F, ^{11}C, ^{13}N, and ^{15}O are useful in PET studies.

11.3.2 Reactor-produced Radionuclides

A large number of radionuclides used in nuclear medicine are produced in the nuclear reactors.[11] The important reactor produced radioisotopes used for medical applications include, ^{32}P, ^{60}Co, ^{99}Mo, ^{153}Sm, ^{177}Lu, ^{131}I, ^{89}Sr, ^{90}Sr, ^{105}Rh, ^{188}W and ^{125}I.

11.3.3 Generator-produced Radionuclides

Radionuclide generators serve as an effective means of production of short-lived radioisotopes at hospital radiopharmacies without the need for on-site nuclear reactors or cyclotrons, and is described in detail in the subsequent sections of this chapter. The important radionuclide generator systems for biomedical applications include ^{99}Mo/^{99m}Tc, ^{68}Ge/^{68}Ga, ^{90}Sr/^{90}Y, ^{188}W/^{188}Re, ^{105}Ru/^{105m}Rh and ^{113}Sn/^{113m}In generators.[5–7,12]

11.4 The Concept of the Radionuclide Generator and the Historical Perspective of its Development

The development of radionuclide generators over the past five decades was primarily motivated by the increasing scope of applications of short-lived radionuclides and their compounds in nuclear medicine.[13] A radionuclide generator is developed on the principle of decay–growth relationship between a long-lived parent and its short-lived daughter, existing in a state of radioactive equilibrium.[7,13] It involves an effective radiochemical separation of the decaying parent and daughter radionuclides, by taking advantage of their difference in chemical properties, such that the daughter activity is obtained with high radionuclidic, radiochemical and chemical purities. Owing to the fact that the daughter and parent are different elements, the daughter activity is obtainable in a no-carrier-added (NCA) form with very high specific activity. The availability of the short-lived radioisotopes from radionuclide generators serves as an inexpensive and convenient alternative to in-house radioisotope production facilities, such as reactors or cyclotrons.[5–7] Radionuclide generators can be conveniently transported to institutions far from the site of reactors or cyclotrons. Radionuclide generators are historically called 'cows' since the daughter activity is 'milked' (*i.e.* separated) from its parent (precursor) and the parent then generates a fresh supply of the daughter.[7] Theoretically, the elution of the daughter activity from the generator can be made repeatedly. Figure 11.1 illustrates the multiple growth and elution of the daughter activity from a radionuclide generator.

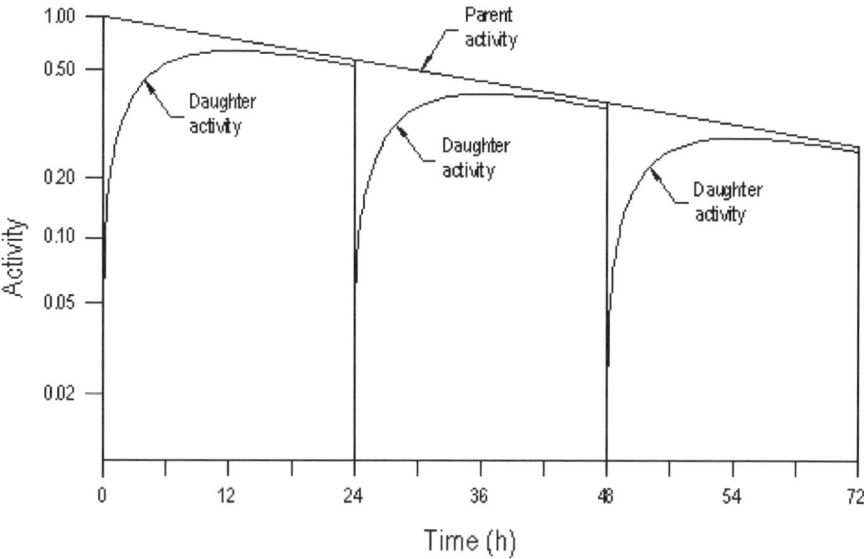

Figure 11.1 Multiple growth and elution of daughter in a radionuclide generator.

The first radionuclide generator for biomedical applications was developed by Fallia in 1920,[7,14] by the use of the ^{226}Rd ($t_{1/2}$ = 1620 y) → ^{222}Rn ($t_{1/2}$ = 3.8 d) parent–daughter pair to obtain ^{222}Rn seeds for radiation therapy. However, the real practical importance of radionuclide generators was realized only in 1951 with the development of the ^{132}Te ($t_{1/2}$ = 78 h)/^{132}I ($t_{1/2}$ = 2.29 h) generator at the Brookhaven National Laboratory (BNL) in USA.[15] A simplified version of this generator was developed in BNL by the middle of 1950s.[15] The similarities between the ^{132}Te/^{132}I and ^{99}Mo/^{99m}Tc parent–daughter pairs led to the pioneering development of ^{99}Mo/^{99m}Tc generator system in 1957 at BNL by Walter Tucker and Margaret Greene.[3] Since then ^{99m}Tc has revolutionized radiopharmaceutical chemistry and nuclear medicine, and numerous clinical uses of ^{99m}Tc based radiopharmaceuticals have been reported.[5,16] Even to date, >80% of diagnostic nuclear medicine procedures are based on ^{99m}Tc based radiopharmaceuticals.[4] Subsequently, several other radionuclide generator systems were developed, some of which have significant practical applications.[5–7]

11.5 The Mathematical Equations of Radioactive Decay and Growth in Radionuclide Generators

In a radionuclide generator, a radioactive nuclide (A) decays into a nuclide (B) that is also radioactive and decays further into the stable nuclide (C). This decay scheme for two successive radionuclides can be represented as in Figure 11.2. In this decay scheme, λ_1 is the decay constant of parent atoms A, having N_1^0 initial number of atoms and λ_2 the decay constant of daughter atoms B. For the sake of simplicity, it is generally assumed that $N_2(t=0) = N_2^0 = 0$, $N_3(t=0) = N_3^0 = 0$ and the grand-daughter product, C, is stable ($\lambda_3 = 0$). N_2 and N_3 are the number of atoms of B and C, respectively, at time t. Then at any given time, t, one can write the following differential equations:[7]

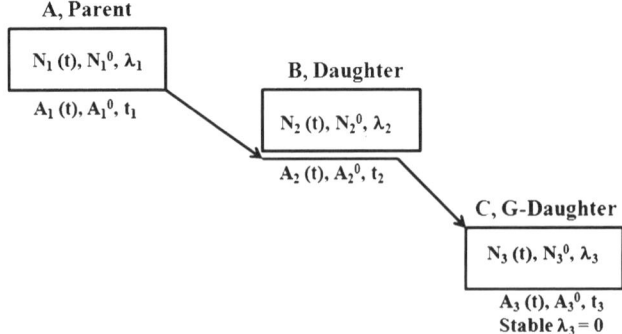

Figure 11.2 The decay scheme of a parent–daughter pair in a radionuclide generator.

$$dN_1(t) = -\lambda_1 N_1(t)dt \quad \text{or} \quad \frac{dN_1(t)}{dt} = A_1(t) = -\lambda_1 N_1(t) \tag{11.1}$$

$$dN_2(t) = +\lambda_1 N_1(t)dt - \lambda_2 N_2(t)dt \quad \text{or} \quad \frac{dN_2(t)}{dt} = -\lambda_1 N_1(t) - \lambda_2 N_2(t) \tag{11.2}$$

$$\text{and} \quad dN_3(t) = +\lambda_2 N_2(t)dt \quad \text{or} \quad \frac{dN_3(t)}{dt} = +\lambda_2 N_2(t) \tag{11.3}$$

The solution of eqn (11.1) leads to the well-known decay equation of a single radioactive nuclide *i.e.*:

$$N_1(t) = N_1^0 e^{-\lambda_1 t} \quad \text{and} \quad A_1(t) = \lambda_1 N_1^0 e^{-\lambda_1 t} = A_1^0 e^{-\lambda_1 t} \tag{11.4}$$

The solution of eqn (11.2) leads to the following expression (assuming, $N_2(0) = N_2^0 = 0$):

$$N_2(t) = N_1^0 \frac{\lambda_1}{\lambda_2 - \lambda_1} \left(e^{-\lambda_1 t} - e^{-\lambda_2 t} \right) \tag{11.5}$$

Here, the activity $A_2(t)$ is given by the general definition of the activity.

$$A_2(t) = \lambda_2 N_2(t) = N_1^0 \frac{\lambda_1 \lambda_2}{\lambda_2 - \lambda_1} \left(e^{-\lambda_1 t} - e^{-\lambda_2 t} \right) \tag{11.6}$$

$$\text{or} \quad A_2(t) = A_1^0 \frac{\lambda_2}{\lambda_2 - \lambda_1} \left(e^{-\lambda_1 t} - e^{-\lambda_2 t} \right) \quad \text{where} \quad A_1^0 = \lambda_1 N_1^0 \tag{11.7}$$

The daughter activity reaches the state of equilibrium or maximum activity, when the feeding of *B* atoms is exactly compensating those which are decaying:

$$\text{when} \quad A_1(t) = A_2(t) \quad \text{or} \quad \lambda_1 N_1(t) = \lambda_2 N_2(t)$$
$$\text{or when} \quad \frac{dA_2(t)}{dt} = \frac{dN_2(t)}{dt} = 0 \tag{11.8}$$

Then one can write:

$$t_{\max} = \frac{\ln(\lambda_2/\lambda_1)}{\lambda_2 - \lambda_1} = \left(\frac{1.44 t_1 t_2}{t_1 - t_2} \right) \ln\left(\frac{t_1}{t_2} \right) \tag{11.9}$$

where, t_1 and t_2 are the half-lives of the parent and daughter radionuclides respectively.

11.6 The Available Options for the Preparation of Radionuclide Generators

The separation of the parent–daughter pairs belonging to the adjacent group of elements is one of the most challenging aspects of radionuclide generator research.[6] Sometimes separation is complicated by the number of oxidation states of the pair and the tendency to form many complexes.[6] Moreover, the requirement of the daughter radioactivity in a form suitable for clinical applications further places stringent requirements on the separation chemistry. This requires selection of the suitable separation procedure, providing high yield of the daughter radionuclide in minimum volumes (high radioactive concentration) and highest purity. Preferably, the daughter activity should be obtained in an uncomplexed form so that it can directly be used for the preparation of radiopharmaceuticals.[6] Additionally, the radionuclide generators for biomedical applications must be simple in design, efficient and easy to operate. Various methods, like column chromatography, solvent extraction, sublimation and gel-type systems are reported for the preparation of radionuclide generators.[17] Out of these, column chromatography and solvent extraction are the most commonly used methods for the preparation of radionuclide generators for biomedical applications. The choice of a particular technique is usually made for technical, economic and logistic reasons, with emphasis on one or other of these factors depending on the circumstances. The relative advantages and disadvantages of using conventional column chromatographic and solvent extraction techniques for the preparation of radionuclide generators are discussed as below

11.6.1 Column Chromatography

This is the most commonly used method for the preparation of radionuclide generators, primarily due to its simplicity of operation. This method relies upon the chromatographic separation of the daughter radioisotope from the parent, by the virtue of the difference in their retention affinity in the column. Generally, the column matrix has very high affinity for retention of the parent radionuclide while the affinity for the daughter radionuclide is substantially low.[13] The long-lived parent is adsorbed on a solid support and the daughter activity is selectively and periodically eluted.[13] Both inorganic and organic sorbents have been evaluated as the chromatographic supports for radionuclide generators.[18] The radiation damage to the organic matrix deteriorates the performance of the generator with respect to the elution yield and parent breakthrough, on prolonged use.[18] Moreover, the radiolytic fragments may come along with the daughter activity in the eluate as undesirable chemical impurities. The limited resistance of these organic resins to radiation also limits the activity of the parent that can be loaded onto the generator column. The inorganic sorbents, on the other hand, possess appreciably high radiation resistance and good ion-exchange properties which make them suitable for the preparation of radionuclide generators.[18] Typically, hydrated metal oxides (such as Al_2O_3, ZrO_2, SnO_2, TiO_2 etc.) are used as inorganic sorbents for the preparation of radionuclide generators.[18] However, a cause of concern with such sorbents is the chemical stability of metal oxides in aqueous solutions. In certain cases, the solubility of metal oxide increases with successive elutions over a prolonged period of time.[18,19] This in turn adds chemical impurities in the daughter product, rendering it unsuitable for the preparation of radiopharmaceuticals.

Another major limitation of the chromatographic approach is the restricted sorption capacity of the column matrices.[18] The required volume of the eluent for the quantitative elution of the daughter radioisotope from a chromatographic generator depends on the size of the column, which in turn is inversely proportional to the specific activity of the parent radioisotope. Consequently, parent radioisotopes with high specific activities are required in order to reduce the size of the column. If the parent activity is obtained with low specific activity, the daughter activity can only be availed with low-radioactive concentration from such systems, thereby requiring post-elution concentration of the daughter activity before utilizing it for the preparation of radiopharmaceuticals.[5] The requirement for the multiple-step post-elution concentration procedure results in a fairly complex system, high dose rates, introduction of chemical impurities in the daughter eluate and overall low reliability. Sometimes column chromatographic generators demonstrate degrading performance on repeated elution, over a prolonged period of time, such that the yield of the daughter activity decreases and the breakthrough of the parent increases with time.[19] This in turn decreases the useful life-time of the generator, thereby rendering it cost-ineffective.

11.6.2 Solvent Extraction

Solvent extraction is another approach which has been widely exploited for the preparation of radionuclide generators.[5,6,17] This method is based on the relative solubility of the parent and the daughter radioelements in two immiscible liquids usually water and an organic solvent. The solvent extraction generators have several economic and technical advantages. Unlike in column chromatographic approach, here, low specific activity parents can be used for the preparation of radionuclide generators and the daughter activity is obtained with high radioactive concentration with low levels of radionuclidic impurities. However, solvent extraction involves cumbersome multi-step separation procedure involving complicated apparatus. Therefore, it requires highly trained personnel for successful operation. The handling of large volumes of organic extractants may pose a possible fire hazard with the volatile organic vapors. Moreover, the organic solvent may undergo radiolysis and introduce organic residues in the daughter product, which may interfere in the subsequent radiolabeling reactions and thus bring undesirable changes in the biological properties of the radiochemical. The detrimental effect of radiolysis may also lower the extraction efficiency of the organic solvent.

The other less commonly adopted procedures like sublimation[20] and 'gel type systems'[21–24] also have their inherent limitations for clinical use. As mentioned earlier, due to the simplicity and ease of operation of column chromatographic radionuclide generators, they can be easily adapted in hospital radiopharmacies. In this chapter, we have focused on the development of radionuclide generators by this approach.

11.7 Essential Components of a Column Chromatographic Radionuclide Generator

A schematic diagram of a typical column chromatographic radionuclide generator system is shown in Figure 11.3. It consists of a borosilicate glass column fitted with a sintered disc at the bottom. The column is packed with sorbent material, which adsorbs the parent radionuclide. This column is kept in a lead shielded container. The sealed vial containing the suitable eluent is connected with the top of the generator column. From the bottom of the column, it is connected with another vial for the collection of the eluate (daughter). The eluate collection vial is either connected with a peristaltic

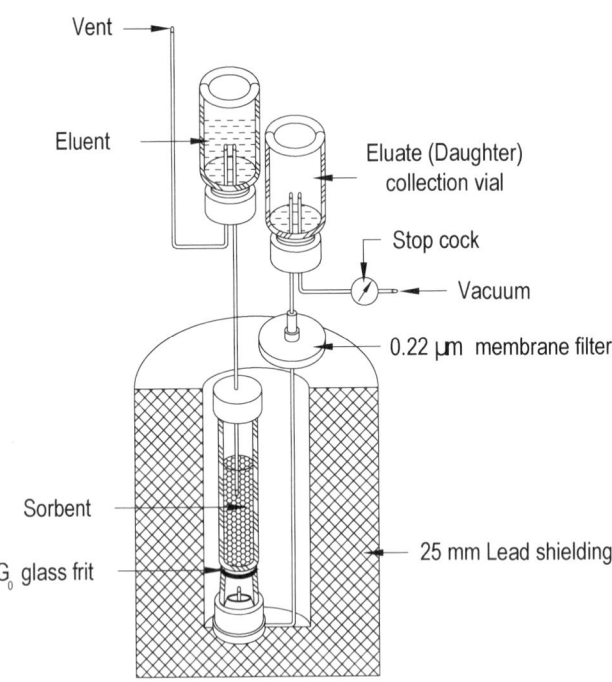

Figure 11.3 Schematic diagram of a typical column chromatographic radionuclide generator.

pump for creation of vacuum or a pre-evacuated vial is used. The eluent passes through the generator column and gets collected in the eluate vial. A disposable 0.22 μm membrane filter is attached to the generator column output by Teflon tubing, to avail the daughter eluate in a sterile form. All the operations are carried out in a 'closed cyclic system' using connecting tubes. The 'closed' system prevents external contaminants from entering the column. Generally, input/output connections are made with standard Teflon tubings of 1 mm inner diameter and connectors. The generator column, connectors and connection tubings are integrated within a small portable lead-shielded unit throughout experimental use for radioprotection purposes. Only the elution vial and output vial are accessible externally.

11.8 Sorbent: The 'Heart' of Column Chromatographic Radionuclide Generator Systems

The most important component of the column chromatographic radionuclide generator system is the sorbent matrix and can be called the 'heart' of the radionuclide generator. The long-lived parent is retained in the sorbent matrix and the short-lived daughter activity can be selectively eluted out at regular intervals with suitable eluting solutions. In all natural nuclear transformation or decay, except isomeric transition, the parent and daughter have different atomic numbers and hence different chemical properties.[6] The retention of a radionuclide in a particular sorbent is dependent on the distribution ratio (K_d) of that radionuclide. Owing to the difference in their K_d values, the parent and daughter radionuclides have different retention affinity towards the sorbent and can thus be effectively separated. Generally, the parent radioisotope has a very high K_d value and that for the daughter radioisotope is negligibly small. Consequently, the parent isotope is quantitatively retained in the sorbent and the daughter isotope can be selectively eluted out. In order to utilize a particular sorbent for the preparation of radionuclide generators it should meet some essential criteria which are enlisted below.

(a) Granularity and free flow characteristics

The sorbent material should be granular in nature and exhibit free flow characteristics. Unagglomerated fine powders are avoided, as column beds prepared with such materials are impervious to the flow of liquid. Additionally, they may block the pores of the sintered disc in the glass column, thereby rendering it unsuitable for use. The mechanical strength of the sorbent granules should be sufficient enough to resist attrition and erosion in chromatographic columns during its application. Preferably, the size of sorbent granules should be in the range 50–100 mesh (149–297 μm), for efficient column operation.

(b) Sorption capacity for the parent radioisotope

The sorbent must possess high sorption capacity for the parent radionuclide. The high sorption capacity of the sorbent reduces the bed volume of the column and hence requires lower volume of eluent for the elution of the daughter radionuclide. This in turn increases the radioactive concentration of the daughter radionuclide. Basic properties of the sorbent that influence this factor are specific surface area, chemical nature of the surface and pore-size, which in turn determine amount of metal ions that can be taken up by per unit mass of sorbent.

(c) Selectivity of sorption and elution

The sorbent must quantitatively adsorb the parent radionuclide and retain it under the conditions of elution. However, the retention of the daughter activity by the sorbent, in the eluting medium, should be negligible. The daughter activity should be selectively eluted, preferably in an uncomplexed form, so that it can be directly complexed with various ligands or biomolecules for the preparation of radiopharmaceuticals. The ease of chemical manipulation of the daughter radionuclide before the preparation of radiopharmaceuticals is an important consideration in making biomedical radionuclide generators commercially viable. The performance of the sorbent must remain consistent on successive elutions over a prolonged period of time.

(d) Sorption kinetics

A fast sorption rate is favored for most radiochemical separations. The accessibility of incoming adsorbates to the reactive surface sites is one of the important properties that decide sorption kinetics. The factors that are responsible for sorption kinetics are particle size, agglomeration behavior and the porosity of the sorbent.

(d) Radiation stability of the sorbent

The radiation stability of the sorbent plays a crucial role in the performance of the radionuclide generator. This in turn directly affects the shelf life of the generator. The sorbent material should be able to withstand continuous exposure to radiation emitted by the parent and the daughter radionuclides, throughout the shelf-life of the generator. The elution efficiency and the radionuclidic purity of the daughter radioisotope must remain consistent over this period of time.

(e) Chemical stability of the sorbent

The sorbent material should be practically insoluble both in the loading as well as the eluting medium. Moreover, the sorbent material should not undergo any chemical modification, thereby affecting the performance of the generator, on multiple elutions over the prolonged period of time.

These stringent requirements can only be met with advances in development of novel sorbent materials for generator fabrication.

11.9 Nanomaterials as New Generation Sorbents for the Preparation of Radionuclide Generators

In recent years, nanomaterials have been the core focus of nanoscience and nanotechnology—which is an ever-growing multidisciplinary field of study attracting tremendous interest and effort in research and development around the world. The present discussion focuses on applications of nanomaterials in radiochemical separations, particularly in the development of radionuclidic generator for medical applications.

Generally, nanomaterials cover objects on the "nano" scale (1–100 nm). The size confinement creates massive changes in physical and chemical properties of substances. The properties that change in the nanoregime include band gaps (for semiconductors), magnetic moments,[25] specific heats,[26] melting points,[27,28] surface chemistry,[29] reactivity and morphology/particle shape.[30–32] The size, surface structures and interparticle interactions of nanomaterials determine their unique properties and the improved performances for their potential applications in many areas.

Nanomaterials are expected to provide unprecedented opportunities in developing a new class of sorbents for column chromatographic applications due to their ability to sorb metal ions on their large active surfaces. The properties of materials with nanometre dimensions are significantly different from those of bulk materials.[33,34] The high surface area to mass ratios of nanomaterials can greatly enhance their sorption capabilities. Nanoparticles also have enhanced surface reactivity due to different distributions of reactive surface sites and disordered surface regions. Most nanomaterials are porous, with varying pore size and pore size distribution, and possess fluid permeability. Moreover, the small size of nanomaterials either renders them free of internal structural imperfections, or impurities present cannot multiply sufficiently to cause mechanical failure.[35,36] This is because the imperfections within the nano dimension are highly energetic and will migrate to the surface to relax themselves under annealing, thus purifying the material and leaving perfect material structures inside the nanomaterials. This phenomenon of increased material perfection favorably affects the properties of nanomaterials. For example, the chemical stability for certain nanomaterials may be enhanced and the mechanical properties of nanomaterials will be better than the corresponding bulk materials.[35] Also it is reported that nanocrystalline materials are sometimes expected to be more radiation tolerant than their bulk counterparts with larger grain sizes.[37]

The properties of synthesized nanomaterials depend on the synthesis protocols and processing conditions and most of them may or may not have all the desirable properties described in Section 11.8. The practical challenges in making useful sorbent materials will be to obtain high sorption capacity in an effective manner to satisfy most of these criteria. The selection of a synthetic route should be based on the creation of highly porous materials with adequate surface area for sorption, a high number of surface active sites and small diffusion resistance.

The utility of nanomaterials as a new generation of sorbents in the chromatographic separation of metal ions[38–44] has been exploited. However, the potential of using such materials as column matrices in radionuclide generators for biomedical applications was not explored before our group ventured into this. For such applications, an appropriate nanomaterial-based sorbent has to be chosen to affect the separation of daughter radionuclide from the parent radionuclide. Owing to their high surface area and intrinsic surface reactivity, nanomaterials-based sorbents are expected to have much higher sorption capacity and selectivity compared to the conventional sorbents. Consequently, relatively lower specific activity parent radioisotopes obtained from medium flux reactors can also be used for the preparation of radionuclide generators, and the daughter activity can be availed with appreciably high radioactive concentration and purity. The enhanced mechanical and chemical stability of such sorbents ensure that successive elutions of the generator over a prolonged period of time does not lead to bleeding of column matrix resulting in the addition of chemical impurities in the daughter eluate. Further, owing to the high radiation stability,[37] nanomaterials can withstand the radiation environment and demonstrate consistently good performance over a prolonged period of time.

11.10 Procedures Involved in Evaluation of a Sorbent Material to Determine its Suitability for the Preparation of Radionuclidic Generators

(a) Granularity, free flow characteristics of the sorbent

In order to evaluate the suitability of the sorbent for column chromatographic applications, it is packed in a generator column and deionized water is passed through the column bed at different flow rates, using a peristaltic pump. The flow characteristic of the effluent is monitored and if the column bed is impervious to the flow of liquid, it cannot be used for chromatographic applications.

(b) Chemical stability of the sorbent

The chemical stability of the sorbent is assessed in various mineral acids and bases, such as HCl, HNO_3, NaOH and NH_4OH (with concentration up to 5 N). For this, generally a fixed weight of the dry sorbent material is placed in fixed volume of solvent of interest and kept for 24 hours with continuous shaking at room temperature. Subsequently it is filtered and the level of metal ions in the solvent is determined by inductively coupled plasma atomic emission spectroscopy (ICP-AES). If the concentration of metal ions corresponding to the sorbent matrix is >10 ppm in the solution, it can be concluded that the sorbent has a tendency for dissolution and hence is unsuitable for use in the preparation of a radionuclide generator.

(c) Distribution ratio (K_d) values of parent and daughter radionuclides

The distribution ratio (K_d) value of a particular metal ion in the sorbent is defined as the ratio of the total amount of ions taken up per gram of the sorbent to the amount of ions remaining in the solution. Generally, the K_d values of the parent and the daughter elements (in their ionic form) are measured in solutions of different pH, using radiotracers of the parent and daughter radioisotopes. In a typical experiment, a known weight of sorbent (200 mg) is suspended in 20 mL solution containing the radiotracer ions in a

50 mL stoppered conical flask, shaken in a wrist arm mechanical shaker for 1 hour at 25 °C and then filtered. The activities of the solution before and after equilibration are measured in a well type NaI(Tl) counter using appropriate window ranges. The distribution ratios are calculated using the following expression:

$$K_d = \frac{(A_i - A_{eq})V}{A_{eq}\, m} \qquad (11.10)$$

where, A_i is the initial radioactivity of 1 mL of the solution, A_{eq} is the unadsorbed activity in 1 mL of the solution at equilibrium, V is the solution volume (mL) and m is the mass (g) of the sorbent.

In order to optimize the time required for maximum sorption of parent radionuclide or the attainment of equilibrium, the K_d values of the parent radionuclide (in its ionic form) is determined at different time intervals, in the solution of optimum pH. The K_d values are taken as an indication of the progress of the sorption process. The time when K_d remained unchanged is taken as the indication of the attainment of equilibrium. The conditions for loading of the parent radionuclide in the generator column and the elution of the daughter radionuclide from it are optimized on the basis of the K_d values. The medium in which the K_d value of the parent radionuclide is maximum is chosen for the loading or sorption of parent radionuclide in the column, while the medium in which the K_d value of the parent radionuclide is appreciably high and that of the daughter radionuclide is significantly low is chosen for the selective elution of the daughter radionuclide.

(d) Zeta potential of the sorbent material

The zeta potential (ζ) of a nanomaterial based sorbent is commonly used to describe the charge developed on the surface of the sorbent when it interacts with its surroundings. It is a very useful parameter in the interpretation of sorption behavior of the nanomaterial. The development of partial positive or negative charge at the surface of the nanoparticle affects the distribution of ions in the surrounding interfacial region, resulting in an increased concentration of counter ions (ions of opposite charge to that of the particle) close to the surface. The zeta-potential values are correlated with the K_d values in order to formulate a mechanism for the sorption of the parent radionuclide and the elution of the daughter radionuclide. For determination of the zeta-potential of the sorbent in a particular solution, ~5 mg of sorbent is suspended in 50 mL of that solution, and the suspension is sonicated. The zeta-potential of the suspensions at different pH is measured by an instrument known as 'Zetasizer', which uses a combination of laser Doppler velocimetry and phase analysis light scattering (PALS). Generally, in aqueous medium, the Smoluchowsky constant $F(K_a)$ of 1.5 is used to calculate the zeta potential values from the electrophoretic mobility.

(e) Determination of sorption capacity of the sorbent for the parent radionuclide

Prior to the fabrication of the generator, it is essential to determine the capacity of the sorbent to sorb the parent radionuclide, in order to optimize the bed size of the generator column. The sorption capacity of the sorbent material is determined under static and dynamic conditions. The static sorption capacity of the material is determined by a batch equilibration method. For this, a known amount of sorbent is taken in a stoppered conical flask and equilibrated with 50 mL solution of the parent radionuclide of a particular concentration (prepared by addition of non-radioactive carrier) for a definite period of time at room temperature. At the end, the contents are filtered and the activities of the parent radionuclide in the solution before and after sorption are estimated. The sorption capacity is calculated using the following expression:

$$\text{Capacity} = \frac{(A_o - A_e)V.C_o}{A_o\,m} \tag{11.11}$$

where A_o and A_e represents the radioactivity of parent radioisotope in 1 mL of supernatant solution before and after sorption, respectively, C_o is the total content of the parent isotope in 1 mL of solution before sorption, V is the volume of solution and m is the mass (g) of the sorbent.

In order to estimate the sorption capacity under dynamic conditions, the generator column is packed with known amount of sorbent. The column is conditioned at the optimum pH for maximal loading of the parent radionuclide. The parent radionuclide solution of a particular concentration (prepared by addition of non-radioactive carrier and spiked with the radionuclide) is allowed to pass through the column at a fixed flow rate. 1 mL of this solution was kept as reference (C_o). The effluent is collected in fractions of 1 mL aliquots (C). The activity in the reference (C_o) and effluent fractions are determined and the ratio of the count rate 'C' of each 1 mL effluent to the count rate 'C_o' of 1 mL of the original feed solution is taken as the parameter to follow the sorption pattern. Theoretically, 'breakthrough point' is said to be reached when $C/C_o > 0$, i.e. the parent radionuclide can be detected in the effluent. For the preparation of the generator, the optimum amount of sorbent must be chosen so that the breakthrough point is not reached in the sorption step and the parent radioactivity can be quantitatively retained.

(f) Fabrication of the radionuclide generator

In order to develop a radionuclide generator, a borosilicate glass column of a particular dimension (G_0) with a sintered disc at the bottom is packed with required amount of sorbent in a lead shield. The sorbent material is preconditioned with the solution of appropriate pH. The column is then loaded with the solution of the parent radionuclide by passing it through the column. Subsequently, the column is washed 100 mL of the eluting solution to remove any unadsorbed or loosely held parent radionuclide from the generator column.

After allowing adequate time for the build up of the daughter, the column is eluted with eluting solution, maintaining a fixed flow rate using a peristaltic pump. In order to examine the elution profile, the eluate is collected as 1 mL aliquots and activity of each fraction is determined. The generator is eluted regularly with the optimum volume of the eluting solution and its performance is evaluated for a period of ~2–3 half-lives of the parent radionuclide, which is normally the expected shelf-life of a radionuclide generator.

11.11 Quality Control of the Generator Produced Radioisotopes

Since generator-produced radioisotopes are intended for clinical applications, it is imperative that the radionuclide generators undergo strict quality control procedures before being handed over to the nuclear medicine physicians for patient applications. Quality control procedures involve specific tests and measurements that ensure the elution efficiency of the generator, product identity, purity and radioactive concentration of the daughter radionuclide, biological safety and the efficacy of the radionuclide for the preparation of radiopharmaceuticals. These methods are described below.

(a) Elution efficiency of the radionuclide generator: The elution efficiency of the radionuclide generator is defined as the proportion of the daughter radioisotope present in the generator system that is separated during the elution process and is usually expressed as a percentage. Theoretically, the activity of the daughter radioisotope present in the generator system at the time of elution is given by eqn (11.7). In practice, the activity of the daughter radioisotope eluted is less than that predicted by the theory. If the measured activity of the separated daughter radioisotope after allowing time 't' for its growth, is denoted by A_s, then the elution efficiency can be defined by the equation:

$$\%Elution\ efficiency = \frac{A_s}{A_2} \times 100 \qquad (11.12)$$

For the radionuclide generator to be cost-effective, it is essential that its elution efficiency should be fairly high (>80%) and it should remain constant during the stipulated period of utilization of the generator.

(b) Radionuclidic purity of the separated daughter radioisotope: In the separated daughter radioisotope obtained from the radionuclide generator, the primary radionuclidic impurity that may be expected is the long-lived parent radioisotope. Sometimes, the parent radioisotope may be associated with other radionuclidic impurities which may also be present in the separated daughter activity. The presence of radionuclidic impurities increases the undue radiation exposure to the patient and also obscures the scintigraphic images, and therefore their level needs to be determined.[13] The determination of radionuclidic impurities in the separated daughter product is mostly done by γ-ray spectrometry using a high purity germanium (HPGe) detector coupled with a multi-channel analyzer.[13] This technique detects the presence of radionuclidic impurities and also determines their level in the daughter activities obtained from the radionuclide generators. However, in the case of generator systems where both parent and daughter radioisotopes are pure β^- emitters (as in the case of $^{90}Sr/^{90}Y$ generator) and no γ-emissions are available to permit γ-analysis, β^- counting has to be done after separating the parent radionuclidic impurity from the daughter unambiguously.[45–47] In such cases, the radionuclidic purity may also be checked and determined by β-spectrometry using a liquid scintillation counter.[47]

(c) Radiochemical purity of the separated daughter radioisotope: Radiochemical impurities may arise in the daughter radionuclide during its separation from the parent or its subsequent storage due to the action of the solvent and the effect of radiolysis, change in temperature or pH, presence of oxidizing or reducing agents. The radiochemical impurities present in the daughter radionuclide may not be suitable for labeling with ligands and biomolecules.[13] This may affect the biological behavior of the radiopharmaceutical, as the agent may not be selectively taken up by the target organ, thus giving rise to unnecessary dose burden to the patient.[13] The presence of radiochemical impurities in generator-produced radioisotopes can be detected and determined by various analytical methods. These include, paper chromatography, thin layer chromatography, paper electrophoresis, high performance liquid chromatography, gel filtration, gel chromatography, ion exchange chromatography, solvent extraction, inverse dilution and precipitation.[13]

(d) Chemical purity: The chemical impurities may be introduced in the generator-derived radionuclides by a variety of ways. This includes the use of impure chemicals and use of radioactive parent solutions containing undesired chemicals introduced into it during its radiochemical processing.[13] Also, in the case of column chromatographic and solvent extraction based generators, the radiolytic or chemical degradation of the column matrix or the organic solvent may lead to the addition of chemical impurities in the daughter radionuclide. The presence of these chemical impurities can be avoided by the use of highly pure chemicals and adoption of appropriate separation methodology. The level of these chemical impurities may be detected and determined by various analytical techniques like colorimetry, spot-tests, spectrophotometry, inductively coupled plasma atomic emission spectroscopy (ICP-AES) *etc*.[48] If required, the chemical impurities present in the radionuclide may be removed by applying simple separation techniques, such as precipitation, solvent extraction, ion-exchange and distillation.[13]

(e) Labeling efficacy: Generally, generator-produced radionuclides are used for clinical applications, only after radiolabeling a suitable ligand or a biomolecule with it. The suitability of a generator-produced radionuclide for radiopharmaceutical applications can be demonstrated by its efficacy to prepare standard radiolabeled agents. This is also an indirect test of the chemical purity of the radionuclide as high chemical purity is required for the preparation of the radiolabeled agent with the 'no-carrier-added' radionuclide.

(f) Biological tests: Biological quality control tests are carried out essentially to examine the sterility and apyrogenicity of the generator produced radionuclide before the preparation of clinical grade radiopharmaceuticals. Sterility testing of a radiochemical is done by incubating the radiochemical either in fluid thioglycollate medium at 30–35 °C or in soyabean–casein digest medium for 7–14 days,[13] and observing if there is any growth of microorganisms, under the microscope. Generally, in order to obtain the generator-produced radionuclide in a sterile form it is filtered through commercially available Millipore filters.[13] The apyrogenicity of a radiochemical must be confirmed before its clinical applications by adopting the limulus amebocyte lysate (LAL) test or bacterial endotoxin test (BET).[13] Pyrogen free radionuclides can be obtained from the generator without much difficulty using high quality chemicals and taking particular care during the preparation and storage of the generator.

Since all the methods for determining the sterility and apyrogenicity of generator-produced radioisotopes are time consuming, these tests are generally done *post-facto*. However, in order to reach a certain level of confidence on the product, several batches have to be produced, tested and repeatedly shown to be complying with the quality requirements, before actual use in humans. The daughter radionuclide obtained from the generator is used for radiopharmaceutical preparation, provided the manufacturer has already established its sterility and apyrogenicity before the generator is commercialized.

11.12 Shelf-life of a Radionuclide Generator

The shelf-life of the radionuclide generator is the period for which the generator can be safely used for the designated clinical applications. The loss of efficacy of a typical radionuclidic generator over a period of time is determined by the physical half-life of the parent radioisotope, yield, radioactive concentration and the purity of the daughter radioisotope. The parent radioisotopes having longer physical half-lives are expected to have longer shelf-life. However, in certain cases,[19] due to radiolysis and chemical degradation, the separation yield of the daughter decreases with time and the daughter is obtained with low radioactive concentration. Moreover, this may also lead to a breakthrough of the parent radioisotope, thereby rendering the daughter radioisotope unsuitable for clinical applications. Sometimes, it may also lead to the addition of chemical impurities in the daughter radioisotope, which may interfere with its complexation chemistry for the preparation of radiopharmaceuticals.[19] Overall, the economics of production of short-lived radioisotopes *via* a radionuclide generator is decided by the shelf-life of the generator and this in turn determines the cost of treatment using radiopharmaceuticals based on generator produced radionuclides.

11.13 Use of Nanomaterial-based Sorbents for the Preparation of ^{99}Mo/^{99m}Tc and ^{188}W/^{188}Re Generators

11.13.1 Current Status and Future Perspectives of ^{99}Mo/^{99m}Tc and ^{188}W/^{188}Re Generators

^{99m}Tc has been the 'work-horse' of diagnostic nuclear medicine and is expected to continue to be so in the foreseeable future. ^{99m}Tc is conveniently available from ^{99}Mo/^{99m}Tc generators. A variety of ^{99}Mo/^{99m}Tc generator systems have been thoroughly investigated during the last 50 years, due to the ever-increasing demand for ^{99m}Tc.[49–61] The column chromatographic generator using a bed of acidic alumina has emerged as the most popular generator system world over.[52,62] The capacity of alumina for taking up molybdate ions is limited (2–20 mg Mo per g of alumina)[52] necessitating the requirement of ^{99}Mo of the highest specific activity available, generally possible only in ^{99}Mo produced through a fission route. The separation of fission ^{99}Mo is an elaborate complex processing technology which is expensive, generates huge quantities of radioactive wastes and requires extensive purification prior to use.[17,53] Worldwide, nearly 95% of fission ^{99}Mo is produced by 5 commercial nuclear reactors — NRU at Chalk River in Canada, HFR at Petten in Netherlands, BR-2 at Fleurus in Belgium, OSIRIS at Saclay in France and SAFARI-1 at Pelindaba in South Africa.[63] The fission ^{99}Mo supply chain has been severely hit hard by the cascade of events in 2008 and this has adversely affected patient services in many countries.[64–69] Although the recent crisis has now passed, this has not only highlighted the fragile nature of fission ^{99}Mo supplies but also raised concerns over its dependency. In order to overcome these unforeseeable problems and to reduce reliance on fission-produced ^{99}Mo, alternative pathways using the $(n,\gamma)^{99}$Mo and their implementation need to be emphasized as a back-up measure and to supplement ^{99m}Tc accessibility to meet the continually growing demand for ^{99m}Tc in nuclear medicine.[68]

Another radionuclide generator of considerable interest is the ^{188}W/^{188}Re generator, which is the source of NCA ^{188}Re, an attractive therapeutic radionuclide.[70–77] Being a congener of ^{99m}Tc, the chemistry of Re is similar to Tc, which is an additional advantage for working with molecules that have shown promising results as '^{99m}Tc-radiopharmaceuticals'.[73] The attractive physical properties of ^{188}Re and its production from a long-lived parent tungsten-188 (^{188}W, $t_{1/2} = 69.2$ d), from a generator with an adequate shelf-life, makes it an interesting option for clinical use. However, the main problem in the use of ^{188}W/^{188}Re generator is the need for high specific activity ^{188}W, which is produced by double neutron capture of ^{186}W in high flux research reactors ($\phi \sim 10^{15}$ n cm^{-2} s^{-1}). Even at high thermal neutron flux, which is only available at the SM reactor in Dimitrovgrad, Russian Federation and at the High Flux Isotope Reactor (HFIR) at the Oak Ridge National Laboratory (ORNL), USA, only relatively low specific activity (148–185 GBq g^{-1} (4–5 Ci

g^{-1})) ^{188}W can be produced because of the low production cross-sections.[74] The commercially available ^{188}W/^{188}Re generators are akin to the ^{99}Mo/^{99m}Tc generators using alumina columns,[70] which have limited W-sorption capacity (maximum 80 mg W g^{-1}).[70,71] Depending on the specific activity of the available ^{188}W, the currently available generators yield ^{188}Re of low specific volume (activity/mL) and consequently often require an additional concentration step,[78] which makes the process cumbersome and leads to addition of chemical impurities. Hence, development of ^{188}W/^{188}Re generators where the concentration step can be avoided and even low specific activity ^{188}W prepared in medium flux research reactors can be used shall be of great advantage.

Several alternate sorbents, such as hydroxyapatite, the hydrous oxides of zirconium, titanium, manganese, tin(IV), and cerium, silica gel, the AG 1-X12 and AG 50 W-X12 ion-exchange resins and activated charcoal were studied to determine their suitability for the preparation of ^{99}Mo/^{99m}Tc and ^{188}W/^{188}Re generators.[79–85] Unfortunately, all these materials showed an unimproved Mo- and/or W-sorption capacity compared to alumina.[79] Other pathways, such as gel generators based on matrices such as zirconium or titanium molybdate or tungstate,[21–24] have also been explored with limited success. Though these matrices could retain higher Mo or W content than alumina, the ^{99m}Tc or ^{188}Re elution performance was still inferior compared to the alumina based systems.

In the last few years, a number of alternative high capacity sorbents for Mo and W, such as gel metal oxide composite, synthetic alumina, polymeric titanium oxychloride and polymeric zirconium compound (PZC), have been developed and exploited for the preparation of ^{188}W/^{188}Re generators.[86–91] Though none of these recent techniques have reached the commercial stage for clinical applications, PZC received considerable attention owing to its remarkably high sorption capacity ($\sim$10 times more than alumina).[86,89] However, the slow kinetics of sorption and appreciably high ^{99}Mo and ^{188}W breakthrough[86,89] are the significant drawbacks that limit its applicability in a clinical context.

In view of the above-described drawbacks, development of nanomaterial based sorbents with high sorption capacity and selectivity for ^{99}Mo and/or ^{188}W, along with appreciable radiation resistance and chemical stability is of considerable importance and deserves a serious consideration. Use of such sorbents would not only facilitate the elution of the daughter with high radioactive concentration but also permits the use of low specific activity parent radioisotopes.

This chapter describes the utilization of two nanomaterial-based sorbents (polymer embedded nanocrystalline titania and nanocrystalline zirconia) developed by our group in the preparation of ^{99}Mo/^{99m}Tc and ^{188}W/^{188}Re generators.[92–95] Experimental parameters that are important in developing radionuclidic generators have been elaborated. The feasibility of the technique both in terms of yield and the purity of the daughter product (^{99m}Tc/^{188}Re) for radiopharmaceuticals application, have been demonstrated and evaluated.

11.13.2 Polymer Embedded Nanocrystalline Titania for the Preparation of ^{99}Mo/^{99m}Tc and ^{188}W/^{188}Re Generators

11.13.2.1 Synthesis of Polymer Embedded Nanocrystalline Titania (TiP)

TiP was synthesized by the *in-situ* reaction of titanium tetrachloride with isopropyl alcohol.[92] In this procedure, TiCl$_4$ was added drop wise to isopropyl alcohol, and the reaction mixture was stirred briskly to avoid vigorous reaction. The viscosity of the reaction mixture increased gradually with time, and after 4–6 hours a semi-solid mass was formed which was difficult to stir. The progress of the reaction could be conveniently monitored by the formation of semi-solid mass. The semi-solid mass was allowed to stand for 5 days at ambient atmosphere in order to allow the reaction to complete. The reaction time and the drying time are crucial for availing sorbent of requisite quality. If the reaction was carried out for shorter time, hetero-structured products were obtained, which on contact with aqueous solution crumbled into fine powder. The powder form was not amenable for use in column operation.

The product obtained after prolonged reaction time was dried under an infrared lamp at a temperature of 80–90 °C for nearly 2 days to remove the generated HCl fumes and excess unreacted isopropyl alcohol. The material obtained after drying was water soluble. In order to make the water soluble precursor into an insoluble sorbent, it was heated at a temperature 160 °C for 2 hours in a furnace. The solid mass obtained was subjected to repeated washings with distilled water to remove traces of remaining HCl and isopropyl alcohol still adhering to the reaction product. The residue was dried, ground manually with a porcelain mortar and sieved using a sieve of 50–100 mesh (149–297 µm). The product thus obtained was granular in texture with adequate mechanical strength and exhibited free flow characteristics when used in fixed-bed column chromatography applications. Investigation of the stability of the material towards various solvents prior to application showed that TiP was stable in water, dilute mineral acids and alkalis.

11.13.2.2 Structural Characterization of TiP

The infrared absorption spectrum of TiP showed a broad absorption peak in the range 3600–3000 cm^{-1}, which was due to the sum of the contributions of hydroxyl groups and water molecules. The absorption peak at 1614 cm^{-1} was due to the bending mode of OH$^-$ group attached to the matrix. The continuous absorption peak at <1000 cm^{-1} was due to Ti–O bonds. The X-ray diffraction (XRD) pattern of the powder is shown in Figure 11.4, which matches well with the rutile modification of TiO$_2$. The average crystallite size of TiO$_2$ was calculated using Scherrer's formula and was found to be ~5 nm. The surface area measurement by standard BET technique was carried out on the prepared sample. The surface area was found to be 30 m^2 g^{-1}. The average pore size of the powder was found to be ~4 Å. The TEM micrograph revealed the network of polymer with dispersed titania phase (Figure 11.5).

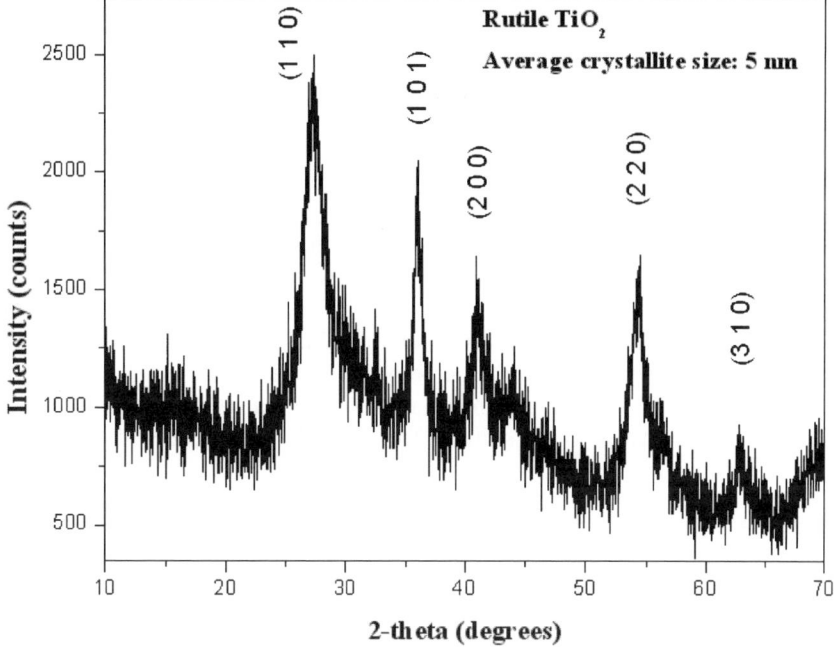

Figure 11.4 XRD pattern of TiP

Figure 11.5 TEM micrograph of TiP

11.13.2.3 Application of TiP in the Preparation of ^{99}Mo/^{99m}Tc and ^{188}W/^{188}Re Generators

11.13.2.3.1 Determination of Distribution Ratio (K_d) for ^{99}Mo/^{99m}Tc ions and ^{188}W/^{188}Re ions on TiP

Although the K_d values of ^{99}Mo/^{99m}Tc ions and ^{188}W/^{188}Re ions are determined under batch equilibrium conditions, it can be used to predict the chromatographic separation of the daughter species from the parent under varied experimental conditions, and therefore was pursued. The K_d values of ^{99}Mo/^{99m}Tc and ^{188}W/^{188}Re ions on TiP, as a function of pH of the solution were determined, and the results are summarized in Table 11.4 and Table 11.5, respectively. It is apparent from the respective tables that the K_d values for both Mo/Tc and W/Re ions attained the maximum value at around pH 3 and hence this medium was found suitable for sorption of ^{99}Mo or ^{188}W. In 0.9%

Table 11.4 Distribution ratios (K_d) of ^{99}MoO$_4^{2-}$ and ^{99m}TcO$_4^-$ ions on TiP.

Medium (pH)	K_d^a	
	^{99}Mo	^{99m}Tc
1	146 ± 2	3 ± 1
2	188 ± 5	14 ± 2
3	489 ± 7	41 ± 4
4	362 ± 3	33 ± 3
5	225 ± 6	3.2 ± 0.8
6	74 ± 2	0.3 ± 0.2
7	14 ± 3	0.2 ± 0.1
8	11 ± 1	0.5 ± 0.3
0.9% NaCl	216 ± 6	0.3 ± 0.1

^{a}n = 3; '±' represents the standard deviation.

Table 11.5 Distribution ratios (K_d) of ^{188}WO$_4^{2-}$ and ^{188}ReO$_4^-$ ions on TiP.

Medium (pH)	K_d^a	
	^{188}W	^{188}Re
1	191 ± 4	23 ± 3
2	244 ± 6	61 ± 6
3	235 ± 3	83 ± 2
4	195 ± 7	56 ± 2
5	189 ± 2	38 ± 1
6	175 ± 8	46 ± 4
7	34 ± 1	15 ± 1
8	22 ± 2	3 ± 1
0.9% NaCl	90 ± 1	0.7 ± 0.3

^{a}n = 3; '±' represents the standard deviation.

NaCl solution, the high K_d values for ^{99}Mo and ^{188}W ions clearly indicate that they can be retained by the sorbent, while the low K_d values for ^{99m}Tc and ^{188}Re ions indicate that they shall not be retained and hence eluted out easily. Therefore, 0.9% NaCl solution was used for the elution of ^{99m}Tc and ^{188}W from the respective generators.

11.13.2.3.2 Determination of Zeta-potential of TiP at Different pH

The zeta potential (ζ), which is commonly used to describe the surface charge of the sorbent that interacts with its surroundings, is a very useful tool to help in the interpretation of the sorption behavior. The surface acid–base properties of nanocrystalline metal oxides are inherent and are primarily responsible for the retention of ^{99}Mo and ^{188}W. The surface association of ^{99}Mo or ^{188}W ions with the sorbent is governed by the pH of the external solution, since it determines the surface charge of the solid particles and the degree of ionization. The dependence of zeta potential (ζ) on the pH of the external solution, helps in finding the sorption mechanism. In aqueous systems, TiP particles are hydrated, and ≡Ti–OH groups cover their surface completely. These amphoteric hydroxyl groups can undergo reaction with either H^+ or OH^- and develop positive or negative charges on the surface depending on the pH. The variation of zeta potential of TiP with pH is illustrated in Figure 11.6. It was observed that the TiP surface has a positive charge at pH < 6.5 and it

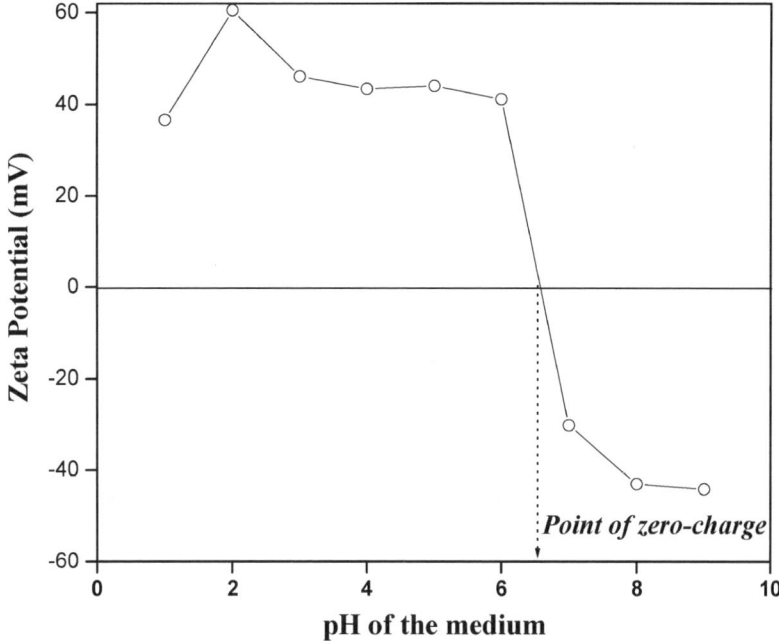

Figure 11.6 Variation of zeta potential of TiP as a function of pH.

reversed at pH > 6.5. The iso-electric point (IEP) was reached at pH ~6.5. At pH < 6.5, a positive charge density builds up on the surface of TiP

$$\equiv \text{Ti-OH}_{\text{surface}} + \text{H}^+_{\text{aq}} \rightleftharpoons \equiv \text{Ti-OH}^+_{2\text{surface}}$$

There is a strong attractive interaction between the positively charged TiP surface and the negatively charged polymolybdate $[\text{Mo}_7\text{O}_{24}]^{6-}$ and polytungstate, $[\text{HW}_6\text{O}_{21}]^{5-}$ anions[55,96] from the liquid solution. The mechanism of ^{99}Mo or ^{188}W uptake therefore may be considered to take place in two steps. The first one may be the exchange of polymolybdate anions or polytungstate anions for hydroxyl ions on the surfaces of the particles. Subsequently, it may form a stable complex of the type $[\text{TiMo}_6\text{O}_{24}]^{8-}$ or $[\text{TiW}_6\text{O}_{24}]^{8-}$, similar to that reported with alumina.[97] The decay of ^{99}Mo to $^{99\text{m}}$Tc and ^{188}W to ^{188}Re are not accompanied by any serious disruption of chemical bonds. ^{99}Mo and ^{188}W are expected to be retained strongly on the sorbent matrix as polymeric molybdate and tungstate ions, respectively. As these molybdate or tungstate ions start transforming into pertechnetate ($^{99\text{m}}\text{TcO}_4^-$) or perrhenate ion ($^{188}\text{ReO}_4^-$), which have only -1 charge, the binding would get weaker and an easy displacement of $^{99\text{m}}\text{TcO}_4^-$ or $^{188}\text{ReO}_4^-$ ions is expected. Therefore, $^{99\text{m}}$Tc or ^{188}Re gets eluted easily with normal saline (0.9% NaCl) solution.

11.13.2.3.3 Determination of the Sorption Capacity of TiP

The results of the capacity determination experiment by the batch equilibrium method indicated that 110 ± 15 mg Mo and 325 ± 10 mg W could be absorbed per g of TiP at pH 3–6. The sorbent, TiP, showed a strong retention of ^{99}Mo and ^{188}W at pH 2 to 3, where optimum K_d was realized. In order to portray the sorption behavior of ^{99}Mo and ^{188}W in the generator column bed containing TiP, the breakthrough curves were developed at pH ~2 and are depicted in Figure 11.7 and Figure 11.8. It was observed that the breakthrough point was reached after ~75 mg Mo and ~102 mg of W were quantitatively retained by 1 g of TiP in the respective column. The saturation capacities of Mo and W were observed to be ~110 mg Mo and ~290 mg of W per g of TiP at $C/C_o = 0.9$. It was observed that the dynamic capacity of the sorbent was less than the static capacity. This was probably due to the mass transfer limitations, such as incomplete external film diffusion and/or to intraparticle transfer. However, the protocol for preparing a chromatographic column generator using sorbent matrix preloaded with ^{99}Mo or ^{188}W under static conditions is not recommended owing to difficulties in handling of radioactive material. This will not only cause associated radioactive contamination problem but also increase the radiation exposure to the personnel involved. Furthermore, even if the column is prepared, the operating performance thereof is likely to deteriorate. Therefore, post loading of ^{188}W activity protocol, was followed despite the lower sorption capacity that could be achieved. However, the dynamic sorption capacity of TiP for Mo and W is much higher than that of

Separation of Medically Useful Radioisotopes: Role of Nano-sorbents 285

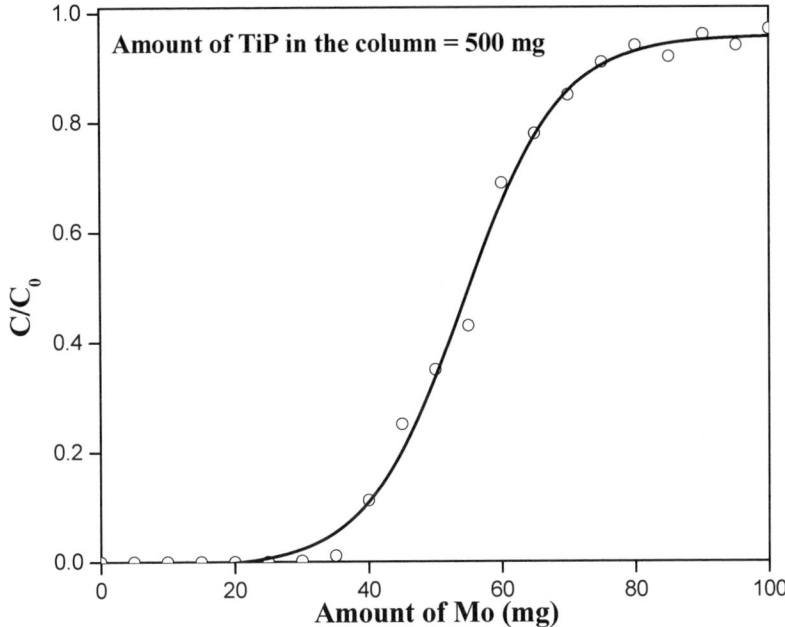

Figure 11.7 Breakthrough profile of ^{99}Mo in TiP column.

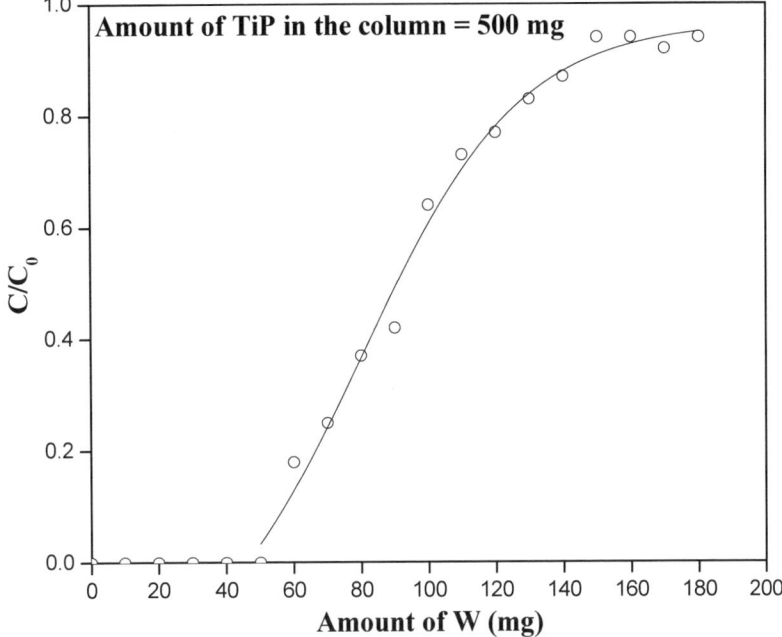

Figure 11.8 Breakthrough profile of ^{188}W in TiP column.

conventionally used alumina, which has a sorption capacity of ~20 mg g^{-1} for Mo,[52] and ~45 mg g^{-1} for W.[70]

11.13.2.3.4 Development of ^{99}Mo/^{99m}Tc and ^{188}W/^{188}Re Generators using TiP as the Sorbent

In order to integrate the experimental findings and to establish the utility of the sorbent, it is necessary to carry out a process demonstration run to evaluate the behavior of the sorbent in the presence of an intense radiation environment and the radiolytic products generated as a result of radioactive parent/daughter radionuclides. The utility of TiP as a column matrix for ^{99}Mo/^{99m}Tc and ^{188}W/^{188}Re generator was exploited by performing process demonstration runs using 1.85 GBq (50 mCi) activities of ^{99}Mo/^{99m}Tc and ^{188}W/^{188}Re and monitoring the elution behaviors of ^{99m}Tc and ^{188}Re, respectively. The respective generator columns were loaded with 1.85 GBq (50 mCi) of ^{99}Mo and ^{188}W solutions, maintained at pH 3. The generator columns were housed in lead shielded assemblies for radioprotection. The ^{99}Mo/^{99m}Tc generator and ^{188}W/^{188}Re generators were regularly eluted with 0.9% NaCl solution. The ^{99}Mo/^{99m}Tc generator was eluted daily for 1 week while the ^{188}W/^{188}Re generator was eluted at 2–3 days intervals for a period of 6 months. After allowing adequate time for build up of the daughter radioisotope, the respective generator columns were eluted with 0.9% NaCl solution at a flow rate of 0.5 mL min^{-1} to study the elution behavior of ^{99m}Tc or ^{188}Re from the column. The practical utility and

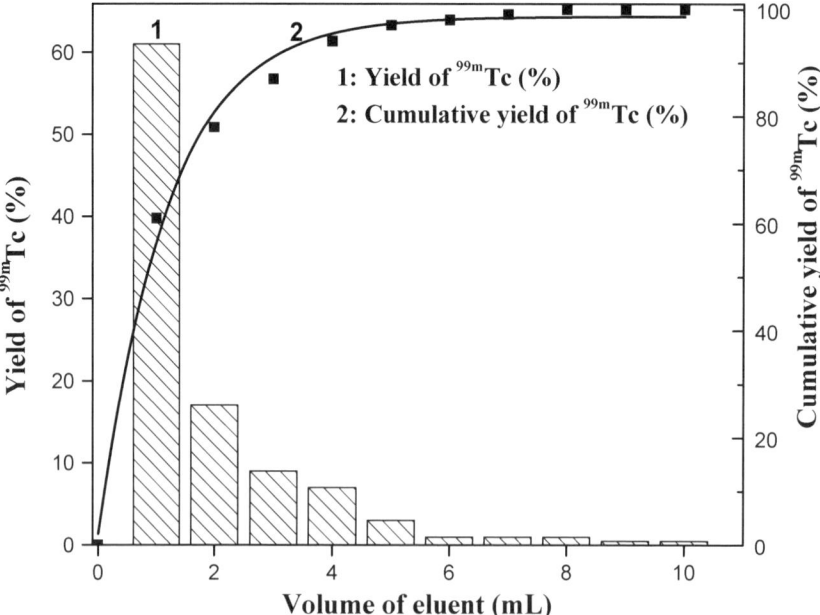

Figure 11.9 Elution profile of the TiP based ^{99}Mo/^{99m}Tc generator.

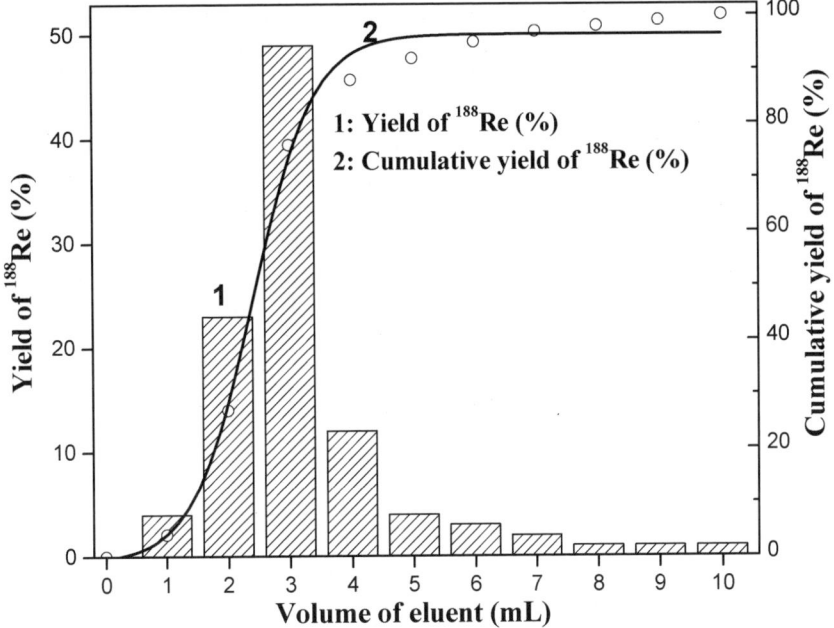

Figure 11.10 Elution profile of the TiP based ^{188}W/^{188}Re generator.

separation capability of ^{99m}Tc and ^{188}Re daughter radioisotopes by the sorbent material were monitored by following the elution profile of the generator. The eluate was collected as 1 mL aliquots, and each fraction was counted for gamma activity. The elution profile of the ^{99}Mo/^{99m}Tc generator is shown in Figure 11.9 and that of the ^{188}W/^{188}Re generator is shown in Figure 11.10. It can be seen from the figures that >85% of ^{99m}TcO$_4^-$ and ^{188}ReO$_4^-$ activity built-up in the generator, could be eluted out with 3–4 mL of 0.9% NaCl solution. Thus, ^{99m}Tc and ^{188}Re could be eluted from these generators with appreciably high radioactive concentrations, suitable for radiopharmaceutical applications.

11.13.2.3.5 Quality Control of ^{188}Re and ^{99m}Tc

The suitability of ^{99m}Tc and ^{188}Re obtained from TiP based ^{99}Mo/^{99m}Tc and ^{188}W/^{188}Re generators for biomedical application were evaluated by adopting the standard quality control procedures. The levels of ^{99m}Tc impurity in ^{99}Mo and ^{188}W impurity in ^{188}Re, as determined by the γ-ray spectrometry of the decayed ^{99m}Tc and ^{188}Re samples were <10^{-3} %. The radiochemical purity of ^{99m}TcO$_4^-$ and ^{188}ReO$_4^-$ was >99%. The level of Ti ions in the ^{188}Re eluate, as determined by ICP-AES analysis of the decayed ^{99m}Tc and ^{188}Re samples was <0.1 ppm. The level of radionuclidic, radiochemical and chemical impurities in ^{99m}Tc and ^{188}Re obtained from both the generators were well within the acceptable limits, as per the British Pharmacopoeia.[98] In order to demonstrate the suitability of ^{99m}Tc and ^{188}Re for the preparation of radiopharmaceuticals, standard ligands like dimercaptosuccinic acid (DMSA), hydroxyethylene diphosphonate (HEDP) and ethylene dicysteine (EC) were labeled with the radionuclides with >98% complexation yield.

The performance of the ^{99}Mo/^{99m}Tc generator was evaluated for 10 days and the ^{188}W/^{188}Re generator was evaluated for a period of 6 months. The generators were eluted regularly with 4 mL of 0.9% NaCl solution. Over this period of time, the ^{99}Mo/^{99m}Tc generator was eluted 9 times and the ^{188}W/^{188}Re generator was eluted >50 times and the performances of both the generators were consistently good. The elution efficiencies of both the generators were always >80% and the levels of ^{99}Mo or ^{188}W impurities present in ^{99m}Tc or ^{188}Re were always <10^{-3}%.

11.13.2.4 Recovery of Enriched ^{98}Mo and ^{186}W from the Exhausted Generator Columns

In order to produce ^{99}Mo and ^{188}W in appreciable quantities, enriched ^{98}Mo and ^{186}W targets are irradiated in the nuclear reactors. Since enriched targets are quite expensive, the scope of increasing the utility of enriched targets should be fully ensured, and this could be realized by the recovery of enriched targets from spent generators for recycling. Moreover, the exhausted generator columns cannot be directly discarded without removing the sorbed ^{99}Mo or ^{188}W from the columns. Therefore, this is an important step from waste disposal point of view, also. Hence, attempts were made to remove Mo and W from the used columns prior to disposal. The spent generator columns were washed with 100 mL of 0.9% NaCl solution. Mo and W were desorbed from the columns by passing 5 M NaOH solution containing H$_2$O$_2$ (15 mL of 5 M NaOH solution + 1 mL of 30% H$_2$O$_2$) through the column as per the method reported for alumina based generators.[99]

The mechanism of desorption of Mo or W from the TiP columns can be explained as follows. The zeta-potential of TiP is negative under alkaline conditions. As the pH of the external solution rises, depolymerisation of polymolybdate and polytungstate anionic species sorbed in the column takes place leading to the transformation of MoO$_4^{2-}$ and WO$_4^{2-}$ species. Owing to electrostatic repulsion, these negatively charged ions gets desorbed from the column and released to the NaOH solution. As a result of radiation and chemical changes caused by the ionizing radiation, Mo or W ions might be sorbed on the nanocrystalline materials in a reduced state. Addition of H$_2$O$_2$ to the NaOH solution promotes the oxidation of the ions and thus facilitates their elution.

11.13.3 Nanocrystalline Zirconia as Sorbent for the Preparation of ^{99}Mo/^{99m}Tc and ^{188}W/^{188}Re Generators

11.13.3.1 Synthesis of Nanocrystalline Zirconia

Various methods are reported for the synthesis of nanocrystalline zirconia,[100–102] out of which a direct precipitation method reported by Yin et al.[100] was chosen due its simplicity, cost-effectiveness and amenability for scale-up. Nanocrystalline zirconia was synthesized by controlled hydrolysis of zirconyl chloride in ammonical medium. 0.17 M zirconyl chloride was added drop-wise into a round bottom flask containing 2.5 M ammonia solution, with careful control of pH (9–11) and with vigorous stirring. It was essential to add zirconium oxychloride solution to the NH_4OH solution in the reaction vessel, with vigorous stirring. This order of addition is important, as formation of zirconia precipitate in alkaline environment helps in stabilization of the precursor to zirconia nanoparticles.[100–102] The vigorous stirring of the reaction mixture prevents agglomeration of the precipitate and hence facilitates the formation of the nanoparticles. The reflux digestion of the basic solution in the glass vessel leads to stabilization of small-size tetragonal crystals at high temperature.[100] The precipitate was washed with de-ionized water until free of chloride ions and then digested under reflux at 96 °C for 24 h in a 1 L round bottom flask that contained an aqueous solution of ammonia (pH ~12). Subsequently, the digested gels were washed extensively with de-ionized water. The washed gel was dried at 100 °C overnight. The dried gel was calcined at 600 °C for 5 h and used for structural studies. Further, it was ground in a porcelain mortar and sieved to get particles of 50–100 mesh (149–297 μm). The product was granular with adequate mechanical strength amenable for fixed-bed column operation.

11.13.3.2 Structural Characterization of Nanocrystalline Zirconia

The XRD pattern of nano zirconia is shown in Figure 11.11. The results suggest that the obtained product has a high degree of crystallinity. The relatively broad peaks are attributed to the very small crystallite size of the material. The XRD studies showed that the major constituent phase is tetragonal. The average crystallite size of nano zirconia as determined from the XRD pattern using Scherrer's formula was ~7 nm. The infrared absorption spectrum of nano zirconia showed a broad absorption peak in the range 3600–3000 cm^{-1}, which was due to the sum of the contributions of hydroxyl groups and water molecules. The absorption peak at 1614 cm^{-1} was due to the bending mode of the OH^- group attached to the matrix. The bands at 1000 cm^{-1} and 490–501 cm^{-1} were attributed to Zr–O–Zr bonds. The surface area measurement and the pore-size distribution in the sorbent were carried out by a standard BET technique. The surface area of nano zirconia was found to be as high as ~340 $m^2 g^{-1}$. The average pore size was determined to be ~4 Å. It was observed that pores sizes are uniform, which facilitate permeation of liquid. The TEM micrograph (Figure 11.12) indicated that the nano zirconia was highly agglomerated. However, the particles were quite uniform in size and shape. The average crystallite size of nano zirconia as determined by TEM measurements was found to be in the range of 6–10 nm, which is in accordance with the results obtained from the XRD method.

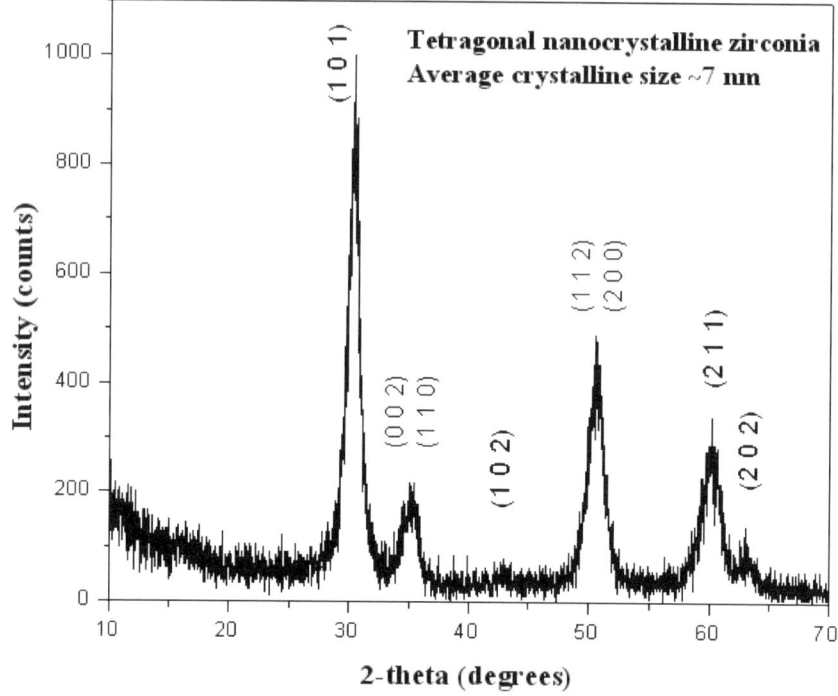

Figure 11.11   XRD pattern of nano zirconia.

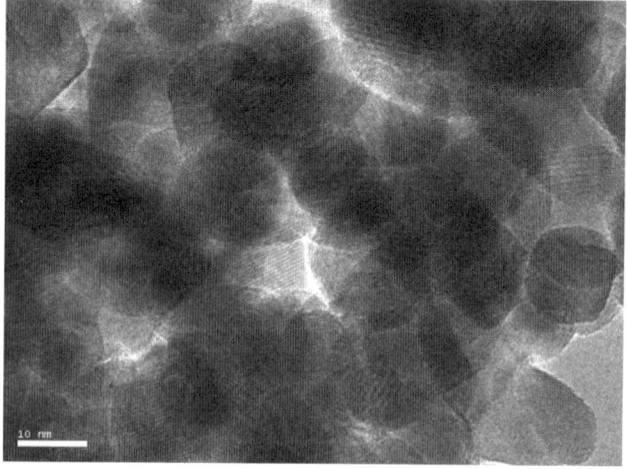

Figure 11.12   TEM micrograph of nano zirconia.

11.13.3.3 Application of Nano Zirconia for the Preparation of $^{99}Mo/^{99m}Tc$ and $^{188}W/^{188}Re$ Generators

11.13.3.3.1 Determination of Distribution Ratio (K_d) Values of $^{99}Mo/^{99m}Tc$ and $^{188}W/^{188}Re$ Ions

The K_d values of $^{99}Mo/^{99m}Tc$ and $^{188}W/^{188}Re$ ions at different pH were evaluated in order to predict the experimental conditions necessary for the optimal loading of parent radionuclides (^{99}Mo or ^{188}W) and efficient elution of daughter radionuclides (^{99m}Tc or ^{188}Re). The results are shown in Table 11.6 and Table 11.7. From the K_d studies, it could be inferred that just as in the case of TiP, maximum sorption of ^{99}Mo and ^{188}W could be achieved at pH 3 and ^{99m}Tc and ^{188}Re could be selectively eluted using 0.9% NaCl solution.

Table 11.6 Distribution ratios (K_d) of $^{99}MoO_4^{2-}$ and $^{99m}TcO_4^{-}$ ions on nano zirconia.

Medium (pH)	K_d^a	
	^{99}Mo	^{99m}Tc
1	129 ± 4	3 ± 1
2	295 ± 9	14 ± 2
3	611 ± 12	35 ± 6
4	560 ± 5	29 ± 3
5	525 ± 6	2.7 ± 0.7
6	174 ± 3	0.3 ± 0.1
7	34 ± 2	0.10 ± 0.07
8	10 ± 1	0.6 ± 0.2
0.9% NaCl	270 ± 7	0.2 ± 0.1

$^a n = 3$; '±' represents the standard deviation

Table 11.7 Distribution ratios (K_d) of $^{188}WO_4^{2-}$ and $^{188}ReO_4^{-}$ ions on nano zirconia

Medium pH	K_d^a	
	^{188}W	^{188}Re
1	143 ± 4	6 ± 3
2	469 ± 10	31 ± 5
3	616 ± 12	61 ± 8
4	315 ± 14	24 ± 12
5	133 ± 7	No uptake
6	122 ± 8	
7	23 ± 3	
8	10 ± 2	
0.9% NaCl	188 ± 9	0.3 ± 0.2

$^a n = 3$; '±' represents the standard deviation.

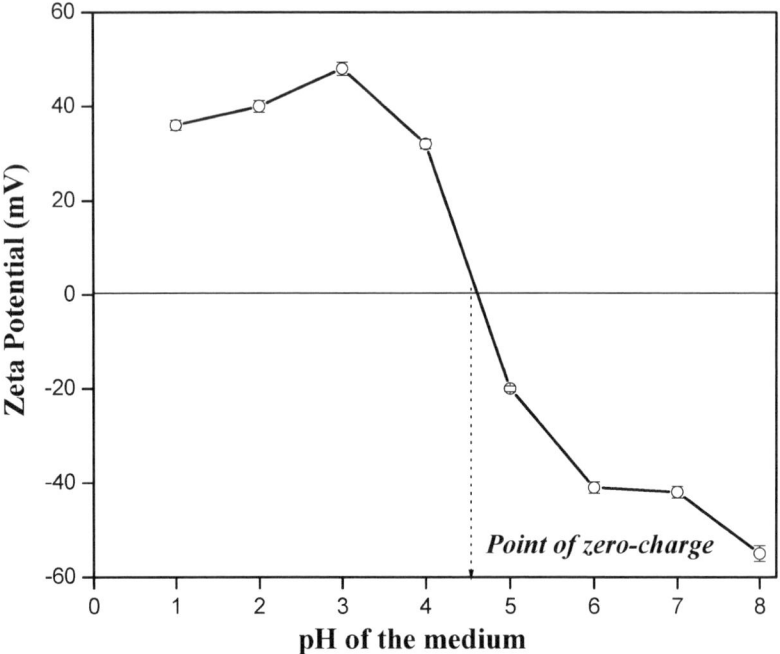

Figure 11.13 Zeta potential of nano zirconia at different pH.

11.13.3.3.2 Determination of Zeta-potential of Nano Zirconia at Different pH

In conjunction with this work that deals with the sorption of ^{99}Mo and ^{188}W, attempts were made to correlate the K_d values with the zeta potential of the sorbent in solutions of different pH. The effect of pH on zeta potential of nano zirconia in aqueous solution is shown in Figure 11.13. In case of nano zirconia, the zeta-potential values are positive in the pH range 1–4. On further increase of pH, the zeta potential values passes through zero (iso-electric point, IEP) and then it develops negative zeta potential. The variation of zeta potential and K_d with pH of the solution show a similar trend. These results give an indication that the electrostatic force is primarily responsible for the sorption of ^{99}Mo and ^{188}W ions in the present study and the mechanism of sorption is similar to that of TiP.

11.13.3.3.3 Determination of the Sorption Capacity of Nano Zirconia

The static sorption capacity of nano zirconia was (250 ± 10) mg Mo g^{-1} and (312 ± 9) mg W g^{-1}. The behavior of nanocrystalline zirconia in a fixed-bed column operation at room temperature was studied to predict the mass transport performance in terms of the breakthrough points of ^{99}Mo and ^{188}W. The breakthrough profiles of ^{99}MoO$_4^{2-}$ and ^{188}WO$_4^{2-}$ ions in nano zirconia are shown in Figure 11.14 and Figure 11.15, respectively. From the breakthrough profiles it could be observed that breakthrough of ^{99}MoO$_4^{2-}$ and ^{188}WO$_4^{2-}$ occurred after 80 mg Mo and 120 mg W were retained by 1 g of

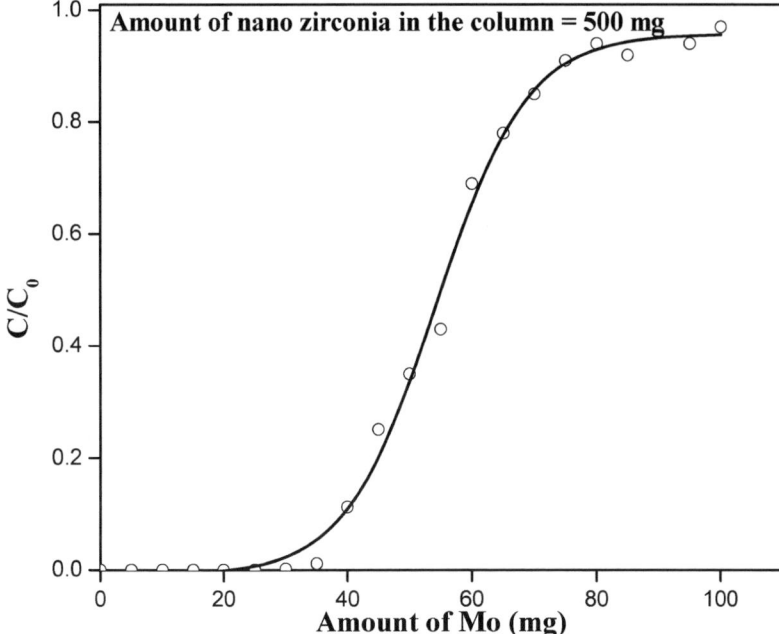

Figure 11.14 Breakthrough profile of ^{99}Mo in nano zirconia column.

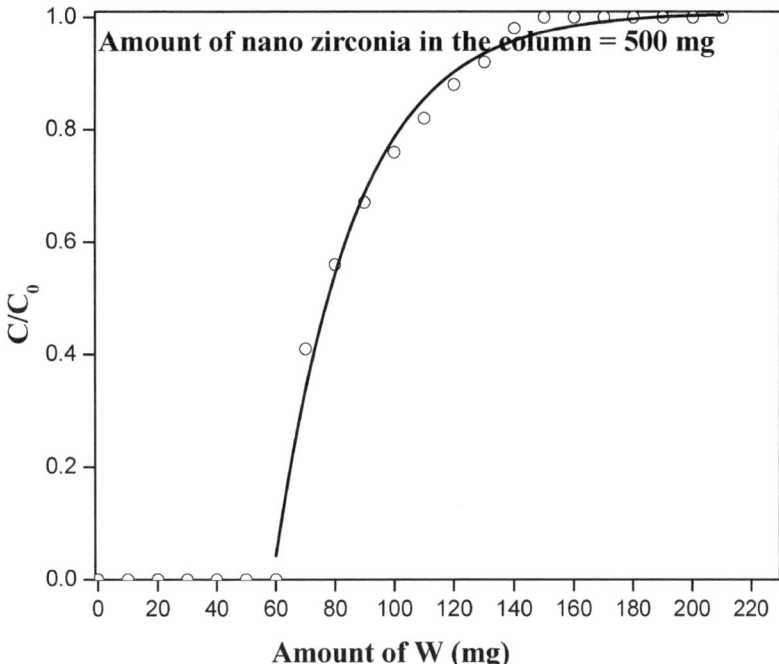

Figure 11.15 Breakthrough profile of ^{188}W in nano zirconia column.

nanocrystalline zirconia in the respective columns. Dynamic sorption capacity observed with this sorbent exhibited marked departure from the static capacity as a result of the impact of mass transfer limitations that occurred during column operation. Even though the breakthrough capacity of nanocrystalline zirconia is much less than the static sorption capacity, still it is much higher than that of alumina.[52,70] Further, in order to achieve higher sorption capacity under column conditions in case of ^{99}Mo, the radioactive solution was allowed to stand in the column for 30 min and then passed through the column maintaining a flow rate of 0.25 mL min^{-1}. By adopting this approach the sorption capacity achieved under dynamic conditions was >140 mg Mo g^{-1}.

11.13.3.3.4 Development of ^{99}Mo/^{99m}Tc and ^{188}W/^{188}Re Generators

In order to establish the utility of the sorbent in radiation environment, it is necessary to carry out a process demonstration run to evaluate the behavior of the adsorbent in the presence of intense radiation environment. The separation processes could be demonstrated by developing 9.25 GBq (250 mCi) ^{99}Mo/^{99m}Tc and 1.85 GBq (50 mCi) ^{188}W/^{188}Re generators. The elution profiles of the ^{99}Mo/^{99m}Tc and ^{188}W/^{188}Re generators are illustrated in Figure 11.16 and Figure 11.17, respectively. The elution profiles of both the generators were quite sharp and >90% of ^{99m}TcO$_4^-$ and ^{188}ReO$_4^-$ could be availed in ~4 mL of 0.9% NaCl solution. Thus, the radioactive concentrations of ^{99m}Tc and ^{188}Re availed from the respective generators were appreciably high for radiopharmaceutical applications.

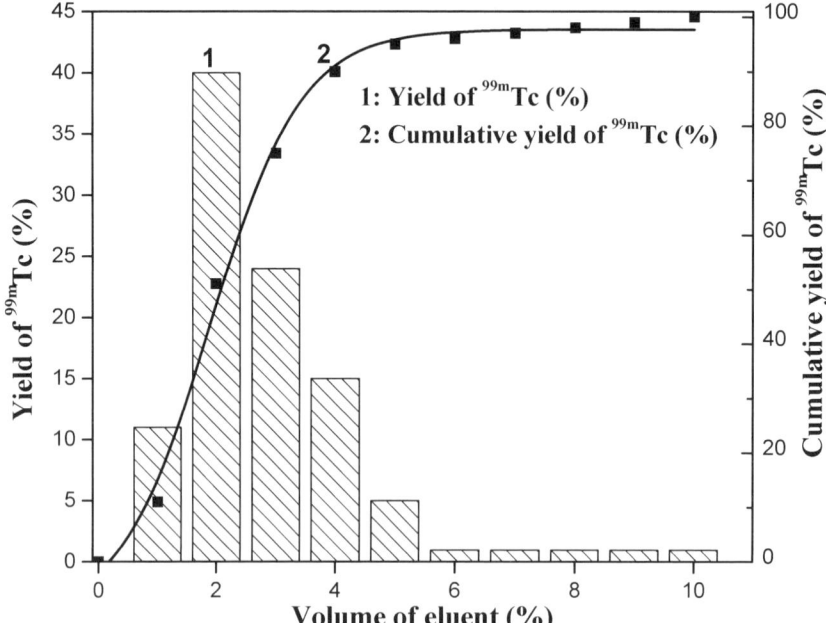

Figure 11.16 Elution profile of nano zirconia based ^{99}Mo/^{99m}Tc generator.

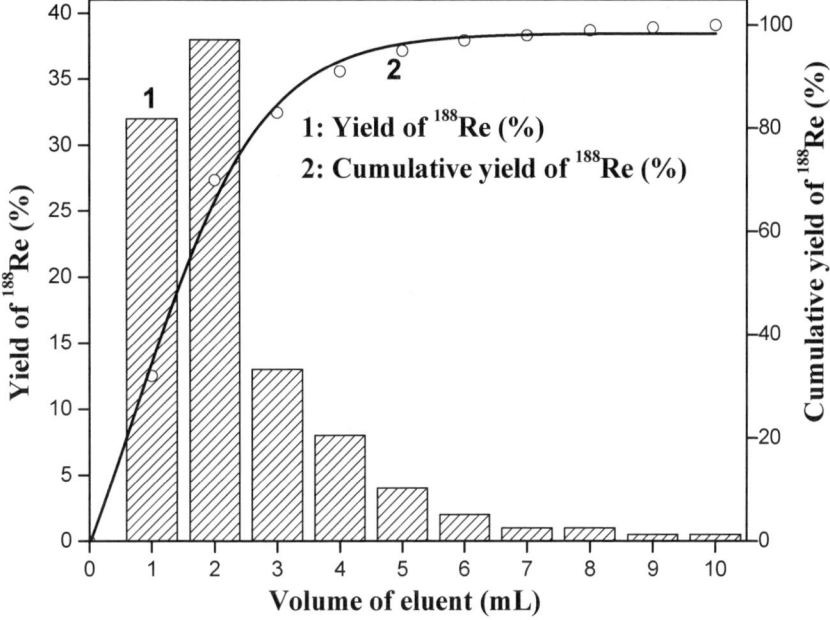

Figure 11.17 Elution profile of nano zirconia based $^{188}W/^{188}Re$ generator.

11.13.3.3.5 Quality Control of ^{99m}Tc and ^{188}Re

The suitability of ^{99m}Tc and ^{188}Re for clinical applications were evaluated adopting standard quality control procedures as described in the case of TiP based generators and the results were found to be satisfactory. The performances of the $^{99}Mo/^{99m}Tc$ and $^{188}W/^{188}Re$ generators remained consistent for a period of 10 days and 6 months respectively, which are normally the shelf-lives of these generators.

11.13.3.3.6 Recovery of Enriched ^{98}Mo and ^{186}W from the Exhausted Generator Columns

After expiry of the shelf-lives of the generators, enriched ^{98}Mo and ^{186}W could be recovered from the generator columns by passing 5 M NaOH solution containing H_2O_2 (15 mL of 5 M NaOH solution + 1 mL of 30% H_2O_2) as per the reported method.[99]

11.14 Conclusions

Perhaps, this is the first time that the potential of nanomaterials have been exploited for the separation of medically useful radioisotopes. The major advantages of using nanomaterial-based sorbents for such applications are the high sorption capacity, excellent selectivity and appreciable radiation and chemical stability. The procedure for the synthesis of these sorbents is simple and economical and can be easily scaled up. Medium specific activity $(n,\gamma)^{99}$Mo and $(2n,\gamma)^{188}$W could be utilized for the preparation of the ^{99}Mo/^{99m}Tc and ^{188}W/^{188}Re generators. This approach retains the simplicity of the alumina-based chromatographic generators and is therefore amenable for automation. ^{99m}Tc and ^{188}Re could be obtained in >85% yields with high radionuclidic, radiochemical and chemical purity and also with appreciably high radioactive concentration. The compatibility of the product in the preparation of labeled formulations was evaluated and found to be satisfactory. Unlike the other non-conventional ^{99}Mo/^{99m}Tc and ^{188}W/^{188}Re generators, these nanomaterial-based column chromatographic generators are expected to be more acceptable to hospital radiopharmacies, as the adaptation of this approach would not require the radiopharmacies to change their operating strategies or increase their personnel resources. The present concept would provide unequivocal support to the nuclear medicine industry, especially in developing countries which neither have neither the technology nor the reactors of adequate flux to produce high specific activity ^{99}Mo and ^{188}W.

11.15 Terminology Used in this Chapter

Nuclear Reactor: A nuclear reactor is an assembly in which fission chain reaction takes place in a controlled manner, for the production of energy and neutrons. A nuclear reactor is constructed with fuel rods made up of fissile materials such as enriched ^{235}U or ^{239}Pu. Nuclear fission is the source of energy in a nuclear reactor.

Cyclotrons: Cyclotrons are good sources of 100–200 µA h^{-1} beam current particles such as p, d, α, ^{3}He *etc.* at fixed as well as variable and energies. In a cyclotron, a high energy beam of accelerated particles is generated by circulating the charged particles by means of an electromagnetic field and targets of stable elements are bombarded with this particle beam.

Radiopharmaceutical: A special class of radiochemical formulation suitable for administration to human patients, either for diagnosis or therapy. Generally, a radiopharmaceutical is a two component system consisting of a radionuclide attached to a carrier moiety. The carrier moiety directs the radiopharmaceutical to the desired target while the radionuclide aids in diagnosis or therapy.

Specific Activity: Specific activity is defined as the radioactivity per unit mass of a radionuclide or a radiolabeled compound.

No-carrier-added (NCA): NCA radionuclides are produced with atomic numbers different from those of the target isotopes or precursor (parent radioisotope) and do not contain any stable ('cold' or 'carrier') isotope.

Radionuclidic purity: Radionuclidic purity is defined as the fraction of the total radioactivity in the form of the desired radionuclide.

Radiochemical purity: The radiochemical purity of a generator-produced radionuclide may be defined as the fraction of the total radioactivity present in the desired chemical form.

Chemical purity: Any unwanted chemical species (organic or inorganic) present in the generator-produced radionuclide is considered as chemical impurity. The presence of these chemical impurities may affect the complexation chemistry of the radionuclide for the preparation of radiopharmaceuticals.

Sterility: The sterility of a radiopharmaceutical indicates the absence of any viable bacteria or micro-organism present in it. Administration of a non-sterile radiopharmaceutical can cause a wide variety of infections leading to several physiological problems, including death.

Apyrogenicity: The absence of polysaccharides or proteins produced by the metabolism of microorganisms in a radiopharmaceutical, which if present can cause a wide variety of physiological problems.

References

1. G. V. S. Rayudu, *Semin. Nucl. Med.*, 1990, **20**, 100.
2. S. J. Adelstein and F. J. Manning, *Isotopes for Medicine and the Life Sciences*, National Academies Press, Washington D.C., 1995.
3. P. Richards, W. D. Tucker and S. C. Srivastava, *Int. J. Appl. Radiat. Isot.*, 1982, **33**, 793.
4. W. C. Eckelman, *J. Am. Coll. Cardiol. Img.*, 2009, **2**, 364.
5. F. F. Knapp Jr. and S. Mirzadeh, *Eur. J. Nucl. Med.*, 1994, **21**, 1151.
6. S. Mirzadeh and F. F. Knapp Jr., *J. Radioanal. Nucl. Chem.*, 1996, **203**, 471.
7. F. Rosch and F. F. Knapp Jr. in *Radiochemistry and radiopharmaceutical chemistry in life sciences: Handbook of Nuclear Chemistry*, ed. A. Vertes, S. Nagy and Z. Klenscar, Kluwer Academic Publisher, Dordrecht, The Netherlands, 2003, p 81.
8. S. M. Qaim, *Radiochim. Acta*, 2001, **89**, 223–231.
9. S. M. Qaim, in *Radiochemistry and radiopharmaceutical chemistry in life sciences: Handbook of Nuclear Chemistry*, ed. A. Vertes, S. Nagy and Z. Klenscar, Kluwer Academic Publisher, Dordrecht, The Netherlands, 2003, p 47.
10. A. P. Wolf and W. B. Jones, *Radiochim. Acta*, 1983, **34**, 1.
11. S. Mirzadeh, L. F. Mausner and M. A. Garland, in *Radiochemistry and radiopharmaceutical chemistry in life sciences: Handbook of Nuclear Chemistry*, ed. A. Vertes, S. Nagy and Z. Klenscar, Kluwer Academic Publisher, Dordrecht, The Netherlands, 2003, p 1.
12. R. M. Lambrecht, *Radiochim Acta*, 1983, **34**, 9.
13. G. B. Saha, *Fundamentals of Nuclear Pharmacy*, Springer-Verlag, New York, 2000.
14. G. Failla., *U.S. Patents*, 1-553-794 and 1-609-614, 1920.
15. W. E. Winsche, L. G. Stang Jr. and W. D. Tucker, *Nucleonics*, 1951, **8**, 14.

16. N. Ramamoorthy, V. Shivarudrappa and A. A. Bhelose, *Radiopharmaceuticals and hospital radiopharmacy practices*, Board of Research in Nuclear Sciences, Department of Atomic Energy, Mumbai, India, 2000.
17. R. E. Boyd, *Int. J. Appl. Radiat. Isot.*, 1982, **33**, 801.
18. A. Mushtaq, *J. Radioanal. Nucl. Chem.*, 2004, **262**, 797.
19. F. Roesch and P. J. Riss, *Curr. Top. Med. Chem.*, 2010, **10**, 1633.
20. L. Zsinka, *Radiochim. Acta*, 1987, **41**, 91.
21. R. E. Boyd, *Appl. Radiat. Isot.*, 1997, **48**, 1027.
22. J. V. Evans, P. W. Moore, M. E. Shying and J. M. Sodeau, *Appl. Radiat. Isot.*, 1987, **38**, 19.
23. P. W. Moore, M. E. Shying, J. M. Sodeau, J. V. Evans, D. J. Maddalena and K. H. Farrington, *Appl. Radiat. Isot.*, 1987, **38**, 25–29.
24. M. S. Dadachov, L. V. So, R. M. Lambrecht and E. Dadachova, *Appl. Radiat. Isot.*, 2002, **57**, 641.
25. G. C. Hadjipanayis and R. W. Siegel, *Nanophase Materials: Synthesis, Properties, Applications*, Kluwer, London, 1994.
26. H J. Fecht, *Nanomaterials: Synthesis, Properties, and Applications*, ed. A. S. Edelstein and R. C. Cammarata, Institute of Physics, Philadelphia, 1996.
27. T. P. Martin, U. Naher, H. Schaber and U. Zimmerman, *J. Chem. Phys.*, 1994, **100**, 2322.
28. A. N. Goldstein, C. M. Euhe and A. P. Alivasatos, *Science*, 1992, **256**, 1425.
29. K. J. Klabunde, J. V. Stark, O. Koper, C. Mohs, D. G. Park, S. Decker, Y. Jiang, I. Lagadic and D. Zhang. *J. Phys. Chem.*, 1996, **100**, 12142.
30. R. Richards, W. Li, S. Decker, C. Davidson, O. Koper, V. Zaikovsk, A. Volodin, T. Rieker and K. J. Klabunde, *J. Am. Chem. Soc.*, 2000, **122**, 4921.
31. J. Karch, R. Birringer and H. Gleiter, *Nature*, 1987, **330**, 556.
32. R. P. Andres, R. S. Averback, W. L. Brown, L. E. Brus, W. A. Goddard, A. Kalder, S. G. Louie, M. Moscovits, P. S. Peercy, S. J. Riley, R. W. Siegel, F. Spaepen and Y. Wang, *J. Mater. Res.*, 1989, **4**, 704.
33. T. Pradeep, *Nano: The essentials Understanding Nanoscience and Nanotechnology*, Tata McGraw Hill Publishing Company Limited, New Delhi, India, 2008.
34. C. P. Poole Jr and F. J. Owens, *Introduction to Nanotechnology*, Wiley India Pvt. Ltd., New Delhi, India, 2007.
35. G. Cao, *Nanostructures & Nanomaterials: Synthesis, Properties & Applications*, Imperial College Press, London, 2004.
36. C. Herring and J. K. Galt, *Phys Rev.*, 1952, **85**, 1060.
37. X. M. Bai, A. F. Voter, R. G. Hoagland, M. Nastasi and B. P. Uberuaga, *Science*, 2010, **327**, 1631.
38. J. Ragai and S.T. Selim, *J. Colloid Interface Sci.*, 1987, **115**, 139.

39. B. A. Manning, E. Scott, S. E. Fendorf and S. Goldberg, *Environ. Sci. Technol.*, 1998, **32**, 2383.
40. M. Hiraide, J. I. Wasawa and H. Kawaguchi, *Talanta*, 1997, **44**, 231.
41. S. Sarkar, P. W. Cara, C. V. Mcneff and A. Subramanian, *J. Chromatogr., B*, 2003, **790**, 143.
42. E. Vassileva and N. Furuta, *Fresenius' J. Anal. Chem.*, 2001, **370**, 52.
43. L. Cumbal, J. Greenleaf, D. Leun and A. K. Sengupta, *React. Funct. Polym.*, 2003, **54**, 167.
44. K. Okuyama and I. W. Lenggoro, *Chem. Eng. Sci.*, 2003, **8**, 537–547.
45. U. Pandey, P. S. Dhami, P. Jagesia, M. Venkatesh and M. R. A. Pillai, *Anal. Chem.* 2008, **80**, 801.
46. M. L. Bonardi, L. Martano, F. Groppi and M. Chinol, *Appl. Radiat. Isot.*, 2009, **67**, 1874.
47. W. J. Skraba, H. Arino and H. H. Kramer, *Int. J. Appl. Radiat. Isot.*, 1978, **29**, 91.
48. H. H. Willard, L. L. Merritt, J. A. Dean and F. A. Settle, *Instrumental Methods of Analysis*, CBH Publishers, India, 1986.
49. R. E. Boyd, *Radiochim Acta*, 1987, **41**, 59.
50. K. H. Bremer and H. Aktiengesellschaft, *Radiochim. Acta*, 1987, **41**, 73.
51. R. E. Boyd, *Radiochim. Acta*, 1982, **30**, 123.
52. V. J. Molinsky, *Int. J. Appl. Radiat. Isot.*, 1982, **33**, 811.
53. T. N. V. Walt and P. P.Coetzee, *Radiochim. Acta*, 2004, **92**, 251.
54. R. J. Baker, *Int. J. Appl. Radiat. Isot.*, 1971, **22**, 483.
55. H. Arino and H. H. Kramer, *Int. J. Appl. Radiat. Isot.*, 1975, **26**, 301.
56. K. Svoboda, *Radiochim. Acta*, 1987, **41**, 83–89.
57. S. Chattopadhyay, S. S. Das and L. Barua, *Appl. Radiat. Isot.*, 2010, **68**, 1.
58. S. Chattopadhyay, S. S. Das and L. Barua, *Nucl. Med. Biol.*, 2010, **37**, 17.
59. S. Chattopadhyay, S. S. Das, L. Barua and N. C. Goomer, *Appl. Radiat. Isot.*, 2008, **66**, 1814.
60. R. Chakravarty, A. Dash and M. Venkatesh, *Nucl. Med. Biol.*, 2010, **37**, 21.
61. R. Ram, R. Chakravarty, Y. Pamale, A. Dash and M. Venkatesh, *Chromatographia*, 2009, **69**, 497.
62. J. J. M. De Goeij, *Trans. Am. Nucl. Soc.*, 1997, 77, 519.
63. D. Cecchin, P. Zucchetta, P. Faggin, E. Bolla and F. Bui, *J. Nucl. Med.*, 2010, **51**, 14N.
64. P. Gould, *Nature*, 2009, **460**, 312.
65. A. C. Perkins and G. Vivian, *Nucl. Med. Commun.*, 2009, **30**, 657.
66. P. Webster, *Lancet*, 2009, **374**,103.
67. J. R. Ballinger, *J. Label. Compds. Radiopharm.*, 2010, **53**,167.
68. M. R. A. Pillai and F. F. Knapp Jr., *J. Nucl. Med.*, 2011, **52**, 15N.
69. F. F. Knapp Jr., *Cancer Biother. Radiopharm.*, 1998, **13**, 337.
70. J. M. Jeong and F. F. Knapp Jr., *Semin. Nucl. Med.*, 2008, **38**, S19.

71. R. Perego, B. Wierczinski, K. Zhernosekov, R. Henkelmann, A. Türler, T. Nikula and O. Buck, *Eur. J. Nucl. Med. Mol. Imaging*, 2007, **34**, S210.
72. E. Dadachova, S. Mirzadeh and R. M. Lambrecht, *J. Phys. Chem.*, 1995, **99**, 10976.
73. U. Abram and R. Alberto, *J. Braz. Chem. Soc.*, 2006, **17**, 1486.
74. H. Kamioki, S. Mirzadeh, R. M. Lambrecht, F. F. Knapp Jr. and K. Dadachova, *Radiochim. Acta*, 1994, **65**, 39.
75. F. F. Knapp Jr., E. C. Lisic and S. Mirzadeh, *US Patent*, 5 275 802, 1994.
76. B. M. Coursey, J. M. Calhoun, J. T. Cessna, D. D. Hoppes, F. J. Schima, M. P. Unterweger, D. B. Golas, A. P. Callahan, S. Mirzadeh and F. F. Knapp Jr., *Radioact. Radiochem.*, 1990, **4**, 38.
77. A. P. Callahan, D. E. Rice and F. F. Knapp Jr., *Nuc. Compact*, 1989, **20**, 3.
78. S. Guhlke, A. L. Beets, K. Oetjen, S. Mirzadeh, H. J. Biersack and F. F. Knapp Jr., *J Nucl Med.*, 2000, **41**, 1271.
79. L. V. So, C. D. Nguyen, P. Pellegrini and V. C. Bui, *Sep. Sci. Technol.*, 2009, **44**, 1074.
80. J. S. Gomez and F. G. Correa, *J. Radioanal. Nucl. Chem.*, 2002, **254**, 625.
81. A. Mushtaq, M. S. Mansoor, H. M. A. Karim and M. A. Khan, *J. Radioanal. Nucl. Chem.*, 1991, **147**, 257.
82. R. E. Lewis, *J. Nucl. Med.*, 1966, **7**, 804.
83. R. L. Hayes, *Oak Ridge Associated Universities. Medical Division Report ORAU*, 1966, **101**, 74.
84. F. Monroy-Guzman, V. E. Badillo-Almaraz, J. A. Flores De La Torre, J. Cosgrove and F. F. Knapp, *Proceedings of the International Symposium on Trends in Radiopharmaceuticals, ISTR-2005, IAEA*, Vienna, Austria, 2007, **1**, 333.
85. F. F. Knapp and S. Mirzadeh, *Oak Ridge National Laboratories, Report ORNL/TM-12222*, 1992.
86. T. Masakazu, T. Katsuyoshi, I. Koji, K. Kiyoyuki, N. Mizuka and H. A. Yoshi, *Appl. Radiat. Isot.*, 1997, **48**, 607.
87. E. Iller, A. Deptula, M. Brykala, M. Sypula and M. Konior, *Eur. J. Nucl. Med. Mol. Imaging*, 2007, **34**, S210.
88. F. Monroy-Guzman, V. E. B. Almaraz, T. R. Gutierrez, L. G. Cohen, J. Cosgrove, F. F. Knapp Jr., P. R. Nava and C. J. Rosales, *Therapeutic Radionuclide Generators: $^{90}Sr/^{90}Y$ and $^{188}W/^{188}Re$ Generators*, IAEA-TRS-470, 2009.
89. H. Matsuoka, K. Hasimoto, Y. Hishinuma, K. Ishikawa, H. Terunuma and K. Tatenuma. *J. Nucl. Radiochem. Sci.*, 2005, **6**, 189.
90. E. Iller, H. Polkowska-Motrenko, W. Lada, D. Wawszczak, M. Sypula, K. Doner, M. Konior, J. Milczarek, J. Zoladek and J. Ralis, *J. Radioanal. Nucl. Chem.*, 2009, **281**, 83.
91. J. S. Lee, J. S. Lee, U. J. Park, K. J. Son and H. S. Han, *Appl. Radiat. Isot.*, 2009, **67**, 1162.

92. R. Chakravarty, R. Shukla, S. Gandhi, R. Ram, A. Dash, M. Venkatesh and A. K. Tyagi, *J. Nanosci. Nanotechnol.*, 2008, **8**, 4447.
93. R. Chakravarty, A. Dash and M. Venkatesh, *Chromatographia*, 2009, **69**, 1363.
94. R. Chakravarty, R. Shukla, R. Ram, A. K. Tyagi, A. Dash and M. Venkatesh, *Appl. Radiat. Isot.*, 2010, **68**, 229–238.
95. R. Chakravarty, R. Shukla, R. Ram, A. K. Tyagi, A. Dash and M. Venkatesh, *Chromatographia*, 2010, **72**, 875.
96. M. Khalid, A. Mushtaq and M. Z. Iqbal, *Sep. Sci.Technol.*, 2001, **36**, 3628.
97. J. Steigman, *Int. J. Appl. Radiat. Isot.*, 1982, **33**, 829.
98. British Pharmacopoeia Commission, British Pharmacopoeia, 2008, The Stationery Office, Norwich, U.K. (www.pharmacopoeia.org.uk)
99. A. Mushtaq, *Appl. Radiat. Isot.*, 1996, **47**, 727.
100. S. F. Yin, B. Q. Xu, *Chemphyschem.*, 2003, **3**, 277.
101. G. K. Chuah, S. Jaenicke, S. A. Cheong and K. S. Chan, *Appl Catal A: Gen.*,1996, **145**, 267.
102. G. Y. Guo and Y. L. Chen, *J. Solid State Chem.*, 2005, **178**, 1675.

CHAPTER 12
Sensing of Biomolecular Charges at Designer Nanointerfaces

TATSURO GODA AND YUJI MIYAHARA*

Institute of Biomaterials and Bioengineering, Tokyo Medical and Dental University, 2-3-10 Kanda-surugadai, Chiyoda, Tokyo 101-0062, Japan
*E-mail: miyahara.bsr@tmd.ac.jp

12.1 Introduction

Biosensors and biosensing protocols are able to detect a wide range of compounds, sensitively and selectively, and have applications in security, point-of-care analyses of diseases and environmental safety. Extensive research is being conducted to develop a series of biosensors for the detection of biomolecules and biorecognition events in more simple and straightforward ways. In recent years, the use of electrical biosensors in clinical diagnostics and in basic research in life sciences has attracted significant interest. Among the proposed concepts for producing such biosensors, the integration of bioactive molecules into a semiconductor device is currently recognized as one of the most attractive. The use of field-effect transistors (FETs) provides a method for label-free biosensing based on the electrostatic interactions of biomolecules in solution with the electrons in a semiconductor device. In other words, FET biosensors directly transform the intrinsic electrical charges in biomolecules into electrical signals at the gate–solution interface, where the electrical diffusion layer of the solution works as a capacitor. Biosensors are composed of a functional interface and an electrical transducer (Figure 12.1). A key issue is therefore the fabrication of a functional layer on the gate surface, which effectively captures target biomolecules and transforms the binding event into

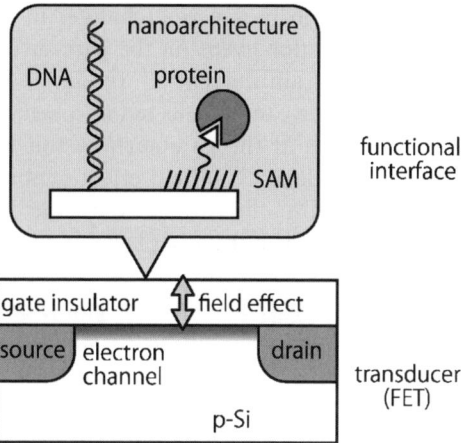

Figure 12.1 Fundamental concept and configuration of the FET biosensor.

an electrical signal. The strategy for achieving this is described in detail in the next section. FET biosensors based on complementary metal oxide semiconductors (CMOS) provide many advantages, such as possible miniaturization, quick signal-tracking, label-free detection, low output-impedance, on-chip integration into array formats, portable microanalysis, parallel detection systems and low-cost mass production. Signal enhancement by tagging an electrical tracer molecule to an analyte in the charge detection system is also becoming important in detecting biomarker species as cancer tracers. In this review, we focus on recent developments in FET-based biosensors achieved using sophisticated nano-/bio-technologies and manipulations at nanoscale interfaces.

12.2 Label-free Biosensing with Semiconductor Devices

12.2.1 FET Designer Nanointerfaces

In principle, FET biosensors can use electrical methods to detect an analyte bound on a gate surface: changes in the charge density on the gate electrode are directly transduced into an electrical signal by the field effect. The metal–insulator–semiconductor structure is useful for explaining the working principle of a FET. The gate insulator is placed between the gate and the semiconductor electrodes; the majority carriers in a p-type semiconductor are positively charged holes. When a sufficient positive bias voltage is applied, negatively charged mobile carriers, namely electrons, are induced at the insulator–silicon interface in p-type silicon, because the number of electrons exceeds that of the holes, and the semiconductor interface is thus inverted to n-type. The n-type source–drain current through the thin layer of electrons at the interface, called the inversion layer, is altered by the applied gate voltage.

When this principle is applied to biosensors, changes in charge density as a result of a biomolecular recognition event on the gate surface can be detected as a variation in the source–drain current.[1-7] In other words, a change in charge density at the gate surface can be detected as a result of an electrostatic interaction between the charges at the gate surface and the electrons in the inversion layer in the p-type silicon. The field effect enables variations in the gate potential, the source–drain current, or the gate conductivity to be monitored. A reference electrode is normally placed in an aqueous solution to control the potential of the gate electrode.

Chemical modification of the gate insulator or gate electrode to endow the semiconductor device with target recognition capability is one of the most interesting aspects of biosensor fabrication. Generally, a functional organosilane is attached to the silicon oxide surface *via* silanization to conjugate with ligands for reaction with biomolecules. The physisorbed organosilane undergoes rapid hydrolysis of the alkoxy groups, forming hydroxyl groups that can covalently interact with the silanol surface. Subsequently, the ligand molecules are immobilized on the organosilane layer using a wide variety of chemical-coupling reactions.[8-12] However, surface modifications by silanization usually lack rigorous arrangement of the molecular orientation of the immobilized ligands. The random orientation of the capturing molecules on the surface reduces affinity and binding efficiency. A more sophisticated method for immobilizing biomolecules on the chip surface uses the interaction between thiol groups and gold as an extended gate. This technique is based on quasi-covalent bonding between gold and thiol groups.[13,14] Alkyl chains of length greater than eight can form well-ordered and densely packed self-assembled monolayers (SAMs) in suitable solvents.[5,15] The use of an appropriate SAM helps in the orientation and controlled immobilization of biomolecules.[16,17] In addition, SAMs can reduce non-specific protein adsorption; this enables highly sensitive measurements with low limits of detection to be performed.[18,19] The other end-group of the alkanethiol is often functionalized so that it reacts with the ligand molecules.[20]

The use of nanomaterials for source–drain channels has attracted significant interest as a platform for potential biosensors; in such biosensors the conductance of a one-dimensional semiconducting nanomaterial is sensitively modulated upon target binding. Although the availability of nanomaterial-based FET biosensors is limited because of the difficulties in manufacturing them by either bottom-up or top-down lithography approaches, the opportunity to develop highly sensitive FET biosensors has attracted many researchers to this developing field. Nanowire (NW)-FETs, single-walled carbon nanotube (SWCNT)-FETs and graphene-FETs have been extensively investigated in recent years.[3,21-27] Semiconducting silicon NW-FETs and indium oxide NW-based FETs are regarded as promising devices with potential applications in the analysis of specific DNA sequences, biomarkers, viruses, enzymatic activities and potential drug molecules.[28-35] There are a number of review articles describing the parameters, such as methods of NW production, device dimensionality, and active measurement conditions, which influence device performance.[24,36,37] Because of their unique single atomic layer structure, charge transport through SWCNTs occurs entirely through the surface that is exposed to the environment.[38–43] SWCNT-FET devices contain a thin, wide Schottky contact region. Graphene is a one-atom-thick material consisting of a honeycomb structure of sp^2-bonded carbon atoms. Because graphene is a low-cost conductive, yet transparent, material with a low environmental impact, it is an ideal material for the construction of biosensor-based devices in various transduction modes, from electrical and electro-chemical transduction to optical transduction.[44] Graphene-FETs have been reported to show potential applications in label-free biosensing.[45–47] Graphene is expected to surpass SWCNTs, because it offers a large detection area, biocompatibility, and unique electrical properties. The development of graphene-based biosensors has become practical with the recent advent of chemical vapour deposition (CVD) of large graphene films.[48] Chemically modified graphenes, such as graphene oxide, are also excellent sensor materials, because they are sensitive, inexpensive, easy to work with, and amenable to wafer-scale production.[49,50] The ultrahigh sensitivity of nanomaterial-based FETs is a result of (i) their small size, which is comparable to that of biomolecules; (ii) their large surface-to-volume ratios; (iii) their ability to facilitate local charge transfers, leading to a current change as a result of the field effect; and (iv) additional signal generation at the Schottky contact region formed between the two different types of material. Experimental observations and theoretical modelling suggest that the sensitivity of NW-based FETs is substantially influenced by the diameter of the NW and its doping level.[51] A computational study showed that the field-effect strength exerted by the binding molecules has a significant impact on scaling behaviours.[52,53] On the other hand, careful control of the solution Debye length is critical for unambiguous selective detection of biomolecules.[54] A single biomolecule present on the surface or in close proximity to a SWCNT can alter its electronic properties by (1) charge transfer between the biomolecule and the nanotube; (2) a scattering potential for mobile charges; and (3) charge scattering caused by local deformation.[24,55] For SWCNT-FETs, a significant contribution to signal generation is attributed to the Schottky contact as well as to the field effect along the exposed lengths of the nanotubes.[56–61]

12.2.2 Nucleic Acids Sensing

Extensive research has been conducted with the aim of developing a series of biosensors for the detection of DNA recognition events, including hybridization, sequencing and single-nucleotide polymorphism (SNP) genotyping. In particular, following the completion of the human genome project, strong competition exists for developing a new technology and platform that can sequence a complete human genome for under US$1000. Attractive features of FET biosensors in genetic analysis include label-free, real-time, low-cost and sensitive detection of oligonucleotides on a solid surface in parallel. Because nucleic acids possess abundant negative charges, the FET can directly readout their intrinsic charges at a gate–solution nanointerface (Figure 12.2a). Generally, oligonucleotide probes are physically or chemically immobilized on the gate surface to capture the target DNA sequences. The target DNA molecule is selectively captured by a hydrogen-bonding-oriented affinity of Watson–Crick base-pairing. A wide variety of FET biosensors with different configurations have already been developed for DNA detection.[11,12,62–75] These DNA-biosensors are particularly attractive since they can be fabricated in very small sizes and massively integrated in lab-on-a-chip devices using silicon-manufacturing technologies that could also integrate many other electronic elements. DNA sequencing can be achieved using CMOS technology for the fabrication of FETs in combination with a stepwise single-base primer extension approach.[76]

Change in the proton concentration (*i.e.* proton activity) and other ion concentrations in solution are also detected as an induction of electrical potential using ion-sensitive FETs (ISFETs) (Figure 12.2b). A voltage between the substrate and the oxide surface is caused by an ion sheath. The surface hydrolysis of the hydroxyl groups of gate materials such as SiO_2, Si_3N_4 and Ta_2O_5 varies in a buffer solution because the concentration of protons is -59 mV per decade under ideal conditions (Nernstian responses).[77] Very recently, Ion Torrent (Guilford, CT, USA) has developed a platform system for label-free electrical DNA sequencing by monitoring the time-course of proton

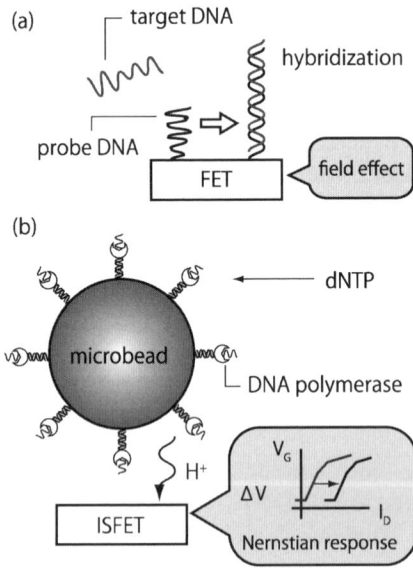

Figure 12.2 DNA sensing using FET: (a) detection of intrinsic charges on target DNA upon hybridization by the field effect and (b) detection of protons released by base-extension using ISFET.

activity using ISFET biosensors.[78] An array chip of 1.2 million CMOS–ISFET sensors in a flow cell can analyse approximately 25 million bases in parallel per run. Each sensor is topped with a small well designed to hold a single-stranded fragment of DNA. To sequence a strand of DNA, the machine synthesizes a complementary strand, sequentially attempting to add each of the four bases that make up DNA to the well one-by-one. When the correct base is incorporated into the growing sequence, it triggers a chemical reaction that releases a proton; this is detected by the ISFET sensor as an electrical signal. A computer stitches together the sequence by integrating these signals with knowledge of when each base flowed through the chip. Ion Torrent, which has recently been acquired by Life Technologies (Carlsbad, CA, USA),[79,80] sells these machines and kits for US$50 000 and US$500, respectively. This is the first example of commercialized genetic FETs for high-speed and high-throughput analysis of DNA.

12.2.3 Protein Sensing

Parallel detection of proteins using FET sensor chips is in huge demand in proteomic analysis. Embedding an antibody on the gate surface as a ligand may be effective for recognizing target proteins with high sensitivity and selectivity.[81–89] However, because of the predominant electrostatic charge-screening effect by mobile counter-ions in solution, caused by the large dimensions of the antibody molecule, immuno-FETs often lack generation of electrical signals in response to target recognition events.[90] Analytes located outside the diffusion layer at the gate/solution interface fail to generate an interfacial potential because of the counter-ion screening-effect by external electrolytes in the buffer solution. Hence, FET biosensors can selectively detect target molecules that are closely located to the gate surface. The characteristic Debye screening length of the solution (κ^{-1}) is generally described as a function of ionic strength (I)[91] as $\propto I^{-1/2}$, and is no more than 10 nm, even at the low ionic strength of 1 mM. The use of an antibody fragment as a ligand molecule may be effective in avoiding the limitations caused by charge-screening in a buffer solution.[92]

To address this issue, immobilization of small nucleic acid aptamers on the FET gate is effective in detecting a wide variety of biological entities with minimum effects from counter-ion screening in the solution.[93] Aptamers are made of oligomeric DNA, RNA, or peptides and show high affinities, specificities, and selectivities to given target molecules.[94–98] More interestingly, these aptamers represent a new type of molecular recognition element; they can be used as biosensors by immobilization on the sensing surface of FET devices because they are very small (ca. 2–4 nm) to be able to operate target recognition events within the diffusion layer at the interface. This can minimize electrical charge-screening from mobile counter-ions in solution. Aptamer-based FET biosensors allow sensitive and selective detection of a wide variety of proteins, based on charge read-outs.[99–107]

One of the attractive features of FET biosensors is the charge-based monitoring of intra- and inter-molecular conformational switching of biomolecules upon biorecognition events.[108] Substrate-binding-induced conformational displacement of proteins can alter distributions of local charge densities at the gate/solution interface, thereby inducing measureable signals with a FET device. Label-free monitoring of protein conformations will be an important toolbox for gaining a deeper understanding of biological science, since conformational transitions of proteins play major roles in biorecognition, cell signalling, gene regulation and biocatalytic reactions.

12.3 Manipulation of Charges for Sensitive Biosensing

12.3.1 DNA Binders and Intercalators

For genetic analysis in clinical diagnostics, SNPs are the most common form of DNA variation in humans and are important markers in "tailored medicine". An extensive collection of SNPs would serve as a valuable resource for the discovery of genetic factors affecting disease susceptibility and resistance. The development of SNP genotyping based on this information has advanced in recent years using a variety of methods,[109–125] and will enable clinicians to determine which pharmacological agent is most effective for treating a given patient's condition.[126] Among these SNP genotyping methods, fluorescent detection is widely used, and chemiluminescence and the mass of the reaction product are detected for pyrosequencing and mass spectrometry, respectively.[114,124] Several FET devices have been used for SNP genotyping in combination with allele-specific oligonucleotide hybridization or the primer extension reaction.[127–129] The use of DNA binders following hybridization events is an effective approach to discriminating between the signals of double-stranded DNA and those of non-specifically adsorbed single-stranded DNA at the gate surface. This is because some DNA binders specifically interact with double-stranded DNA. After hybridization, introduction of a DNA groove-binder and a DNA intercalator such as DAPI (4′,6-diamidino-2-phenylindole), Hoechst33258, EB (ethidium bromide), and PI (propidium iodide) can induce an additional V_T (threshold voltage) shift in the negative direction (Figure 12.3). The negative shift of the V_T is correlated with an increase in the positive charges on the gate surface as a result of the specific interaction of the DNA binder with the double-stranded DNA. This is the inverse phenomenon to the positive shift of the V_T caused by DNA hybridization. Thus, changes in the charge density on the gate surface after each molecular recognition event can be successfully detected using genetic FETs. The binding abilities of DNA binders and intercalators depend not only on their charges and chemical structures, but also on the base sequence of the DNA strands. Hoechst 33258, Hoechst 33342, and DAPI have affinities for the minor grooves of AT-rich sites of DNA molecules,[130,131] but show weak binding to GC sites.[132] In contrast, the intercalators EB and PI show high affinities for

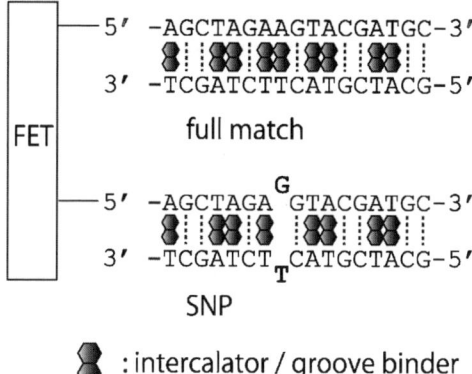

Figure 12.3 SNP genotyping based on charge manipulation using cationic intercalators/DNA binders.

double-stranded DNA and generally exhibit modest base-pair selectivity.[133,134] Thus, the ability to discriminate single base changes is enhanced by the use of a DNA binder in combination with genetic FETs. The use of a DNA binder after hybridization is more effective in distinguishing genotypes of the target DNA than is hybridization alone. Notably, the directions of the V_T shifts were positive for hybridization with target DNA and negative for the DNA binder. This is because of the intrinsic charges on the DNA (negative) and the DNA binder (positive).

12.3.2 Site-selective Charge Conversion of Proteins

In the context of charge detection, proteins represent more challenging targets in the following two ways. First, in contrast to the case with DNA biosensing, not all amino acids are charged, and only five amino acids are primarily responsible for electrical properties (arginine, lysine, and histidine are cationic, and aspartic acid and glutamic acid are anionic). Second, proteins are polyampholytes and form zwitterions in a certain pH range, so the magnitude of the change in the charge density in response to the target-binding event is insufficient. Indeed, the estimated charge density of human serum albumin is −5733 Da per charge at pH 7.4. The substantially lower charge density of proteins compared with that of DNA limits the detection of proteins at the subnanomolar level using typical FET-based protein biosensors.[99,101] To address this issue, we attempted to tag additional and stable negative charges to proteins using site-selective chemical modifications of the cationic amino acids, using conventional chemical reactions (Figure 12.4).[135] Succinic anhydride (SA) reacts specifically with the ε-amino groups of lysine and the α-amino groups of N-termini, converting the residues into anionic moieties.[136] The succinylation of each amino group therefore changes the net charge on a protein by up to two units. 2,3-Butanedione (BD) reacts solely with arginine, converting the cationic group to a neutral group.[137–139] Methylglyoxal (MG) can react with both lysine and arginine to reduce their cationic charges.[140–143] The site-selectivity of the charge conversions maintains the structural integrities and bioactivities of the proteins. The estimated negative charge-density of albumin at physiological pH increased by five-fold as a result of the formation of stable succinic lysine. The potentiometric signal upon albumin adsorption increased three-fold as a result of succinylation, and a more significant amplification of the signal (eleven-fold) was achieved by lysosome modification. Furthermore, *in-situ* modification of amino acids during potentiometry was possible for lysosomes adsorbed on the gate surface. Thus, manipulation of the protein charges provides a new type of molecular "label" for biosensing, with preserved conformational integrity of target proteins. Charge manipulation in nanomaterials finds many applications not only in protein diagnostics but also in drug-delivery systems using smart nanocarriers. Recently, charge-converting techniques have been used for efficient delivery of bioactive proteins with polyionic complex micelles into the cytoplasm of living cells *via* endocytic pathways.[144,145] The temporary introduction of additional electrical charges to the selected side-chains of proteins using reversible charge-converting chemistry can stabilize polyionic complex micelles by increasing the charge density under physiological salt conditions. When the surrounding pH drops from 7.4 to 5.5 at the endosomal-digestion stage, the charge-converted proteins degrade to their original form. The loss of the artificial charges on the protein induces disruption of micelles and efficient endosomal escape of the cargo biomolecules. Charge-converting techniques are also useful for manipulation of the electrical properties of synthetic non-viral vectors, such as lipoplexes and polyplexes, to facilitate intracellular delivery of nucleotides.[146–153] The encapsulation of nucleic acids/polycation polyplexes with smart charge-converting polymers decreases the cytotoxicity of the therapeutic nanocarriers and enhances endosomal escape and transfection efficiency. Gold nanoparticles coated with a charge-reversing

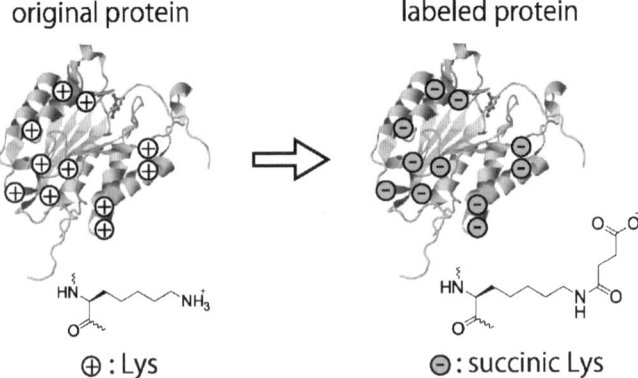

Figure 12.4 Charge conversion of specific amino acid residues in protein by chemical reactions.

polyelectrolyte also facilitate endosomal escape of bioactive nucleotides.[154] Charge manipulation of nanomaterials is a promising method for designing drug- and gene-delivery systems.

12.4 Summary

We described the importance of the preparation of nanoscale organic functional interfaces for FET biosensors. Such designer nanoarchitectures at the gate/solution interface can achieve highly sensitive and specific label-free detection of bioanalytes. Manipulations of the electrical charges on DNA and proteins at the molecular level are also essential for advanced biosensing such as SNP genotyping, as described in this chapter. Given the many inherent advantages of CMOS-based semiconductor devices, the FET biosensor is a strong candidate for use in lab-on-a-chip devices for point-of-care diagnostics.

Acknowledgments

Part of this work was financially supported by JST-CREST.

References

1. A. Poghossian, A. Cherstvy, S. Ingebrandt, A. Offenhausser and M. J. Schoning, *Sens. Actuators, B*, 2005, **111**, 470–480.
2. M. J. Schoning and A. Poghossian, *Electroanalysis*, 2006, **18**, 1893–1900.
3. B. L. Allen, P. D. Kichambare and A. Star, *Adv. Mater.*, 2007, **19**, 1439–1451.
4. F. R. R. Teles and L. R. Fonseca, *Talanta*, 2008, **77**, 606–623.
5. S. K. Arya, P. R. Solanki, M. Datta and B. D. Malhotra, *Biosens. Bioelectron.*, 2009, **24**, 2810–2817.
6. M. Mir, A. Homs and J. Samitier, *Electrophoresis*, 2009, **30**, 3386–3397.
7. C. S. Lee, S. K. Kim and M. Kim, *Sensors*, 2009, **9**, 7111–7131.
8. S. Ingebrandt and A. Offenhausser, *Phys. Status Solidi A*, 2006, **203**, 3399–3411.
9. E. Katz and I. Willner, *Electroanalysis*, 2003, **15**, 913–947.
10. T. Ohtake, C. Hamai, T. Uno, H. Tabata and T. Kawai, *Jpn. J. Appl. Phys., Part 2*, 2004, **43**, L1137–L1139.
11. T. Sakata, M. Kamahori and Y. Miyahara, *Mater. Sci. Eng., C*, 2004, **24**, 827–832.
12. Y. Han, A. Offenhausser and S. Ingebrandt, *Surf. Interface Anal.*, 2006, **38**, 176–181.
13. A. L. Crumbliss, S. C. Perine, J. Stonehuerner, K. R. Tubergen, J. G. Zhao and R. W. Henkens, *Biotechnol. Bioeng.*, 1992, **40**, 483–490.
14. M. C. Parker, N. Patel, M. C. Davies, C. J. Roberts, S. J. B. Tendler and P. M. Williams, *Protein Sci.*, 1996, **5**, 2329–2332.

15. M. Mrksich, C. S. Chen, Y. N. Xia, L. E. Dike, D. E. Ingber and G. M. Whitesides, *Proc. Natl. Acad. Sci. U. S. A.*, 1996, **93**, 10775–10778.
16. S. H. Jung, H. Y. Son, J. S. Yuk, J. W. Jung, K. H. Kim, C. H. Lee, H. Hwang and K. S. Ha, *Colloids Surf. B*, 2006, **47**, 107–111.
17. S. J. Kim, K. V. Gobi, H. Tanaka, Y. Shoyama and N. Miura, *Sens. Actuators, B*, 2008, **130**, 281–289.
18. E. Klein, P. Kerth and L. Lebeau, *Biomaterials*, 2008, **29**, 204–214.
19. O. R. Bolduc and J. F. Masson, *Langmuir*, 2008, **24**, 12085–12091.
20. A. B. Kharitonov, M. Zayats, A. Lichtenstein, E. Katz and I. Willner, *Sens. Actuators, B*, 2000, **70**, 222–231.
21. E. Katz and I. Willner, *ChemPhysChem*, 2004, **5**, 1085–1104.
22. K. Balasubramanian and M. Burghard, *Anal. Bioanal. Chem.*, 2006, **385**, 452–468.
23. S. N. Kim, J. F. Rusling and F. Papadimitrakopoulos, *Adv. Mater.*, 2007, **19**, 3214–3228.
24. S. Roy and Z. Q. Gao, *Nano Today*, 2009, **4**, 318–334.
25. R. A. Villamizar, A. Maroto and F. X. Rius, *Sens. Actuators, B*, 2009, **136**, 451–457.
26. C. B. Jacobs, M. J. Peairs and B. J. Venton, *Anal. Chim. Acta*, 2010, **662**, 105–127.
27. K.-I. Chen, B.-R. Li and Y.-T. Chen, *Nano Today*, 2011, **6**, 131–154.
28. J. Hahm and C. M. Lieber, *Nano Lett.*, 2004, **4**, 51–54.
29. Y. L. Bunimovich, Y. S. Shin, W. S. Yeo, M. Amori, G. Kwong and J. R. Heath, *J. Am. Chem. Soc.*, 2006, **128**, 16323–16331.
30. Z. Q. Gao, A. Agarwal, A. D. Trigg, N. Singh, C. Fang, C. H. Tung, Y. Fan, K. D. Buddharaju and J. M. Kong, *Anal. Chem.*, 2007, **79**, 3291–3297.
31. W. C. Maki, N. N. Mishra, E. G. Cameron, B. Filanoski, S. K. Rastogi and G. K. Maki, *Biosens. Bioelectron.*, 2008, **23**, 780–787.
32. C. H. Lin, C. H. Hung, C. Y. Hsiao, H. C. Lin, F. H. Ko and Y. S. Yang, *Biosens. Bioelectron.*, 2009, **24**, 3019–3024.
33. T. Kudo, T. Kasama, T. Ikeda, Y. Hata, S. Tokonami, S. Yokoyama, T. Kikkawa, H. Sunami, T. Ishikawa, M. Suzuki, K. Okuyama, T. Tabei, K. Ohkura, Y. Kayaba, Y. Tanushi, Y. Amemiya, Y. Cho, T. Monzen, Y. Murakami, A. Kuroda and A. Nakajima, *Jpn. J. Appl. Phys.*, 2009, **48**, 06FJ04.
34. K. S. Chang, C. C. Chen, J. T. Sheu and Y. K. Li, *Sens. Actuators, B*, 2009, **138**, 148–153.
35. G. F. Zheng, X. P. A. Gao and C. M. Lieber, *Nano Lett.*, 2010, **10**, 3179–3183.
36. M. Curreli, R. Zhang, F. N. Ishikawa, H. K. Chang, R. J. Cote, C. Zhou and M. E. Thompson, *IEEE Trans. Nanotechnol.*, 2008, **7**, 651–667.
37. D. Wei, M. J. A. Bailey, P. Andrew and T. Ryhanen, *Lab Chip*, 2009, **9**, 2123–2131.
38. A. Star, E. Tu, J. Niemann, J. C. P. Gabriel, C. S. Joiner and C. Valcke, *Proc. Natl. Acad. Sci. U. S. A.*, 2006, **103**, 921–926.

39. T. Dastagir, E. S. Forzani, R. Zhang, I. Amlani, L. A. Nagahara, R. Tsui and N. Tao, *Analyst*, 2007, **132**, 738–740.
40. S. Kim, T. G. Kim, H. R. Byon, H. J. Shin, C. Ban and H. C. Choi, *J. Phys. Chem. B*, 2009, **113**, 12164–12168.
41. S. Sorgenfrei, C. Y. Chiu, R. L. Gonzalez, Y. J. Yu, P. Kim, C. Nuckolls and K. L. Shepard, *Nat. Nanotechnol.*, 2011, **6**, 125–131.
42. J. W. Ko, J.-M. Woo, J. Ahn, J. H. Cheon, J. H. Lim, S. H. Kim, H. Chun, E. Kim and Y. J. Park, *ACS Nano*, 2011, **5**, 4365–4372.
43. R. A. Villamizar, A. Maroto and F. Xavier Rius, *Anal. Bioanal. Chem.*, 2011, **399**, 119–126.
44. A. K. Geim, *Science*, 2009, **324**, 1530–1534.
45. X. C. Dong, Y. M. Shi, W. Huang, P. Chen and L. J. Li, *Adv. Mater.*, 2010, **22**, 1649–1652.
46. Y. Ohno, K. Maehashi and K. Matsumoto, *J. Am. Chem. Soc.*, 2010, **132**, 18012–18013.
47. S.-R. Guo, J. Lin, M. Penchev, E. Yengel, M. Ghazinejad, C. S. Ozkan and M. Ozkan, *J. Nanosci. Nanotechnol.*, 2011, **11**, 5258–5263.
48. X. S. Li, W. W. Cai, J. H. An, S. Kim, J. Nah, D. X. Yang, R. Piner, A. Velamakanni, I. Jung, E. Tutuc, S. K. Banerjee, L. Colombo and R. S. Ruoff, *Science*, 2009, **324**, 1312–1314.
49. R. Stine, J. T. Robinson, P. E. Sheehan and C. R. Tamanaha, *Adv. Mater.*, 2010, **22**, 5297–5300.
50. S. Mao, G. H. Lu, K. H. Yu, Z. Bo and J. H. Chen, *Adv. Mater.*, 2010, **22**, 3521–3526.
51. N. Elfstrom, R. Juhasz, I. Sychugov, T. Engfeldt, A. E. Karlstrom and J. Linnros, *Nano Lett.*, 2007, **7**, 2608–2612.
52. F. S. Zhou and Q. H. Wei, *Nanotechnology*, 2008, **19**, 015504.
53. X. P. A. Gao, G. F. Zheng and C. M. Lieber, *Nano Lett.*, 2010, **10**, 547–552.
54. E. Stern, R. Wagner, F. J. Sigworth, R. Breaker, T. M. Fahmy and M. A. Reed, *Nano Lett.*, 2007, **7**, 3405–3409.
55. A. Star, J. C. P. Gabriel, K. Bradley and G. Gruner, *Nano Lett.*, 2003, **3**, 459–463.
56. X. D. Cui, M. Freitag, R. Martel, L. Brus and P. Avouris, *Nano Lett.*, 2003, **3**, 783–787.
57. R. J. Chen, H. C. Choi, S. Bangsaruntip, E. Yenilmez, X. W. Tang, Q. Wang, Y. L. Chang and H. J. Dai, *J. Am. Chem. Soc.*, 2004, **126**, 1563–1568.
58. A. B. Artyukhin, M. Stadermann, R. W. Friddle, P. Stroeve, O. Bakajin and A. Noy, *Nano Lett.*, 2006, **6**, 2080–2085.
59. L. Zhang, S. Zaric, X. M. Tu, X. R. Wang, W. Zhao and H. J. Dai, *J. Am. Chem. Soc.*, 2008, **130**, 2686–2691.
60. M. Cha, S. Jung, M. H. Cha, G. Kim, J. Ihm and J. Lee, *Nano Lett.*, 2009, **9**, 1345–1349.
61. R. Kalantari-Nejad, M. Bahrami, H. Rafii-Tabar, I. Rungger and S. Sanvito, *Nanotechnology*, 2010, **21**, 445501.

62. E. Souteyrand, J. P. Cloarec, J. R. Martin, C. Wilson, I. Lawrence, S. Mikkelsen and M. F. Lawrence, *J. Phys. Chem. B*, 1997, **101**, 2980–2985.
63. H. Berney, J. West, E. Haefele, J. Alderman, W. Lane and J. K. Collins, *Sens. Actuators, B*, 2000, **68**, 100–108.
64. J. Fritz, E. B. Cooper, S. Gaudet, P. K. Sorger and S. R. Manalis, *Proc. Natl. Acad. Sci. U. S. A.*, 2002, **99**, 14142–14146.
65. F. Wei, B. Sun, Y. Guo and X. S. Zhao, *Biosensors Bioelectron.*, 2003, **18**, 1157–1163.
66. D. S. Kim, Y. T. Jeong, H. K. Lyu, H. J. Park, H. S. Kim, J. K. Shin, P. Choi, J. H. Lee, G. Lim and M. Ishida, *Jpn. J. Appl. Phys., Part 1*, 2003, **42**, 4111–4115.
67. D. S. Kim, Y. T. Jeong, H. J. Park, J. K. Shin, P. Choi, J. H. Lee and G. Lim, *Biosens. Bioelectron.*, 2004, **20**, 69–74.
68. D. S. Kim, H. J. Park, H. M. Jung, J. K. Shin, P. Choi, J. H. Lee and G. Lim, *Jpn. J. Appl. Phys., Part 1*, 2004, **43**, 3855–3859.
69. F. Uslu, S. Ingebrandt, D. Mayer, S. Bocker-Meffert, M. Odenthal and A. Offenhausser, *Biosens. Bioelectron.*, 2004, **19**, 1723–1731.
70. F. Pouthas, C. Gentil, D. Cote and U. Bockelmann, *Appl. Phys. Lett.*, 2004, **84**, 1594–1596.
71. T. Sakata, S. Matsumoto, Y. Nakajima and Y. Miyahara, *Jpn. J. Appl. Phys., Part 1*, 2005, **44**, 2860–2863.
72. T. Sakata, M. Kamahori and Y. Miyahara, *Jpn. J. Appl. Phys., Part 1*, 2005, **44**, 2854–2859.
73. T. Sakata and Y. Miyahara, *Biosens. Bioelectron.*, 2005, **21**, 827–832.
74. M. Barbaro, A. Bonfiglio, L. Raffo, A. Alessandrini, P. Facci and I. Barak, *Sens. Actuators, B*, 2006, **118**, 41–46.
75. H. S. Im, X. J. Huang, B. Gu and Y. K. Choi, *Nat. Nanotechnol.*, 2007, **2**, 430–434.
76. T. Sakata and Y. Miyahara, *Angew. Chem., Int. Edn.*, 2006, **45**, 2225–2228.
77. E. Bakker, P. Buhlmann and E. Pretsch, *Talanta*, 2004, **63**, 3–20.
78. J. M. Rothberg, W. Hinz, T. M. Rearick, J. Schultz, W. Mileski, M. Davey, J. H. Leamon, K. Johnson, M. J. Milgrew, M. Edwards, J. Hoon, J. F. Simons, D. Marran, J. W. Myers, J. F. Davidson, A. Branting, J. R. Nobile, B. P. Puc, D. Light, T. A. Clark, M. Huber, J. T. Branciforte, I. B. Stoner, S. E. Cawley, M. Lyons, Y. T. Fu, N. Homer, M. Sedova, X. Miao, B. Reed, J. Sabina, E. Feierstein, M. Schorn, M. Alanjary, E. Dimalanta, D. Dressman, R. Kasinskas, T. Sokolsky, J. A. Fidanza, E. Namsaraev, K. J. McKernan, A. Williams, G. T. Roth and J. Bustillo, *Nature*, 2011, **475**, 348–352.
79. *Chem Eng. News*, 2010, **88**, 18.
80. M. Eisenstein, *Nat. Biotechnol.*, 2010, **28**, 544–546.
81. M. Gotoh, M. Suzuki, I. Kubo, E. Tamiya and I. Karube, *J. Mol. Catal.*, 1989, **53**, 285–292.
82. P. Bergveld, *Biosens. Bioelectron.*, 1991, **6**, 55–72.

83. B. S. Kang, H. T. Wang, T. P. Lele, Y. Tseng, F. Ren, S. J. Pearton, J. W. Johnson, P. Rajagopal, J. C. Roberts, E. L. Piner and K. J. Linthicum, *Appl. Phys. Lett.*, 2007, **91**, 112106.
84. S. Gupta, M. Elias, X. J. Wen, J. Shapiro, L. Brillson, W. Lu and S. C. Lee, *Biosens. Bioelectron.*, 2008, **24**, 505–511.
85. J. P. Kim, B. Y. Lee, J. Lee, S. Hong and S. J. Sim, *Biosens. Bioelectron.*, 2009, **24**, 3372–3378.
86. A. Kim, C. S. Ah, C. W. Park, J. H. Yang, T. Kim, C. G. Ahn, S. H. Park and G. Y. Sung, *Biosens. Bioelectron.*, 2010, **25**, 1767–1773.
87. H. J. Park, S. K. Kim, K. Park, S. Y. Yil, J. W. Chung, B. H. Chung and M. Kim, *Sens. Lett.*, 2010, **8**, 233–237.
88. S. Hideshima, R. Sato, S. Kuroiwa and T. Osaka, *Biosens. Bioelectron.*, 2011, **26**, 2419–2425.
89. S. M. Kwon, G. B. Kang, Y. T. Kim, Y.-H. Kim and B.-K. Ju, *J. Nanosci. Nanotechnol.*, 2011, **11**, 1511–1514.
90. P. R. Nair and M. A. Alam, *Nano Lett.*, 2008, **8**, 1281–1285.
91. D. C. Brydges and P. A. Martin, *J. Stat. Phys.*, 1999, **96**, 1163–1330.
92. T. Sakata, M. Ihara, I. Makino, Y. Miyahara and H. Ueda, *Anal. Chem.*, 2009, **81**, 7532–7537.
93. H. R. Byon, S. Kim and H. C. Choi, *Nano*, 2008, **3**, 415–431.
94. L. Bock, L. Griffin, J. Latham, E. Vermaas and J. Toole, *Nature*, 1992, 564–566.
95. R. Macaya, P. Schultze, F. Smith, J. Roe and J. Feigon, *Proc. Natl. Acad. Sci. U. S. A.*, 1993, **90**, 3745–3749.
96. D. Wilson and J. Szostak, *Annu. Rev. Biochem.*, 1999, **68**, 611–647.
97. S. Jayasena, *Clin. Chem.*, 1999, **45**, 1628–1650.
98. T. Hermann and D. Patel, *Science*, 2000, **287**, 820–825.
99. K. Maehashi, T. Katsura, K. Kerman, Y. Takamura, K. Matsumoto and E. Tamiya, *Anal. Chem.*, 2007, **79**, 782–787.
100. H. Yoon, J. H. Kim, N. Lee, B. G. Kim and J. Jang, *ChemBioChem*, 2008, **9**, 634–641.
101. K. S. Kim, H. S. Lee, J. A. Yang, M. H. Jo and S. K. Hahn, *Nanotechnology*, 2009, **20**, 235501.
102. H. S. Lee, K. S. Kim, C. J. Kim, S. K. Hahn and M. H. Jo, *Biosens. Bioelectron.*, 2009, **24**, 1801–1805.
103. T. An, K. S. Kim, S. K. Hahn and G. Lim, *Lab Chip*, 2010, **10**, 2052–2056.
104. O. S. Kwon, S. J. Park and J. Jang, *Biomaterials*, 2010, **31**, 4740–4747.
105. P. Estrela, D. Paul, Q. F. Song, L. K. J. Stadler, L. Wang, E. Huq, J. J. Davis, P. K. Ferrigno and P. Migliorato, *Anal. Chem.*, 2010, **82**, 3531–3536.
106. S. H. Han, S. K. Kim, K. Park, S. Y. Yi, H. J. Park, H. K. Lyu, M. Kim and B. H. Chung, *Anal. Chim. Acta*, 2010, **665**, 79–83.
107. J. P. Kim, S. Hong and S. J. Sim, *J. Nanosci. Nanotechnol.*, 2011, **11**, 4182–4187.
108. H. J. Park, S. K. Kim, K. Park, H. K. Lyu, C. S. Lee, S. J. Chung, W. S. Yun, M. Kim and B. H. Chung, *FEBS Lett.*, 2009, **583**, 157–162.

109. U. Landegren, R. Kaiser, J. Sanders and L. Hood, *Science*, 1988, **241**, 1077–1080.
110. M. Orita, H. Iwahana, H. Kanazawa, K. Hayashi and T. Sekiya, *Proc. Natl. Acad. Sci. U. S. A.*, 1989, **86**, 2766–2770.
111. R. K. Saiki, P. S. Walsh, C. H. Levenson and H. A. Erlich, *Proc. Natl. Acad. Sci. U. S. A.*, 1989, **86**, 6230–6234.
112. M. Chee, R. Yang, E. Hubbell, A. Berno, X. C. Huang, D. Stern, J. Winkler, D. J. Lockhart, M. S. Morris and S. P. A. Fodor, *Science*, 1996, **274**, 610–614.
113. T. Pastinen, J. Partanen and A. C. Syvanen, *Clin. Chem.*, 1996, **42**, 1391–1397.
114. D. J. Lockhart, H. L. Dong, M. C. Byrne, M. T. Follettie, M. V. Gallo, M. S. Chee, M. Mittmann, C. W. Wang, M. Kobayashi, H. Horton and E. L. Brown, *Nat. Biotechnol.*, 1996, **14**, 1675–1680.
115. B. L. Parsons and R. H. Heflich, *Mutat. Res., Rev. Mutat. Res.*, 1997, **387**, 97–121.
116. T. Pastinen, A. Kurg, A. Metspalu, L. Peltonen and A. C. Syvanen, *Genome Res.*, 1997, **7**, 606–614.
117. L. A. Haff and I. P. Smirnov, *Genome Res.*, 1997, **7**, 378–388.
118. S. Tyagi, D. Bratu and F. Kramer, *Nat. Biotechnol.*, 1998, **16**, 49–53.
119. M. Ronaghi, M. Uhlen and P. Nyren, *Science*, 1998, **281**, 363–365.
120. W. M. Howell, M. Jobs, U. Gyllensten and A. J. Brookes, *Nat. Biotechnol.*, 1999, **17**, 87–88.
121. P. N. Gilles, D. J. Wu, C. B. Foster, P. J. Dillon and S. J. Chanock, *Nat. Biotechnol.*, 1999, **17**, 365–370.
122. T. Pastinen, M. Raitio, K. Lindroos, P. Tainola, L. Peltonen and A. C. Syvanen, *Genome Res.*, 2000, **10**, 1031–1042.
123. A. C. Syvanen, *Nat. Rev. Genet.*, 2001, **2**, 930–942.
124. M. A. Cooper, F. N. Dultsev, T. Minson, V. P. Ostanin, C. Abell and D. Klenerman, *Nat. Biotechnol.*, 2001, **19**, 833–837.
125. M. Jobs, W. M. Howell, L. Stromqvist, T. Mayr and A. J. Brookes, *Genome Res.*, 2003, **13**, 916–924.
126. M. Schena, D. Shalon, R. W. Davis and P. O. Brown, *Science*, 1995, **270**, 467–470.
127. T. Kajiyama, Y. Miyahara, L. J. Kricka, P. Wilding, D. J. Graves, S. Surrey and P. Fortina, *Genome Res.*, 2003, **13**, 467–475.
128. T. Sakata and Y. Miyahara, *ChemBioChem*, 2005, **6**, 703–710.
129. S. Ingebrandt, Y. Han, F. Nakamura, A. Poghossian, M. J. Schoning and A. Offenhausser, *Biosens. Bioelectron.*, 2007, **22**, 2834–2840.
130. M. Sriram, G. A. Vandermarel, H. Roelen, J. H. Vanboom and A. H. J. Wang, *Biochemistry*, 1992, **31**, 11823–11834.
131. D. L. Boger, B. E. Fink, S. R. Brunette, W. C. Tse and M. P. Hedrick, *J. Am. Chem. Soc.*, 2001, **123**, 5878–5891.
132. W. D. Wilson, F. A. Tanious, H. J. Barton, L. Strekowski, D. W. Boykin and R. L. Jones, *J. Am. Chem. Soc.*, 1989, **111**, 5008–5010.

133. H. S. Rye, S. Yue, M. A. Quesada, R. P. Haugland, R. A. Mathies and A. N. Glazer, *Methods Enzymol.*, 1993, **217**, 414–431.
134. W. A. Dengler, J. Schulte, D. P. Berger, R. Mertelsmann and H. H. Fiebig, *Anti-Cancer Drugs*, 1995, **6**, 522–532.
135. T. Goda and Y. Miyahara, *Anal. Chem.*, 2010, **82**, 8946–8953.
136. A. D. Gounaris and G. E. Perlmann, *J. Biol. Chem.*, 1967, **242**, 2739–2746.
137. J. A. Yankeelov Jr, *Biochemistry*, 1970, **9**, 2433–2439.
138. J. F. Riordan, K. D. McElvany and C. L. Borders, *Science*, 1977, **195**, 884–886.
139. A. Leitner and W. Lindner, *Anal. Chem.*, 2005, **77**, 4481–4488.
140. T. W. C. Lo, M. E. Westwood, A. C. McLellan, T. Selwood and P. J. Thornalley, *J. Biol. Chem.*, 1994, **269**, 32299–32305.
141. M. E. Westwood, A. C. McLellan and P. J. Thornalley, *J. Biol. Chem.*, 1994, **269**, 32293–32298.
142. M. U. Ahmed, E. B. Frye, T. P. Degenhardt, S. R. Thorpe and J. W. Baynes, *Biochem. J.*, 1997, **324**, 565–570.
143. Y. Gao and Y. Wang, *Biochemistry*, 2006, **45**, 15654–15660.
144. Y. Lee, T. Ishii, H. Cabral, H. J. Kim, J.-H. Seo, N. Nishiyama, H. Oshima, K. Osada and K. Kataoka, *Angew. Chem., Int. Ed.*, 2009, **48**, 5309–5312.
145. Y. Lee, T. Ishii, H. J. Kim, N. Nishiyama, Y. Hayakawa, K. Itaka and K. Kataoka, *Angew. Chem., Int. Ed.*, 2010, **49**, 2552–2555.
146. Y. Lee, S. Fukushima, Y. Bae, S. Hiki, T. Ishii and K. Kataoka, *J. Am. Chem. Soc.*, 2007, **129**, 5362–5363.
147. Y. Lee, K. Miyata, M. Oba, T. Ishii, S. Fukushima, M. Han, H. Koyama, N. Nishiyama and K. Kataoka, *Angew. Chem., Int. Ed.*, 2008, **47**, 5163–5166.
148. K. Miyata, M. Oba, M. Nakanishi, S. Fukushima, Y. Yamasaki, H. Koyama, N. Nishiyama and K. Kataoka, *J. Am. Chem. Soc.*, 2008, **130**, 16287–16294.
149. Z. Zhou, Y. Shen, J. Tang, M. Fan, E. A. Van Kirk, W. J. Murdoch and M. Radosz, *Adv. Funct. Mater.*, 2009, **19**, 3580–3589.
150. M. Sanjoh, S. Hiki, Y. Lee, M. Oba, K. Miyata, T. Ishii and K. Kataoka, *Macromol. Rapid Commun.*, 2010, **31**, 1181–1186.
151. Y. Q. Shen, Z. X. Zhuo, M. H. Sui, J. B. Tang, P. S. Xu, E. A. Van Kirk, W. J. Murdoch, M. H. Fan and M. Radosz, *Nanomedicine*, 2010, **5**, 1205–1217.
152. J.-Z. Du, T.-M. Sun, W.-J. Song, J. Wu and J. Wang, *Angew. Chem., Int. Ed.*, 2010, **49**, 3621–3626.
153. F. Pittella, M. Zhang, Y. Lee, H. J. Kim, T. Tockary, K. Osada, T. Ishii, K. Miyata, N. Nishiyama and K. Kataoka, *Biomaterials*, 2011, **32**, 3106–3114.
154. S. Guo, Y. Huang, Q. Jiang, Y. Sun, L. Deng, Z. Liang, Q. Du, J. Xing, Y. Zhao, P. C. Wang, A. Dong and X.-J. Liang, *ACS Nano*, 2010, **4**, 5505–5511.

CHAPTER 13
Nanostructured Materials for Biosensor Applications: Comparative Review of Preparation Methods

V. I. CHEGEL

V. E. Lashkaryov Institute of Semiconductor Physics, National Academy of Sciences of Ukraine, Department of Functional Optoelectronic Transducers, 41, prospect Nauki, Kyiv, 03028, Ukraine
E-mail: vche111@yahoo.com

13.1 Introduction

Manufacturing of nanostructures for biosensing is a promising and interdisciplinary research area involving biology, chemistry, physics and medicine.[1,2] Today, most of the traditional and new experimental methods are evolving to develop nanomaterials, and are used to fabricate different nanostructured systems that can be used in biosensors. It is crucially important to develop and utilize novel functional nanomaterials with well-defined structure and properties for biosensor applications.[3–5] During the last decade, a great number of nanomaterials and nanofabrication techniques have been developed, opening up the new possibilities in biological and biomedical studies.[1,2,6,7] Up to now, a number of novel nanomaterials, such as carbon and polymeric nanotubes,[5] quantum dots[3] and magnetic nanoparticles[4] have been studied and utilized in biotechnology applications. Their unique physicochemical properties provide a new basis for researching complicated biological processes that are difficult to study using conventional methods.[3–7] Nevertheless, most of widely applied methods in the nano-manufacturing of biosensors at present time are based on utilization of materials with well-known properties, such as gold, silver, carbon and silicon. On the other hand, application of these materials for creation of nanostructures reveals new problems related to nanoscale sizes, where properties of even well-studied materials can differ widely.[8] From this point of view, one of the problems in the manufacturing of nanostructures for biosensors is to control their parameters during the fabrication process. Usually used characterization techniques, such as AFM, STM or TEM, are expensive, whereas even specialized spectral methods do not provide clear picture for a full range of parameters. The modeling of structural and physical characteristics, provided by finite-difference time domain (FDTD), discrete dipole approximation (DDA) or other similar methods, offers additional possibilities in this direction,[9] especially with rapid development of powerful computing facilities, but does not offer a complete resolution of the problem. These problems probably explain the slower appearance of commercially available nanobiosensors on the world market in comparison with conventional biosensors,[10] despite the huge number of scientific papers in the field of manufacturing of nanostructures for biosensors.[11–15]

Because the topic of nano-manufacturing for biosensing is quite extensive for a limited publication, this review highlights the current state of manufacturing of the most widely used so-called "plasmonic" (made of Au and Ag) nanomaterials and nanosystems only. Different techniques used for the fabrication and assembly alignment of nanostructures from Au and Ag for biosensor applications are presented here. The advantages and disadvantages of various nano-patterning techniques are discussed. The evaluation of such technologies for improving biosensor performance in terms of high sensitivity and low detection limits is presented. Starting from simple methods, such as preparation of nanostructures from colloidal solutions, to more complicated technologies, like nanoimprint lithography or multistage synthesis methods are highlighted.

13.2 Nanostructures Manufactured Using Plasmonic Colloidal Nanoparticles

13.2.1 Utilization of Noble Metal Nanostructures in Plasmonic Biosensors

Highly conductive Au or Ag (so-called plasmonic) colloidal nanoparticles (NPs) are widely used in manufacturing of nanostructures for biosensors,[16–18] starting from pioneer work (Turkevich *et al.*)[19] with preparation of citrate-stabilized colloidal gold solutions. Due to their small size (1–150 nm) these NPs have high surface area to volume ratio and high adsorption activity[20] that provide good conditions for chemical modification of their surface and stable immobilization for a large amount of biomolecules. The strong ability of Au and Ag nanoparticles for electric field enhancing and light scattering are widely used for amplification of signals at manufactured nanostructured surfaces for surface enhanced Raman spectroscopy (SERS)[21–23] and surface enhanced fluorimetry (SEF).[24–26] Because gold and silver nanoparticles are excellent objects for observation of phenomenon of localized surface plasmon resonance (LSPR), this technology offers wide possibilities to be utilized in developing of optoelectronic LSPR-biosensors.[17,27–30] Development of biosensors that make use of LSPR for transducing the biomolecular processes near the metal surface into the detectable optical response has been carried out widely during the last decade.[17,27–30] This considerable attention is explained by weighty advantages provided by this technique, among which are real-time label-free biomolecular detection[31–33] and low-cost measuring equipment – the measured response can be registered using a compact UV-Vis spectrophotometer. Among the advantages of LSPR-based biosensors is a possibility to work in light transmission configuration that provides a simpler optical arrangement, and the possibility to minimize the probed area in comparison with conventional Kretschmann-type surface plasmon polariton resonance systems. Because LSPR will be often mentioned later on, we consider this phenomenon in more detail. The physical properties of LSPR are defined by the collective properties of conduction electrons of noble metal nanoparticles. A high concentration of free electron gas leads to its collective oscillations being resonantly excited by incident light, with wavelengths ranging from UV to NIR depending on the material properties[34,35] and the geometrical parameters of the nanoparticle.[34,36] Such a specific electronic response appears as an unusual optical property of nanosized metal formations – a light extinction peak appears in contrast to the peakless extinction spectrum for the bulk material. Both the shape and spectral position of the nanoparticle LSPR extinction peak depend on the refractive index of the ambient medium, which also applies to any possible coatings on the surface of the nanoparticle.[37,38] This specific peculiarity is a basis for the development of a surrounding medium refractive index sensor or a biosensor that can detect the presence of biomolecules covering the nanoparticle surface. For plasmonic nanosystems, optical absorbance strongly depends on the distance between nanoparticles and, for instance, during aggregation of colloidal Ag or Au nanoparticles, the absorbance spectrum changes significantly, which results in color transformation of solution.[39] An identical color transformation could be detected upon addition of chemical or biological substances, which promotes the aggregation of nanoparticles, and this peculiarity can be a basis for future chemical and biological sensors against corresponding molecules.[40–42]

13.2.2 Preparation of Colloidal Plasmonic Nanostructures

The basic technology for preparation of colloidal metal nanoparticles is reduction of metals from their ion solution using different types of reductants. In commonly a used method (developed by Turkevich et al. in 1951) the synthesis of gold nanoparticles (AuNPs) is performed by reduction from the oxidized state of tetrachloroaurate ions (AuCl$_4^-$) by boiling in sodium citrate (Na$_3$C$_6$H$_5$O$_7$) aqueous solution.[19] AuNPs obtained using this method appear as spheroids with a diameter about of 10–15 nm, which due to weakly bound citrate-ion cover are charged negatively and characterized by a plasmonic absorbance band approximately at 520 nm.[43] In 1998, Murray used dodecanethiol as a stabilizer instead of citrate and NaBH$_4$ as oxidant for synthesis of colloidal NPs with diameters varied between of 1.5 and 5.2 nm.[44] It should be noted that the size and monodispersity of plasmonic NPs is of importance for LSPR biosensors, because, theoretically, bigger NP size and more narrow peak of absorbance reveal better sensitivity,[30,45,46] whereas the narrow peak of the absorbance spectrum is provided by the high monodispersity of NPs. In practice, different objects for biosensing require NPs with different size, so it is mandatory to develop techniques, which allow preparation of nanoparticles with high monodispersity and wide size range. Several approaches exist to prepare particles with required dimensions using protocols with controlled nanoparticle formation.[47,48,49] For instance, a method in which a ratio between stabilizing and reducing reagents is varied was proposed by Frens et al.[48] This allows the synthesis of particles within a wide range of sizes, having diameters from 10 up to 140 nm. The disadvantage of this approach is poor monodispersity for NPs with a diameter more than 20 nm. It is possible to prepare smaller gold nanoparticles with diameters from 1.5 to 20 nm using a similar approach proposed by Leff et al., in which thiol was used as stabilizing agent.[49] This method requires approximately 15 h to reach equilibrium in the system. To obtain the AuNPs of different size and good monodispersity, Sau and coauthors[47] developed a fast technique of size control, in which gold nanoparticles with diameters ranging from 5 to 110 nm were prepared in two stages using seed-based technology. In the first stage, seed nanoparticles with a diameter between 5 and 20 nm were prepared by reducing gold ions by a TX-100 non-ionic surfactant with UV irradiation as the reaction amplifier. The bigger size nanoparticles were prepared on the second stage, where seed nanoparticles were placed in a fresh gold ion solution and the subsequent process of NP growth was provided by adsorption of Au ions on the surface of seed particles and subsequent reduction of adsorbed ions by ascorbic acid. Commonly, multiple factors of the gold salt reductive reaction, including concentration of agents, type of stabilizer, pH, ionic strength, temperature and lighting level, are used to influence the size, morphology and monodispersity of nanoparticles. Using stronger reductive agents, like NaBH$_4$, results in smaller NPs. The nanoparticles with a bigger size can be also be effectively prepared using electrochemical reduction of gold ions in presence of a surfactant.[50–52]

13.2.3 Core–Shell Nanoparticles

There has been increased interest in the bionanotechnology research for core–shell nanoparticles due to their potential in biosensing, diagnostic and imaging technologies.[53,54] Core–shell nanoparticles containing both magnetic material and gold have been used for bioseparation applications.[55] The LSPR of Au NPs made it possible to follow the positions of individual particles with size smaller than the optical diffraction limit.[56] The recent progress in microfluidic systems opens a way to apply nanoparticles with biomodified surfaces in biologic assays that allow selective detection of pathogens or antibodies to these pathogens from biological fluids.[57,58] The important property of core–shell nanoclusters is the tunability of their properties by changing of chemical composition and relative geometry of core and shell. The compositions of core–shell particles change from metal surrounding metal[59,60] to dielectric surrounding metal[61] and metal surrounding dielectric.[62] For the case of metal surrounding metal, particle synthesis starts from reduction of metal ions by one of methods mentioned above to create a metal core, with subsequent addition of dimensional confinements/stabilizing agents. In the following stage, a reduction of metal shell ions is performed using another reducing agent. Nash and colleagues[63] have proposed a new method for synthesis of temperature-responsive γ-Fe$_2$O$_3$-core/Au-shell magnetic nanoparticles (Au-mNPs) from an amphiphilic diblock copolymer. The synthesis of diblock copolymer was performed using poly(N,N-dimethylaminoethylacrylamide)

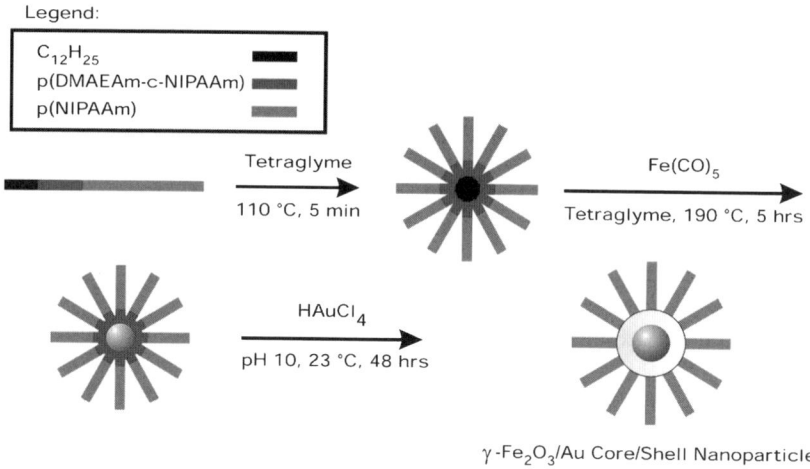

Figure 13.1 Magnetic-core–gold-shell nanoparticle synthesis scheme. Amphiphilic diblock copolymers were designed with a dodecyl hydrocarbon tail (black), an amine-containing reducer (ACR) block (dark grey), and a thermally responsive homo-pNIPAAm block (light grey). These polymers form micelles, which are used as nanoscale reactors for magnetic nanoparticle (mNP) synthesis from iron pentacarbonyl precursors. The mNPs retain the DMAEAm reducing moieties at the nanoparticle surface, allowing for subsequent reduction and formation of a gold shell on the mNP core. Reprinted with permission from M. A. Nash, J. J. Lai, A. S. Hoffman, P. Yager and P. S. Stayton, *Nano Lett.*, 2010, **10**(1), 85. Copyright 2010 American Chemical Society.

(DMAEAm) block as reductant and poly(*N*-isopropylacrylamide) (pNIPAAm) as a thermally sensitive block (Figure 13.1). An important advantage is that mNPs synthesized on the first stage retain the DMAEAm reducing moieties near the mNP surface, allowing the subsequent reduction of ionic gold salt (*e.g.*, $HAuCl_4$ or $KAuCl_4$) onto the Fe_2O_3-core/Au-shell nanoparticles, forming a gold shell.

Another approach for synthesis of multifunctional core–shell microspheres using the layer by layer (LBL) technique in aqueous solution was proposed by Spasova and colleagues.[64] For this research they used a polystyrene 640 nm diameter core covered with a selectable number of layers of magnetite and silica-coated gold 15 nm nanoparticles. Among the advantages of this method is a possibility of varying the shell thickness and layer composition with independent control of the magnetic and optical properties and the diameter of colloids. A well-resolved LSPR peak was observed which exhibits red shift related to the number of adsorbed Au nanoparticle layers. The linear increase of the magnetic moment per sphere corresponds with the number of adsorbed magnetite nanoparticle layers. Interestingly, that deposition of prepared spheres in a magnetic field results in assembling of mNPs into chains of up to 1 mm length.

The tunability of plasmonic properties of Au mNP was also reached with another synthesis method realized by Xu and colleagues.[65] Here, preparation of Au- and Ag-coated Fe_3O_4 nanoparticles starts from synthesis of magnetic nanoparticles *via* thermal decomposition of iron(III) oleate in the mixture of oleylamine and oleic acid.[66] The next stage was synthesis of Fe_3O_4/Au nanoparticles by reducing $HAuCl_4$ in a chloroform solution of oleylamine. The interesting finding of the authors is a possibility to dissolve the Au-coated Fe_3O_4 nanoparticles in water by previously mixing the Au mNPs with sodium citrate and cetyltrimethylammonium bromide (CTAB). After the solubilization procedure, Fe_3O_4/Au nanoparticles were well-stabilized due to the formation of a CTAB double layer on the nanoparticle surface.[67,68] The water-soluble nanoparticles were then used as seeds for growth of a thicker Au coating, which was performed by adding more $HAuCl_4$ in the reducing mixture. The preparation of Fe_3O_4/Au/Ag nanoparticles was realized by adding $AgNO_3$ to the reaction solution. The regulation of the coating thickness and material allows the tuning of the plasmonic properties of these core/shell nanoparticles. For this kind of Au mNPs, the tuning range of plasmonic properties was found to extend from 501 nm to 560 nm. An advantage of the proposed method of synthesis is the uniform coating of Au and Ag and good monodispersity of Fe_3O_4 nanoparticles.

Another type of core-shell NPs have a dielectric core covered by metal. Zhu *et al.*[69] have presented an interesting approach based on coating of metallic nanolayers onto soft polystyrene (PS) nanotemplates through simple electro- less methods. Firstly, the aggregates of nanospheres or nanowires in aqueous solution were prepared by dissolving the triblock copolymer polystyrene/ poly(4-vinyl pyridine) (P4VP-b-PS-b-P4VP) in the mixture of solvents

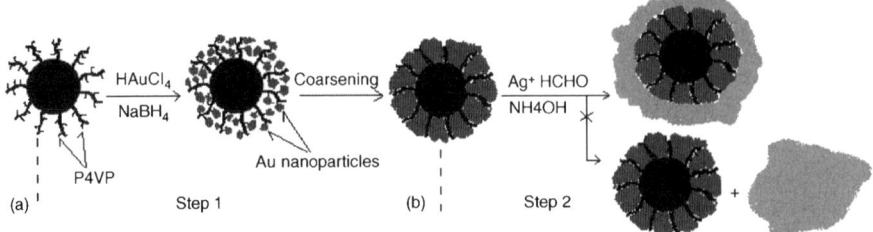

Figure 13.2 Schematic illustration of the metallization route (cross-section of the nanostructure, all of the aggregates have the similar cross-section structure), starting from PS nanostructure covered with P4VP corona (hairy shell) (a) complexation and reduction of Au ions with P4VP on the Si substrates and (b) electroless plating of Ag (or Au) on the precoated Au surface. Reprinted from J. Zhu and W. Jiang, Fabrication of conductive metallized nanostructures from self-assembled amphiphilic triblock copolymer templates: Nanospheres, nanowires, nanorings, *Materials Chemistry and Physics*, 2007, **101**, 56–62, with permission from Elsevier.

containing DMF, dioxane and dioxane–tetrahydrofuran. Then, segregation of PS blocks was reached by adding large amount (*ca.* 500 wt%) of deionized water. The removal of the organic solvent from resulting solution was performed by dialysis against distilled water (or methanol) for a several days with continuous adjusting of the pH of solution to 4 to prevent the colloidal solution from precipitating. The reduction of metal salts or electroless plating was used to cover the copolymer nanotemplates with Au, Ag, Pd and Au@Ag layers. To prepare the Au or Ag metallic nanostructures, the nanosphere aggregates, spin-coated on the Si-wafer, were wiped with a 0.1 wt% $HAuCl_4$ or 0.15 mM solution of fresh $AgNO_3$, correspondingly, for 10–15 min. After rinsing with water, $HAuCl_4$ was reduced by 0.01% $NaBH_4$ aqueous solution. Correspondingly, Ag^+ ions were reduced by 4 mM aqueous solution of NH_2OH. The advantage of this method is that it allows preparation of metal covered nanostructures with different forms, including nanorings, nanospheres and nanowires with variable sizes (Figure 13.2).

13.2.4 Surface Nano-patterning of Random-fashion Nanostructures Using Colloidal Au

Obviously, colloidal Au and Ag NPs can be used for preparation of chip-based structures in array-format by immobilization of NPs on the solid support. A number of articles from different groups has presented chip-based format optical biosensors, in which gold or silver nanostructures are immobilized in a random fashion on an optically transparent substrate.[31,70–72] This approach opens a way for high-throughput biosensor control and multiplexed analysis of biomolecular interactions with minimal consumption of reagents. There are several different approaches to fabrication of chip-based structures based on colloidal plasmonic NPs. Liu and colleagues[71] demonstrated simple fabrication of large-area gold nanostructures using thiol-stabilized gold nanoparticles without complicated lithography and vacuum evaporation techniques being involved in the fabrication process. Gold nanoparticles having a mean size of about 5 nm with a distribution range from 2 to 8 nm were obtained using modernized method of Murray et al.[44] In this work, hexanethiol was used as the stabilizing functional ligand to cover the gold nanoparticles to reach good dispersion in organic solvents like xylene and toluene. The Au NP colloidal solution was spin-coated onto the 10 mm × 10 mm indium–tin-oxide glass substrate at a speed of 2000 rpm for 30 s and then annealed at temperatures in the range of 200–550 °C in air. As a result of the Au NPs melting, nanostructures with different morphologies were formed (Figure 13.3). The high limit of temperature (550 °C) was found due to observation of undesired structural changes in the transparent glass support. The regulation of temperature during the annealing process and changing the concentration of the Au NP colloids allowed the tunability of the optical response. In addition, a concentration ranging from 40 to 120 mg ml^{-1} was recommended by authors for the large-area fabrication of size-optimized gold nanoisland structures. It should be noted that a disadvantage of this method is a relatively wide LSPR absorbance peak of prepared structures, which is explained by the large size dispersity of nanostructures.

S. Malynych and G. Chumanov[72] proposed the method of forming stable random-fashion monolayers of colloidal Ag NPs on the glass substrate with a layer of polymer maintaining only the lower part of the nanoparticles. This enables LSPR oscillations at the top part of particles, which are free from the polymer. Importantly, arrays of Ag NPs prepared in such a way exhibit extra narrow peaks in extinction spectra, which can be used to develop high-sensitive biosensors. To produce such a nanoparticle array, self-assembly of 100 nm silver particles on glass or silicon substrates coated with poly(vinylpyridine) (PVP) was performed. PVP acts as an efficient surface modifier for immobilization of metal nanoparticles due to its capability of simultaneous attachment to different substrates via hydrogen bonding and interaction with metal particles due to metal-ligand interactions of the nitrogen atom of the pyridyl group.[73] The PVP-treated substrates were immersed into a colloidal solution of silver nanoparticles in deionized water at low ionic strength. Low ionic strength is needed to support long-range electrostatic repulsion between silver nanoparticles to yield two-dimensional arrays of non-touching particles. By adjusting the exposure time, arrays with various surface densities can be produced. If the average interparticle distance for such arrays becomes of the same order of magnitude as the nanoparticle diameter, the light extinction spectrum changes drastically, with appearance of a new sharp peak located at 436 nm (Figure 13.4, black curve).

In several papers, techniques of colloidal NPs utilization were extended to obtain nanorod chip-based structures. Two chip-based methods (slow and rapid) have been proposed by the Geddes group[74] for the deposition of silver

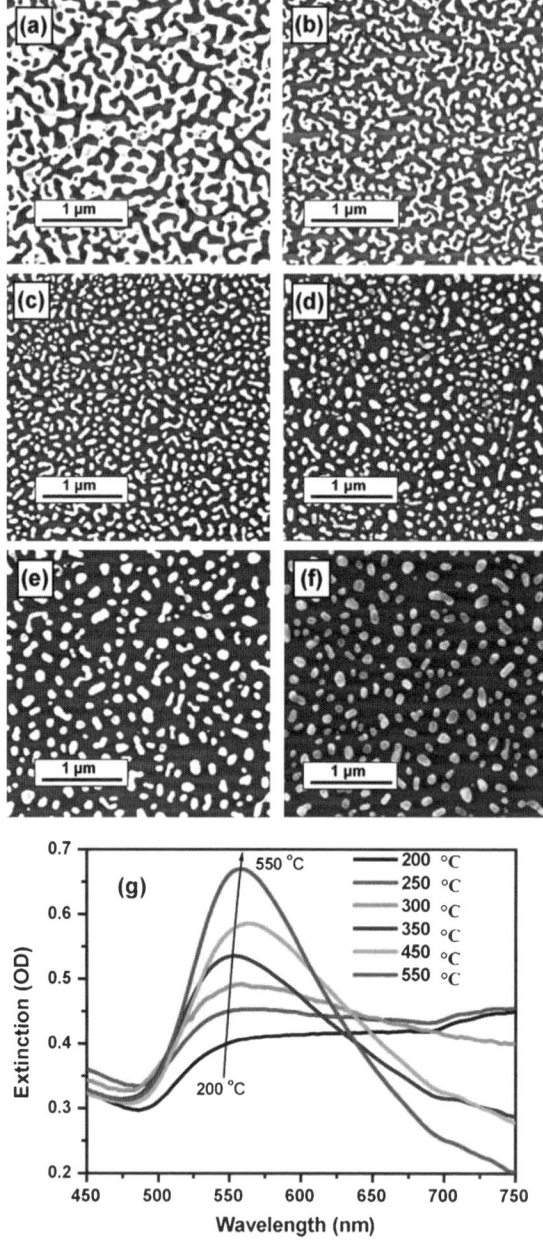

Figure 13.3 (a)–(f) SEM images of the samples of gold nanostructures that have been fabricated using an annealing temperature of 200, 250, 300, 350, 450, and 550 °C, respectively. (g) The corresponding optical extinction spectra. Reprinted from Photonics and Nanostructures – Fundamentals and Applications, 8, H. Liu, X. Zhang and Z. Gao, Lithography-free fabrication of large-area plasmonic nanostructures using colloidal gold nanoparticles, 131–139, Copyright (2010), with permission from Elsevier.

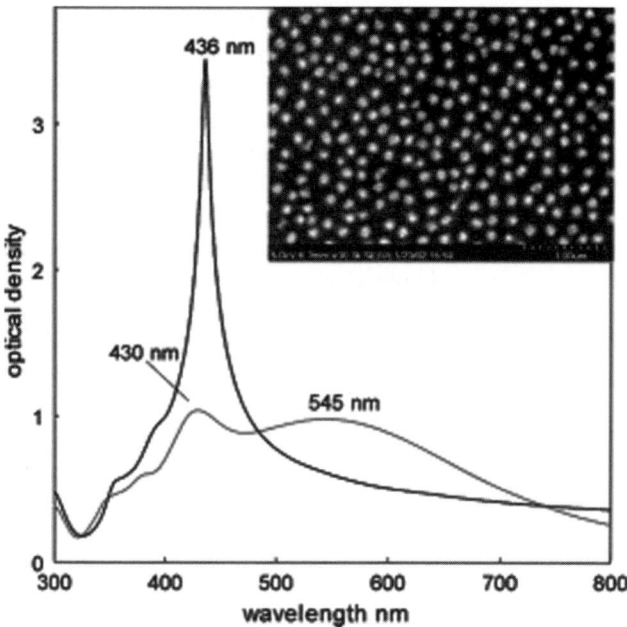

Figure 13.4 Extinction spectra of 100 nm Ag particles in water (grey curve) and the same particles assembled into a closely spaced 2D array imbedded in poly(dimethylsiloxane) (PDMS) (black curve). Inset: electron microscopy image of this 2D array. Reprinted with permission from S. Malynych and G. Chumanov, *J. Am. Chem. Soc.*, 2003, **125**(10), 2896. Copyright 2003 American Chemical Society.

nanorods onto glass substrates using colloidal Ag NPs. Before nanorod deposition, the surface of glass slides was modified in several steps. At first, glass slides were treated with "piranha solution" (3 : 7 30% hydrogen peroxide/concentrated sulfuric acid) for at least 2 h. After that, the substrates were rinsed copiously with deionized water and dried under a stream of dry N_2 gas. The pretreated slides were silanized by dipping into a 2% ethanolic solution of 3-(aminopropyl)triethoxysilane (APS) for 2 h. Then, the APS-coated glass substrates were rinsed in ethanol and water with further sonication in ethanol for 30 s. Subsequently, the glass slides were rinsed with water and dried under a stream of dry N_2 gas. Furthermore, the technology developed by Murphy and colleagues[75] for preparation of silver nanorods in solution was modernized for a chip-based variant. Briefly, Murphy proposed to use a previously prepared silver seed to stimulate silver nanorods growth by chemical reduction of a silver salt. To fabricate nanorods and nanowires of different aspect ratio, $AgNO_3$ salt was reduced by ascorbic acid in the presence of the silver seed, CTAB surfactant and NaOH. The rod-like surfactant micelles in solution promoted silver nanorod growth. As a result, it became able to reproducibly fabricate silver nanorods having varying aspect ratios of 2.5–15 (with 10–15 nm short axes), and nanowires 1–4 micrometres long with 12–18 nm short axes. Subsequently, in the slow method, silver nanorods were precipitated onto APS-coated glass slides by ordinary immersion into the silver nanorod solution. The adsorption of silver nanorods on the surface of glass slides from the solution continued for a few days and the light absorption at 550 nm reached only 20% that of the silver nanorods solution. In the rapid method, spherical silver seeds chemically bound to the glass surface were grown into silver nanorods due to a cationic surfactant and silver ions present in the solution. The length of nanorods was determined by the number and duration of immersions of silver seed-coated glass slides into a growth solution, and ranged from tens of nanometers to a several micrometers. The silver nanorod formation on the glass substrates was evident after 10 min of immersion due to the color change (clear to green) on the glass slide and in the solution. To increase the concentration of silver nanorods on the surface, the silver

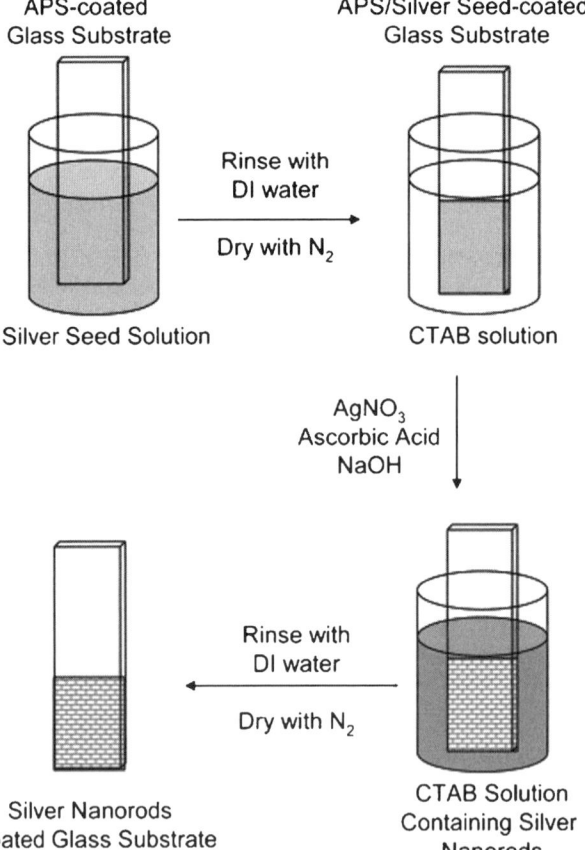

Figure 13.5 Rapid deposition of silver nanorods on a glass substrate. Reprinted with permission from K. Aslan, Z. Leonenko, J. R. Lakowicz and C. D. Geddes, *J. Phys. Chem. B*, 2005, 109(8), 3157. Copyright 2005 American Chemical Society.

nanorod-coated glass slides were immersed in a similar fresh growth solution again containing CTAB, AgNO$_3$, ascorbic acid and NaOH. This procedure can be repeated until the needed loading of the silver nanorods onto the glass substrate is reached (Figure 13.5). Interestingly, that minimal change in the process of preparation (immersing the silver-seed-coated glass slides in 40 mL of 0.80 M CTAB solution for 1–3 min,[76] instead of 5 min) results in the growth of triangle structures. These properties of surfactant-based technology are of importance, because the shape of nanostructures sometimes plays a crucial role in their further biosensor's applications. For instance, silver triangles are more suitable for SERS applications, whereas silver nanorods are preferable for biomolecular sensing.[30] The major limitations of the abovementioned technologies are related to NP shape, disordering, monodispersity and reproducibility which negatively influence their extensive application.

13.2.5 Surface Nano-patterning of Periodic Ordered Nanostructures Using Colloidal Au

Two-dimensional patterned substrates with nanosize features have attracted great interest due to a necessity for high-resolution devices in biosensor research. The manufacturing of periodically ordered Au and Ag nanostructures has obvious advantages in comparison with disordered nanoarrays due to higher possibility for cooperative oscillation of free electrons in the ordered nanoparticle array, so-called "cooperative plasmon resonance". Different methods have been demonstrated to fabricate periodic plasmonic nanostructures. Electron beam lithography with subsequent evaporation and lift-off is commonly used to produce one- or two-dimensional ordered nanoarrays. The obvious disadvantages of this method are the small working range (<200 μm), low speed and high costs. Interference lithography is a high-speed method for fabrication of large-area nanostructures with possibilities to change the periodicity of array. A drawback is a necessity to perform the lift-off procedure that is complicated for nano-manufacturing.

Zhang and colleagues[77] propose interesting, and one of the simplest, methods in manufacturing of similar structures based on high-speed combination of colloidal gold synthesis and interference lithography. In this interesting work, the authors propose an alternative method for fabricating periodic ordered nanostructures demonstrating the advantages of simplicity, high throughput and low cost. In this technique, a one-dimensional photoresist (PR) grating on top of an indium tin oxide (ITO) glass substrate is produced initially by interference lithography. Subsequently, a gold nanoparticle colloidal suspension is spincoated onto the PR mask. The average size of the nanoparticles was about 1.5 nm in diameter, and the ITO layer has a thickness of about 210 nm. Bearing in mind the low melting temperature of the gold nanoparticles, heating the sample to above 200 °C for about 5 min was performed with subsequent cooling down to room temperature. With a temperature higher than melting point, the gold colloidal solution aggregates into the grating grooves of the ITO substrate due to the strong wetting on the ITO surface and the constraint of the groove structures. The processes of annealing of gold nanoparticles and their accumulation in the grooves only result in formation of nanowires, which are distributed periodically over an area of up to 20 mm^2. The mask-determined parameters of gold nanowires structure such as a period below 500 nm and a width from 100 to 300 nm were obtained (Figure 13.6). The height of the nanowires was dependent on the

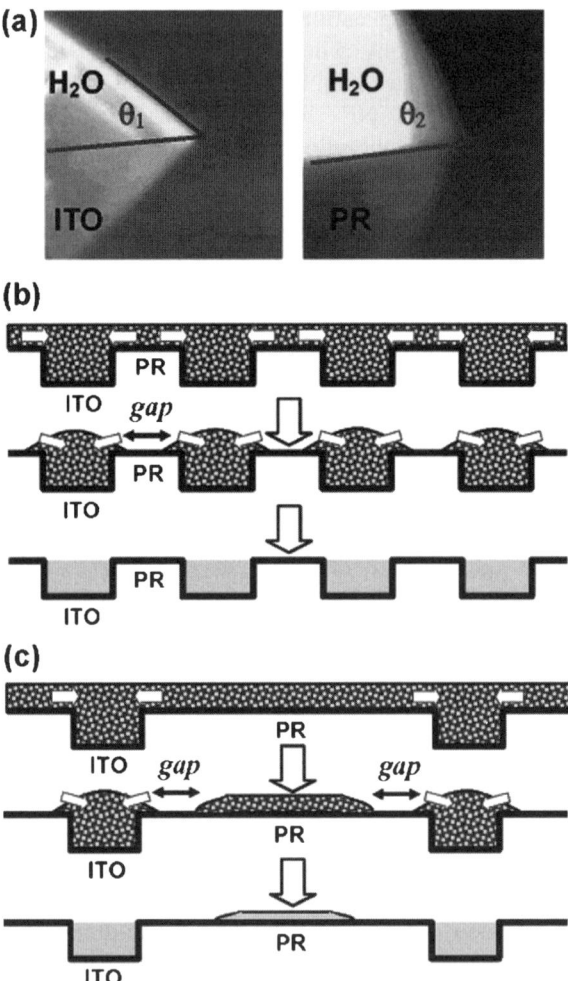

Figure 13.6 (a) Measurements of contact angles of water on the ITO and on the PR surfaces: $\theta_1 = 44\text{--}45°$, $\theta_2 = 73\text{--}75°$. (b) Mechanisms for the confinement of the gold nanoparticles into the grating grooves when the PR channel is small enough. (c) Two gaps will form structures with large PR channels. Reprinted with permission from X. Zhang, B. Sun, R. H. Friend, H. Guo, D. Nau and H. Giessen, *Nano Lett.*, 2006, **6**(4), 651. Copyright 2006 American Chemical Society.

concentration of the gold colloid and the depth of the mask grating. The main advantage of this promising method is ordering of nanostructures and absence of a lift-off process, however, the rough surface of the prepared nanowires due to the nanocrystal size distribution broadens the plasmon resonance spectrum.

13.3 Lithography-free Methods

13.3.1 Oblique Angle Deposition Method

There are substantial difficulties in easily preparing robust, metal-coated substrates with the desired surface morphology that provide repeatability and high sensitivity for biosensor applications. Substrates patterned by silver nanorod arrays for biosensing, and, particularly, for SERS and SEF, can be prepared by the method of oblique angle deposition (OAD) recently developed by Dluhy and colleagues.[78,79] This nanofabrication technique offers a flexible, easy and inexpensive method for fabrication of integrated nanostructured substrates for high-sensitive biological spectral applications. The substrates with nanorods produced by OAD have the advantage of a large area and allow preparation of nanopatterned structures for plasmonic biosensors rapidly, accurately and cost-effectively to detect, for instance, extremely low levels of viruses.[80] In this work, SERS substrates were prepared by electron beam/sputtering evaporation. In the OAD technique, the angle between the metal vapor and the normal of the substrate surface is fixed at 86°, and about a 500 nm Ag thin film is deposited. During deposition, randomly located but uniformly aligned nanorod arrays rise on the substrate. The length of the nanorods is proportional to the deposition time and the nanorods are inclined with respect to the normal of the substrate surface. SEM micrographs (Figure 13.7) show the average rod length and diameter of the nanorod arrays to be (868 ± 95) nm and (99 ± 29) nm, respectively. The surface density of the nanorods was estimated to be (13.3 ± 0.5) rods μm^{-2} with an average tilt angle of (71.3 ± 4.0)°.

13.3.2 Synthesis of Hybrid Nanostructures of Gold Nanoparticles and Carbon Nanotubes

Carbon nanotubes (CNTs) are very strong materials with excellent properties for use in molecular nanoelectronics, which have already been incorporated into biosensor systems.[81–83] On the other hand, plasmonic nanostructures, namely, gold nanoparticles, are of increasing interest because of their unique electronic and optical properties. The hybrid structures consisting of these two kinds of materials can be applied to extend the possible applications of both CNTs and Au NPs. Promising fields for exploitation of such hybrid materials include catalysis, solar energy conversion and gas, electrochemical and biosensing, which have attracted many investigators during last few years.[84,85] Tello and colleagues[86] presented a method of synthesis of hybrid structures

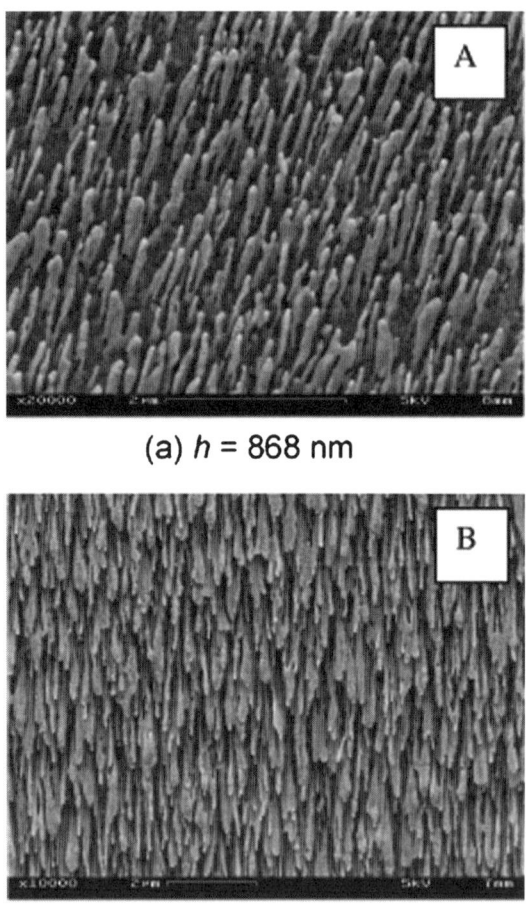

(a) h = 868 nm

(b) h = 2080 nm

Figure 13.7 Representative scanning electron micrographs of the Ag nanorod arrays deposited with different lengths, (a) $h = 868$ nm and (b) $h = 2080$ nm. The typical SERS substrate used for virus detection is represented in (a), *i.e.* 870 nm. Reprinted with permission from S. Shanmukh, L. Jones, J. Driskell, Y. Zhao, R. Dluhy and R. A. Tripp, *Nano Lett.*, 2006, **6**(11), 2630. Copyright 2006 American Chemical Society.

formed by Au NPs and multiwall carbon nanotubes (MWCNT)s (Figure 13.8). MWCNTs were fabricated by chemical vapor deposition (CVD), using the decomposition of acetylene at 800 °C over a Pd/γ-Al$_2$O$_3$ catalyst.[87] Synthesized MWCNTs were purified by standard methods that include air oxidation, alkali and acid treatments.[88,89] Au NPs@MWCNTs hybrids were obtained by the solvated metal atom dispersion method (SMAD).[90,91] Briefly, this technique consists of the resistive evaporation of bulk gold into acetone atmosphere, which is subsequently embodied into a frozen matrix at liquid nitrogen temperature. Warming up this structure to

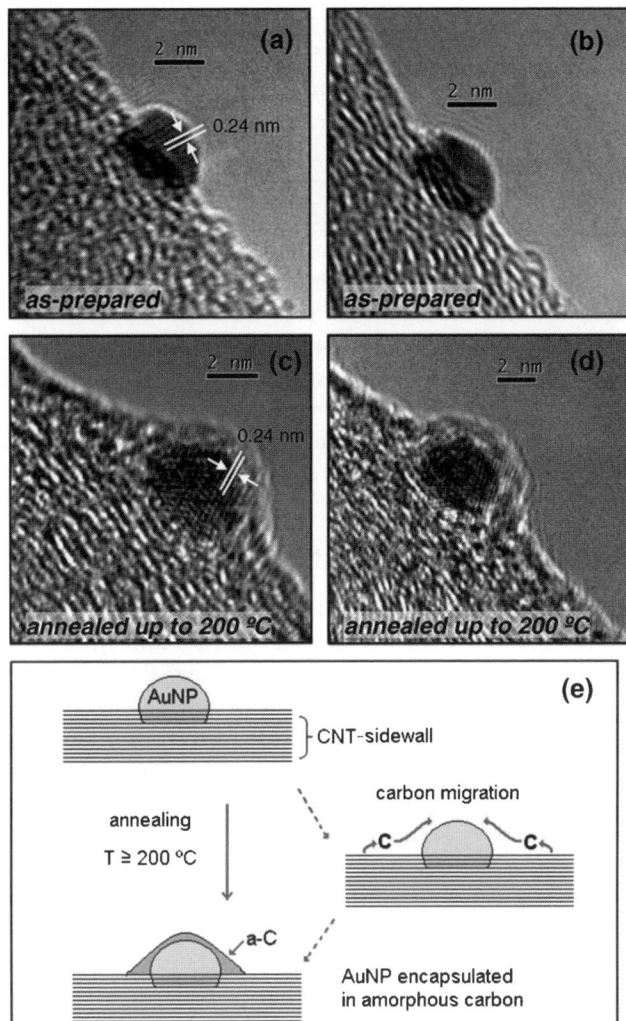

Figure 13.8 HRTEM micrographs of Au NPs anchored on the CNTs-walls. (a) and (b) are the as prepared AuCNT hybrids; (c) and (d) are the Au-CNT hybrids annealed up to 200 °C. A carbon layer is encapsulating the AuNPs after the thermal process. Lattice fringes on gold particles are consistent with (111) fcc orientation (0.235 nm) and (e) is a schematic representation of the model of this process. Reprinted from A. Tello, G. Cárdenas, P. Häberle and R. A. Segura, The synthesis of hybrid nanostructures of gold nanoparticles and carbon nanotubes and their transformation to solid carbon nanorods, *Carbon*, **46** 884–889, Copyright (2008), with permission from Elsevier.

room temperature yields a colloidal solution with tiny and chemically reactive gold clusters. These clusters subsequently interact with the CNTs, which have been incorporated into the reactor. The highly reactive Au NPs attach to the

CNTs side walls and form the hybrid nanostructures. Following synthesis, the freshly prepared hybrids were purified by filtration with a nitrocellulose membrane (pore 3 μm), washed and dried. After purification, the hybrids were annealed for 1 h at various temperatures in an Ar atmosphere. The analysis of HRTEM results shows that thermal treatment to temperatures as low as 200 °C leads to the generation of a thin layer of amorphous carbon, encapsulating the attached gold nanoparticles (Figure 13.8). This C-layer prevents subsequent sintering of the Au NPs, rendering the hybrid structures very stable up to temperatures near to 400 °C. For temperatures higher than 600 °C the Au NPs disconnect from the tubes and undergo a sintering process. An important particle size increase occurs (from 4 nm to almost 20 nm in average), but no damage is induced on the CNTs. The situation diverges totally if the samples are annealed up to 800 °C. At these conditions the CNTs transform into solid carbon nanorods. Apparently, this process takes place because of the presence of the Au NPs in the hybrids.[86]

13.3.3 Anodic Porous Alumina Membranes

Usage of well-known anodic porous alumina (Al_2O_3) nanomembranes was established five decades ago as a convenient and powerful tool for nanomanufacturing, starting from pioneer works of Possin with preparation of nanowires within a mica template.[92] The subsequent works of Martin[93] and Moscowits[94] using anodized aluminum oxide (AAO) as a template for metal nanowires (NW) fabrication initiated wide utilization of this method for different applications. Among the advantages of utilization of such AAO templates are simplicity, possibility to use cheap equipment, tunability in preparation of nanostructures with different size and interparticle distance, and the possibility to work with different materials.[95,96] The AAO membrane method allows better control of NW parameters than other protocols for synthesis of similar nanostructures. Because AAO is also highly suitable for manufacturing of nanostructure arrays that provide high surface to volume ratio and good access for biomolecules due to vertical alignment,[97–100] we will consider this method in more detail.

The main approaches for preparation and utilization of AAO were presented in an excellent review by Lei with coauthors.[101] The nano-template from anodic porous alumina is a highly ordered nanopore array with controllable geometry. Using different acid solutions (sulfuric for small (10–30 nm), oxalic for medium (30–80 nm) and phosphoric for larger than 80 nm pore sizes) in electrochemical treatment of thin Al foil (or evaporated film), careful control of pore diameter is possible.[101,102] The size of pores can be regulated also by anodization voltage in the range of 10–200 nm.[101] It should be noted that good tuning of the nanopore diameter is observed for small and medium pore size, whereas difficulties occur for large pores. Usually, the quality of nanopore arrays ordering is better on the bottom side of the membrane, so Masuda and Satoh[103] proposed an anodization process performed in two steps, in the which top alumina layer is removed from the Al foil, leaving highly ordered bottom layer. The next stage of the anodization process results in a pore array with high regularity for both sides of AAO membrane. Subsequent annealing and polishing improves pore regularity and membrane surface smoothness. The area of pores free of defects typically does not exceed several square micrometers, so "pretexturing" procedure was proposed by Masuda et al.,[103] in which a single silicon carbide (SiC) mold was used as a stamp before the anodization procedure, that allows the extension of the defect-free ordered areas up to millimetre sizes. A disadvantage of thick (more than 1 micrometre) AAO membranes are difficulties in manufacturing of zero-dimensional nanostructures and preparation of nanostructures directly on the substrate. To overcome this problem, a novel nano-patterning technology for preparation of nanostructures using ultra-thin alumina mask (UTAM) was developed.[104,105] The preparation of nanoparticles or nanohole arrays using AAO can be performed on almost any smooth substrate (Figure 13.9). Two different types of membranes – attached (which is placed on the surface) and connected (fabricated on the surface) – can be used. An advantage of attached membranes is higher regularity of pore arrays; however, if attachment with substrate is not in good conditions, detachments are possible. A thin (5 nm) Au layer covering the back side of the AAO usually serves as a working electrode and as an initial substrate for deposition of materials within the pore array. Due to direct contact of electrolyte solutions with gold, the deposition process takes place mainly in the pores.[106] The geometry of AAO template defines the diameter of the nanowires and the distance between them, whereas the charge passed during deposition procedure is proportional to the length of prepared nanowires. The process of preparation of NW is finished by dissolving the AAO template in 1 M acidic or basic solution for several hours.[106] The total area of this kind of a mask can be up to a few square centimetres. An advantage of the AAO method, especially in biosensing, is a possibility to prepare nanostructures with a regulated interparticle gap that is of importance due to the role of so-called "hot spots" (interspaces in the vicinity of plasmonic NPs, where a strong electric field is generated). Decreasing the distance between nanoparticles results in increasing of the electric field, and, mainly, in a higher response of a biosensor. Using AAO membranes, Mirkin and coworkers[107] developed an on-wire-lithography (OWL) technique, which allows producing segmented gold nanowires with unique narrow regulated gaps (from 2.5 nm). By periodically changing the composition of electrolyte and time of deposition, segmented nanowires were prepared. After membrane dissolution, an oxide (e.g. SiO_2) or metal backing was deposited on the wires by thermal evaporation. An etching procedure was performed to dissolve the segments between the gold segments, whereas the backing maintains the spacing between them. Similar structures allow enhancment, for instance, of a SERS signal up to 10^8 times. Using OWL technology it is possible to produce nanostructures with distance between segments especially for biomolecules with known length. Besides the fabrication of one-dimensional nanostructures, AAO membranes can be used in preparation of metal replicated membranes,[108,109] and for the present time is an excellent low cost alternative method for manufacturing of large scale ordered arrays.

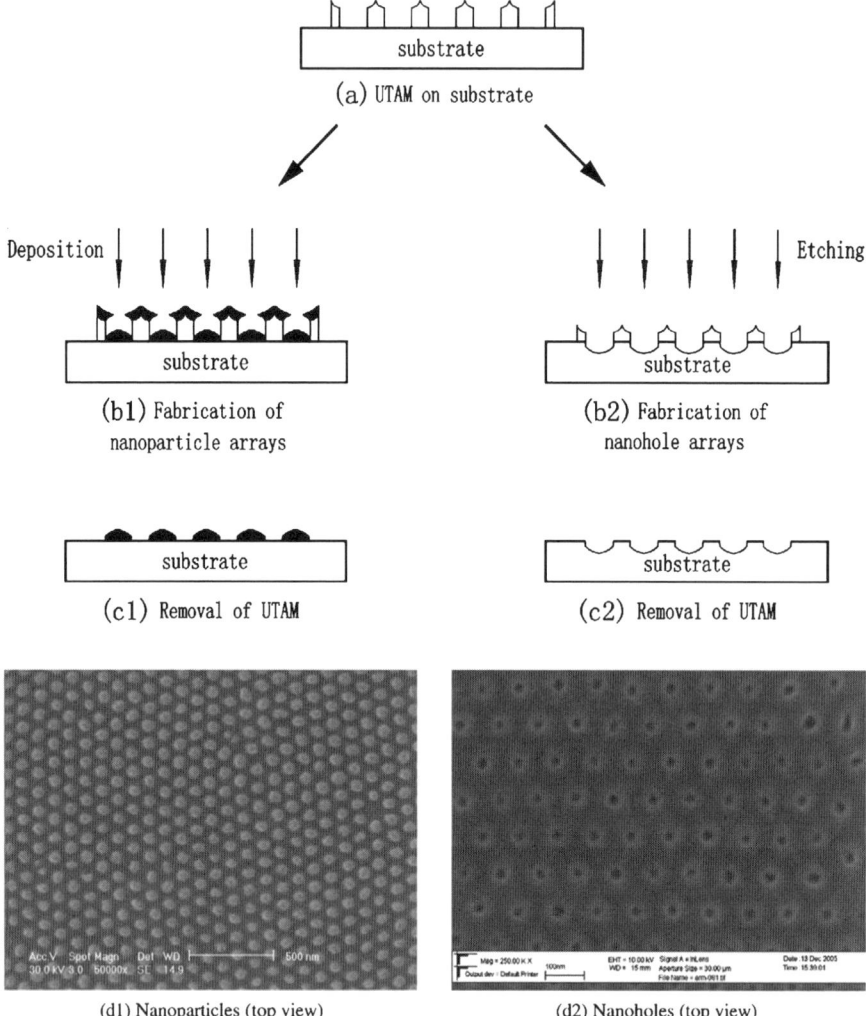

Figure 13.9 Schematic outline of the general fabrication processes of nano-particles (a–b1–c1) and nano-holes (a–b2–c2). d1 and d2 are highly ordered nano-particle (Pd) and nano-hole arrays (etched using focused ion beam) on Si substrates, respectively. Reprinted from Progress in Materials Science, 52, Y. Lei, W. Cai and G. Wilde, Highly ordered nanostructures with tunable size, shape and properties: A new way to surface nano-patterning using ultra-thin alumina masks, 465–539, Copyright (2007), with permission from Elsevier.

13.4 Lithographic Methods

Nanolithographic techniques, which are top-down fabrication methods, exploit various nanoscale processing methods to form highly reproducible noble metal nanostructures through the application of masking, patterning, imprinting and treatment procedures to the lithographic resists and thin metal films on the flat substrate surface. Photolithography as an approach that is widely used for micropatterning purposes possesses the inherent resolution limitations related to the light wavelength, and therefore this technique is rarely applied for the fabrication of nanosized plasmonic objects. However, state-of-the-art photolithographic method allows production of features with an average half-pitch in semiconductor devices of 32 nm using 193 nm laser light combined with sophisticated immersion optics, phase-shifting masks and multiple exposures,[110] and this limit permanently diminishes, paving the way for the fabrication of plasmonic nanostructures in the future. In the field of modern lithography-based methods for production of nanostructured biosensor chips such techniques as scanning beam, colloidal and nanoimprint lithographies are prevailing.

13.4.1 Scanning Beam Lithographies

Scanning beam techniques, such as electron beam lithography (EBL) and focused ion beam (FIB) lithography, are the most common techniques for the production of metal nanostructures on the surfaces. The main advantage of these methods is the ability to fabricate nearly arbitrary one- and two-dimensional patterns with precise control of the size, shape and interparticle distance of surface-bound plasmonic nanoparticles. This type of control is needed to meet the requirements of sensor chip reproducibility and optical properties tunability for biosensing, SERS and SEF applications.

EBL uses a highly focused electron beam to scan across a thin layer of radiation-sensitive resist that is deposited on a substrate, which is subsequently used as a mask for etching and metal deposition procedures to yield the desired metallic nanopattern. EBL provides outstanding resolution down to the sub-10 nm range, which is attainable through the exploitation of specialized resists, such as hydrogen silsesquioxane,[111,112] or traditional poly(methyl methacrylate) resist with ultrasonically assisted development.[113] The EBL technique is often used when dense gold and silver nanoparticle arrays of a particular geometry and size are needed. This method has been used to fabricate highly uniform arrays of monomers,[114–116] dimers[117] and oligomers[118] of discs (Figure 13.10), elliptical discs,[119] trigonal prisms,[114,120] spheroids,[121] half spheroids,[122] rods,[123] split rings,[124] L-shaped[125] and almost spherical particles[126] with varying size and interparticle distance. EBL has been also applied for creation of gold nanogratings[127–129] and subwavelength apertures[130] in thin gold films. Such nanoparticle arrays have been exploited to achieve low detection limits for anti-biotin and streptavidin in a LSPR biosensor (3 nM and 7 pM, respectively),[116] to create a multianalyte DNA sensor based on surface enhanced resonance Raman spectroscopy[124] and to enhance the dye fluorescence both on periodic nanoantenna[119] and subwavelength aperture[130] arrays. Another important application of EBL is the preparation of reusable masks and master stamps for other lithographic techniques such as nanostencil lithography[131] and nanoimprint lithography.[132–134] However, limitations of fabrication area ($<200 \times 200$ μm^2), low speed due to serial processing and high costs entail the low throughput sample preparation with EBL method.

Another scanning beam lithographic technique, FIB, allows writing of both subtractive and additive patterns into a resist or directly onto the thin metal film by a highly focused beam of ions through milling, ion-assisted etching and ion-induced deposition processes.[135,136] The attainable resolution of FIB lithography method is <10 nm,[137,138] which is comparable to that of EBL. FIB techniques usually employ Ga$^+$ ions to form a scanning beam;[139–141] however, the use of other ions such as Ar$^+$ for ion-assisted lithography has also been reported.[142] In principle, FIB can be used to fabricate arbitrary two-dimensional metal nanostructures, but due to its relatively slow processing capabilities in common with EBL, it is mostly used for the production of holes in metal films rather than isolated metal nanoparticles. Various metallic (i.e. Au, Ag, Al) nanostructures such as single and arrayed circular[139–141,143–147]

Figure 13.10 SEM images of a typical gold heptamer sample fabricated by EBL. (top) A normal view of the sample. The interparticle gap distance is 20 nm. The thickness of the gold nanoparticles is 80 nm. Reprinted with permission from M. Hentschel, M. Saliba, R. Vogelgesang, H. Giessen, A. P. Alivisatos and N. Liu, *Nano Lett.*, 2010, **10**(7), 2721. Copyright 2010 American Chemical Society.

and rectangular[145] nanoholes, circular double-holes,[148] slits,[127,149,150] circular slits[151] and V-grooves[152] with varying size and periodicity have been fabricated by means of FIB lithography for research in the field of plasmonics and, especially, for biosensor applications.[139–141,143,144,146,148] Namely, biosensors based on nanohole arrays demonstrated detection of attomolar streptavidin concentrations,[143] 6-fold improvement in response time for surface adsorption of mercaptoundecanoic acid when operated in flow-through format[140] and up to 6.5-fold enhancement of rhodamine 6G single-molecule fluorescence detection as compared to open solution.[139]

13.4.2 Colloidal Lithographies

Colloidal lithography, an economical alternative to the common scanning beam lithography techniques, utilizes self-assembly of colloidal particles (usually polystyrene) on the solid substrate surface to form a mask for subsequent evaporation and/or etching processes in order to fabricate metallic nanostructure arrays. The main advantage of this group of techniques is the possibility to produce both ordered and random nanostructured patterns over large areas, not achievable by conventional nanofabrication methods, resulting in relatively low fabrication costs and high throughput. In addition, narrow nanostructure size distribution (less than 5%) can be achieved if polystyrene colloids with high monodispersity are used. Depending on the structure and composition of the colloidal mask formed, colloidal lithography methods are subdivided into sparse colloidal lithography (SCL), hole-mask colloidal lithography (HCL) and nanosphere lithography (NSL).

The SCL technique involves adsorbing colloidal polystyrene spheres onto an oppositely-charged substrate *via* electrostatic self-organization into a sparse random pattern with interparticle distance defined by the particle–particle repulsion, which can be adjusted by the concentration and ionic strength of colloidal solution.[153] Adsorbed polystyrene particles are further used as a mask to produce different types of nanostructured metal surface with feature sizes down to 20 nm,[153] which are defined by the size of polystyrene spheres used. Single nanoholes and random nanohole arrays in thin metal films are produced by metal evaporation on the colloidal mask and subsequent lift-off of the polystyrene particles.[146,154–157] Biosensing potential of such nanostructured thin gold and silver films has been demonstrated for selective sensing of antigens such as cancer antigen 19-9 (of less than 1 pg on a 0.1 mm^2 probing area, after surface functionalization with respective antibodies),[154] detection of NeutrAvidin with a detection limit of <0.1 ng cm^{-2},[155] sensing of lipid-membrane-mediated biorecognition reactions in the zeptomole regime[156] and for probing the formation of macroscopic and laterally mobile supported lipid bilayers and their binding with proteins.[157] Fabrication of metallic nanorings with a similar technique involves an additional technological step – Ar$^+$ ion-beam etching is used to deposit a metal shell on the polystyrene spheres due to secondary sputtering process. After the removal of polystyrene spheres, a random array of nanorings is formed.[158,159] Gold nanorings were shown to exhibit a redshifted LSPR that can be tuned over a wide wavelength range in the NIR region by varying the ratio of the ring thickness to its radius[158] and had "bulk" refractive index sensitivities of up to 880 nm per refractive index unit (RIU) (observed for 150 nm diameter rings formed with a 20 nm sacrificial gold layer), which is substantially (>5 times) larger than those of nanodiscs with similar diameters,[159] suggesting nanorings as potential ultrasensitive refractive index biosensor platforms. The SCL technique can be also used to produce nanodiscs in an approach when a colloidal mask is deposited on top of metal film and acts as a protecting layer during the subsequent etching of the metal. After the removal of colloidal mask, a random array of metallic nanodiscs is formed.[160] More oblate disc shapes were demonstrated to have higher refractive index sensitivity, making them of interest as substrates for optimizing optical biosensing methods at the nanometre scale.[160] Common drawbacks of the SCL approach are limitations in producing nanostructures composed of materials with low etching selectivity, necessity of the reactive oxygen treatment for the polystyrene mask removal and the restriction of nanostructure dimensions to the polystyrene sphere size.[161] In case of biosensor application, one may face an issue with insufficient sensor chip reproducibility of SCL-fabricated nanopatterned metallic films due to the random nature of these nanostructure arrays.

A related technique based on SCL combined with UV-lithography and reactive ion etching (RIE) steps has been developed to fabricate short-range ordered nanoplasmonic pores penetrating through a thin (around 250 nm) multilayer membrane composed of gold and silicon nitride (SiN) that is supported on a Si wafer (Figure 13.11).[162] At first, a thin metal film with nanoholes is fabricated by colloidal lithography on a Si wafer coated with SiN (Figure 13.11, i-iv). In the subsequent step, the whole nanostructure is covered with a second SiN layer (Figure 13.11, v) for the purpose of protection during wet etching of Si wafer. UV-lithography is applied to prepare areas where Si removal will take place (Figure 13.11, vi), and the Si wafer is subsequently etched in tetramethyl-ammonium hydroxide (TMAH) (Figure 13.11, vii). During the final step, the holes in the metal film are converted into pores by etching the SiN with RIE. Here, gold film acts as an etch mask. It is possible to apply RIE from the front side (Figure 13.11, viii a) or the backside (Figure 13.11, viii b) of the sample. In the latter case, the gold will be accessible only in the regions of the membranes. Flow-through nanoplasmonic sensing of specific biorecognition reactions has been demonstrated using this nanohole membrane with a signal-to-noise ratio of around 50 at a temporal resolution below 190 ms and molecular uptake at least 1 order of magnitude faster than under stagnant conditions. Additionally, a high-throughput fabrication scheme that enables parallel production of multiple (more than 50) separate sensor chips or more than 1000 separate nanoplasmonic membranes on a single wafer has been presented.

The HCL method extends the approach of SCL technique by introducing a sacrificial resist layer combined with a thin film mask with nanoholes (a so-

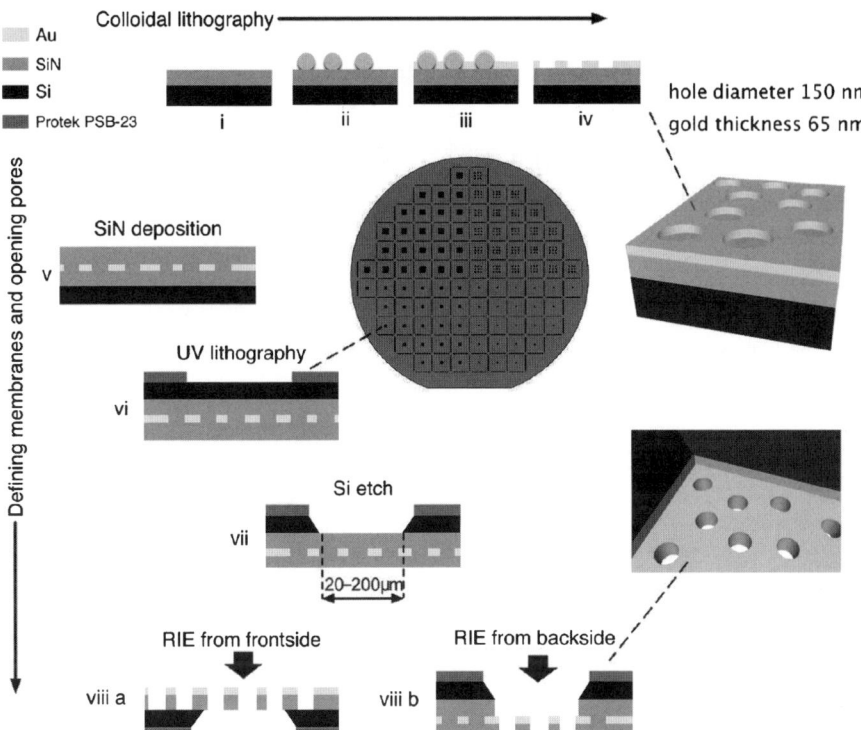

Figure 13.11 Schematic image illustrating the fabrication of sensor chips with membranes penetrated by short-range ordered nanoplasmonic pores. Note that the schematic illustration is not to scale. Adapted with permission from M. P. Jonsson, A. B. Dahlin, L. Feuz, S. Petronis and F. Höök, *Anal. Chem.*, 2010, 82(5), 2087. Copyright 2010 American Chemical Society.

called "hole-mask".[161] Figure 13.12 shows the common fabrication steps of the HCL. A sacrificial film (usually poly(methyl methacrylate)) is deposited onto a substrate. Similarly to SCL, a colloidal solution of polystyrene beads is deposited onto the charged-substance pretreated sacrificial surface, and a short-ordered polystyrene nanoparticle array is formed due to electrostatic interaction between a colloid and a surface. After quick stimulated drying of the surface, a thin film, which is resistant to sacrificial film etchant, is deposited, and polystyrene particles are removed, thus forming a hole-mask in the etching-resistant mask. Subsequently, the sacrificial resist layer is selectively etched through the hole-mask. After the abovementioned preparation stages, the hole-mask can then either be used as a deposition- or etch-mask, or both, to produce a variety of metallic nanostructures on the substrate surface. The HCL technique has been used to fabricate Au and Ag nanodiscs,[161,163–165] embedded nanodiscs,[161] metal-dielectric nanodiscs[166–168] and nanocones.[161] Au nanodiscs have been employed for plasmon-enhanced

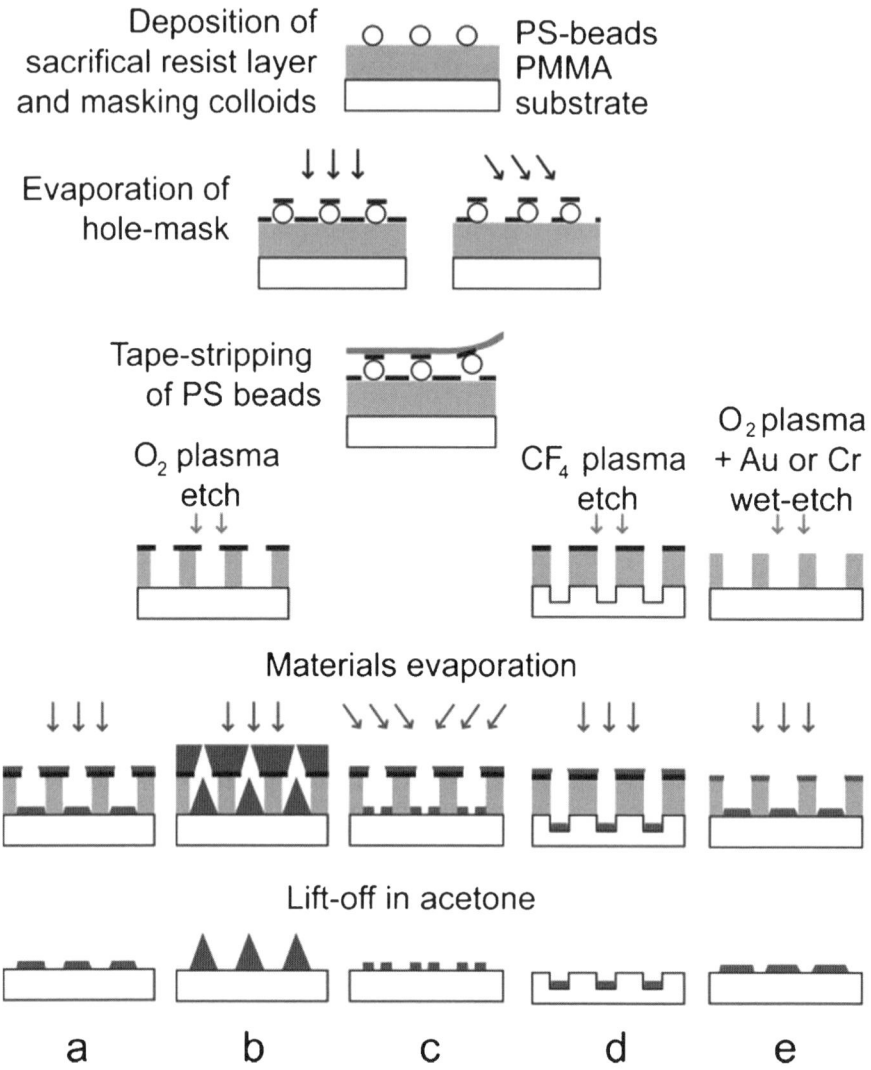

Figure 13.12 Diagram illustrating the basic process steps and resulting structures produced with HCL nanofabrication. Resulting structures are (a) arrays of nanodiscs and oriented elliptical nanostructures, (b) nanocone arrays, (c) (binary) arrays of nanodisc pairs, (d) embedded nanodiscs, and (e) discs with fine-tunable diameters. Reprinted with permission from H. Fredriksson, Y. Alaverdyan, A. Dmitriev, C. Langhammer, D. S. Sutherland, M. Zäch and B. Kasemo, Hole–mask colloidal lithography, *Adv. Mater.*, **19**(23), 4297–4302. Copyright 2007 WILEY-VCH Verlag GmbH & Co. KGaA, Weinheim.

colorimetric enzyme-linked immunosorbent assay with single molecule sensitivity against horseradish peroxidase[163] and to achieve extremely low limit of detection for bacterial and cancer diagnostics (down to several pg cm^{-2})[164] based on LSPR optical label-free biodetection. An interesting nanoplasmonic biosensor chip consisting of gold nanodiscs on a silicon solar cell with integrated electrical detection was successfully used to monitor a specific biorecognition reaction in real-time.[165] Enhanced refractive index sensitivity of metal nanodiscs when placed on dielectric nanopillars has been demonstrated.[168] Ag nanocones have been shown as suitable substrates for SERS applications with even single nanocones yielding appropriate signal intensity and showing excellent reproducibility of the SERS features.[161] Though the HCL technique removes certain restrictions of SCL method, it is still prone to sensor chip reproducibility issues due to unordered fashion of produced metallic nanoparticle arrays.

NSL is a type of colloidal lithography that employs a densely packed two-dimensional colloidal crystal as a deposition/etching mask. As opposed to SCL and HCL techniques, NSL is capable of producing ordered nanoparticle arrays with well-controlled shape, size and interparticle spacing.[169] Such a possibility is achieved by charging the substrate surface with the same charge sign that is carried by polystyrene nanospheres. When polystyrene colloidal solution is deposited on such a pretreated substrate and dried, capillary forces drag the nanospheres together and they crystallize in a hexagonally-packed pattern (Figure 13.13). This colloidal mask can be readily used for both additive and subtractive lithographic processes to produce metallic nanostructures with sizes down to 20 nm.[170] In one of the NSL approaches, a metal film is deposited through the mask to produce a metallic "film over nanosphere"

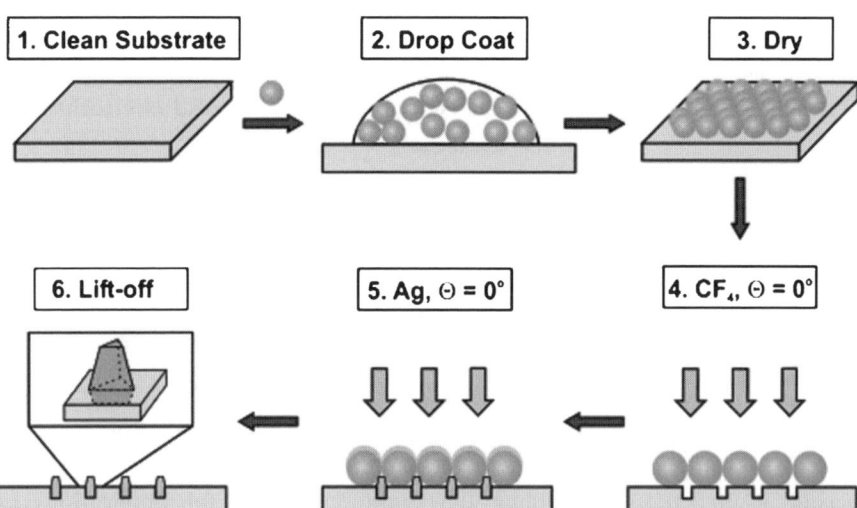

Figure 13.13 Schematic representation of the anchored Ag nanoparticle array fabrication using NSL. Reprinted with permission from E. M. Hicks, O. Lyandres, W. P. Hall, S. Zou, M. R. Glucksberg and R. P. Van Duyne, *J. Phys. Chem. C*, 2007, **111**(11), 4116. Copyright 2007 American Chemical Society.

(FON) structure. Enhanced electromagnetic fields generated by the surface roughness of the Ag FON structure have been employed to excite strong SERS signals for detection of *Bacillus subtilis* (a harmless analog of *Bacillus anthracis* – an anthrax causative agent) with limit of detection of 2.1×10^{-14} M,[171] for *in vivo* glucose detection using an implantable SERS sensor[171] and for quantitative lactate detection in the clinically relevant range 0.5–22 mM.[172] Alternatively, if the nanosphere mask is removed after the metal deposition, typically by sonicating the sample in a solvent, a substrate can be obtained that is covered with surface-confined metallic nanoparticles having triangular cross-section, which corresponds to the shape of voids between polystyrene nanospheres. Such nanostructures, mostly made of Ag, have been widely exploited in LSPR and SERS biosensing.[173–176] Ordered arrays of nanovoids in gold films have also been prepared by electrochemical deposition of Au through the colloidal mask and subsequent removal of polystyrene nanospheres[177] and have been used as substrates for NIR-SERS.[178] Other types of nanostructures that have been prepared using NSL-based techniques include rhombic nanoparticles,[179] nanopillars,[180] anchored nanoparticles (Figure 13.13),[181] "film over nanowell" structures,[182] nanodiscs,[183,184] nanoholes,[183] split rings[185] and micropatterned metallic films[186] for applications in LSPR biosensing and surface-enhanced spectroscopies. The main disadvantage of NSL is a variety of defects that arise in the process of crystallization as a result of nanosphere polydispersity, site randomness, point and line defects; typical defect-free domains have sizes in the 10–100 μm range. Dense nanosphere mask also imposes geometric constraints on the shape of produced nanoparticles – specifically, only triangular or hexagonally shaped metal nanoparticles can be fabricated by metal deposition through the stacked layers of close-packed nanospheres. In order to tune the nanoparticle shape, additional processing is required such as moving the sample or changing an angle during metal deposition, annealing and electrochemical growth.

13.4.3 Nanoimprint Lithography

Nanoimprint lithography (NIL) is a nanopatterning technique that provides a tradeoff between fabrication throughput, precision, costs and sample area, which eliminates drawbacks of scanning beam and colloidal lithographies for specific applications. This fast-developing technology provides a means to replicate features on a hard or soft stamps in a thermoplastic or photocurable resist by embossing or molding. Subsequent deposition of a metal film on such a replica with or without further resist lift-off can be used to produce plasmonic structures resembling the stamp nanorelief for applications in chemical and biosensing. Among the advantages of NIL is the capability to create nanoreliefs over large sample areas in parallel mode with high pattern uniformity and low defect densities, which is crucial for mass production of reproducible sensor chips for biosensing instruments. Namely, NIL is capable of molding a variety of materials and pattern features with a sub-10 nm resolution on the cm² area scale.[187–189] Another advantage of NIL is a wide range of nanoparticle shapes that can be fabricated using specially prepared reusable NIL stamps (Figure 13.14),[190] which is important in biosensing applications to optimize the performance of plasmonic nanoparticle arrays.

NIL has been employed to produce uniformly oriented and homogenous noble metal nanoparticle arrays using "nanoblock" molds fabricated from a collection of one-dimensional grating molds with different profiles.[190] Such anisotropic Ag and Au nanoparticle arrays have shown light polarization- and dimension-dependent plasmonic properties.[190] Similar approach has been applied to produce arrays of Au nanorectangles, which were shown to be a suitable basis for LSPR biosensor by detecting BSA-anti-BSA specific interaction.[191] High-density square arrays of free-standing cylindrical gold-coated polymer "nanofingers" fabricated using NIL have been demonstrated to be an efficient SERS substrate providing enhancement factors up to 2×10^{10}.[192] A two-dimensional array of cavities fabricated by NIL using a soft elastomeric mold in polyurethane, coated by 50 nm Au film, has demonstrated its sensing potential in light transmission experiments during the formation of self-assembled monolayer of hexadecanethiol.[193] Similar quasi-3D plasmonic crystals have shown a refractive index sensitivity of 700–800 nm/RIU[194] and enabled full multi-wavelength spectroscopic and spatially resolved detection of biomolecular binding events with sensitivities that correspond to small fractions of a monolayer,[194–197] e.g. 400 pM limit of detection for anti-IgG/IgG binding.[197] Au dot and ring arrays produced by NIL were used to detect the binding of streptavidin to biotin and it was shown that the sensitivity of the LSPR spectra to the binding of the biomolecules was enhanced as the ring width of Au rings was decreased.[198] Arrays of gold-coated nanodomes fabricated on glass substrates using a soft NIL revealed complex plasmonic resonances highly sensitive to the array dimensions, the thickness of the gold layer, and the refractive index of the surrounding medium.[199] Other metallic nanostructures fabricated using NIL, which are promising for biosensing applications, include arrays of Pt nanowires,[200] Ag island films deposited on grated and pillared Si substrates,[201] elliptical Au nanodisc arrays[202] and Au nanodot and nanowire arrays.[203,204]

However, this attractive nanolithographic technique is still under development and has several challenges to overcome. One of the important problems to be solved is the useful lifetime of the mold – at present, nanoimprint stamps require replacement after about 50 imprints made[205] due to wear produced by temperature variations and high pressures applied during the imprinting process. A possible way to partially solve this issue is to avoid high temperatures by using special resist treatment, which allows imprinting at room temperature.[206] Thermally-assisted imprinting has also limited throughput caused by time requirements of thermal cycling; however, employment of photocurable resist instead of thermoplastic or imprinting at room temperature and high pressures[207] can speed up the nanofabrication process. Other conditions that should be maintained in order to enable uniform pattern

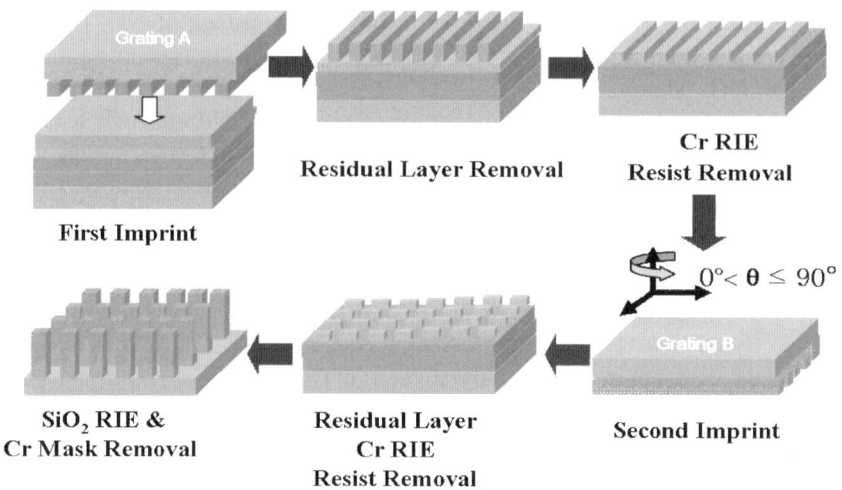

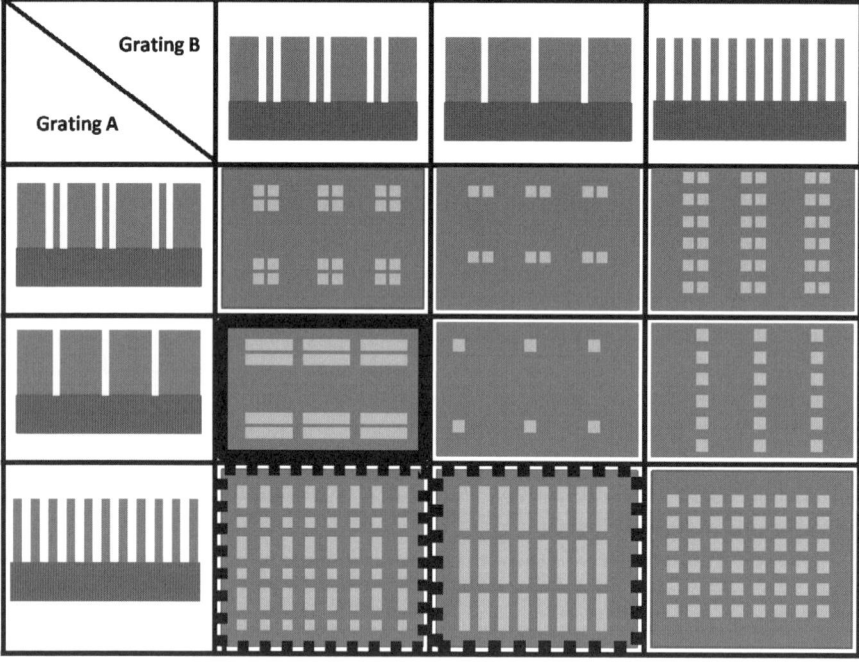

Figure 13.14 (Top) Illustration of the general process used to fabricate "nanoblock" NIL molds possessing different lattice and particle geometries. (Bottom) Matrix illustrating possible nanoparticle array configurations from nanoblock molds produced using a collection of one-dimensional grating molds with different profiles. The array bordered by a solid line is produced using the inverse profile of grating A. Similarly, the arrays bordered by broken lines are achieved using the inverse of grating B. Reprinted with permission from B. D. Lucas, J.-S. Kim, C. Chin and L. J. Guo, Nanoimprint lithography based approach for the fabrication of large-area, uniformly oriented plasmonic arrays, *Adv. Mater.*, **20**(6), 1129–1134. Copyright 2008 WILEY-VCH Verlag GmbH & Co. KGaA, Weinheim.

transfer over large areas include absence of air bubbles between the mold and resist, parallelism of rigid mold and resist surfaces and applied pressure uniformity.

References

1. K. Riehemann, S. W. Schneider, T. A. Luger, B. Godin, M. Ferrari and H. Fuchs, *Angew. Chem., Int. Ed. Engl.*, 2009, **48**(5), 872.
2. H. Hong, Y. Zhang, J. Sun and W. Cai, *Nano Today*, 2009, **4**(5), 399.
3. X. Michalet, F. F. Pinaud, L. A. Bentolila, J. M. Tsay, S. Doose, J. J. Li, G. Sundaresan, A. M. Wu, S. S. Gambhir and S. Weiss, *Science*, 2005, **307**(5709), 538.
4. J. Won, M. Kim, Y.-W. Yi, Y. H. Kim, N. Jung and T. K. Kim, *Science*, 2005, **309**(5731), 121.
5. K. Kostarelos, A. Bianco and M. Prato, *Nat. Nanotechnol.*, 2009, **4**, 627.
6. D. A. Rothenfluh, H. Bermudez, C. P. O'Neil and J. A. Hubbell, *Nat. Mater.*, 2008, **7**, 248.
7. M. De, P. S. Ghosh and V. M. Rotello, *Adv. Mater.*, 2008, **20**(22), 4225.
8. L. M. Liz-Marzán and P. V. Kamat in *Nanoscale Materials*, ed. L. M. Liz-Marzán and P. V. Kamat, Kluwer Academic Publishers, Boston, 2003, p. 1.
9. N. Harris, L. K. Ausman, J. M. McMahon, D. J. Masiello and G. C. Schatz, in *Computational Nanoscience*, ed. E. Bichoutskaia, The Royal Society of Chemistry, Cambridge, 2011, p. 147.
10. L. Huang, Y. Guo, Z. Peng and A. L. Porter, *Technol. Anal. Strateg.*, 2011, **23**(5), 527.
11. V. Vamvakaki and N. A. Chaniotakis, *Sens. Actuators, B*, 2007, **126**(1), 193.
12. J. Wang, *Analyst*, 2005, **130**, 421.
13. L. Xia, Z. Wei and M. Wan, *J. Colloid Interface Sci.*, 2010, **341**(1), 1.
14. R. J. Martín-Palma, M. Manso and V. Torres-Costa, *Sensors*, 2009, **9**(7), 5149.
15. L. Guo, G. Chen and D.-H. Kim, *Anal. Chem.*, 2010, **82**(12), 5147.
16. D. J. Maxwell, J. R. Taylor and S. Nie, *J. Am. Chem. Soc.*, 2002, **124**(32), 9606.
17. B. Sepúlveda, P. C. Angelomé, L. M. Lechuga and L. M. Liz-Marzán, *Nano Today*, 2009, **4**, 244.
18. J. Satija, R. Bharadwaj, V. V. R. Sai and S. Mukherji, *Nanotechnol., Sci. Appl.*, 2010, **3**, 171.
19. J. Turkevich, P. C. Stevenson and J. Hillier, *Discuss. Faraday Soc.*, 1951, **11**, 55.
20. X. Ren, X. Meng and F. Tang, *Sens. Actuators, B*, **110**(2), 358.
21. J. Zhang, X. Li, X. Sun and Y. Li, *J. Phys. Chem. B*, 2005, **109**(25), 12544.
22. X.-M. Qian and S. M. Nie, *Chem. Soc. Rev.*, 2008, **37**, 912.

23. Y. Yang, J. Shi, G. Kawamura and M. Nogami, *Scr. Mater.*, 2008, **58**(10), 862.
24. K. Sokolov, G. Chumanov and T. M. Cotton, *Anal. Chem.*, 1998, **70**(18), 3898.
25. E. G. Matveeva, I. Gryczynski, A. Barnett, Z. Leonenko, J. R. Lakowicz and Z. Gryczynski, *Anal. Biochem.*, 2007, **363**(2), 239.
26. J. R. Lakowicz, K. Ray, M. Chowdhury, H. Szmacinski, Y. Fu, J. Zhang and K. Nowaczyk, *Analyst*, 2008, **133**, 1308.
27. C. R. Yonzon, E. Jeoung, S. Zou, G. C. Schatz, M. Mrksich and R. P. Van Duyne, *J. Am. Chem. Soc.*, 2004, **126**(39), 12669.
28. A. J. Haes and R. P. Van Duyne, *Anal. Bioanal. Chem.*, 2004, **379**(7–8), 920.
29. P. Englebienne, A. Van Hoonacker, M. Verhas and N. G. Khlebtsov, *Comb. Chem. High Throughput Screening*, 2003, **6**(8), 777.
30. J. N. Anker, W. P. Hall, O. Lyandres, N. C. Shah, J. Zhao and R. P. Van Duyne, *Nat. Mater.*, 2008, **7**(6), 442.
31. N. Nath and A. Chilkoti, *Anal. Chem.*, 2002, **74**(3), 504.
32. A. J. Haes and R. P. Van Duyne, *Mater. Res. Soc. Symp. Proc.*, 2002, **723**, O3.1.1.
33. T. Endo, S. Yamamura, N. Nagatani, Y. Morita, Y. Takamura and E. Tamiya, *Sci. Technol. Adv. Mater.*, 2005, **6**(5), 491.
34. N. G. Khlebtsov, *Anal. Chem.*, 2008, **80**(17), 6620.
35. K.-S. Lee and M. A. El-Sayed, *J. Phys. Chem. B*, 2006, **110**(39), 19220.
36. X. Liu, M. Atwater, J. Wang and Q. Huo, *Colloids Surf., B*, 2007, **58**(1), 3.
37. M. D. Malinsky, K. L. Kelly, G. C. Schatz and R. P. Van Duyne, *J. Am. Chem. Soc.*, 2001, **123**(7), 1471.
38. W. A. Murray, B. Auguie and W. L. Barnes, *J. Phys. Chem. C*, 2009, **113**(13), 5120.
39. M. Faraday, *Philos. Trans. R. Soc. London*, 1857, **147**, 145.
40. P. K. Jain, K. S. Lee, I. H. El-Sayed and M. A. El-Sayed, *J. Phys. Chem. B.*, 2006, **110**, 7238.
41. C. A. Mirkin, R. L. Letsinger, R. C. Mucic and J. J. Storhoff, *Nature*, 1996, **382**, 607.
42. F. Wei, R. Lam, S. Cheng, S. Lu, D. Ho and N. Li, *Appl. Phys. Lett.*, 2010, **96**, 133702.
43. A. S. de Dios and M. E. Díaz-García, *Anal. Chim. Acta*, 2010, **666**(1–2), 1.
44. M. J. Hostetler, J. E. Wingate, C.-J. Zhong, J. E. Harris, R. W. Vachet, M. R. Clark, J. D. Londono, S. J. Green, J. J. Stokes, G. D. Wignall, G. L. Glish, M. D. Porter, N. D. Evans and R. W. Murray, *Langmuir*, 1998, **14**(1), 17.
45. Y. B. Zheng, B. K. Juluri, X. Mao, T. R. Walker and T. J. Huang, *J. Appl. Phys.*, 2008, **103**(1), 014308.
46. A. Lopatynskyi, O. Lopatynska and V. Chegel, *Semicond. Phys., Quantum Electron. Optoelectron.*, 2011, **14**(1), 114.

47. T. K. Sau, A. Pal, N. R. Jana, Z. L. Wang and T. Pal, *J. Nanopart. Res.*, 2001, **3**(4), 257.
48. G. Frens, *Nature*, 1973, **241**, 20.
49. D. V. Leff, P. C. Ohara, J. R. Heath and W. M. Gelbart, *J. Phys. Chem.*, 1995, **99**(18), 7036.
50. M. Yun, N. V. Myung, R. P. Vasquez, C. Lee, E. Menke and R. M. Penner, *Nano Lett.*, 2004, **4**(3), 419.
51. E. C. Walter, M. P. Zach, F. Favier, B. J. Murray, K. Inazu, J. C. Hemminger and R. M. Penner, *ChemPhysChem*, 2003, **4**(2), 131.
52. P. Kohli, M. Wirtz and C. R. Martin, *Electroanalysis*, 2004, **16**(1–2), 9.
53. D. G. Georganopoulou, L. Chang, J. M. Nam, C. S. Thaxton, E. J. Mufson, W. L. Klein and C. A. Mirkin, *Proc. Natl. Acad. Sci. U. S. A.*, 2005, **102**(7), 2273.
54. A. W. H. Lin, N. A. Lewinski, J. L. West, N. J. Halas and R. A. Drezek, *J. Biomed. Opt.*, 2005, **10**(6), 064035.
55. F. Bao, J.-L. Yao and R.-A. Gu, *Langmuir*, 2009, **25**(18), 10782.
56. J. K. Lim, R. D. Tilton, A. Eggeman and S. A. Majetich, *J. Magn. Magn. Mater.*, 2007, **311**(1), 78.
57. Y. K. Hahn, Z. Jin, J. H. Kang, E. Oh, M. K. Han, H. S. Kim, J. T. Jang, J. H. Lee, J. Cheon, S. H. Kim, H. S. Park and J. K. Park, *Anal. Chem.*, 2007, **79**(6), 2214.
58. F. Lacharme, C. Vandevyver and M. A. M. Gijs, *Anal. Chem.*, 2008, **80**(8), 2905.
59. D.-K. Lim, I.-J. Kim and J.-M. Nam, *Chem. Commun.*, 2008, **42**, 5312.
60. M. Schierhorn and L. M. Liz-Marzán, *Nano Lett.*, 2002, **2**, 13.
61. T. Ung, L. M. Liz-Marzán and P. Mulvaney, *Langmuir*, 1998, **14**(14), 3740.
62. J. B. Jackson and N. J. Halas, *J. Phys. Chem. B*, 2001, **105**(14), 2743.
63. M. A. Nash, J. J. Lai, A. S. Hoffman, P. Yager and P. S. Stayton, *Nano Lett.*, 2010, **10**(1), 85.
64. M. Spasova, V. Salgueiriño-Maceira, A. Schlachter, M. Hilgendorff, M. Giersig, L. M. Liz-Marzán and M. Farle, *J. Mater. Chem.*, 2005, **15**(21), 2095.
65. Z. Xu, Y. Hou and S. Sun, *J. Am. Chem. Soc.*, 2007, **129**(28), 8698.
66. J. N. Park, K. J. An, Y. S. Hwang, J. G. Park, H. J. Noh, J. Y. Kim, J. H. Park, N. M. Hwang and T. W. Hyeon, *Nat. Mater.*, 2004, **3**, 891.
67. B. Nikoobakht and M. A. El-Sayed, *Chem. Mater.*, 2003, **15**, 1957.
68. J. X. Gao, C. M. Bender and C. J. Murphy, *Langmuir*, 2003, **19**, 9065.
69. J. Zhu and W. Jiang, *Mater. Chem. Phys.*, 2007, **101**(1), 56.
70. N. Nath and A. Chilkoti, *Proc. SPIE*, 2002, **4626**, 441.
71. H. Liu, X. Zhang and Z. Gao, *Photonics and Nanostruct.*, 2010, **8**(3), 131.
72. S. Malynych and G. Chumanov, *J. Am. Chem. Soc.*, 2003, **125**(10), 2896.
73. S. Malynych, I. Luzinov and G. Chumanov, *J. Phys. Chem. B*, 2002, **106**, 1280.

74. K. Aslan, Z. Leonenko, J. R. Lakowicz and C. D. Geddes, *J. Phys. Chem. B*, 2005, **109**(8), 3157.
75. N. R. Jana, L. Gearheart and C. J. Murphy, *Chem. Commun.*, 2001, **7**, 617.
76. K. Aslan, J. R. Lakowicz and C. D. Geddes, *J. Phys. Chem. B*, 2005, **109**(13), 6247.
77. X. Zhang, B. Sun, R. H. Friend, H. Guo, D. Nau and H. Giessen, *Nano Lett.*, 2006, **6**(4), 651.
78. S. B. Chaney, S. Shanmukh, Y.-P. Zhao and R. A. Dluhy, *Appl. Phys. Lett.*, 2005, **87**, 31908.
79. Y.-P. Zhao, S. B. Chaney, S. Shanmukh and R. A. Dluhy, *J. Phys. Chem. B*, 2006, **110**, 3135.
80. S. Shanmukh, L. Jones, J. Driskell, Y. Zhao, R. Dluhy and R. A. Tripp, *Nano Lett.*, 2006, **6**(11), 2630.
81. B. Kim and W. M. Sigmund, *Langmuir*, 2004, **20**(19), 8239.
82. Y.-Y. Ou and M. H. Huang, *J. Phys. Chem. B*, 2006, **110**(5), 2031.
83. X. Li, Y. Liu, L. Fu, L. Cao, D. Wei, G. Yu and D. Zhu, *Carbon*, 2006, **44**(14), 3139.
84. S. Fullam, D. Cottell, H. Rensmo and D. Fitzmaurice, *Adv. Mater.*, 2000, **12**, 1430.
85. H. C. Choi, M. Shim, S. Bangsaruntip and H. Dai, *J. Am. Chem. Soc.*, 2002, **124**, 9058.
86. A. Tello, G. Cárdenas, P. Häberle and R. A. Segura, *Carbon*, 2008, **46**(6), 884.
87. R. Segura, A. Tello, G. Cardenas and P. Häberle, *Phys. Status Solidi A*, 2007, **204**(2), 513.
88. K. Balasubramanian and M. Burghard, *Small*, 2005, **1**(2), 180.
89. K. Niesz, A. Siska, I. Vesselényi, K. Hernadi, D. Méhn, G. Galbács, Z. Kónya and I. Kiricsi, *Catal. Today*, 2002, **76**(1), 3.
90. G. Cárdenas, R. Oliva, P. Reyes and B. L. Rivas, *J. Mol. Catal. A: Chem.*, 2003, **191**(1), 75.
91. G. Cárdenas-Triviño, R. A. Segura and J. Reyes-Gasga, *Colloid Polym. Sci.*, 2004, **282**(11), 1206.
92. G. E. Possin, *Rev. Sci. Instrum.*, 1970, **41**, 772.
93. C. R. Martin, *Science*, 1994, **266**, 1961.
94. C. K. Preston and M. Moskovits, *J. Phys. Chem.*, 1993, **97**, 8495.
95. N. Kwon, N. Kim, J. Yeon, G. Yeom and I. Chung, *J. Vac. Sci. Technol., B*, 2011, **29**, 031805.
96. C. Iida, M. Sato, M. Nakayama and A. Sanada, *Int. J. Electrochem. Sci.*, 2011, **6**, 4730.
97. D. Routkevitch, A. A. Tager, J. Haruyama, D. Almawlawi, M. Moskovits and J. M. Xu, *IEEE Trans. Electron Devices*, 1996, **43**(10), 1646.
98. C. Mu, Y.-X. Yu, R. M. Wang, K. Wu, D. S. Xu and G.-L. Guo, *Adv. Mater.*, 2004, **16**(17), 1550.

99. G. Meng, Y. J. Jung, A. Cao, R. Vajtai and P. M. Ajayan, *Proc. Natl. Acad. Sci. U. S. A.*, 2005, **102**(20), 7074.
100. F. Berti, S. Todros, D. Lakshmi, M. J. Whitcombe, I. Chianella, M. Ferroni, S. A. Piletsky, A. P. F. Turner and G. Marrazza, *Biosens. Bioelectron.*, 2010, **26**(2), 497.
101. Y. Lei, W. Cai and G. Wilde, *Prog. Mater. Sci.*, 2007, **52**(4), 465.
102. G. A. Wurtz, P. R. Evans, W. Hendren, R. Atkinson, W. Dickson, R. J. Pollard and A. V. Zayats, *Nano Lett.*, 2007, **7**(5), 1297.
103. H. Masuda and M. Satoh, *Jpn. J. Appl. Phys.*, 1996, **35**, L126.
104. H. Masuda, K. Yasui and K. Nishio, *Adv. Mater.*, 2000, **12**(14), 1031.
105. Y. Lei and W.-K. Chim, *Chem. Mater.*, 2005, **17**(3), 580.
106. R. Inguanta, S. Piazza and C. Sunseri, *Electrochem. Commun.*, 2008, **10**(4), 506.
107. A. B. Braunschweig, A. L. Schmucker, W. D. Wie and C. A. Mirkin, *Chem. Phys. Lett.*, 2010, **486**(4–6), 89.
108. H. Masuda and K. Fukuda, *Science*, 1995, **268**(5216), 1466.
109. Y. Lei, W.-K. Chim, Z. Zhang, T. Zhou, L. Zhang, G. Meng and F. Phillipp, *Chem. Phys. Lett.*, 2003, **380**(3–4), 313.
110. D. J. Lipomi, R. V. Martinez and G. M. Whitesides, *Angew. Chem., Int. Ed.*, 2011, **50**, 8566.
111. K. Yamazaki and H. Namatsu, *Jpn. J. Appl. Phys.*, 2004, **43**, 3767.
112. M. J. Word, I. Adesida and P. R. Berg, *J. Vac. Sci. Technol. B*, 2003, **21**, L12.
113. W. Hinsberg, F. Houle, M. Sanchez, J. Hoffnagle, G. Wallraff, D. Medeiros, G. Gallatin and J. Cobbc, *Proc. SPIE*, 2003, **5039**, 1.
114. C. L. Haynes, A. D. McFarland, L. Zhao, R. P. Van Duyne, G. C. Schatz, L. Gunnarsson, J. Prikulis, B. Kasemo and M. Käll, *J. Phys. Chem. B*, 2003, **107**, 7337.
115. R. Hillenbrand, F. Keilmann, P. Hanarp, D. S. Sutherland and J. Aizpurua, *Appl. Phys. Lett.*, 2003, **83**, 368.
116. G. Barbillon, J.-L. Bijeon, J. Plain and P. Royer, *Thin Solid Films*, 2009, **517**, 2997.
117. W. Rechberger, A. Hohenau, A. Leitner, J. R. Krenn, B. Lamprecht and F. R. Aussenegg, *Opt. Commun.*, 2003, **220**, 137.
118. M. Hentschel, M. Saliba, R. Vogelgesang, H. Giessen, A. P. Alivisatos and N. Liu, *Nano Lett.*, 2010, **10**(7), 2721.
119. R. M. Bakker, H.-K. Yuan, Z. Liu, V. P. Drachev, A. V. Kildishev, V. M. Shalaev, R. H. Pedersen, S. Gresillon and A. Boltasseva, *Appl. Phys. Lett.*, 2008, **92**, 043101.
120. D. P. Fromm, A. Sundaramurthy, P. J. Schuck, G. Kino and W. E. Moerner, *Nano Lett.*, 2004, **4**, 957.
121. J. R. Krenn, W. Gotschy, D. Somitsch, A. Leitner and F. R. Aussenegg, *Appl. Phys. A: Mater. Sci. Process.*, 1995, **61**(5), 541.

122. J. R. Krenn, A. Dereux, J. C. Weeber, E. Bourillot, Y. Lacroute, J. P. Goudonnet, G. Schider, W. Gotschy, A. Leitner, F. R. Aussenegg and C. Girard, *Phys. Rev. Lett.*, 1999, **82**, 2590.
123. B. Auguié and W. L. Barnes, *Phys. Rev. Lett.*, 2008, **101**, 143902.
124. A. W. Clark, A. Glidle, D. R. S. Cumming and J. M. Cooper, *J. Am. Chem. Soc.*, 2009, **131**(48), 17615.
125. B. K. Canfield, S. Kujala, K. Laiho, K. Jefimovs, J. Turunen and M. Kauranen, *Opt. Express*, 2006, **14**(2), 950.
126. S. A. Maier, P. G. Kik and H. A. Atwater, *Appl. Phys. Lett.*, 2002, **81**, 1714.
127. M. U. González, J.-C. Weeber, A.-L. Baudrion, A. Dereux, A. L. Stepanov, J. R. Krenn, E. Devaux and T. W. Ebbesen, *Phys. Rev. B*, 2006, **73**, 155416.
128. N.-F. Chiu, C. Yu, S.-Y. Nien, J.-H. Lee, C.-H. Kuan, K.-C. Wu, C.-K. Lee and C.-W. Lin, *Opt. Express*, 2007, **15**(18), 11608.
129. K. Kim, D. J. Kim, S. Moon, D. Kim and K. M. Byun, *Nanotechnology*, 2009, **20**, 315501.
130. Y. Liu and S. Blair, *Opt. Lett.*, 2003, **28**(7), 507.
131. S. Aksu, A. A. Yanik, R. Adato, A. Artar, M. Huang and H. Altug, *Nano Lett.*, 2010, **10**(7), 2511.
132. http://www.nilt.com/ (accessed Oct 2011).
133. D.-G. Choi, J. Jeong, Y. Sim, E. Lee, W.-S. Kim and B.-S. Bae, *Langmuir*, 2005, **21**(21), 9390.
134. J. Chu, F. Meng, Z. Han and Q. Guo, *Sci. China: Technol. Sci.*, 2010, **53**(1), 248.
135. K. Gamo, *Microelectron. Eng.*, 1996, **32**, 159.
136. T. Morita, R. Kometani, K. Watanabe, K. Kanda, Y. Haruyama, T. Hoshino, K. Kondo, T. Kaito, T. Ichihashi, J.-i. Fujita, M. Ishida, Y. Ochiai, T. Tajima and S. Matsui, *J. Vac. Sci. Technol., B*, 2003, **21**, 2737.
137. R. L. Kubena, J. W. Ward, F. P. Stratton, R. L. Joyce and G. M. Atkinson, *J. Vac. Sci. Technol., B*, 1991, **9**, 3079.
138. J. Gierak, C.Vieu, M. Schneider, H. Launois, G. Ben Assayag and A. Septier, *J. Vac. Sci. Technol., B*, 1997, **15**, 2373.
139. H. Rigneault, J. Capoulade, J. Dintinger, J. Wenger, N. Bonod, E. Popov, T. W. Ebbesen and P. F. Lenne, *Phys. Rev. Lett.*, 2005, **95**(11), 117401.
140. F. Eftekhari, C. Escobedo, J. Ferreira, X. Duan, E. M. Girotto, A. G. Brolo, R. Gordon and D. Sinton, *Anal. Chem.*, 2009, **81**(11), 4308.
141. A. G. Brolo, R. Gordon, B. Leathem and K. L. Kavanagh, *Langmuir*, 2004, **20**, 4813.
142. H. J. Jeon, K. H. Kim, Y. K. Baek, D. W. Kim and H. T. Jung, *Nano Lett.*, 2010, **10**(9), 3604.
143. J. Ferreira, M. J. Santos, M. M. Rahman, A. G. Brolo, R. Gordon, D. Sinton and E. M. Girotto, *J. Am. Chem. Soc.*, 2009, **131**(2), 436.

144. A. Lesuffleur, H. Im, N. C. Lindquist, K. S. Lim and S. H. Oh, *Opt. Express*, 2008, **16**(1), 219.
145. A. Degiron, H. J. Lezec, N. Yamamoto and T. W. Ebbesen, *Opt. Commun.*, 2004, **239**, 61.
146. B. Brian, B. Sepúlveda, Y. Alaverdyan, L. M. Lechuga and M. Käll, *Opt. Express*, 2009, **17**(3), 2015.
147. H. F. Ghaemi, T. Thio, D. E. Grupp, T. W. Ebbesen and H. J. Lezec, *Phys. Rev. B*, 1998, **58**(11), 6779.
148. A. Lesuffleur, H. Im, N. C. Lindquist and S.-H. Oh, *Appl. Phys. Lett.*, 2007, **90**, 243110.
149. Z. Liu, Y. Wang, J. Yao, H. Lee, W. Srituravanich and X. Zhang, *Nano Lett.*, 2009, **9**(1), 462.
150. F. López-Tejeira, S. G. Rodrigo, L. Martín-Moreno, F. J. García-Vidal, E. Devaux, T. W. Ebbesen, J. R. Krenn, I. P. Radko, S. I. Bozhevolnyi, M. U. González, J. C. Weeber and A. Dereux, *Nat. Phys.*, 2007, **3**, 324.
151. C.-K. Chang, D.-Z. Lin, C.-S. Yeh, C.-K. Lee, Y.-C. Chang, M.-W. Lin, J.-T. Yeh and J.-M. Liu, *Appl. Phys. Lett.*, 2007, **90**, 061113.
152. V. S. Volkov, S. I. Bozhevolnyi, E. Devaux, J.-Y. Laluet and T. W. Ebbesen, *Nano Lett.*, 2007, **7**(4), 880.
153. P. Hanarp, D. S. Sutherland, J. Gold and B. Kasemo, *Colloids Surf., A*, 2003, **214**, 23.
154. D. Gao, W. Chen, A. Mulchandani and J. S. Schultz, *Appl. Phys. Lett.*, 2007, **90**, 073901.
155. A. B. Dahlin, J. O. Tegenfeldt and F. Höök, *Anal. Chem.*, 2006, **78**(13), 4416.
156. A. Dahlin, M. Zäch, T. Rindzevicius, M. Käll, D. S. Sutherland and F. Höök, *J. Am. Chem. Soc.*, 2005, **127**(14), 5043.
157. M. P. Jonsson, P. Jönsson, A. B. Dahlin and F. Höök, *Nano Lett.*, 2007, **7**(11), 3462.
158. J. Aizpurua, P. Hanarp, D. S. Sutherland, M. Käll, G. W. Bryant and F. J. García de Abajo, *Phys. Rev. Lett.*, 2003, **90**(5), 057401.
159. E. M. Larsson, J. Alegret, M. Käll and D. S. Sutherland, *Nano Lett.*, 2007, **7**(5), 1256.
160. P. Hanarp, M. Käll and D. S. Sutherland, *J. Phys. Chem. B*, 2003, **107**(24), 5768.
161. H. Fredriksson, Y. Alaverdyan, A. Dmitriev, C. Langhammer, D. S. Sutherland, M. Zäch and B. Kasemo, *Adv. Mater.*, 2007, **19**(23), 4297.
162. M. P. Jonsson, A. B. Dahlin, L. Feuz, S. Petronis and F. Höök, *Anal. Chem.*, 2010, **82**(5), 2087.
163. S. Chen, M. Svedendahl, R. P. Van Duyne and M. Käll, *Nano Lett.*, 2011, **11**(4), 1826.
164. S. Chen, M. Svedendahl, M. Käll, L. Gunnarsson and A. Dmitriev, *Nanotechnology*, 2009, **20**, 434015.
165. F. Mazzotta, G. Wang, C. Hägglund, F. Höök and M. P. Jonsson, *Biosens. Bioelectron.*, 2010, **26**(4), 1131.

166. A. Dmitriev, T. Pakizeh, M. Käll and D. S. Sutherland, *Small*, 2007, **3**, 294.
167. T. Pakizeh, A. Dmitriev, M. S. Abrishamian, N. Granpayeh and M. Käll, *J. Opt. Soc. Am. B*, 2008, **25**(4), 659.
168. A. Dmitriev, C. Hägglund, S. Chen, H. Fredriksson, T. Pakizeh, M. Käll and D. S. Sutherland, *Nano Lett.*, 2008, **8**(11), 3893.
169. J. C. Hulteen and R. P. Van Duyne, *J. Vac. Sci. Technol., A*, 1995, **13**, 1553.
170. C. L. Haynes and R. P. Van Duyne, *J. Phys. Chem. B*, 2001, **105**(24), 5599.
171. N. C. Shah, O. Lyandres, C. Yonzon, X. Zhang and R. P. Van Duyne in *New Approaches in Biomedical Spectroscopy*, ed. KKneipp, RAroca, HKneipp and EWentrup-Byrne, American Chemical Society, Washington, DC, 2007, p. 107.
172. N. C. Shah, O. Lyandres, J. T. Walsh Jr., M. R. Glucksberg and R. P. Van Duyne, *Anal. Chem.*, 2007, **79**, 6927.
173. W. P. Hall, J. Modica, J. N. Anker, Y. Lin, M. Mrksich and R. P. Van Duyne, *Nano Lett.*, 2011, **11**, 1098.
174. W. Zhou, Y. Ma, H. Yang, Y. Ding and X. Luo, *Int. J. Nanomed.*, 2011, **6**, 381.
175. A. J. Haes and R. P. Van Duyne, *J. Am. Chem. Soc.*, 2002, **124**(35), 10596.
176. A. J. Haes, W. P. Hall, L. Chang, W. L. Klein and R. P. Van Duyne, *Nano Lett.*, 2004, **4**(6), 1029.
177. T. A. Kelf, Y. Sugawara, R. M. Cole, J. J. Baumberg, M. E. Abdelsalam, S. Cintra, S. Mahajan, A. E. Russell and P. N. Bartlett, *Phys. Rev. B*, 2006, **74**, 245415.
178. S. Mahajan, M. Abdelsalam, Y. Suguwara, S. Cintra, A. Russell, J. Baumberg and P. Bartlett, *Phys. Chem. Chem. Phys.*, 2007, **9**, 104.
179. S. Zhu, C. Dua and Y. Fu, *Opt. Mater.*, 2009, **31**, 769.
180. C. L. Cheung, R. J. Nikolić, C. E. Reinhardt and T. F. Wang, *Nanotechnology*, 2006, **17**(5), 1339.
181. E. M. Hicks, O. Lyandres, W. P. Hall, S. Zou, M. R. Glucksberg and R. P. Van Duyne, *J. Phys. Chem. C*, 2007, **111**(11), 4116.
182. E. M. Hicks, X. Zhang, S. Zou, O. Lyandres, K. G. Spears, G. C. Schatz and R. P. Van Duyne, *J. Phys. Chem. B*, 2005, **109**(47), 22351.
183. Y. B. Zheng and T. J. Huang, *JALA*, 2008, **13**(4), 215.
184. Y. B. Zheng, B. K. Juluri, X. Mao, T. R. Walker and T. J. Huang, *J. Appl. Phys.*, 2008, **103**, 014308.
185. M. C. Gwinner, E. Koroknay, L. Fu, P. Patoka, W. Kandulski, M. Giersig and H. Giessen, *Small*, 2009, **5**(3), 400.
186. W. Ruan, Z. Lu, T. Zhou, B. Zhao and L. Niu, *Anal. Methods*, 2010, **2**, 684.
187. S. Y. Chou, P. R. Krauss, W. Zhang, L. Guo and L. Zhuang, *J. Vac. Sci. Technol., B*, 1997, **15**, 2897.

188. S. Y. Chou and P. R. Krauss, *Microelectron. Eng.*, 1997, **35**(1–4), 237.
189. M. Li, L. Chen, W. Zhang and S. Y. Chou, *Nanotechnology*, 2003, **14**(1), 33.
190. B. D. Lucas, J.-S. Kim, C. Chin and L. J. Guo, *Adv. Mater.*, 2008, **20**(6), 1129.
191. V. Chegel, B. Lucas, J. Guo, A. Lopatynskyi, O. Lopatynska and L. Poperenko, *Semicond. Phys., Quantum Electron. Optoelectron.*, 2009, **12**(1), 91.
192. M. Hu, F. S. Ou, W. Wu, I. Naumov, X. Li, A. M. Bratkovsky, R. S. Williams and Z. Li, *J. Am. Chem. Soc.*, 2010, **132**(37), 12820.
193. V. Malyarchuk, F. Hua, N. Mack, V. Velasquez, J. White, R. Nuzzo and J. Rogers, *Opt. Express*, 2005, **13**(15), 5669.
194. M. E. Stewart, N. H. Mack, V. Malyarchuk, J. A. N. T. Soares, T.-W. Lee, S. K. Gray, R. G. Nuzzo and J. A. Rogers, *Proc. Natl. Acad. Sci. U. S. A.*, 2006, **103**(46), 17143.
195. V. Malyarchuk, M. E. Stewart, R. G. Nuzzo and J. A. Rogers, *Appl. Phys. Lett.*, 2007, **90**, 203113.
196. J. Yao, M. E. Stewart, J. Maria, T. W. Lee, S. K. Gray, J. A. Rogers and R. G. Nuzzo, *Angew. Chem., Int. Ed. Engl.*, 2008, **47**(27), 5013.
197. M. E. Stewart, J. Yao, J. Maria, S. K. Gray, J. A. Rogers and R. G. Nuzzo, *Anal. Chem.*, 2009, **81**(15), 5980.
198. S. Kim, J. M. Jung, D. G. Choi, H. T. Jung and S. M. Yang, *Langmuir*, 2006, **22**(17), 7109.
199. J. McPhillips, C. McClatchey, T. Kelly, A. Murphy, M. P. Jonsson, G. A. Wurtz, R. J. Winfield and R. J. Pollard, *J. Phys. Chem. C*, 2011, **115**(31), 15234.
200. X. M. Yan, S. Kwon, A. M. Contreras, J. Bokor and G. A. Somorjai, *Nano Lett.*, 2005, **5**(4), 745.
201. R. Alvarez-Puebla, B. Cui, J.-P. Bravo-Vasquez, T. Veres and H. Fenniri, *J. Phys. Chem. C*, 2007, **111**(18), 6720.
202. S.-W. Lee, K.-S. Lee, J. Ahn, J.-J. Lee, M.-G. Kim and Y.-B. Shin, *ACS Nano*, 2011, **5**(2), 897.
203. S. H. Ko, I. Park, H. Pan, C. P. Grigoropoulos, A. P. Pisano, C. K. Luscombe and J. M. J. Frechet, *Nano Lett.*, 2007, **7**, 1869.
204. I. Park, S. H. Ko, H. Pan, C. P. Grigoropoulos, A. P. Pisano, J. M. J. Fréchet, E.-S. Lee and J.-H. Jeong, *Adv. Mater.*, 2008, **20**(3), 489.
205. C. M. Sotomayor Torres, S. Zankovych, J. Seekamp, A. P. Kam, C. Clavijo Cedeño, T. Hoffmann, J. Ahopelto, F. Reuther, K. Pfeiffer, G. Bleidiessel, G. Gruetzner, M. V. Maximov and B. Heidari, *Mater. Sci. Eng., C*, 2003, **23**(1–2), 23.
206. D.-Y. Khang and H. H. Lee, *Appl. Phys. Lett.*, 2000, **76**, 870.
207. D.-Y. Khang, H. Yoon and H. H. Lee, *Adv. Mater.*, 2001, **13**(10), 749.

Materials Nanoarchitechtonics for Advanced Devices

CHAPTER 14
Nanostructure Manipulation in Organic Solar Cells

PINYI YANG, SHANE D. BOYD AND
CHRISTINE K. LUSCOMBE

Department of Materials Science and Engineering, University of Washington, Box 352120, Seattle, WA 98195, USA

14.1 Introduction

In order to meet the planet's future energy needs, it will become increasingly urgent to look for inexpensive, clean, and renewable sources of energy. Because of the overwhelming abundance of natural sunlight, solar energy has long been recognized as an important future source of power. Conventional solar cells are most often made from inorganic semiconductor materials such as crystalline Si or CdTe, which can offer high energy conversion efficiencies but at the price of requiring cleanroom and high-vacuum equipment. These technologies are expensive and relatively difficult to scale up, resulting in final installed energy prices too high to compete with current conventional energy sources (~$2–$4 per Watt).[1–4] As a result, organic solar cells have attracted much attention for their potential for facile processing *via* solution methods and the possibility of easy scale-up to roll-to-roll printing of thin, light, large-area, flexible units.[5–8]

Organic solar cells (OSCs) are based on conjugated polymers, which are characterized by alternating single and double carbon–carbon bonds along the polymer backbone (Figure 14.1).[9,10] Within such a polymer, a delocalized π electron system forms, much the same as in an inorganic semiconductor.[11–15]

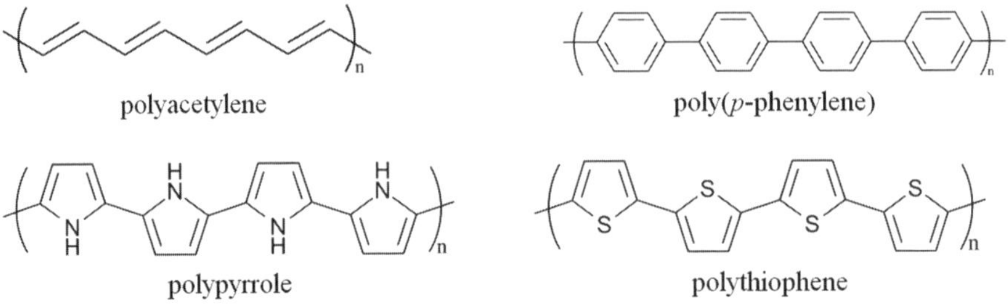

polyacetylene

poly(p-phenylene)

polypyrrole

polythiophene

Figure 14.1 Examples of conjugated polymers.

However, due to the lower dielectric constant of organic materials, excitation with light forms a strongly localized and electrostatically bound electron–hole pair known as an exciton instead of a free electron and hole (Figure 14.2 step 1).[16–21] One way to overcome the exciton binding energy and separate electron from hole is to use an electron-accepting material in the OSCs; the conjugated polymer acts as an electron donor and transfers an electron at a donor–acceptor interface (Figure 14.2 step 2, 3).[17,18,22,23] However, the lifetime of an exciton is on the order of 1 ns, which is equivalent to a diffusion length of approximately 10 nm.[7,24–27] Consequently, the morphology of the conjugated polymer donor and acceptor film (active layer) must be fine enough to allow excitons to diffuse to a donor–acceptor interface before they decay. In addition, separated free electrons and holes can only be collected if they are able to conduct to their corresponding electrodes (cathode and anode, respectively), which requires the nanostructure of each component phase to be continuous throughout the thickness of the active layer. The requirements of exciton dissociation and charge collection on organic solar cell active layer morphology make manipulation of the nanostructure crucial to the performance of the final device.[28–31] Many of the manipulation methods devised by researchers in the past decade will be reviewed in the following sections.

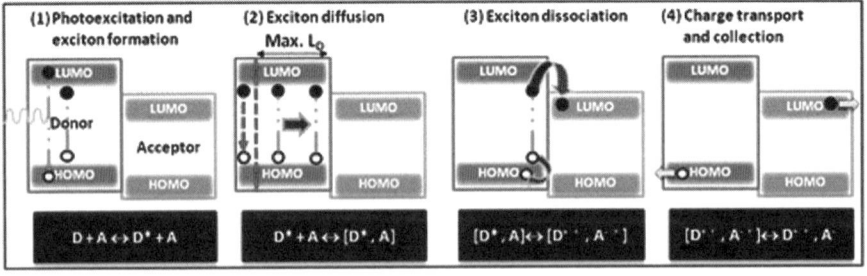

- electron
- hole

Figure 14.2 Important processes towards formation of photocurrent in organic solar cells.[32]

14.2 Nanostructure Manipulation In Conjugated Polymer–Fullerene Organic Solar Cells

As discussed previously, the primary goal of organic solar cell nanostructure manipulation is to simultaneously satisfy the requirements of both exciton dissociation and charge collection by creating a finely-textured (20 nm feature size), bicontinuous donor–acceptor network. An ideal OSC morphology would be as shown in Figure 14.3 (left), with interdigitating conjugated polymer and electron acceptor "fingers," each connected to their respective electrodes. With each of the conjugated polymer phase "fingers" approximately 20 nm in thickness, every exciton generated in the conjugated polymer should in theory be able to diffuse to the interface and transfer an electron to the acceptor phase,[28,30,33,34] thereby minimizing loss by exciton decay. In addition, every "finger" of each phase possesses a contiguous conduction pathway to its corresponding electrode, thus also minimizing charge collection losses.

However, it has proven extremely challenging to make an OSC with the ideal morphology shown above. Instead, an active layer structure known as a bulk heterojunction (Figure 14.3, middle and left) is most often used. Such a device can be achieved by simply spin-coating a blend solution of electron donor (conjugated polymer) and electron acceptor materials (usually a small-molecule fullerene derivative);[28,30] because of the two species' different chemical structures, they naturally separate into two phases within the film. This type of morphology can easily satisfy either one of the exciton separation or charge conduction conditions by discouraging or encouraging phase separation (also increasing or decreasing donor–acceptor interfacial surface area), respectively. However, if the phase separation is excessive (Figure 14.3, middle), most excitons will decay before they reach an interface; while if the phase separation or crystallinity is insufficient (Figure 14.3, left), free charges will not be collected for lack of a continuous or sufficiently conductive pathway to electrodes. Thus, the focus of any treatment is to optimize the balance between a large donor–acceptor material interfacial area—thus maximizing exciton separation—and a high degree of crystallinity and

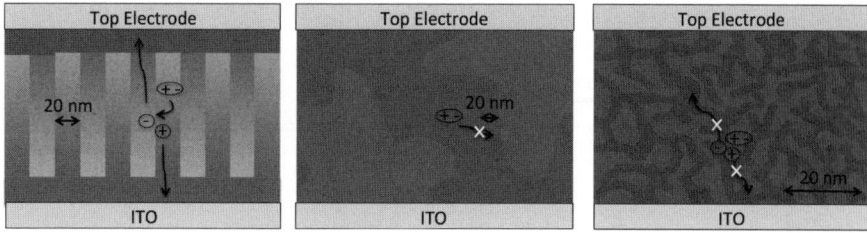

■ Conjugated Polymer; ■ N type organic material

Figure 14.3 Organic solar cell morphology: ideal interdigitated morphology (left); bulk heterojunction active layer with oversized (middle) and undersized (right) phase separation.

mesoscopic order—thus maximizing charge carrier mobility and charge collection. It should be noted that the effects of active layer nanostructure on the final device performance are very complex. The discussion above merely illustrates one aspect that is relatively well understood. Experiments are still being conducted to further our knowledge of different systems.

The degree of phase separation, feature size, and crystallinity of the active layer can be manipulated by various treatments during different stages of device making. Generally speaking, there are 2 main opportunities to influence the active layer morphology of an OSC: (1) during solution preparation and film deposition, and (2) post-film deposition. Section 2.1 will discuss techniques used to affect active layer morphology during solution preparation and film deposition such as the use of different solvent or material systems, the addition of high boiling point additives, and solvent annealing. The morphology can be further tuned after film deposition by thermal treatment near the polymer's glass transition temperature or solvent vapour annealing, and this will be the subject of section 4.2.2.

14.2.1 During Solution Preparation and Film Formation

During the early years of OSC study, research was focused on devices based on blends of poly[2-methoxy-5-(3′,7′-dimethyloctyloxy)-1,4-phenylene vinylene] (MDMO-PPV, Figure 14.4 a)) and 6,6-phenyl C_{61} butyric acid methyl ester (PCBM, Figure 14.4 b)). After initial reports of the superior device performance and properties of poly(3-hexylthiophene) (P3HT, Figure 14.4c)) began to surface in 2002,[35] studies on the various aspects of P3HT:PCBM film morphology manipulation and characterization dominated the OPV field. Because MDMO-PPV and P3HT have been the subject of a great many systematic studies, the following discussion of nanostructure manipulation will be based primarily on examples from research on these two polymers.

The most common way to form an active layer film is to spin-coat a blend solution of conjugated polymer and PCBM (electron acceptor). Researchers have been able to manipulate the nanostructure of the active layer during this important step by manipulating factors such as: (1) chemical structure of the materials, (2) solvent mixture, (3) polymer : PCBM weight ratio, (4) blend

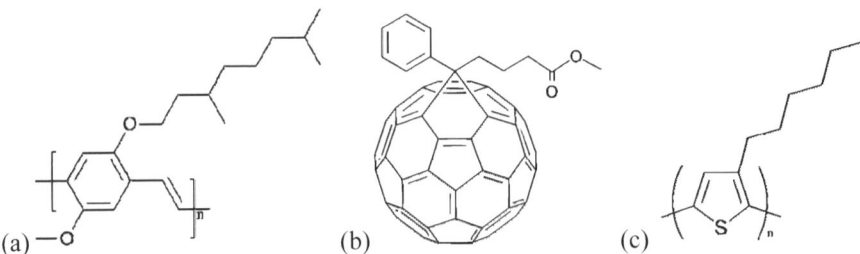

Figure 14.4 Chemical structure of (a) MDMO-PPV, (b) PCBM, and (c) P3HT.

solution concentration, and (5) film drying speed.[27,30,36] The molecular structures of polymer and fullerene determine their solubility in organic solvents, packing behaviour, and the miscibility between the two compounds. Solvent choice not only determines solubility, but also influences the film drying time during spin-coating.[28,34,37] If the film dries over a long period of time, it allows polymers more time to crystallize, self-organize, and diffuse; by controlling the drying time, the degree of crystallization and organization can be controlled, and hence so can the nanostructure of film.[38–40] The diffusion of one or both components in the blend leads to a modification of the phase separation.

14.2.1.1 Solvent effect

Figure 14.5 shows an example of the influence of the solvent on the nanostructure of the spin-coated MDMO-PPV:PCBM (80 wt% PCBM) active layer. In both images, darker structures are attributed to PCBM-rich regions (PCBM phase).[41] The surrounding lighter-coloured phase (matrix) was a homogenous phase of 50 : 50 wt% MDMO-PPV:PCBM.[42] When the solvent was changed from toluene to chlorobenzene (CB), it was observed that the

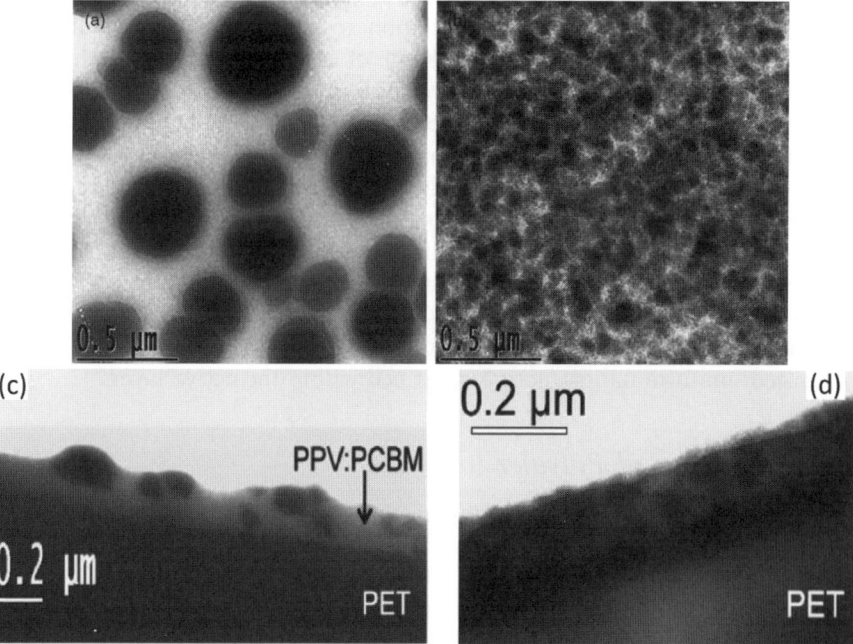

Figure 14.5 Transmission electron microscope (TEM) images of 80 wt% MDMO-PPV:PCBM thin films on PET substrates spin-coated from toluene (a) and chlorobenzene (b). Cross-sectional view of the same toluene (c) and chlorobenzene (d) films.[41]

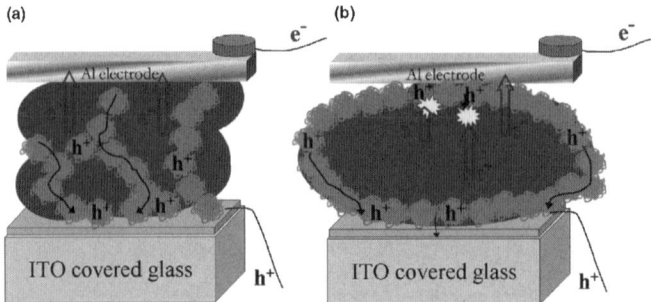

Figure 14.6 Schematic of MDMO-PPV:PCBM blend films in solar cells cast from (a) chlorobenzene and (b) toluene. In (a), holes and electrons find percolated pathways to reach their respective electrodes. In (b), electrons and holes suffer recombination due to incomplete charge percolation pathways.[46]

average radius of PCBM-rich domains was reduced from 100 nm to 15 nm.[41,43] It is suggested that these larger PCBM-rich domains are due to the lower solubility of PCBM in toluene; while the film is drying, PCBM in toluene precipitates sooner, allowing more time for formation of larger crystallites.[44,45]

Films cast from toluene exhibit relatively large PCBM clusters (100 nm), which are considered over-size (in relation to the 10 nm exciton diffusion length) and result in a reduction in the rate of exciton separation.[41] However, these PCBM clusters also caused an insufficient electron percolation network in toluene-cast MDMO-PPV:PCBM blend films, which is believed to be the main photocurrent loss mechanism in these devices (Figure 14.6.).[46] Direct comparison of MDMO-PPV:PCBM (80 wt% PCBM) blend OPV devices spin-cast from either toluene or chlorobenzene (CB) by S. E. Shaheen *et al.*,[47] showed that the power conversion efficiency (PCE) of devices cast from CB was nearly triple that of those from toluene (from toluene: PCE = 0.9 %, from CB:PCE = 2.5 %). The tendency of PCBM molecules to phase segregate into clusters is suppressed when CB is used as the solvent, and thus a more optimized, uniform nanostructure is formed within the active layer.[47]

14.2.1.2 Ratio of Polymer to PCBM

One way to more precisely manipulate the size of both matrix and PCBM phases is to change the MDMO-PPV:PCBM weight ratio. In atomic force microscopy (AFM) images of MDMO-PPV:PCBM blend films of varying weight ratios (Figure 14.7) spin-cast from CB, nanoscale phase separation is not observed in films with lower than 50 wt% PCBM (Figure 14.7 (d,h)). Nanoscale phase separation throughout the film sets in for concentrations of more than 67 wt% PCBM (MDMO-PPV : PCBM = 1 : 3) (Figure 14.7 (c,g)).[42] Also, the film surface becomes increasingly uneven with increasing PCBM content (for 50–90 wt%): peak-to-valley roughness increased from 3.3 nm to 22 nm throughout this range. It was found that the virtually homogeneous 50 wt% PCBM films (Figure 14.7 (d,h)) was a perfect morphology for efficient charge separation; nearly every exciton generated in the MDMO-PPV was quenched by charge transfer to PCBM. However, little device efficiency was shown in these films due to the lack of efficient charge conduction pathways to the electrodes. Devices with 80 wt% PCBM actually exhibited the highest device efficiencies; despite a lesser charge separation efficiency, an interpenetrating percolation network with pathways long enough to facilitate charge collection to the electrodes proved more important.[42]

By further studying the vertical composition profile of these samples with time-of-flight secondary ion mass spectrometry (TOF-SIMS), it was shown that the phase distribution present at the top surface is also present throughout the depth, although with a slightly increasing domain size with increasing depth.[42] Similar conclusions were made in a scanning electron microscope (SEM) cross-section study of MDMO-PPV:PCBM blend films in which 10–20 nm-sized PCBM nanospheres were found to be evenly dispersed throughout the films.[48] It is important to note that the effect of PCBM content on the nanostructure of MDMO-PPV:PCBM blend films is dampened when spin-cast from chlorobenzene. The size of PCBM crystallites increases from 0 to 500 nm when spin-cast from toluene over the range 50 to 80 wt% PCBM (Figure 14.8), compared to 0 to 23 nm when spin-cast from CB.[48] This allows more precise control over a fine-textured film nanostructure; considering the short exciton diffusion length (~ 20 nm), more precise nanostructure control is highly desirable for the optimization of MDMO-PPV:PCBM blend solar cells.

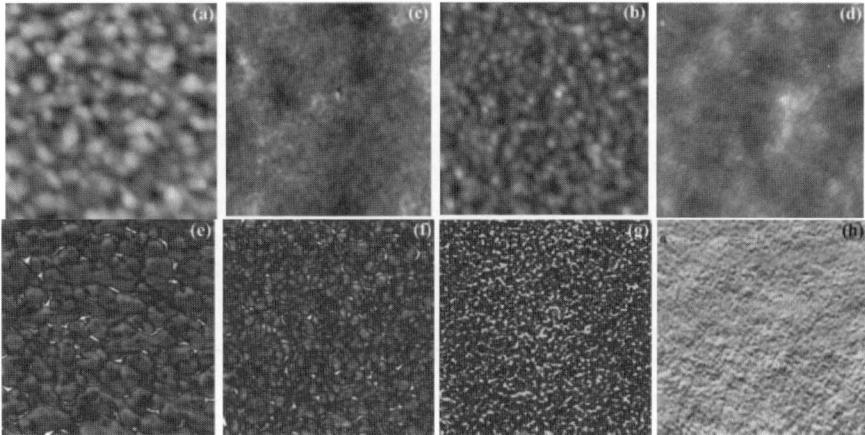

Figure 14.7 The AFM height (a–d) and phase (e–h) images of MDMO-PPV:PCBM composite films of 90 wt% (a,e), 80 wt% (b,f), 67 wt% (c,g), and 50 wt% (d,h) PCBM. Maximum peak-to-valley z-range values were 20 nm (a), 10 nm (b), 3 nm (c), and 3 nm (d). The size of the images is 2.0 μm × 2.0 μm.[42]

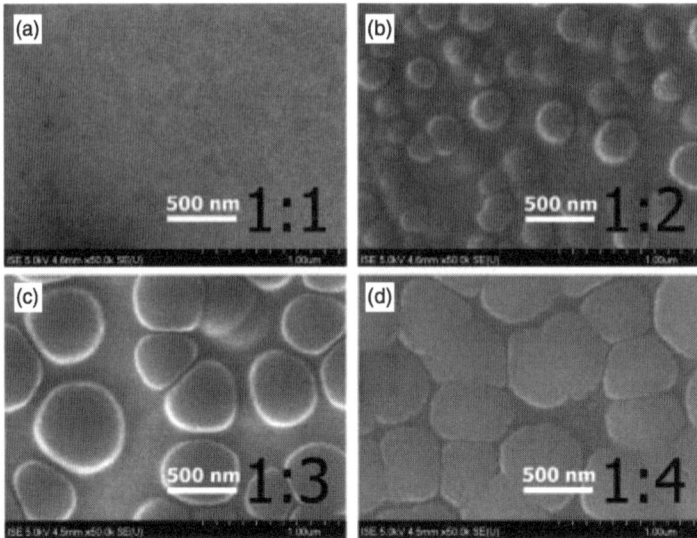

Figure 14.8 SEM micrographs of MDMO-PPV:PCBM blend films cast from toluene with varying PCBM weight fraction. The surface of the 50 wt% (1 : 1) PCBM film (a) looks flat and shows no features. For the films containing a larger percentage of PCBM (b–d), nanoclusters are visible, increasing in size with increasing PCBM content.[48]

14.2.1.3 Molecular Structure and Morphology

The molecular structure of materials determines their solubility in different solvents, as well as their self-organizing and solid-state packing behaviours. In general, the donor and acceptor materials used in active layer blends are immiscible due to their different chemical structures; this is the cause of phase separation in active layer blends. The materials' different surface energies also cause vertical phase separation—the lower surface energy component tends to diffuse to the solid–air interface to reduce the energy there; creating a concentration gradient across the thickness of the film.[49]

As described in section 14.2.1.1, use of a better solvent for PCBM—chlorobenzene instead of toluene—hindered phase separation between MDMO-PPV and PCBM, resulting in a more finely-textured bulk heterojunction film.[41] The same principle holds for the solubility of the materials themselves: by comparing C_{60} with its more soluble derivative PCBM, Yu *et al.* showed that a better network was formed in the PCBM devices, exhibiting improved charge collection.[50] Conjugated polymers also contain side-chains to improve their solubility; it has been shown that increased polymer solubility also engenders increased charge carrier network interpenetration.[51,52] Thus it is clear that material solubility is an important parameter for achieving nanotextured bulk heterojunctions.[28]

The structure of a conjugated polymer is also known to affect its aggregation behaviour. MDMO-PPV has been shown to take a coiled nanosphere conformation when deposited from solution (Figure 14.9).[46] Assuming a molecular mass of 10^6 Da and a density of 910 kg m^{-3} for MDMO-PPV, the volume of a single polymer chain becomes approximately 1.8×10^{-24} m^3 (1800 nm^3); a sphere of said volume would have a diameter of approximately 15 nm, which is in good agreement with SEM data.[46] The resulting nanosphere diameter can be varied between 10 and 20 nm by changing the molecular weight through the range 2.5×10^5 Da to 2.5×10^6 Da.[28] This is an example of how the polymer aggregation behaviour can be affected by the conjugated polymer molecular property; however, the coiled conformation is not preferred in solar cells because it requires more inter-molecular jumps for charge conduction.

One way to affect conjugated polymer conformation and aggregate shape is to change the crystallinity of the polymer. The most successful examples are regioregular poly(3-alkylthiophene)s (rr-P3ATs or simply P3ATs), which were first reported as a promising type of conjugated polymer by Brabec *et al.* in 2002.[35] Depending on the relative position of the side chain, a pair of coupled thiophene rings can have head-to-head (HH), head-to-tail (HT), or tail-to-tail (TT) conformations (Figure 14.10).[53] A regioregular polymer is one made up of almost entirely HT couplings, resulting in a polymer with side chains extending in a single direction. This regular side chain structure is the basis for one of the most fascinating and useful physical properties of rr-P3AT: its supramolecular self-assembly, which is not found in regiorandom poly(alkylthiophene).[9,549,549,559,549,559,54] The most commonly-studied of the P3ATs is poly(3-hexylthiophene) (P3HT), which was first shown *via* X-ray diffraction studies by the McCullough group to self-assemble into crystalline structures.[55,56] The strong intermolecular association of rr-P3HT encourage self-association and crystallization, even in good solvents.[57,58]

Because of the self-organizing properties of rr-P3ATs, they tend to pack with other chains to form highly-ordered semi-crystalline domains as illustrated in Figure 14.11.[55,56,59] The size and shape of such domains can be varied by processing rr-P3ATs with different side chain length,[60,61] degree of regioregularity,[55,62] and molecular weight.[63–65] Atomic force microscopy and X-ray diffraction show that low-mobility, low-MW films have a highly ordered structure composed of nanorods and high-mobility, high-MW films have a less ordered, isotropic nodule structure (Figure 14.12 left).[63] Moreover, by careful

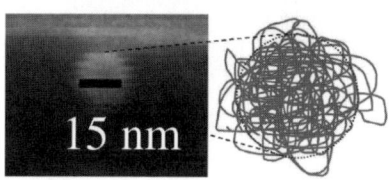

Figure 14.9 Magnification of an SEM cross-sectional measurement: a MDMO-PPV nanosphere has a typical radius of 15 nm in chlorobenzene based blends. The conformation of the polymer chain is illustrated in the right side.[46]

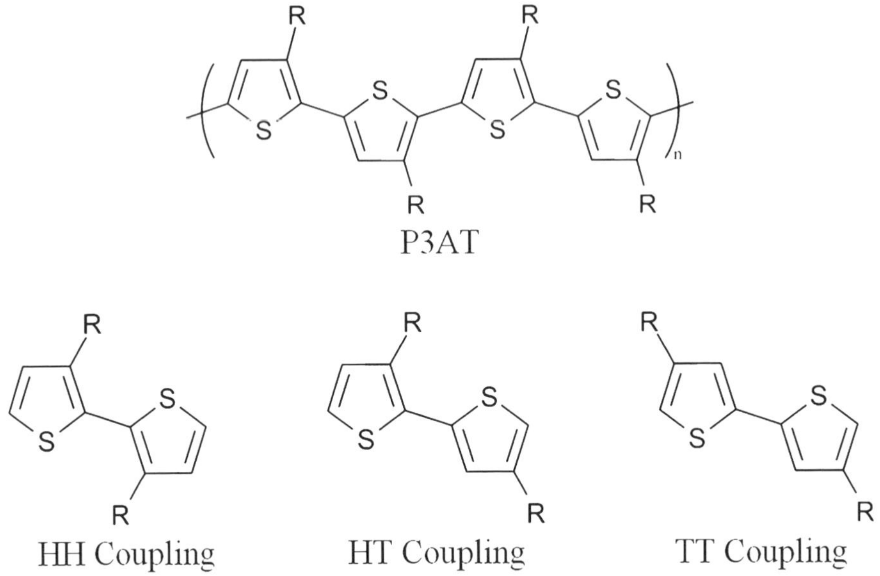

Figure 14.10 Structure of poly(3-alkylthiophene) and examples of head-to-head (HH), head-to-tail (HT), and tail-to-tail (TT) coupling pairs.

choice of processing conditions, rr-P3HT with a narrow polydispersity index (PDI) can form a very well-defined nanofibrillar morphology in which the width of the nanofibrils corresponds closely to the chain length of the smallest polymer chains (Figure 14.12 right).[65]

It should be pointed out that in most cases, the nanostructure of spin-cast active layers is far from its thermodynamic equilibrium due to fast solvent quenching, which results in amorphous or disorder regions within the film.[38,66,67] In such films, donor and acceptor materials are well-mixed, which

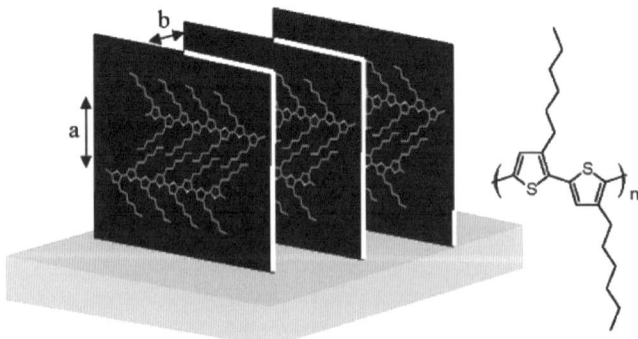

Figure 14.11 (left) Schematic of the proposed structural orientation of P3HT with respect to the horizontal plane (air/water or solid support). (right) Molecular structure of P3HT.[59]

Materials Nanoarchitechtonics for Advanced Devices 369

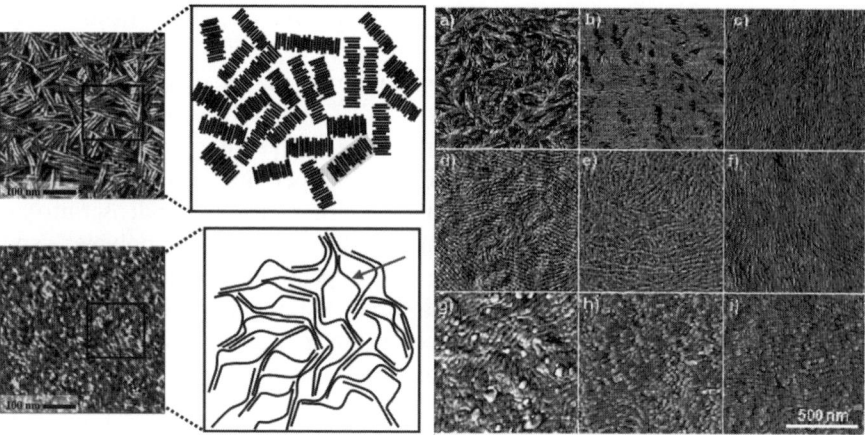

Figure 14.12 (left) Model for transport in low-MW (top) and high-MW (bottom) films. Charge carriers are trapped on nanorods in low-MW films. Long chains in high-MW films bridge the ordered regions and soften the boundaries (marked with an arrow).[63] (right) Tapping mode AFM images (phase contrast, 1 μm × 1 μm) of thin films of rr-P3HTs of various molecular weights in field effect transistor devices prepared by drop casting from toluene. Corrected weight average molecular weights in (a–i) were equal to: 2.4, 4.8, 5.1, 7.0, 7.5, 11.8, 15.7, 17.3, and 18.4 kDa, respectively.[65]

as discussed previously is not favoured for charge conduction. Trapping regiorandom PAT in a nonequilibrium state leads to simple amorphous films; nonequilibrium rr-P3AT films, on the other hand, contain large internal tensile strains because of the spontaneous formation of semi-crystalline domains.[68] One way to prevent a nonequilibrium state is to lengthen the film drying time to give polymers enough time to self-organize, which will be discussed in the next section. Another method to relieve these strains is to apply some kind of post-film-deposition treatment, which will be discussed in Section 4.2.2.

14.2.1.4 Film Drying Condition

During the film drying process, the conjugated polymer and fullerene first precipitate from solution, then, due to their different surface energies, self-organize into their respective domains, determining the nanostructure of as-cast film. In the previous section, it was shown that the film morphology can be manipulated by modifying the material's solubility, polymer molecular weight, or regioregularity. It has also been shown that active layer morphology manipulation is possible by control of drying conditions such as solvent evaporation rate, airflow on the solution surface, and casting temperature.[36,69–75] Among these methods, changing the solvent evaporation rate has been the most widely studied for its comparably better morphology control and improvement on the final device performance, but airflow and temperature can also play a role.

Spin-coating has long been the dominant method for quickly and easily depositing uniformly thin polymer films.[30,76,33,36] When the substrate begins to spin, excess solution is propelled off the edge of the substrate and solvent rapidly evaporates, concentrating the solution. Depending on the evaporation rate of the solvent, the two components will either quickly condense into an amorphous glassy state or slowly phase-separation to form polymer- and PCBM-rich domains on a length scale that depends on the layer formation rate.[33] If the solvent quench occurs faster than the polymer self-organization, the polymer conformational state formed during spin-coating will be relatively amorphous and thermodynamically unstable.[69] Exchanging a higher boiling point solvent for a lower boiling point one slows down the solvent quenching and allows the film time to self-organize. P3HT films spin-coated from xylenes (boiling point: 138 °C) are locally-ordered and appear to be better connected with neighbours, whereas the small needle-like aggregates of P3HT in films spin-cast from chloroform (boiling point: 61 °C) film appear loosely connected and randomly oriented (Figure 14.13). Additionally, average domain size in the xylenes-cast film is also much larger than that of the chloroform-cast film (Figure 14.13 b,d).[63] This is expected because the higher boiling point of xylene should allow the spin-cast films more time to reach their equilibrium

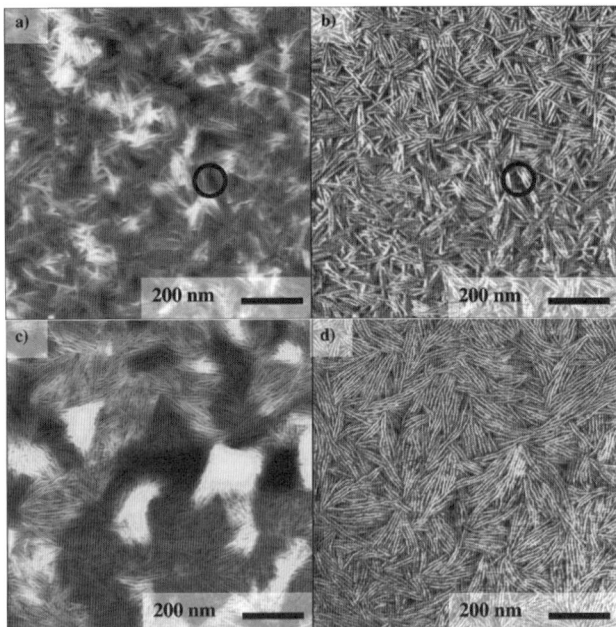

Figure 14.13 AFM images comparing low-MW films spin-cast from chloroform (boiling point: 61 °C) [(a) topography and (b) phase] and xylenes (boiling point: 138 °C) [(c) topography and (d) phase].[63]

morphology during film formation.[63,77] Because quenching time determines domain size to a first approximation, the boiling point of the solvent determines the initial morphology of the film.[78,79] Similar observations have also been made from P3HT:PCBM OPV device experiments. Solar cells with 100 nm thick active layers spin-cast from 1, 2, and 3 % solutions of P3HT:PCBM (50 wt% PCBM) in chlorobenzene were characterized. From FE-TEM (field emission transmission electron microscope) images (Figure 14.14), nanoscale interpenetrating networks of donor–acceptor domains with feature sizes of ~10 nm were observed in all devices, but fibrillar structures were observed only in the 1% device. This indicates that slow evaporation of solvent from a low concentration P3HT:PCBM solution leads to a high degree of polymer ordering which is efficient for carrier generation and transportation.[80] Consequently, the 1% concentration device showed a PCE of 2.11%, while 2 and 3% devices exhibited 1.58% and 1.35%, respectively.[80]

The self-organization of P3HT under different drying time is summarized in Figure 14.15. In drop-cast P3HT films (slow-drying), well-defined "crystalline" nanorods are formed,[81] and little internal strain was shown inside.[82] Slow drying (called "solvent annealing") was first used to manipulate the active layer morphology by Yang's group.[83] P3HT:PCBM (50 wt% PCBM) dissolved in 1,2-dichlorobenzene (DCB, b.p.=180 °C) was spin-cast at low speed (600 rpm) for 60 s. Spin-coating at 600 rpm left the films wet, which were then allowed to dry in covered glass petri dishes ("slow growth") or were baked at 70 °C for 30 s ("fast growth"). Excess solvent after spin-coating allows the components to remain partially dissolved and diffuse at a higher rate to arrange into a more energetically favourable ordered structure.[82,29] It was found that solvent annealing significantly improved the efficiency of P3HT:PCBM based solar cells to a value as high as 3.5%.[81,38] Controlling the active layer growth rate results in an increased hole mobility and balanced charge transport. Together with increased active layer absorption due to greater P3HT packing, this results in much-improved device performance, particularly in external quantum efficiency.[81]

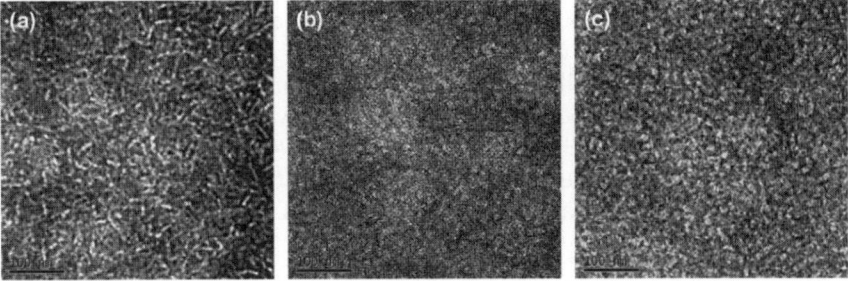

Figure 14.14 FE-TEM micrographs of P3HT:PCBM (50 wt% PCBM) blend films from CB solutions of different concentration: (a) 1%, (b) 2% and (c) 3%.[80]

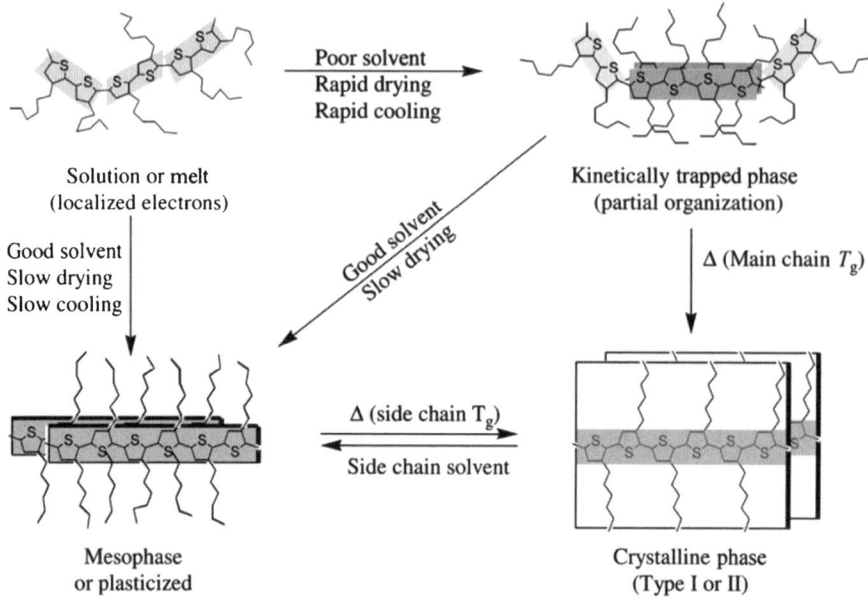

Figure 14.15 Polymer self-assembly flow chart according to different treatments.[58]

Figure 14.16 shows AFM topography and phase images for rr-P3HT:PCBM (1 : 1 weight ratio) blend films after spin-coating at 600 rpm for 30 and 80 s. After 30 s of spin-coating, the sample remained wet and was allowed to solvent anneal afterward; 80 s of spin-coating resulted in films that were almost dry already, a much shorter drying time. The 80 s spun film images show fibrils that are quite distinct from each other, indicative of a longer disordered zone between semi-crystalline domains. This is in contrast to the 30 s film images, in which domains appear indistinct; these shorter inter-fibril disordered zones contribute to lower energy barriers to conduction.[85–87] Grazing incidence X-ray diffraction (GI-XRD) showed that even 80 s spin-cast films exhibited highly ordered fibrils with rr-P3HT rings edge-on to the substrate, resulting in both π-conjugated planes and π–π stacking parallel with respect to the substrate; 30 s spin-cast films, however, showed a much higher crystallinity than the film cast for 80 s spin-cast with longer solvent annealing time.[39] The higher crystallinity and longer-range order of the 30 s spin-cast devices result in both higher absorption and mobility, and a PCE of about 3%, over double the value of the 80 s spin-cast device (PCE = 1.17%). The effects of solvent annealing on the morphology of polymer films has also been further studied and applied by many other research groups.[40,88–90]

14.2.1.5 Solvent Additives

The solvent additive method, also sometimes known as the solvent mixture or co-solvent method,[91–96] is yet another method to adjust the drying time of a

Materials Nanoarchitechtonics for Advanced Devices

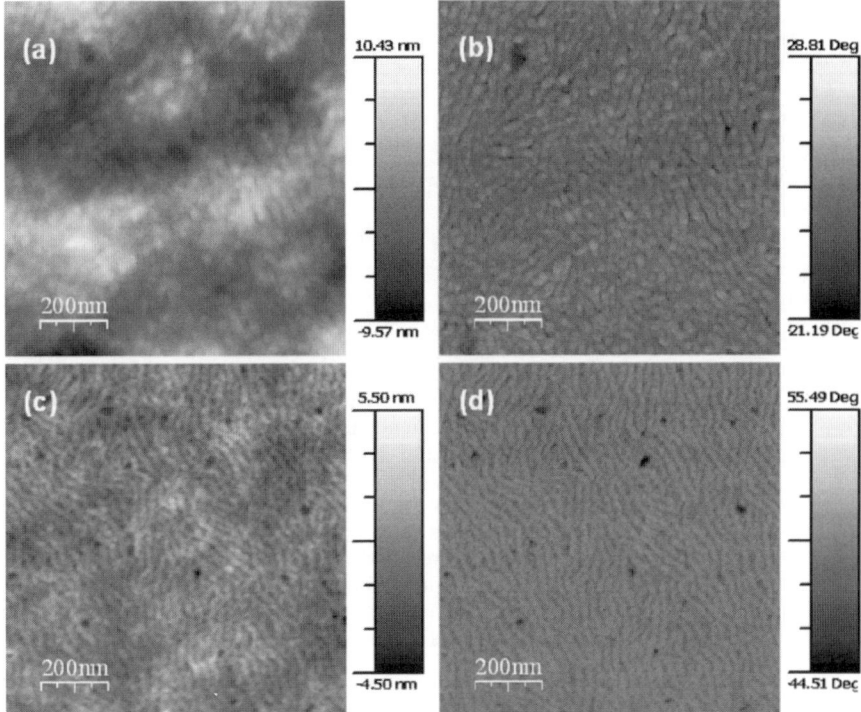

Figure 14.16 TM-AFM topography (left) and phase (right) images for rr-P3HT:PCBM films fabricated using different processing conditions. The spin-coating times are (a,b) 30 s and (c,d) 80 s.[39]

deposited film. From previous studies, we have seen that film crystallinity is strongly related to drying time (in turn related to boiling point),[97] and that degree of phase separation is strongly influenced by material solubility.[98] The solvent additive method adds a few % of a solvent that is selectively soluble with one material and has a higher boiling point than the host solvent to affect the film morphology during solvent evaporation (Figure 14.17).[91]

In a study of [2,6-(4,4-bis (2-ethylhexyl)-4H-cyclopenta[2,1-b;3,4-b′]-dithiophene)-4,7-(2,1,3-benzothiadiazole)] (PCPDTBT) and C_{71}-PCBM by the Heeger group,[95,99,100] the power conversion efficiency was increased from 2.8% to 5.5% by incorporation of a few per cent of an additive such as 1,8-diiodooctane or 1,8-octanedithiol into the chlorobenzene.[99,100] The formation of a phase-separated BHJ was clearly observed, and no additive remained in thoroughly dried films, despite the additive's higher boiling point.[100] During drying, fullerene molecules are selectively dissolved in the additive, resulting in the formation of three separate phases (Figure 14.17): a fullerene-additive phase, a polymer aggregate phase, and a polymer-fullerene phase. Because of the additional C_{71}-PCBM-additive phase, most of the C_{71}-PCBM remains in

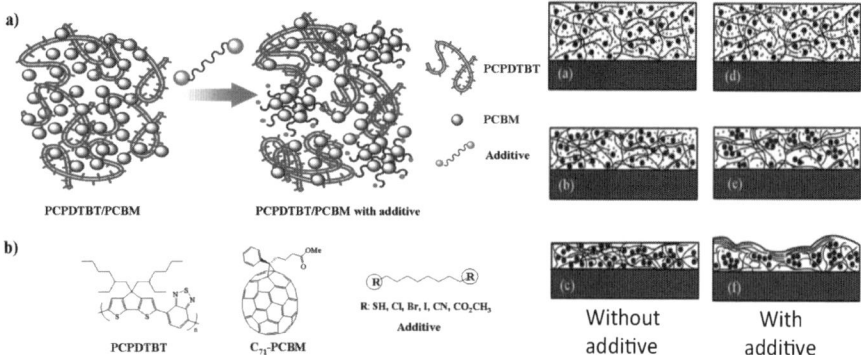

Figure 14.17 (Left) Schematic of the role of processing additive in the self-assembly of bulk heterojunction blend materials (a) and structures of PCPDTBT, C_{71}-PCBM, and additives (b).[95] (Right) Black wire: conjugated polymer chain; large black dots: PCBM; (a–c) small dots: solvent molecules; (d–f) small dots: additive molecules. (a–c) Correspond to three stages in the spin-coating process when DCB is the sole solvent; (d–f) correspond to three stages in the spin-coating process when additive is added in host solvent. Note the difference of PCBM distribution in the final stage of each case, (c) and (f). The total number of large black dots is the same in all images.[91]

solution longer, allowing the PCPDTBT to self-assemble more fully and avoiding excessive C_{71}-PCBM aggregation (Figure 14.18).[95]

Other compounds, such as other alkanedithiols, 1-chloronaphthalene (CN),[94] and nitrobenzene (NB),[96] have also been studied as solvent additives. CN is a good solvent for both P3HT and PCBM, and allows more time for P3HT chains to self-organize into highly ordered molecular structures for higher hole mobility.[94] NB follows the opposite logic; as a non-solvent for both P3HT and PCBM, the addition of NB to the casting solvent causes the solvent quality for P3HT to gradually decline due to evaporation. This encourages polymer aggregation in the early stages of film drying, thus increasing crystallinity and order within the P3HT domains.[96]

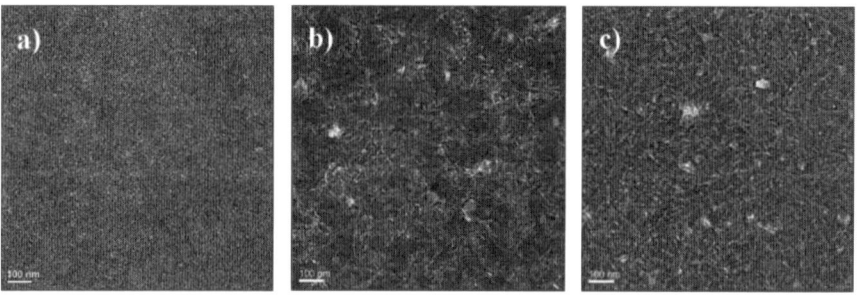

Figure 14.18 TEM images of films cast from PCPCTBT/C_{71}-PCBM with: (a) no additives, 2.5 vol% of (b) 1,8-octanedithiol and (c) 1,8-diiodooctane.[95]

14.2.2 Post-Film Formation Treatment

After the active layer has been formed, its nanostructure can be modified with post-film formation treatment, further improving device performance in many cases. For example, P3HT:PCBM OSC devices without any morphology manipulation during film formation show a dramatic increase in efficiency after post-film formation thermal annealing (from ~1% to 3.5~5%).[101–104] Even for devices optimized during film formation, thermal annealing is still often able to further improve the device performance. For example, the P3HT:PCBM (50 wt% PCBM) solvent annealed devices discussed in section 2.1.4, exhibited additional improvement from 3.5% to 4.4% by thermal annealing at 110 °C for 10 min.[83] Solvent vapour annealing is another common post-film-formation treatment used in organic solar cell research.[105–107] The method, mechanism, and effect of these two post-film-formation treatments on microstructure and device performance will be discussed in the following sections.

14.2.2.1 Thermal Annealing.

Thermal annealing was developed very early on in solar cell research, before most other morphology modification methods. Even so, it has remained an effective method to optimize device performance for most active layer material systems. Thermal annealing can both enhance donor–acceptor phase separation and increase individual phase crystallinity,[102,108] hence forming a well-ordered morphology with better transportation pathways and higher mobility.[109–111]

By annealing MDMO-PPV:PCBM solar cell devices at relatively high temperatures (150 °C) for sufficiently long times, micron-sized PCBM crystals nucleate from MDMO-PPV:PCBM blends and phase separation between the two constituents is completed.[42,48,46] The morphology of MDMO-PPV:PCBM films annealed at different temperatures is shown in Figure 14.19. Above

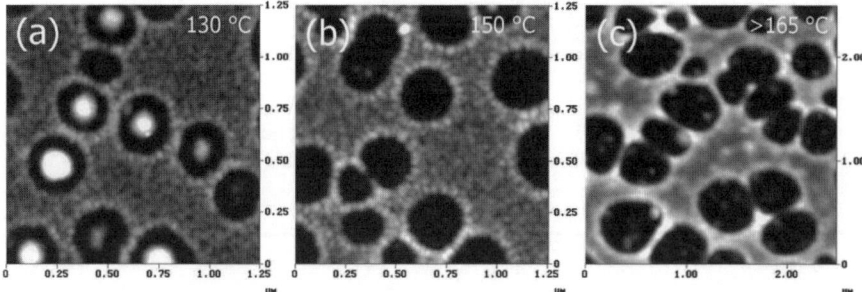

Figure 14.19 AFM images of the polymer-rich region between the PCBM crystallites (dark circles) of MDMO-PPV:PCBM 1:4 (80 wt. % PCBM) films cast from toluene after thermal annealing for 4 hours at different temperatures. While up to 150 °C polymer nanospheres (light dots) can be detected, for temperatures above 165 °C these spheres are no longer visible and the polymer appears molten.[48]

165 °C, the spherical polymer phase are no longer visible and the previously rigid polymer matrix is molten.[28,48]

As mentioned in section 2.1.3, due to the high crystallinity of rr-P3HT, internal strain develops in P3HT:PCBM films if the as-cast film is far from its thermodynamic equilibrium.[82] This strain can be released by post-film formation thermal annealing, which also increases the size and crystallinity of both P3HT and PCBM phases (Figure 14.20),[104,114–117] increasing performance from 1.3% to 4.0%.[112] Films of neat P3HT exhibit a natural tendency to self-organize into crystalline lamellae but this is inhibited by the addition of PCBM.[115] However, heating of P3HT:PCBM films to temperatures above the glass transition temperature of the as-deposited film facilitates polymer backbone flexing, which allows P3HT chains to crystallize and forces phase segregation, creating an interpenetrating network with better charge conduction properties. As a result, films annealed over 100 °C show a 4 to 5 order of magnitude increase in hole mobility (Figure 14.21).[120]

Thermal annealing leads to microstructural coarsening and mobility improvement,[110–112,117] which both benefit charge transportation and collection in organic solar cells. However, formation of very large (*ca.* 100 nm) crystallites is undesirable in solar cells, as the large domain size increases losses through geminate recombination and photoluminescence.[103,121,122] Sudden drops in P3HT:PCBM device efficiency have been reported after annealing either for only a few seconds at 130 °C or a few minutes at 75 °C.[101,123] This drop in efficiency was attributed to a degradation of the morphology resulting

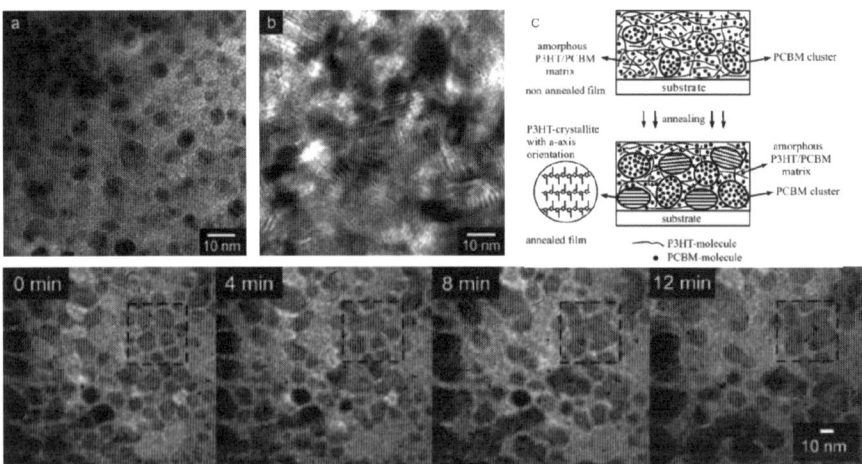

Figure 14.20 (Top) Low voltage, high resolution TEM micrographs of P3HT:PCBM blend active layers (a) as-deposited and (b) after annealing at 150 °C for 10 min. (c) Schematic showing structural changes of P3HT/PCBM films upon annealing. (Bottom) Time series of a P3HT:PCBM blend during the annealing process. The highlighted area shows one area of PCBM domain agglomeration.[112,113]

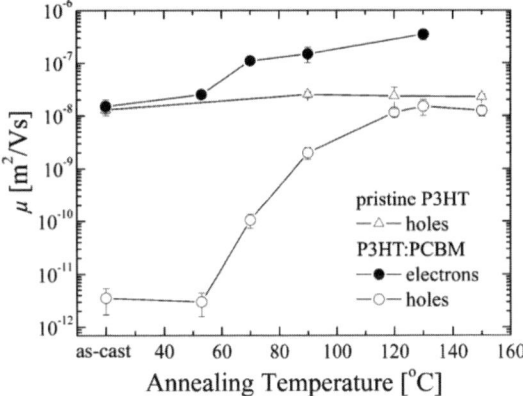

Figure 14.21 Room temperature electron and hole zero-field mobilities in 50 wt% PCBM blends of P3HT:PCBM as a function of post film deposition annealing temperature. For comparison, the hole mobility measured in pristine P3HT devices is also shown. The mobilities were calculated from the space charge-limited current measured using electron- and hole-only device configurations.[111]

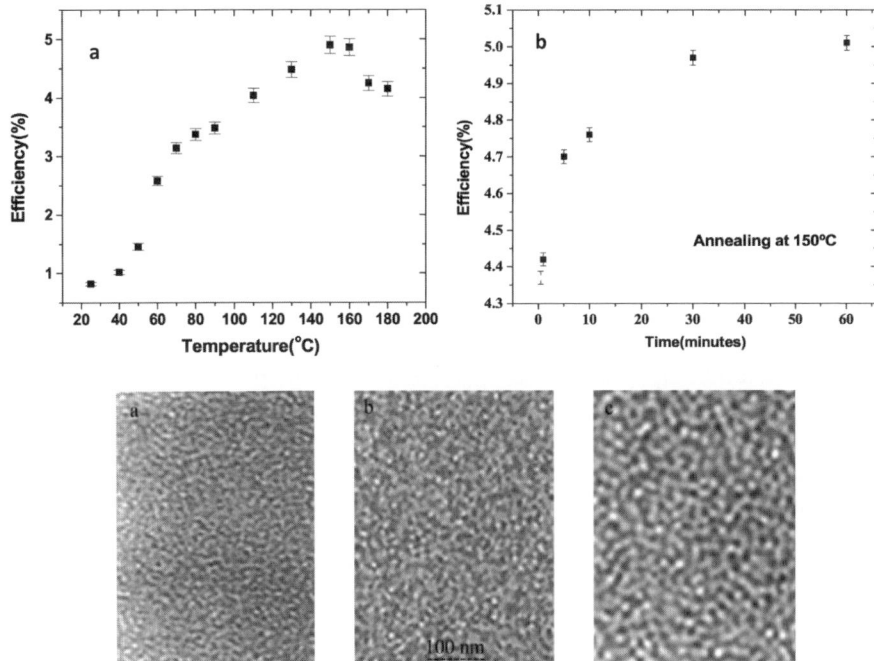

Figure 14.22 (Top) (a) Plot of device efficiency *versus* annealing temperature (15 min anneal time). (b) Evolution of device efficiency *versus* annealing time at 150 °C. (Bottom) TEM images of P3HT:PCBM film bulk morphology (a) before annealing, after annealing at 150 °C for (b) 30 min, and (c) 2 h.[114]

from the formation of oversized PCBM aggregates.[114] As controlling and limiting phase separation to the proper size ($\sim$5–20 nm) is important for optimization of the overall device efficiency, it is necessary to define suitable annealing temperatures and times.[124] Through a detailed study of both annealing temperature and time, the Heeger group determined that the optimum annealing temperature for P3HT:PCBM films was 150 °C (Figure 14.22); even after annealing at 150 °C for hours, device performance remained stable.[114] Morphologically, without any heat treatment, the interpenetrating networks are not well developed and the donor and acceptor domains are difficult to distinguish (Figure 14.22 bottom (a)). After annealing at 150 °C for 30 min and 2 h (Figure 14.22 bottom (b), (c)), respectively, the morphology of the interpenetrating DA networks becomes clearer and more easily visible with a feature size of $\sim$10 nm.[114] In a more recent study, it was found that at 150 °C, the crystallization rate of P3HT and PCBM phases is slower than a temperature around 125 °C.[124] This is because at higher temperatures close to the melting point, the crystallization rate decreases despite an increase in diffusion rate due to the reduction of the thermodynamic driving force for crystallization.[124] These slower crystallization rates allow better control of the nanostructure of P3HT:PCBM films.

14.2.2.2 Solvent Vapour Annealing

Another technique used to manipulate active layer nanostructure after film deposition is known as solvent vapour annealing (SVA) or vapour annealing (VA),[105,106,125–127] later developed into controlled solvent vapour annealing (C-SVA).[107,128] In this technique, as-cast films are placed into a closed vessel filled with solvent vapour for a period of time;[126] once the vapour is absorbed by the polymer film, it can wet and reorganize one or both components inside the active layer film.[106,126,127] It was first reported by Zhao *et al.* in 2007 that 1,2-dichlorobenzene (DCB) vapour treatment can induce P3HT self-organization into ordered structures resulting in enhanced absorption, hole transport properties, and improved device performance from 0.8% (as-cast film) to 3.4%[121]. After that, a more complete study on solvent choice showed that the nanostructure of active layers could be controlled by changing the boiling point of the annealing solvent.[104] It is shown by TEM (Figure 14.23 top), AFM and XRD (Figure 14.23 bottom) that a good solvent vapour (chloroform, CB and DCB) induces better self-organization of P3HT than a poor solvent vapour (acetone, methylene chloride). However, the result of the increased self-organization induced by good solvent vapours was lower performance due to excessive phase separation and decreased exciton separation rates.[104]

Controlled solvent vapour annealing (C-SVA) uses vapour pressure as an additional parameter to more precisely control film exposure to vapour.[123,124] Various types of glassware can be used, but the most simple consists of a long vertical glass tube open to atmosphere at the top with solvent in the bottom

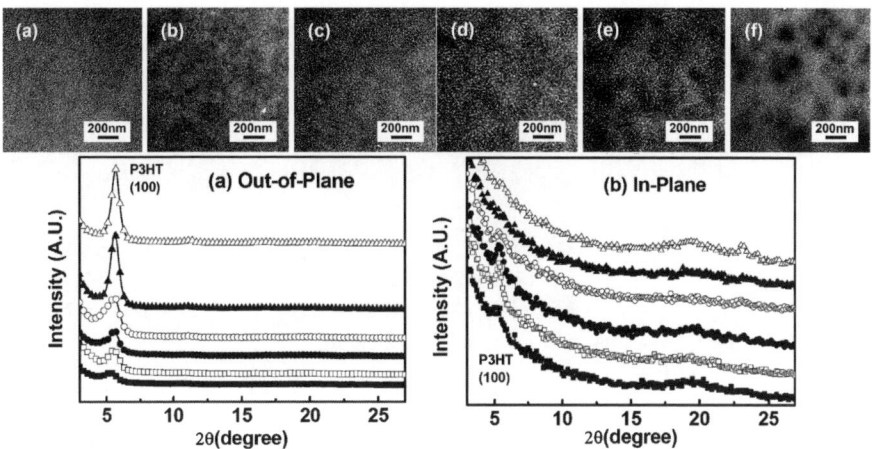

Figure 14.23 (Top) TEM images of P3HT:PCBM films (50 wt% PCBM, 10 mg mL^{-1}) after solvent annealing (PCE) with various solvent vapours: (a) as-prepared (0.94%), (b) acetone (3.29%), (c) methylene chloride (3.27%), (d) chloroform (2.66%), (e) chlorobenzene (2.88%), and (f) 1,2-dichlorobenzene (2.86%). (Bottom) Out-of-plane (a) and in-plane (b) GI-XRD spectra for P3HT:PCBM films before and after solvent annealing. In ascending order in (a): as-prepared (solid square), acetone (open square), methylene chloride (solid circle), chloroform (open circle), chlorobenzene (solid triangle), and 1,2-dichlorobenzene (open triangle).[104]

(Figure 14.24 left).[123] After some time, a stable solvent vapour pressure gradient is established, and the treatment vapour pressure can be determined by precise vertical positioning of the film inside the glass tube. Treatment of P3HT:C$_{60}$ films by carbon disulphide (CS$_2$) C-SVA resulted in formation of C$_{60}$ nanorods inside the films (Figure 14.24 left). The size and aspect ratio of C$_{60}$ nanorods were found to be tuneable by varying the vapour pressure and treatment time. The final device efficiency was improved from 0.4% for untreated films to 2.5% after treatment.[128]

It should be noted that in general P3HT:C$_{60}$ blend devices exhibit lower performance than P3HT:PCBM devices as mentioned in section 14.1.3. However, the optimization of active layer nanostructure through the growth of uniform one-dimensional C$_{60}$ nanocrystals within the film is a novel method of morphology design for photovoltaic device applications.[128] Still, the P3HT phase is not fully optimized by CS$_2$ C-SVA, which leaves opportunity for further device optimization.

More recently, a two-step C-SVA with THF followed by CS$_2$ on P3HT:PCBM film showed that not only could an optimized active layer morphology be achieved *via* this method, but also the nanostructures of P3HT and PCBM could be manipulated separately (Figure 14.25 right).[107] THF vapour was first employed to form PCBM clusters. CS$_2$ vapour was then used

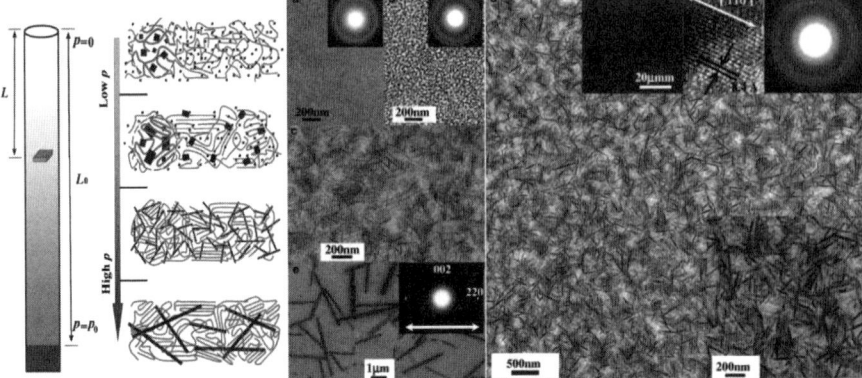

Figure 14.24 (Left) the setup used for C-SVA, with schematic illustrations of the shape and sizes of C_{60} fullerene nanorods grown at the corresponding solvent vapour pressure.[128] (Right) Bright-field TEM images and corresponding selected area electron diffraction (SAED) patterns showing the morphology of P3HT/C_{60} composite films upon CS_2 C-SVA (% saturated vapour pressure, exposure time) (PCE) with (a) pristine film (0.40%), (b) (0.5, 30 s) (1.57%), (c) (0.77, 5 s), (d) (0.83, 5 s) (2.5 %), (e) (0.96, 30 s).[128]

to simultaneously increase the crystallinity of P3HT and to partially dissolve the large PCBM clusters, resulting in an optimized morphology with appropriately sized PCBM aggregates as well as improved P3HT crystallization (Figure 14.25 left). After being treated with CS_2, the size of PCBM-rich domains reduced from 29 nm (on the top of film) and 23 nm (on the bottom of the film) to 19 nm and 12 nm respectively (Figure 14.26). The final devices showed a PCE as high as 3.9%, in contrast to 3.2% for the thermally annealed device under the same characterization conditions.[107]

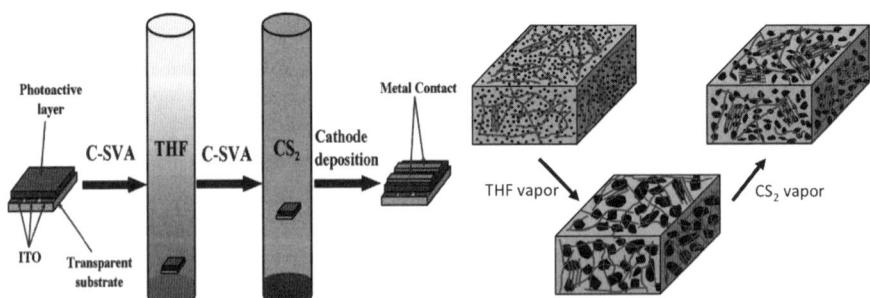

Figure 14.25 (Left) Scheme of the fabrication of P3HT/PCBM solar cells using two-step C-SVA approach. (Right) Schematic illustrations of the morphology evolution of P3HT/PCBM composite film upon the two-step C-SVA.[107]

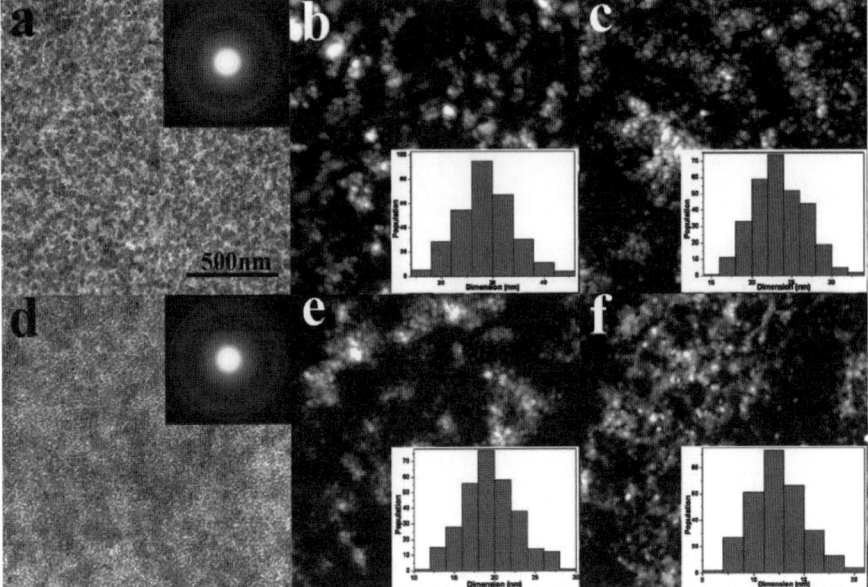

Figure 14.26 TEM micrographs and corresponding SAED patterns of P3HT/PCBM composite films spin-coated from ODCB. (a) one-step treatment: THF C-SVA for 5 s; (d) two-step treatment: THF C-SVA for 5 s then CS_2 C-SVA for 15 s. AFM topography images of composite film and corresponding size distribution of PCBM clusters: one-step top (b) and bottom (c) faces; two-step top (e) and bottom (f) faces.[107]

14.2.2.3 Vertical Phase Separation

Vertical phase segregation with increased concentrations of electron donor toward the anode and electron acceptor toward the cathode have been shown to benefit solar cell performance.[112,130–133] In organic solar cells, vertical phase-segregation has been observed in films of P3HT blended with isotactic semi-crystalline polystyrene,[134] blends of two different polyfluorene copolymers,[135] and blends of polyfluorene with PCBM[136,137] by high resolution TEM (HR-TEM) (Figure 14.27 top),[112] variable-angle spectroscopic ellipsometry (VASE),[131] and by characterizing the top and bottom of the released active layer with various microscopes (Figure 14.27 bottom).[107,112,130] The root cause of this phenomenon has been attributed variously to differences in the solubility or surface energy of the components[137] or the dynamics of the spin-coating process.[131,138] In the HR-TEM images in Figure 14.27, the evidence of vertical segregation is clear: images taken at focal depths of 0 and 28 nm revealing PCBM crystals, while the image taken at a depth of 84 nm reveals P3HT lamellae.[112] Higher roughness and AFM phase contrast in images of the bottom surface in Figure 14.27 also indicate that more fullerene phase exists at the bottom surface. These results indicate that a negative vertical composition gradient exists inside the P3HT:PCBM film (a gradient with increasing

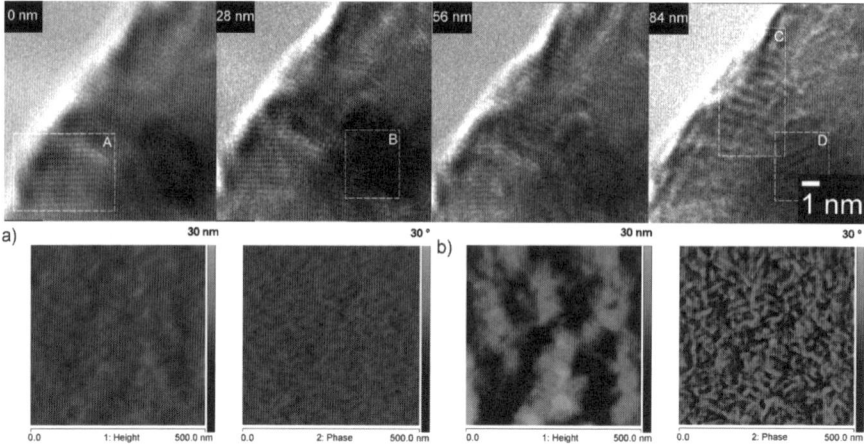

Figure 14.27 (Top) HR-TEM focal series of P3HT:PCBM blend films (1 : 0.8 weight ratio, 18 mg mL^{-1}). The distances quoted refer to microscope focal depth.[112] Regions A and B are PCBM crystals. Parts C and D are P3HT lamella. (Bottom) AFM topography and phase images of (a) top and (b) bottom surfaces of the exposed P3HT networks. The PCBM in the fast grown blend films spin-coated on glass was selectively removed using OT.[130]

acceptor concentration closer to the cathode is defined as positive),[131] which is disfavoured for devices with Al cathodes. Following thermal annealing, the negative composition gradient value was reduced from −25 to −5.9, while the PCE increased from 1.13% to 3.92% (Table 14.1).[131]

Vertical phase segregation with increased concentrations of electron donor toward the anode and electron acceptor toward the cathode have been shown to benefit solar cell performance.[112,130–133] In organic solar cells, vertical phase-segregation has been observed in films of P3HT blended with isotactic semi-crystalline polystyrene,[134] blends of two different polyfluorene copolymers,[135] and blends of polyfluorene with PCBM[136,137] by high resolution TEM (HR-TEM) (Figure 14.27 top),[112] variable-angle spectroscopic ellipsometry (VASE),[131] and by characterizing the top and bottom of the released active layer with various microscopes (Figure 14.27 bottom).[107,112,130] The root cause

Table 14.1 Optical density (OD) and slope of the PCBM vertical profile for P3HT:PCBM (1 : 1) blend films deposited on PEDOT:PSS, and power conversion efficiency (PCE) for solar cells as a function of annealing time (at 140 °C).[131]

Annealing time/min	OD(520 nm) (arb. units)	Slope (on PEDOT) (%)	PCE (%)
0	0.709	−25	1.13
5	0.879	−6.8	3.64
30	0.879	−5.9	3.92

Materials Nanoarchitechtonics for Advanced Devices 383

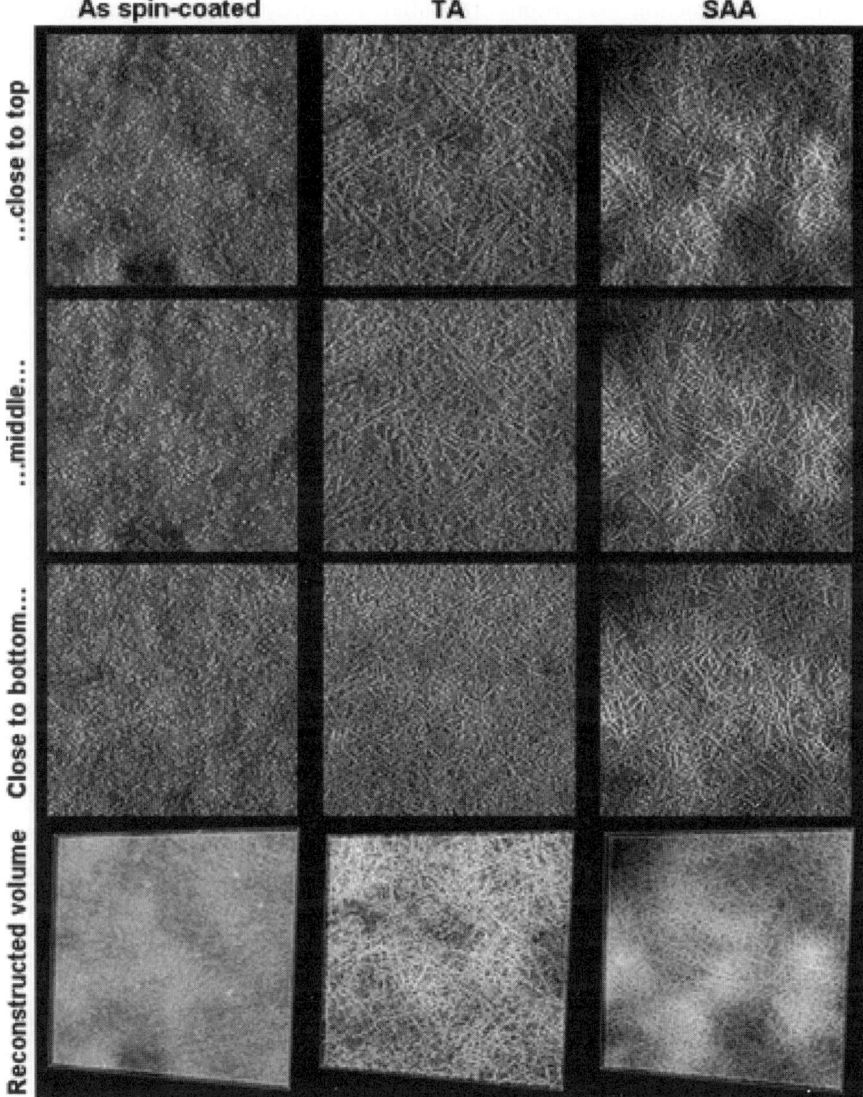

Figure 14.28 Electron tomography slices of P3HT/PCBM photoactive layers with various treatments: as spin-coated, thermally annealed at 130 °C for 20 min (TA) and solvent-assisted annealed for 3 h (SAA). The first three rows contain horizontal slices taken from the reconstructed film volume at different depths: one slice close to the top of the film (near the cathode), another one in the middle of the film, and the third one close to the bottom of the film (near the anode). The dimensions of the slices are around 1700 nm × 1700 nm. Images in the fourth row are snapshots of the whole reconstructed film volume, with dimensions of around 1700 nm × 1700 nm × 100 nm.[139]

of this phenomenon has been attributed variously to differences in the solubility or surface energy of the components[137] or the dynamics of the spin-coating process.[131,138] In the HR-TEM images in Figure 14.27, the evidence of vertical segregation is clear: images taken at focal depths of 0 and 28 nm show PCBM crystals, while the image taken at a depth of 84 nm shows P3HT lamellae.[112] Higher roughness and AFM phase contrast in images of the bottom surface in Figure 14.27 also indicate that more fullerene phase exists at the bottom surface. These results indicate that a vertical composition gradient with more acceptor at the anode and more donor at the cathode exists inside as-deposited P3HT:PCBM films, which disfavours devices with Al cathodes. Following thermal annealing, the negative composition gradient was reduced, increasing PCE from 1.13% to 3.92% (Table 14.1).[131]

From the previous sections, it can be seen that in nearly all cases, the main effect of nanostructure manipulation is the enhancement of charge mobility and charge collection through enabling spatial rearrangement of the polymer chains and fullerene molecules and thus self-assembly of P3HT chains into crystallites.[134] It is also important to keep in mind that the morphological changes in organic solar cell induced by various treatments are three-dimensional in nature. Studying their effects on vertical phase segregation of the active layer is as important as other studies on the lateral nanostructure. From the TEM profiles of thermal annealed and solvent vapour annealed films at different depths (Figure 14.28), both annealing techniques were found to enrich crystalline P3HT nanorods in the lower part of the photoactive layer close to the anode and PCBM close to the top cathode.[139] This partial reversal of the original as-deposited vertical concentration gradient benefits OPV charge collection because the quality of the charge percolation network for holes increases in the vicinity of the anode, and similarly for electrons in the vicinity of the cathode. The profile of PCBM concentration *versus* film depth can be quantitatively profiled using variable-angle spectroscopic ellipsometry (VASE) (Figure 14.29); using this method, it was shown that PCBM

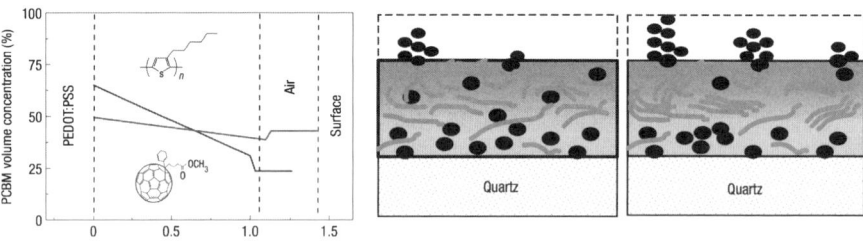

Figure 14.29 (Left) PCBM concentration *vs.* depth in films spin-coated on PEDOT:PSS-coated fused silica before (black) and after (grey) solvent vapour annealing. Schematic representations of the model used to fit the ellipsometry data, showing typical PCBM distributions before (middle) and after (right) vapour or thermal annealing. Note that PCBM crystals in real samples are often much larger than illustrated here.[131]

concentration does indeed decrease with increasing distance from the top electrode after solvent vapour annealing.[131]

14.3 Conclusion

It can be seen that organic solar cell active layer morphology manipulation is crucial to balance the contrasting morphology requirements of exciton separation and charge collection in the final devices. Various factors and methods that affect the morphology of the active layer during device fabrication were reviewed with examples largely drawn from the most-studied MDMO-PPV:PCBM and P3HT:PCBM systems. Nanostructure manipulation methods, their mechanisms, and especially their effects on device performance were discussed and compared. Although some methods showed very similar effects on both lateral and vertical morphology of organic photovoltaic active layer films, application of these techniques to different systems yields different effects. However, familiarity with and understanding of the mechanisms of these nanostructure modification techniques will help researchers in the field to optimize new active layer systems in order to achieve the highly efficient, inexpensive, large area devices necessary to fulfil the promise of this technology.

References

1. J. Zhao, A. Wang and M. A. Green, *Prog. Photovoltaics*, 1999, **7**, 471–474.
2. R. R. King, D. C. Law, K. M. Edmondson, C. M. Fetzer, G. S. Kinsey, H. Yoon, R. A. Sherif and N. H. Karam, *Appl. Phys. Lett.*, 2007, **90**, 183516-183516-3.
3. M. Gr\ätzel, *Nature*, 2001, **414**, 338–344.
4. E. Maruyama, A. Terakawa, Y. Yoshimine, D. Idel, T. Baba, T. Shima, H. Sakata and M. Tanaka, *4th World Conference on Photovoltaic Energy Conversion (WCEP-4)*, 2006.
5. C. J. Brabec, N. S. Sariciftci and J. C. Hummelen, *Adv. Funct. Mater.*, 2001, **11**, 15–26.
6. L. Chen, Z. Hong, G. Li and Y. Yang, *Adv. Mater.*, 2009, **21**, 1434–1449.
7. G. Dennler, M. C. Scharber and C. J. Brabec, *Adv. Mater.*, 2009, **21**, 1323–1338.
8. C. Winder and N. S. Sariciftci, *J. Mater. Chem.*, 2004, **14**, 1077.
9. T. A. Skotheim, *Handbook of Conducting Polymers*, CRC Press, 2nd edn, 1997.
10. G. Hadziioannou and P. F. van Hutten, *Semiconducting Polymers: Chemistry, Physics, and Engineering*, Wiley-VCH, New York, 2000.
11. B. Kippelen and J.-L. Brédas, *Energy Environ. Sci.*, 2009, **2**, 251.
12. B. J. Schwartz, *Nat. Mater.*, 2008, **7**, 427–428.
13. E. Collini and G. D. Scholes, *Science*, 2009, **323**, 369–373.

14. J.-L. Brédas and R. Silbey, *Science*, 2009, **323**, 348–349.
15. R. Österbacka, C. P. An, X. M. Jiang and Z. V. Vardeny, *Science*, 2000, **287**, 839–842.
16. B. C. Thompson and J. M. J. Fréchet, *Angew. Chem., Int. Ed.*, 2008, **47**, 58–77.
17. V. I. Arkhipov, E. V. Emelianova and H. Bässler, *Phy. Rev. Lett.*, 1999, **82**, 1321–1324.
18. I. H. Campbell, T. W. Hagler, D. L. Smith and J. P. Ferraris, *Phy. Rev. Lett.*, 1996, **76**, 1900–1903.
19. A. Salleo, M. L. Chabinyc, M. S. Yang and R. A. Street, *Appl. Phys. Lett.*, 2002, **81**, 4383.
20. V. D. Mihailetchi, L. J. A. Koster, J. C. Hummelen and P. W. M. Blom, *Phys. Rev. Lett.*, 2004, **93**, 216601.
21. M. Hallermann, S. Haneder and E. Da Como, *Appl. Phys. Lett.*, 2008, **93**, 053307.
22. M. C. Scharber, D. Mühlbacher, M. Koppe, P. Denk, C. Waldauf, A. J. Heeger and C. J. Brabec, *Adv. Mater.*, 2006, **18**, 789–794.
23. G. D. Scholes and G. Rumbles, *Nat. Mater.*, 2006, **5**, 683–696.
24. M. Yan, L. J. Rothberg, F. Papadimitrakopoulos, M. E. Galvin and T. M. Miller, *Phys. Rev. Lett.*, 1994, **73**, 744–747.
25. T. Q. Nguyen, V. Doan and B. J. Schwartz, *J. Chem. Phys.*, 1999, **110**, 4068.
26. N. C. Greenham, X. Peng and A. P. Alivisatos, *Phys. Rev. B*, 1996, **54**, 17628.
27. S. Günes, H. Neugebauer and N. S. Sariciftci, *Chem. Rev.*, 2007, **107**, 1324–1338.
28. H. Hoppe and N. S. Sariciftci, *J. Mater. Chem.*, 2006, **16**, 45.
29. G. Li, V. Shrotriya, Y. Yao, J. Huang and Y. Yang, *J. Mater. Chem.*, 2007, **17**, 3126.
30. C. J. Brabec, S. Gowrisanker, J. J. M. Halls, D. Laird, S. Jia and S. P. Williams, *Adv. Mater.*, 2010, **22**, 3839–3856.
31. F. Yang and S. R. Forrest, *ACS Nano*, 2008, **2**, 1022–1032.
32. Y.-J. Cheng, S.-H. Yang and C.-S. Hsu, *Chem. Rev.*, 2009, **109**, 5868–5923.
33. A. J. Moulé and K. Meerholz, *Adv. Funct. Mater.*, 2009, **19**, 3028–3036.
34. H. Hoppe and N. S. Sariciftci, *J. Mater. Res.*, 2011, **19**, 1924–1945.
35. P. Schilinsky, C. Waldauf and C. J. Brabec, *Appl. Phys. Lett.*, 2002, **81**, 3885.
36. K. Norrman, A. Ghanbari-Siahkali and N. B. Larsen, *Annu. Rep. Prog. Chem., Sect. C*, 2005, **101**, 174.
37. S. Günes, H. Neugebauer and N. S. Sariciftci, *Chem. Rev.*, 2007, **107**, 1324–1338.
38. L. Cui, Y. Ding, X. Li, Z. Wang and Y. Han, *Thin Solid Films*, 2006, **515**, 2038–2048.

39. G. Li, Y. Yao, H. Yang, V. Shrotriya, G. Yang and Y. Yang, *Adv. Funct. Mater.*, 2007, **17**, 1636–1644.
40. J.-H. Huang, K.-C. Li, F.-C. Chien, Y.-S. Hsiao, D. Kekuda, P. Chen, H.-C. Lin, K.-C. Ho and C.-W. Chu, *J Phys. Chem. C*, 2010, **114**, 9062–9069.
41. T. Martens, *Synth. Methods*, 2003, **138**, 243–247.
42. J. K. J. van Duren, X. Yang, J. Loos, C. W. T. Bulle-Lieuwma, A. B. Sieval, J. C. Hummelen and R. A. J. Janssen, *Adv. Funct. Mater.*, 2004, **14**, 425–434.
43. J.-M. Nunzi, *C. R. Phys.*, 2002, **3**, 523–542.
44. N. S. Sariciftci, D. Braun, C. Zhang, V. I. Srdanov, A. J. Heeger, G. Stucky and F. Wudl, *Appl. Phys. Lett.*, 1993, **62**, 585.
45. J. J. M. Halls, K. Pichler, R. H. Friend, S. C. Moratti and A. B. Holmes, *Appl. Phys. Lett.*, 1996, **68**, 3120.
46. H. Hoppe, T. Glatzel, M. Niggemann, W. Schwinger, F. Schaeffler, A. Hinsch, M. C. Lux-Steiner and N. S. Sariciftci, *Thin Solid Films*, 2006, **511**, 587–592.
47. S. E. Shaheen, C. J. Brabec, N. S. Sariciftci, F. Padinger, T. Fromherz and J. C. Hummelen, *Appl. Phys. Lett.*, 2001, **78**, 841.
48. H. Hoppe, M. Niggemann, C. Winder, J. Kraut, R. Hiesgen, A. Hinsch, D. Meissner and N. S. Sariciftci, *Adv. Funct. Mater.*, 2004, **14**, 1005–1011.
49. L.-M. Chen, Z. Xu, Z. Hong and Y. Yang, *J. Mater. Chem.*, 2010, **20**, 2575.
50. G. Yu, J. Gao, J. C. Hummelen, F. Wudl and A. J. Heeger, *Science*, 1995, **270**, 1789–1791.
51. R. D. McCullough, P. C. Ewbank and R. S. Loewe, *J. Am. Chem. Soc.*, 1997, **119**, 633–634.
52. Y. Li and Y. Zou, *Adv. Mater.*, 2008, **20**, 2952–2958.
53. E. Bundgaard and F. C. Krebs, *Sol. Energy Mater. Sol. Cells*, 2007, **91**, 954–985.
54. T. A. Skotheim and J. Reynolds, *Conjugated Polymers: Theory, Synthesis, Properties, and Characterization*, CRC Press, 3rd edn., 2006.
55. M. J. Winokur, R. D. McCullough and T. J. Prosa, *Macromol.*, 1996, **29**, 3654–3656.
56. R. D. McCullough, S. Tristram-Nagle, S. P. Williams, R. D. Lowe and M. Jayaraman, *J. Am. Chem. Soc.*, 1993, **115**, 4910–4911.
57. S. Yue, G. C. Berry and R. D. McCullough, *Macromolecules*, 1996, **29**, 933–939.
58. C. J. Brabec, V. Dyakonov and U. Scherf, *Organic photovoltaics: materials, device physics, and manufacturing technologies*, Wiley-VCH, 2008.
59. V. Ruiz, P. G. Nicholson, S. Jollands, P. A. Thomas, J. V. Macpherson and P. R. Unwin, *J. Phys. Chem. B*, 2005, **109**, 19335–19344.

60. A. Gadisa, W. D. Oosterbaan, K. Vandewal, J. Bolsée, S. Bertho, J. D'Haen, L. Lutsen, D. Vanderzande and J. V. Manca, *Adv. Funct. Mater.*, 2009, **19**, 3300–3306.
61. C. H. Woo, B. C. Thompson, B. J. Kim, M. F. Toney and J. M. J. Fréchet, *J. Am. Chem. Soc.*, 2008, **130**, 16324–16329.
62. S. Pal and A. K. Nandi, *Macromolecules*, 2003, **36**, 8426–8432.
63. R. J. Kline, M. D. McGehee, E. N. Kadnikova, J. Liu, J. M. J. Fréchet and M. F. Toney, *Macromolecules*, 2005, **38**, 3312–3319.
64. R. J. Kline, M. D. McGehee, E. N. Kadnikova, J. Liu and J. M. Fréchet, *Adv. Mater.*, 2003, **15**, 1519–1522.
65. R. Zhang, B. Li, M. C. Iovu, M. Jeffries-EL, G. Sauvé, J. Cooper, S. Jia, S. Tristram-Nagle, D. M. Smilgies, D. N. Lambeth, R. D. McCullough and T. Kowalewski, *J. Am. Chem. Soc.*, 2006, **128**, 3480–3481.
66. J. Liu, T.-F. Guo and Y. Yang, *J. Appl. Phys.*, 2002, **91**, 1595.
67. S. Walheim, M. Böltau, J. Mlynek, G. Krausch and U. Steiner, *Macromolecules*, 1997, **30**, 4995–5003.
68. A. G. Jones, C. Balocco, R. King and A. M. Song, *Appl. Phys. Lett.*, 2006, **89**, 013119.
69. L. L. Spangler, J. M. Torkelson and J. S. Royal, *Polym. Eng. Sci.*, 1990, **30**, 644–653.
70. H. Sirringhaus, N. Tessler and R. H. Friend, *Science*, 1998, **280**, 1741–1744.
71. H. Sirringhaus, P. J. Brown, R. H. Friend, M. M. Nielsen, K. Bechgaard, B. M. W. Langeveld-Voss, A. J. H. Spiering, R. A. J. Janssen, E. W. Meijer, P. Herwig and D. M. de Leeuw, *Nature*, 1999, **401**, 685–688.
72. J. J. Apperloo, R. A. J. Janssen, M. M. Nielsen and K. Bechgaard, *Adv. Mater.*, 2000, **12**, 1594–1597.
73. R. J. Kline, M. D. McGehee, E. N. Kadnikova, J. Liu and J. M. Fréchet, *Adv. Mater.*, 2003, **15**, 1519–1522.
74. A. C. Arias, J. D. MacKenzie, R. Stevenson, J. J. M. Halls, M. Inbasekaran, E. P. Woo, D. Richards and R. H. Friend, *Macromolecules*, 2001, **34**, 6005–6013.
75. N. Corcoran, A. C. Arias, J. S. Kim, J. D. MacKenzie and R. H. Friend, *Appl. Phys. Lett.*, 2003, **82**, 299.
76. N. Sahu, B. Parija and S. Panigrahi, *Indian J. Phys.*, 2009, **83**, 493–502.
77. J.-F. Chang, B. Sun, D. W. Breiby, M. M. Nielsen, T. I. Sölling, M. Giles, I. McCulloch and H. Sirringhaus, *Chem. Mater.*, 2004, **16**, 4772–4776.
78. L. L. Spangler, J. M. Torkelson and J. S. Royal, *Polym. Eng. Sci.*, 1990, **30**, 644–653.
79. C.-Y. Lin, A. Garcia, P. Zalar, J. Z. Brzezinski and T.-Q. Nguyen, *J Phys. Chem. C*, 2010, **114**, 15786–15790.
80. W.-H. Baek, H. Yang, T.-S. Yoon, C. J. Kang, H. H. Lee and Y.-S. Kim, *Sol. Energy Mater. Sol. Cells*, 2009, **93**, 1263–1267.
81. D. H. Kim, Y. Jang, Y. D. Park and K. Cho, *J. Phys. Chem. B*, 2006, **110**, 15763–15768.

82. A. G. Jones, C. Balocco, R. King and A. M. Song, *Appl. Phys. Lett.*, 2006, **89**, 013119.
83. G. Li, V. Shrotriya, J. Huang, Y. Yao, T. Moriarty, K. Emery and Y. Yang, *Nat. Mater.*, 2005, **4**, 864–868.
84. T. Erb, U. Zhokhavets, G. Gobsch, S. Raleva, B. Stühn, P. Schilinsky, C. Waldauf and C. J. Brabec, *Adv. Funct. Mater.*, 2005, **15**, 1193–1196.
85. R. Zhang, B. Li, M. C. Iovu, M. Jeffries-EL, G. Sauvé, J. Cooper, S. Jia, S. Tristram-Nagle, D. M. Smilgies, D. N. Lambeth, R. D. McCullough and T. Kowalewski, *J. Am. Chem. Soc.*, 2006, **128**, 3480–3481.
86. R. J. Kline, M. D. McGehee, E. N. Kadnikova, J. Liu and J. M. Fréchet, *Adv. Mater.*, 2003, **15**, 1519–1522.
87. T. Shimomura, H. Sato, H. Furusawa, Y. Kimura, H. Okumoto, K. Ito, R. Hayakawa and S. Hotta, *Phys. Rev. Lett.*, 1994, **72**, 2073.
88. T. H. Kim, J. Hwang, W. S. Hwang, J. Huh, H. -C Kim, S. H. Kim, J. M. Hong, E. L. Thomas and C. Park, *Adv. Mater.*, 2008, **20**, 522–527.
89. G. Wei, R. R. Lunt, K. Sun, S. Wang, M. E. Thompson and S. R. Forrest, *Nano Lett.*, 2010, **10**, 3555–3559.
90. W. Yin and M. Dadmun, *ACS Nano*, 2011, **5**, 4756–4768.
91. Y. Yao, J. Hou, Z. Xu, G. Li and Y. Yang, *Adv. Funct. Mater.*, 2008, **18**, 1783–1789.
92. F. Zhang, K. G. Jespersen, C. Björström, M. Svensson, M. R. Andersson, V. Sundström, K. Magnusson, E. Moons, A. Yartsev and O. Inganäs, *Adv. Funct. Mater.*, 2006, **16**, 667–674.
93. J. Peet, C. Soci, R. C. Coffin, T. Q. Nguyen, A. Mikhailovsky, D. Moses and G. C. Bazan, *Appl. Phys. Lett.*, 2006, **89**, 252105.
94. F.-C. Chen, H.-C. Tseng and C.-J. Ko, *Appl. Phys. Lett.*, 2008, **92**, 103316.
95. J. K. Lee, W. L. Ma, C. J. Brabec, J. Yuen, J. S. Moon, J. Y. Kim, K. Lee, G. C. Bazan and A. J. Heeger, *J. Am. Chem. Soc.*, 2008, **130**, 3619–3623.
96. A. J. Moulé and K. Meerholz, *Adv. Mater.*, 2008, **20**, 240–245.
97. C. Y. Kwong, A. B. Djurisic, P. C. Chui, K. W. Cheng and W. K. Chan, *Chem. Phys. Lett.*, 2004, **384**, 372–375.
98. S. Walheim, M. Böltau, J. Mlynek, G. Krausch and U. Steiner, *Macromolecules*, 1997, **30**, 4995–5003.
99. J. Peet, C. Soci, R. C. Coffin, T. Q. Nguyen, A. Mikhailovsky, D. Moses and G. C. Bazan, *Appl. Phys. Lett.*, 2006, **89**, 252105.
100. J. Peet, J. Y. Kim, N. E. Coates, W. L. Ma, D. Moses, A. J. Heeger and G. C. Bazan, *Nat Mater*, 2007, **6**, 497–500.
101. F. Padinger, R. S. Rittberger and N. S. Sariciftci, *Adv. Funct. Mater.*, 2003, **13**, 85–88.
102. G. Li, V. Shrotriya, Y. Yao and Y. Yang, *J. Appl. Phys.*, 2005, **98**, 043704.
103. U. Zhokhavets, T. Erb, H. Hoppe, G. Gobsch and N. Serdar Sariciftci, *Thin Solid Films*, 2006, **496**, 679–682.

104. P. Vanlaeke, A. Swinnen, I. Haeldermans, G. Vanhoyland, T. Aernouts, D. Cheyns, C. Deibel, J. D'Haen, P. Heremans, J. Poortmans and J. V. Manca, *Sol. Energy Mater. Sol. Cells*, 2006, **90**, 2150–2158.
105. J. Jo, S. Kim, S. Na, B. Yu and D. Kim, *Adv. Funct. Mater.*, 2009, **19**, 866–874.
106. J. H. Park, J. S. Kim, J. H. Lee, W. H. Lee and K. Cho, *J Phys. Chem. C*, 2009, **113**, 17579–17584.
107. H. Tang, G. Lu, L. Li, J. Li, Y. Wang and X. Yang, *J. Mater. Chem.*, 2010, **20**, 683.
108. T. Erb, U. Zhokhavets, G. Gobsch, S. Raleva, B. Stühn, P. Schilinsky, C. Waldauf and C. J. Brabec, *Adv. Funct. Mater.*, 2005, **15**, 1193–1196.
109. F. Padinger, R. S. Rittberger and N. S. Sariciftci, *Adv. Funct. Mater.*, 2003, **13**, 85–88.
110. L. H. Nguyen, H. Hoppe, T. Erb, S. Günes, G. Gobsch and N. S. Sariciftci, *Adv. Funct. Mater.*, 2007, **17**, 1071–1078.
111. V. D. Mihailetchi, H. X. Xie, B. de Boer, L. J. A. Koster and P. W. M. Blom, *Adv. Funct. Mater.*, 2006, **16**, 699–708.
112. R. M. Beal, A. Stavrinadis, J. H. Warner, J. M. Smith, H. E. Assender and A. A. R. Watt, *Macromolecules*, 2010, **43**, 2343–2348.
113. T. Erb, U. Zhokhavets, G. Gobsch, S. Raleva, B. Stühn, P. Schilinsky, C. Waldauf and C. J. Brabec, *Adv. Funct. Mater.*, 2005, **15**, 1193–1196.
114. W. Ma, C. Yang, X. Gong, K. Lee and A. J. Heeger, *Adv. Funct. Mater.*, 2005, **15**, 1617–1622.
115. Youngkyoo Kim, S. Cook, S. M. Tuladhar, S. A. Choulis, J. Nelson, J. R. Durrant, D. D. C. Bradley, M. Giles, I. McCulloch, Chang-Sik Ha and M. Ree, *Nat. Mater.*, 2006, **5**, 197–203.
116. R. M. Beal, A. Stavrinadis, J. H. Warner, J. M. Smith, H. E. Assender and A. A. R. Watt, *Macromolecules*, 2010, **43**, 2343–2348.
117. X. Yang, J. Loos, S. C. Veenstra, W. J. H. Verhees, M. M. Wienk, J. M. Kroon, M. A. J. Michels and R. A. J. Janssen, *Nano Lett.*, 2005, **5**, 579–583.
118. J. Zhao, A. Swinnen, G. Van Assche, J. Manca, D. Vanderzande and B. V. Mele, *J. Phys. Chem. B*, 2009, **113**, 1587–1591.
119. P. E. Hopkinson, P. A. Staniec, A. J. Pearson, A. D. F. Dunbar, T. Wang, A. J. Ryan, R. A. L. Jones, D. G. Lidzey and A. M. Donald, *Macromolecules*, 2011, **44**, 2908–2917.
120. V. D. Mihailetchi, H. X. Xie, B. de Boer, L. J. A. Koster and P. W. M. Blom, *Adv. Funct. Mater.*, 2006, **16**, 699–708.
121. M. Reyes-Reyes, K. Kim, J. Dewald, R. López-Sandoval, A. Avadhanula, S. Curran and D. L. Carroll, *Org. Lett.*, 2005, **7**, 5749–5752.
122. M. Campoy-Quiles, T. Ferenczi, T. Agostinelli, P. G. Etchegoin, Y. Kim, T. D. Anthopoulos, P. N. Stavrinou, D. D. C. Bradley and J. Nelson, *Nat. Mater.*, 2008, **7**, 158–164.
123. D. Chirvase, J. Parisi, J. C. Hummelen and V. Dyakonov, *Nanotechnology*, 2004, **15**, 1317–1323.

124. F. Demir, N. Brande, B. Mele, S. Bertho, D. Vanderzande, J. Manca and G. Assche, *J. Therm. Anal. Calorim.*, 2011, **105**, 845–849.
125. J. C. Conboy, E. J. C. Olson, D. M. Adams, J. Kerimo, A. Zaban, B. A. Gregg and P. F. Barbara, *J. Phys. Chem. B*, 1998, **102**, 4516–4525.
126. Y. Zhao, Z. Xie, Y. Qu, Y. Geng and L. Wang, *Appl. Phys. Lett.*, 2007, **90**, 043504.
127. S. Miller, G. Fanchini, Y.-Y. Lin, C. Li, C.-W. Chen, W.-F. Su and M. Chhowalla, *J. Mater. Chem.*, 2008, **18**, 306.
128. G. Lu, L. Li and X. Yang, *Small*, 2008, **4**, 601–606.
129. H. Tang, G. Lu, L. Li, J. Li, Y. Wang and X. Yang, *J. Mater. Chem.*, 2010, **20**, 683.
130. A. C. Arias, N. Corcoran, M. Banach, R. H. Friend, J. D. MacKenzie and W. T. S. Huck, *Appl. Phys. Lett.*, 2002, **80**, 1695.
131. M. Campoy-Quiles, T. Ferenczi, T. Agostinelli, P. G. Etchegoin, Y. Kim, T. D. Anthopoulos, P. N. Stavrinou, D. D. C. Bradley and J. Nelson, *Nat Mater*, 2008, **7**, 158–164.
132. J. Xue, B. P. Rand, S. Uchida and S. R. Forrest, *J. Appl. Phys.*, 2005, **98**, 124903.
133. H. J. Snaith, N. C. Greenham and R. H. Friend, *Adv. Mater.*, 2004, **16**, 1640–1645.
134. S. Goffri, C. Muller, N. Stingelin-Stutzmann, D. W. Breiby, C. P. Radano, J. W. Andreasen, R. Thompson, R. A. J. Janssen, M. M. Nielsen, P. Smith and H. Sirringhaus, *Nat. Mater.*, 2006, **5**, 950–956.
135. A. C. Arias, N. Corcoran, M. Banach, R. H. Friend, J. D. MacKenzie and W. T. S. Huck, *Appl. Phys. Lett.*, 2002, **80**, 1695.
136. C. M. Björström, A. Bernasik, J. Rysz, A. Budkowski, S. Nilsson, M. Svensson, M. R. Andersson, K. O. Magnusson and E. Moons, *J. Phys.: Condens. Matter*, 2005, **17**, L529–L534.
137. C. M. Björström, S. Nilsson, A. Bernasik, A. Budkowski, M. Andersson, K. O. Magnusson and E. Moons, *Appl. Surf. Sci.*, 2007, **253**, 3906–3912.
138. S. Y. Heriot and R. A. L. Jones, *Nat. Mater.*, 2005, **4**, 782–786.
139. S. S. van Bavel, E. Sourty, G. de With and J. Loos, *Nano Lett.*, 2009, **9**, 507–513.

CHAPTER 15
Substrate Alignment with Scanning Probe Lithography

MICHAEL V. LEE

International Center for Materials Nanoarchitectonics (MANA), National Institute for Materials Science (NIMS), 1-1 Namiki, Tsukuba 305-0044, Japan
E-mail: LEE.Michael@nims.go.jp

15.1 Introduction

The electrical, mechanical, or photonic response of addressable individual atoms that compose molecules or that are embedded in a crystal represents the limit of imagination for intentional manipulation of matter to accomplish computation. Single molecules can still conduct electrons in a manner that can be modeled by Ohm's Law although the current is generally small, but they are also in the range where quantum effects are dominant. Individual atoms can move across a surface between different bound states, or atoms—notably hydrogen—can move by tunneling as part of chemical reactions. The spin of electrons or nuclei can maintain a superposition of all possible quantum states. Atoms or molecules can adsorb or chemisorb onto a surface and when molecules are stimulated they can rotate or revolve around an axis. This unique intersection of the classical and quantum at the atomic and molecular scale will someday enable a revolution in information processing.

The limits of photolithography are in some part due to this unique interaction of the classical and quantum, as well as the transition from patterning a structure in a jumble of random molecules to placing each individual molecule. At the scale of molecular electronics, each molecule, atom, or spin is precisely positioned, separately stimulated or manipulated, and observed. This is the finest precision for which spatial lithography will be useful. Even after molecules can be placed on a surface, electrodes or molecules or some other form of interface is required to feed information to the system and extract the result. The interface must connect the molecule to microelectronics, by which we can use them. This requires the molecules to be aligned on a substrate relative to a point of reference.

This chapter discusses the scientific progress towards patterning and addressing individual molecules and atoms relative to the macroscale world. The chemistry for anchoring of molecules to substrates or electrodes, as well as a basic overview of relevant molecules, will be presented. Much of the chapter will focus on the application of surface probe microscopy (SPM) techniques, specifically scanning tunneling microscopy (STM) and atomic force microscopy (AFM), for molecule or atom placement and functional pattern creation to interface between the molecules and electrodes fabricated by traditional methods. The chapter concludes with an evaluation of the current situation and directions for focused research.

15.2 Calculation Using Single Molecules or Atoms

Hyperconjugation in higher-order carbon bonds provides for overlapping π orbitals that can extend over the whole molecule. These overlapping π bonds are the source of conductivity in some common polymers such as polyaniline or polyacetylene. Conducting polymers can act as molecular wires.[1] The same phenomenon is observed in polyaromatic molecules. A π orbital that extends over a large, flat molecule leads to high mobility like that observed in graphene. Incorporating electron donating and electron withdrawing character by adding functional groups was suggested in 1974 to create the analog of p-type and n-type regions in a single molecule.[2] This effect has been experimentally demonstrated.[3] A more recent alternate route to create rectification uses cross conjugation in hyperconjugated molecules. When a voltage is applied to a molecule with multiple cross conjugations, their interference induces a p–n junction as shown in Figure 15.1.[4] Such p–n junctions are an essential part of diodes, transistors, and much of silicon and inorganic semiconductor electronics. Conducting organic molecules aligned on electrodes or multicomponent arrays of molecules have been suggested as routes to form electronic circuits out of molecules. An array of such donor and acceptor molecules is pictured in Figure 15.2, where the intersection of the two components acts as a p–n junction.[5,6] Such devices face many challenges. A recent review cited some of the glaring problems, including the challenge of placement of molecules on a surface with the required precision, the lack of a method to connect active components from solution to pre-placed electrodes, and the instability of noncovalent bonds.[7] However, a more significant problem is the low conductivity in conductive organic molecules,[8] which leads to an exponential decay of signals for molecules connected in series.[9] This problem has increased the focus on using quantum states for molecular or atomic scale calculation.

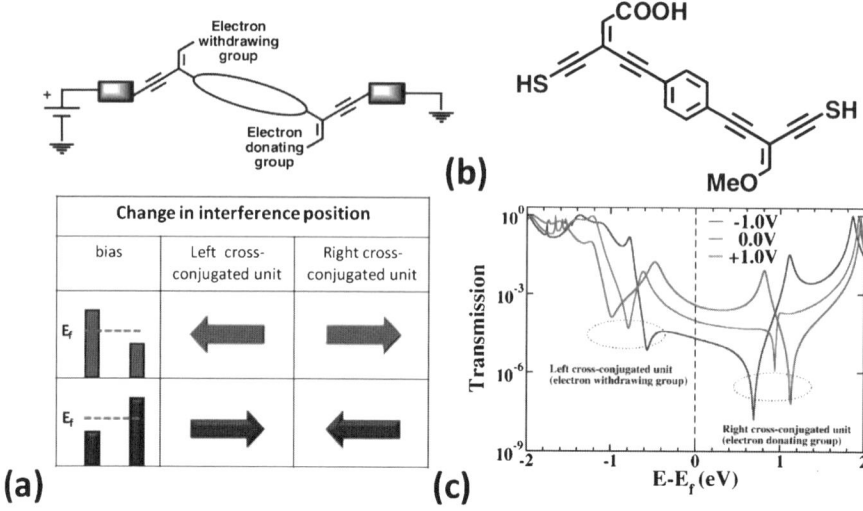

Figure 15.1 In molecules with multiple cross-conjugated units, an applied bias voltage will split the interference features. This splitting occurs because an applied bias has an electron-donating or -withdrawing effect that moves the interference position. (a) The more positive an electrode is, the more it moves the interference feature of the closest cross-conjugated group to lower energy, and conversely, the more negative an electrode, the more it moves the interference feature to higher energy. (b) An asymmetric molecule and (c) the corresponding transmission plots. The asymmetry causes two antiresonance features at different energies. At negative bias these antiresonance features move together, and at positive applied bias they move apart. Reprinted with permission from D. Q. Andrews, G. C. Solomon, R. P. Van Duyne, M. A. Ratner, *J. Am. Chem. Soc.* 2008, **130**, 17309. Copyright 2008 American Chemical Society.

Figure 15.2 STM image of a donor and acceptor adsorbed prostrate on the surface and next to each other. Reprinted with permission from K. Müllen, J. P. Rabe, *Acc. Chem. Res.*, 2008, **41**, 511. Copyright 2008 American Chemical Society.

15.2.1 Quantum Qubits

The term "qubit" is used in the general science literature as the quantum version of a computer bit. By superposition of states, a single qubit can represent either "1" or "0" or both of these options. By making the quantum states of an array of qubits all interdependent, *i.e.*, entangling the quantum states, all possible solutions can be represented and manipulated simultaneously. When prepared, stimulated, and measured appropriately, a qubit system can provide the solution to problems that are intractable by classical computation, such as Shor's algorithm that factors larger numbers.[10] Many groups have been working to implementing such qubits in a real system.[11] The first successful systems used nuclear magnetic resonance (NMR) of spins on seven individual atoms on a single molecule to represent an array of qubits. The average spins of the atoms in a solution of these molecules was used to confirm Shor's algorithm by factoring the number "15".[12]

One of the most desired platforms for realizing a qubit device is a solid-state implementation. This has been attempted with dopant atoms embedded in an isotopically pure silicon crystal. Fabrication of such geometries has been realized by utilizing the nuclear spin of two ^{29}Si atoms embedded in isotopically pure ^{28}Si.[13] The useful implementation also requires gates to stimulate each dopant atoms individually; such geometries have been presented including that by Los Alamos National Laboratory which placed ^{28}P atoms in a regular array on a silicon crystal, growing Si crystal over the top of them, and placed nanoscale metallic gates over each phosphorus atom.[14] Creating the basic geometry is still very far from creating a working device, since impurities and imperfections will increase the effect of decoherence which will be one of the greatest challenges to qubit calculations, especially with a useful number of qubits. However, by using both the electron and nuclear spin of ^{31}P atoms instead of just the nuclear spin of ^{29}Si, both spins can be manipulated independently and both exhibit long coherence times.[15,16] Additionally in this system, if electron spin is measured repeatedly, the state of the nuclear spin can be refocused and maintain coherence longer.[17–19]

Significant progress is being made toward fabrication of a solid state device, but there is still much work to do. Quantum calculations with entangled spins have been experimentally demonstrated, the coherence times of spins are approaching the time length required for a reasonable calculation, and refocusing of the spins has been demonstrated. Technology is also at the point where fabrication of an isotopically pure silicon crystal embedded with phosphorus atoms qubits can possibly be created at significant effort and expense, although crystal defects could still pose a significant problem. Recently, entangled solid-state qubits were demonstrated with phosphorus atoms doped in a isotopically purified silicon crystal.[20] While this represents a dramatic step toward single atom quantum electronics, selectively entangling large arrays of them, and more importantly the challenge of moving from averaging ensembles of spins to manipulating, refocusing, and measuring individual spins in a reasonable manner, are still beyond current technology.

15.2.2 Single Molecule Quantum Logic

An alternate approach, which is the focus of this chapter, harnesses the power of quantum mechanics for computation. This approach uses classical binary inputs to a single molecule, where the valence state of the molecule can be represented by a Hamiltonian equation. The inputs can be the close interaction, or absence of interaction, between a metal atom and a part of a molecule. These inputs alter the quantum state of the molecule, which state can then be measured in a way that will produce an output value. The molecule will generally be polyaromatic so that orbital overlap allows a single quantum state. A Boolean truth table then can represent the state change at different points on the molecule. With the appropriate molecule, an appropriate output

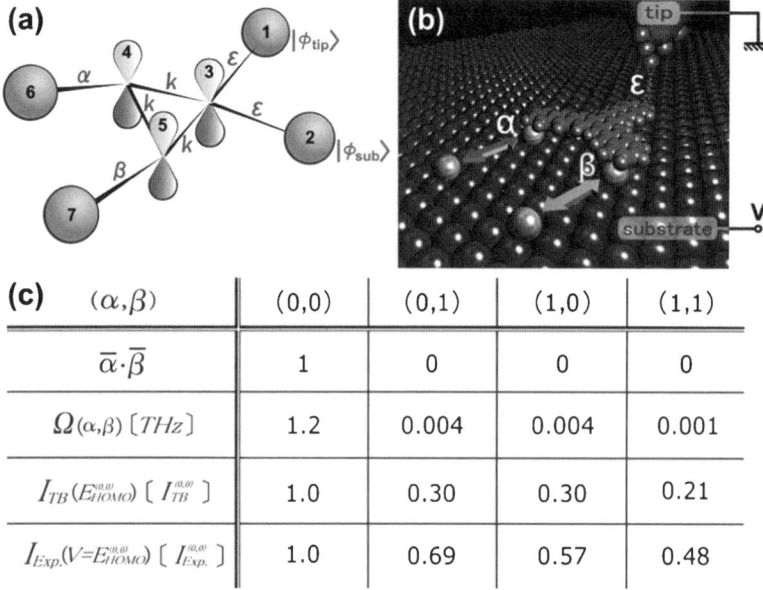

(c) (α,β)	(0,0)	(0,1)	(1,0)	(1,1)
$\overline{\alpha}\cdot\overline{\beta}$	1	0	0	0
$\Omega(\alpha,\beta)\ [THz]$	1.2	0.004	0.004	0.001
$I_{TB}(E_{HOMO}^{(0,0)})\ [I_{TB}^{(0,0)}]$	1.0	0.30	0.30	0.21
$I_{Exp.}(V=E_{HOMO}^{(0,0)})\ [I_{Exp.}^{(0,0)}]$	1.0	0.69	0.57	0.48

Figure 15.3 Schematic view of a quantum Hamiltonian computing system for a NOR logic gate, the corresponding experimental setup, and the resulting NOR truth table. (a) A seven valence bond-like quantum states model of a quantum logic gate that acts as a NOR gate. (b) The experimental implementation with a trinaphthylnene molecule NOR gate where the intensity of the tunnel current through the measurement end of the molecule corresponds to the logic status. (c) NOR truth table: the calculated Rabi-like oscillation frequency $\Omega(\alpha,\beta)$ for a weak measurement coupling $\varepsilon = 0.001$ eV, the normalized calculated current intensity is in the third line, and the measured intensity, using the configuration in (b), is in the fourth line. Reprinted with permission from W.-H. Soe, C. Manzano, N. Renaud, P. de Mendoza, A. De Sarkar, F. Ample, M. Hliwa, A. M. Echavarren, N. Chandrasekhar and C. Joachim, *ACS Nano*, 2011, **5**, 1436–1440. Copyright (2011) American Chemical Society.

measurement method, and great enough signal-to-noise in the output measurement, any truth table could theoretically be possible.[21] Interfacing with traditional electronics is more difficult since the logical "1" or "0" need to be transformed into the form of a wavefunction, which makes it difficult to control the input.[22]

An example system of this is a NOR gate produced by the interaction of two Au atoms with a triphenylene molecule. The two Au atoms act as the Boolean inputs, shifting the level of the HOMO levels of the molecule depending on whether (1) or not (0) they are close enough to interact with two arms of the triphenylene. The output state can then be measured as tunneling bias in STM at the third arm. When either or both Au atoms are present the HOMO shifts to a more negative bias voltage. This shift reduces the current intensity and represents the "0" state of the output, while the typical shift represents the "1" output This NOR gate was realized and the results published, as shown in Figure 15.3.[23] Simulation of other 3-arm molecules suggested XOR gates are also be possible,[24] or alternatively, different arms could produce both AND or XOR outputs, resulting in half-adder function.[25]

The demonstration of logic function with a single molecule rather than an ensemble of molecules shows that logic with single molecules is considerably closer to realization than quantum calculation by manipulating spins. If a single molecule can be manipulated by STM to perform logic calculations, the same operation is possible embedded in a solid-state device. The challenge is to align the molecule to the substrate, anchor the molecule in a stable, reliable, and reproducible manner, and connect the molecule for input and output. These are significant challenges, but are within the reach of technology today.

15.3 Anchoring Molecules to Substrates

Molecules binding to a surface to modify surface energy or impart function is often traced back to a classic publication by Bigelow, Pickett, and Zisman. These authors found that eicosyl alcohol molecules adsorbed as a thin oleophobic and hydrophobic layer on glass and other oxidized surfaces.[26] Since that time, similar self-assembly or self-organization has been observed to form reproducible and stable single molecule coatings on metal, oxide, and even plastic surfaces. Whether covalently bound or simply adsorbed, these single molecular layers form the interfaces that are used to investigate the binding and properties of molecules bound to a solid surface. These monolayers have been extensively reviewed in the literature,[27,28] so here we will focus primarily on the aspects that pertain to realizing molecular electronics.

15.3.1 Gold

Gold has been the most commonly used substrate in research on molecular electronics. The lack of oxidation and corrosion of gold is a significant advantage for investigations of molecular interactions just as in many other applications. The surface is chemically stable and inert, which improves reproducibility in experiments that are highly sensitive to contamination. STM, which is one of the main tools for single molecule investigations, depends on conduction of small currents and benefits from the high conductivity of gold. Additionally, self-assembled monolayers (SAMs) on gold can tailor the surface energy,[29] and form what are possibly the most complete and most crystalline monolayers.[30–33]

Organic monolayers on gold are part of a promising platform for molecular electronics. A theoretical study about the electrical connection between organic polyaromatic molecules and gold electrodes concluded that, of the functional groups investigated, S, Se, and Te connections to gold carried the lowest impedance.[34] Organic sulfhydryls or thiols, especially those with a "hard" end group opposite the thiol group, form well-ordered monolayers. This is one of the reasons why they are most often used.[35] One simple experiment to investigate conductivity of molecules bound to gold was to bring an STM tip slowly towards a gold substrate coated with a monolayer. The conductivity was measured at each point as the tip approached, contacted, and moved through the monolayer until the tip reached the gold substrate. As shown in Figure 15.4, the current increases as the tip approaches, but the current increases the most as the tunnel current directly to the substrate becomes

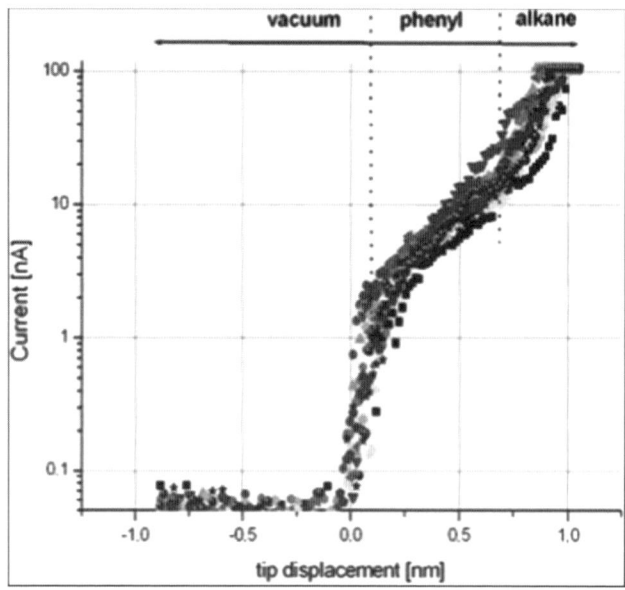

Figure 15.4 Dependence of the tunneling current on the distance. The origin of the x-axis is near the outer edge of the monolayer. Reprinted with permission from B. Lüssem, L. Müller-Meskamp, S. Karthäuser, M. Homberger, U. Simon and R. Waser, *J. Phys. Chem. C*, 2007, **111**, 6392. Copyright 2007 American Chemical Society.

dominant.[36] Other experiments have trapped thiols between electrodes and measured the conductivity.[5,37–39]

Sulfhydryl molecules reacted with gold are stable for experimental devices, but may not bind molecules strongly enough for practical devices. Sulfur–gold binding is used often for molecular machines and not just molecular electronics. This is principally because STM, which requires a conductive surface, is the main tool for observing and manipulating molecular machines. The sulfur–gold bond provides a stable interface at low temperature for molecules to remain bound to the surface even when they rotate and move.[40,41] In contrast, in the presence of a thiol solution at room temperature, bound thiol molecules will slowly desorb and be replaced by other thiol molecules.[42] Or, when heated the molecules will quickly desorb,[43] which shows their instability. Additionally gold atoms themselves tend to migrate.[44,45] This will certainly degrade devices that are dependent on the scale of single molecules.

15.3.2 Oxides

A few different chemical groups will self-assemble in a monolayer on a surface. Indeed, the original Zisman monolayers of alcohols were first observed on surface of glassware.[26] Langmuir-Blodgett monolayers have become popular as well,[46,47] but self-assembled monolayers may be most common at present. A variety of groups including alcohols, carboxylic acids, and alkylsilanes can form monolayers on oxide surfaces, however annealing is required to drive off water and convert the hydrogen bonding to covalent bonding.[48] Alkylsilanes are most commonly used because they will also form crosslinks between the molecules and form more stable monolayers. Alkylchlorosilanes, will react with hydroxyl groups on an oxide surface to form a very stable Si–O–Si linkages. The chlorides can also be replaced with methoxy or ethoxy groups to make the molecules less reactive and easier to work with, but these will still produce the same strong bonding structure.[49] The monolayers that silanes form on oxide surfaces are not as well ordered as monolayers of thiols on gold, but they are still well ordered.[50] One of the principle challenges, however, is that the silane molecules prefer to crosslink rather than bind to the surface. Relatively few molecules actually bind the monolayer to the surface.[51] Another question with conflicting answers in the literature is the effect of amine functional groups in a silane monolayer. One group reported, based on angle-resolved XPS, that the amine group will abstract a hydrogen and electrostatically bind to the substrate in the absence of thermal curing.[52] Other groups used similar characterization methods, except angle-resolved XPS and reflection IR spectroscopy, on similar monolayers, but claimed optimized monolayer formation.[53,54]

Similar chemistry has been used on many surfaces for different applications. Some examples are a silane monolayer on a thin oxide film on a polymer substrate,[55] directly on PDMS,[56] for electropolymerization of aniline on an ITO electrode,[57] or on aluminium metal.[58] Similar monolayers have been used to act as a handle for polymerization of a hydrogel,[59] for lowering the surface free energy to improve deposition of organic semiconductors,[60] as a handle for binding biological molecules,[61] or for reducing friction.[62] However, the direct application of monolayers on oxide surfaces are generally limited because the most common oxides are not conducting. That generally limits electrical conduction to hopping between molecules, which significantly lowers the useful current for molecular electronics.

15.3.3 Silicon

The third principal class of covalent self-assembled monolayers was first introduced by Linford and Chidsey in two papers in 1993 and 1995.[63,64] Initially they introduced monolayers of peroxide molecules bound to a silicon surface through oxygen molecules,[63] and later monolayers bound by a strong Si–C bond produced by a reaction between hydrogen-terminated silicon and 1-alkenes.[64] Since that time the hydrogen-terminated silicon surface has proven reactive to many other functional groups, including alkynes, alcohols, amines, epoxides, and acid chlorides.[65] Many of the same reactions have been demonstrated on other silicon-containing surfaces and on germanium as well.[66,67] Mechanical, electrical, or optical stimulation have catalyzed similar monolayer formation with the same strong Si–C bonds.[68–70] These stable monolayers are the most versatile of the covalent monolayers.

Silicon would be advantageous for molecular electronics based on the compatibility of silicon with traditional semiconductor processes and based on the stability of the covalent Si–C bond. The principal substrate used in commercial electronics is silicon, from individual transistors or logic gates to complex integrated circuits like microprocessors, microcontrollers, and memory. A common substrate for traditional electronics and molecular electronics would bring molecular scale computing one step closer to realization. The covalent Si–C bond that directly anchors carbon-based molecules to a silicon substrate is significantly more stable than Au-thiol bonds.[71] Organic monolayers bound by a Si–C bond to a silicon surface are stable over long periods of time, even in the presence of acid or solvents. Stability is of utmost importance when calculation relies on a single molecule and when small molecular rotations or the interaction of a single atom can modify the output.

The chemical reactions of organic molecules with silicon are significantly more complicated than on either gold or oxides. This has led to a wide variety of proposed pathways and bonding structures for individual mechanisms.[72–77] Organic molecules with a terminal alkene or alkyne group binding to either hydrogen-terminated silicon or bare crystalline silicon are the most common for binding single organic molecules to silicon surfaces. Organic monolayers can be patterned on hydrogen-terminated silicon with atomic precision, when the reaction is initiated by an STM tip as will be discussed later, or when initiated thermally, they can protect the silicon surface in preparation for mechanical patterning. The mechanical patterning method is commonly called scribing, when performed on the macroscale with a sharp diamond-coated implement, or AFM chemomechanical nanografting, when nanoscale patterns are mechanically etched into a passivated surface with an AFM tip. Macroscale patterns completely change the properties of a surface from hydrophilic to hydrophobic; as demonstrated by pinning water drops on a shard of silicon in Figure 15.5. The hydrophobic lines will hold the water on the surface even when it is turned vertical.[68] Nanoscale patterns can disturb less than one layer of silicon atoms, but provide chemical handles for further functionalization.[78]

Particular interest has focused on the chain reaction of alkenes and alkynes with hydrogen-terminated silicon. After a single radical has been formed, a double bond from the unsaturated hydrocarbon groups will react with the radical to form a Si–C bond and leave a carbon-centered radical. The carbon centered radical can then pull a hydrogen from a neighboring silicon to regenerate a silicon radical. On a Si(100)-2×1 surface, the reaction preferentially reacts along the dimer rows creating one-dimensional patterns on the substrate.[79] Patterning with this method will be discussed further later in the chapter.

There are still several questions regarding the reaction of these particular reactants, namely hydrogen-terminated silicon and unsaturated hydrocarbons. One of the most trying is reconciling complete monolayer formation by the radical chain reaction with the low natural occurrence of silicon radicals or reconciling the formation of monolayers with the high barriers to known direct reaction pathways without radical mediation. Trace oxygen has been suggested as a possible catalyst for cleaving Si–H bonds,[74] but even with high concentration of oxygen in air, UV stimulation is required for efficient loss of hydrogen.[67] Other groups have proposed a direct reaction catalyzed by trace

Figure 15.5 Water droplets corralled by monolayer lines of 1-hexadecene on an otherwise hydrophilic surface. The lines were formed by scribing Si(100) in the presence of 1-hexadecene. Reprinted with permission from T. L. Niederhauser, G. Jiang, Y.-Y. Lua, M. J. Dorff, A. T. Woolley, M. C. Asplund, D. A. Berges and M. R. Linford, *Langmuir* 2001, **17**, 5889. Copyright 2001 American Chemical Society.

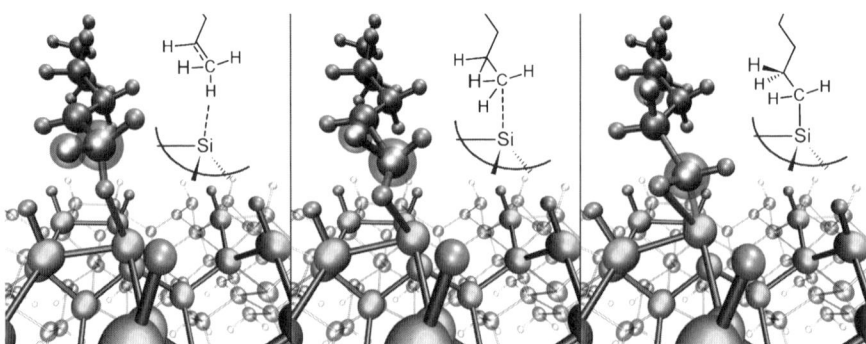

Figure 15.6 Direct reaction of a 1-hexene molecule with hydrogen-terminated Si(100). A hydrogen migrates from the α-carbon to the β-carbon as the α-carbon inserts itself into the surface Si–H bond. From M. V. Lee, R. Scipioni, M. Boero, P. L. Silvestrelli and K. Ariga, *Phys. Chem. Chem. Phys.*, 2011, **13**, 4862–4867, reproduced by permission of the PCCP Owner Societies.

fluorine or excitons in the silicon substrate.[77,75] Either some species catalyzes the formation of silicon radicals on the surface or other mechanisms are also at work.[72] A recent work, suggested a more straightforward possibility, visible light could stimulate the alkenes themselves to abstract hydrogen atoms from the silicon surface to produce radicals. In a surprise twist, they also found that a 1-alkene molecule near the hydrogen-terminated silicon surface can transfer a hydrogen from the α- to the β- carbon and insert the terminal carbon into the Si–H bond with a lower reaction barrier than other known reaction mechanisms. This mechanism is illustrated in Figure 15.6.[72] Alkene abstraction of hydrogen together with α- to β-transfer and direct reaction provide a thermodynamically reasonable explanation for complete monolayer formation on silicon by 1-alkenes. However, it still doesn't answer an open question as to the source of the unexpectedly high double bond content in monolayers formed by thermal reaction in addition to those formed by scribing.[80]

15.4 Surface Patterning

Intentionally forming a design on a surface at the molecular level is the key to fabrication for molecular electronics. Surface probe microscopes, such as the scanning-tunneling microscope (STM) and atomic force microscope (AFM) are able to image surfaces near the atomic scale and, under certain conditions, also alter surfaces at that scale. These are the principle instruments that have been used in attempts to manipulate molecules on an individual basis. Other than these surface probe instruments, we are generally left with only the self-assembly forces of nature to align molecules on a surface. Below, some examples from the literature are presented.

15.4.1 Scanning Tunneling Microscope

The discovery of the STM in 1983 by Binnig and Rohrer was one of the key events that enabled modern nanotechnology.[81] Relatively quickly researchers found straightforward ways to scan surfaces and translate the charge density into images.[82,83] The power of the STM was clear with publication of an image of the <7×7> reconstruction of the Si(111) surface, which settled one of the great scientific questions of the time and produced one of the earliest images of individual atoms on a surface.[84] Although it is not as conductive as metal surfaces, crystalline silicon can still be imaged by STM reasonably well. Silicon surfaces were some of the first to be imaged by STM because they are stable, well-characterized, and technologically important. Many excellent reviews have been written about STM lithography.[85–88]

STM is still one of the few and one of the most well-known analytical instruments that can image and also manipulate individual atoms and molecules. Just as reading and writing are complementary, imaging with an STM quickly led to experiments demonstrating lithography using thin resist layers to create electrical patterns down to 15 nm.[89–91] Alternatively, surface modification of individual surface atoms demonstrates the finest possible patterning.[92] Refinements over the next decade enabled improved control over manipulation and patterning even down to the single atom level.[93–95] Hydrogen atoms can be desorbed from a hydrogen-terminated silicon surface to leave "dangling bonds" or partially filled orbitals that are prominently visible in STM imaging.[96–98] These bonds can then be functionalized by introducing other atoms, such as the common silicon dopants nitrogen or phosphorous, into the vacuum chamber.

Another approach is the formation of oxidative patterns on silicon and GaAs. By combining characterization of the patterns by STM with that by scanning electron microscopy (SEM), time-of-flight secondary ion mass spectrometry (TOF-SIMS), reflection high energy electron diffraction (RHEED), and X-ray photoelectron spectroscopy (XPS), researchers demonstrated that the oxide patterns are stable during the high temperature conditions typical of semiconductor processing. By using oxide patterns as an etch resist, patterns could be transferred into the underlying substrate. Features are limited to around 10 nm laterally by this method and depend on crystalline surfaces, or at least those with subnanometre roughness.[99–101] Metallic lines down to 35 nm wide can be patterned by decomposing inorganic precursors by current from the STM tip.[102]

STM provides not only the most powerful method for surface alignment of atoms, but also for alignment of individual molecules. One notable example of this takes advantage of a chain reaction that occurs when alkenes react with a silicon radical on a hydrogenated Si(100) surface, as shown in Figure 15.7.[79] Reaction of one alkene prepares the next site for reaction so only one radical must be created by STM to form a row of covalently bound molecules. This chain reaction allows long rows of styrene molecules to react in a manner such that the aromatic rings are aligned to create a molecular wire. This represents an example of alignment not just of one molecule, but of a whole ensemble.

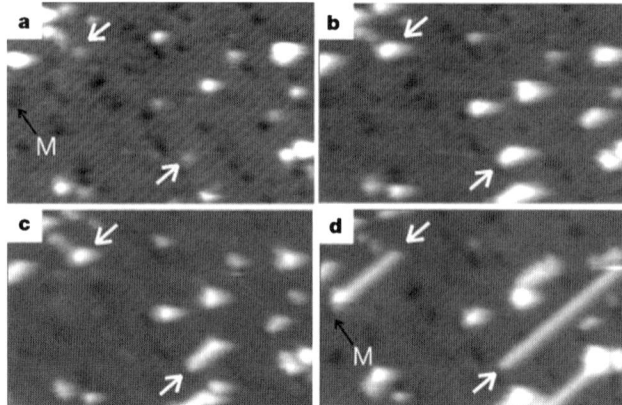

Figure 15.7 Sequence of images showing growth of styrene lines on a H-terminated Si(100). (a–d) correspond to an increasing exposure to styrene: a, 3 L; b, 28 L; c, 50 L; and d, 105 L. The white arrows mark the position of two particular dangling-bond sites that lead to the growth of long styrene lines. The missing dimer defect (M) terminates the growth of the line in the top left-hand corner of the image. Reprinted by permission from Macmillan Publishers Ltd: *Nature*, 2000, **406**, 48–51, copyright 2000.

Patterning by STM has some important advantages over other patterning methods for laboratory experiments, but likely has insurmountable challenges with regard to patterning for molecular electronics. E-beam lithography has been producing linewidths of ≤10 nm for more than a decade and other related methods can achieve these scales as well, but this is still about an order of magnitude greater than the molecular scale. But possibly the greatest advantage of STM over e-beam lithography, X-ray lithography, or ion beam lithography is the lower cost of an STM system.[102] Additionally, during patterning, the STM uses voltages or energies near 4eV; although this is in the range of field emission rather than tunneling, it is still low enough that it won't cause deep or significant damage to the substrate. On the contrary, other methods use voltages or energies significantly higher that reach deep into the substrate. Although the STM is versatile and precise, the serial nature of the STM still poses the greatest challenge. Other significant problems are the requirement of a conducting substrate and thermal drift. One challenge that is particular to the STM is the desorption, migration, or other atomic scale changes in the tip that can distort imaging or patterning during the scan; atomic rearrangements of the atoms of the STM tip can confound images of a surface.[87]

15.4.2 AFM

The atomic force microscope has lower resolution than STM, but it is generally less temperamental and more robust, as well as having a lower cost for upkeep, use, and maintenance. For these reasons commercial AFMs are more widely available and more widely used than STM instruments even though the AFM was invented later.[103] Rather quickly after its invention, the AFM was already being used for patterning, especially by mechanically or electrically modifying a surface, or even combining these forces.[104] As discussed above for STM, surface probe microscopies are inherently serial processes; thus the time requirement for patterning increases by the second order as dimensions are decreased. Immediately researchers began grappling with this huge barrier to practical use of scanning probes for patterning. Rather quickly researchers began reaching for parallelism at the nanoscale for SPM instruments. Within ten years of the AFM instrument's introduction, multi-tip lithography was already underway[100] and now even thousands of tips can work in parallel.[105] Thousands of tips are practical for patterning features on the scale of tens of nanometers and may be useful for fabrication of portions of patterning for practical molecular electronics, but still requires assistance in aligning single molecules for active components.

15.4.2.1 Electrical Patterning

Even within a few years of its development, groups were using the AFM to pattern at the nanoscale. Some of the earliest patterning experiments used electrical stimulation through the AFM tip. Electrical stimulation for patterning is a logical choice for use with an atomic force microscope. The AFM operation uses light reflected off of the back of the cantilever to determine height and the head that holds the cantilever is moved laterally by piezoelectric crystals. Thus, an electrical current can be decoupled and operated independently from the regular AFM operation. For patterning, as well as other analytical analyses, this decoupling gives AFM significant advantage over STM.

The earliest experiments exposed a resist on a surface, oxidized a bare silicon surface, or modified monolayers to achieve features of tens of nanometres.[104,106,91] In more recent advances useful for molecular electronics, the Buriak group demonstrated chemical control in addition to patterning. Initial experiments with porous silicon showed that with porous silicon acting as an anode while submerged in a liquid 1-alkyne, the alkyne molecules bonded as alkyl chains, losing their triple bonds completely. However, when the surface was biased positively, the alkynes bound without losing their triple bonds.[69] Later they added more molecules to the repertoire and transferred the process to the nanoscale with AFM patterning of Si(111)-H surfaces. The method is illustrated in Figure 15.8.[107] Their features were still in the tens of nanometres, but this important work demonstrated that chemical control can be maintained at the nanoscale.[108–111]

A seminal work by the Sagiv group demonstrated an alternate approach by oxidizing organic silane monolayers and building up multilayer structures.[112] Because silicon, rather than a silicon dioxide, was used as the substrate, a voltage could be applied directly across insulating octadecyl trichlorosilane

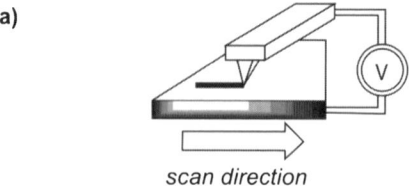

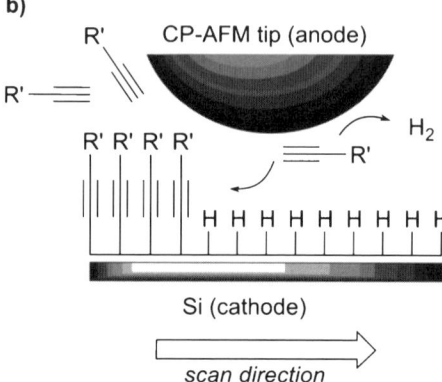

Figure 15.8 Schematic of the method for chemically functionalizing a surface by cathodic alkyne electrografting on a hydrogen-terminated silicon surface. (a) A potential is applied between a tip and the surface. (b) Between the tip and the surface, the alkynes react with Si–H bonds to form alkynyl moieties and evolve hydrogen molecules. Reprinted with permission from P. T. Hurley, A. E. R. Ribbe and J. M. Buriak, *J. Am. Chem. Soc.*, 2003, **125**, 11334. Copyright 2003 American Chemical Society.

(OTS) monolayers to oxidize the terminal group.[113] The oxidized terminus will then react with a subsequent layer of silane molecules, which in turn can be oxidized by applying a potential again (Figure 15.9). This method introduces the ability to create both insulating regions and conductive regions on top. One could imagine a platform being constructed on which a molecular circuit could be fabricated; the OTS foundation could likely be insulating enough to isolate the molecule for low enough potentials.

This sequence of reactions when using an AFM tip is one of the few truly bottom-up strategies with nanoscale control, but it has some significant obstacles. In addition to the problems associated with serial surface probe methods, like processing time and drift, electrical modification requires an electrically conductive substrate and an insulating molecule. A conductive substrate may cause problems for electrical devices. Also, the resistance will grow with each additional layer, which changes the conditions at each additional level and requires higher voltages. Since each octadecyl layer is insulating, the electrochemical oxidation is limited to only a few layers. Additionally, because the pattern resolution is dependent on conduction from

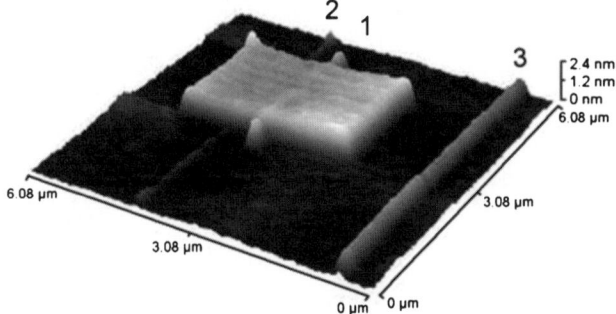

Figure 15.9 Monolayer–bilayer patterns produced by a two step process. (a–d) pairs of topographic (left side) and corresponding friction (right side) contact-mode AFM images of three lines and a rectangle drawn sequentially with a conductive diamond tip. Reprinted with permission from R. Maoz, S. R. Cohen and J. Sagiv, *Adv. Mater.*, 1999, **11**, 55. Copyright 1999 John Wiley and Sons.

the tip rather than being constrained by self-assembly, resolution will degrade with each subsequent layer.

15.4.2.2 Nanografting/Nanoshaving/Mechanical Patterning

An important topic for our purpose of examining patterning and alignment to a substrate requires not just introduction of patterns, but also functionalization of those patterns. A notable example of this is nanografting of monolayers. The molecules of one monolayer can be removed by an AFM tip and the shaved region can then be functionalized by another monolayer-forming molecule. The classic example is Xu and Liu's nanografting thiols on gold (Figure 15.10). C10-thiol molecules can be removed from a monolayer in a 10 nm-wide line and by submersion in a longer chain C18-thiol solution, a raised line forms. Monolayer patterns can be reproducibly written and erased by changing the thiol solution.[114] If a thiol with a functional end group, *e.g.* alcohol, carboxyl, or amine, is grafted in the patterned regions, those functional groups become available for further chemical derivatization.

Formation and patterning of thiol monolayers on gold is simplified by the soft–soft Au–S interaction. This bond is rather weak, as discussed previously, which allows molecules to be easily bound, removed, or dynamically replaced. This allows highly-ordered monolayers to form in a manner that is largely independent of the atomic structure of the gold surface. The weakness of the bond, however, carries the obvious challenge of instability.

The surfaces of silicon and also silicon dioxide, which are well-characterized and common in semiconductor patterning are alternatives. Alkenes have been grafted to silicon surfaces. After hydrogen-terminating a silicon surface and covering it with 1-alkene, an AFM tip was scanned along the surface with an increased contact force to displace hydrogen atoms. The resultant silicon

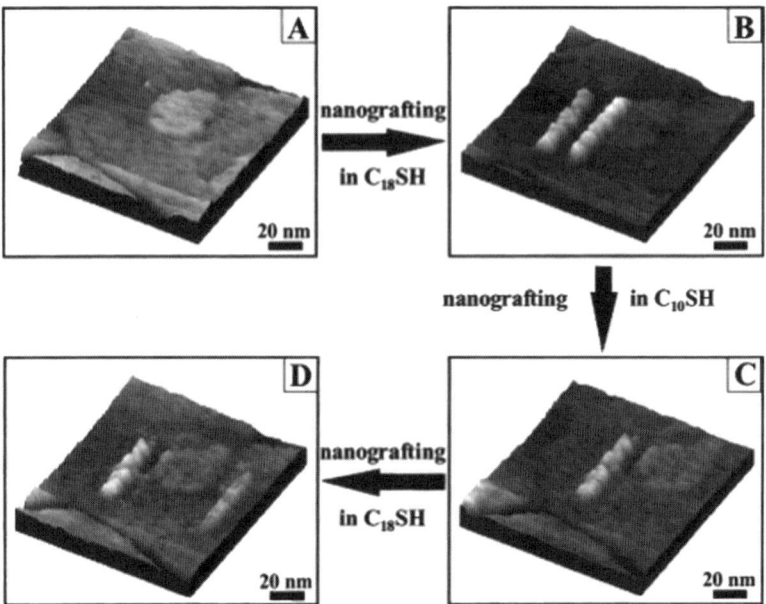

Figure 15.10 *In situ* modification of grafted nanostructures. (a) AFM image of the matrix $C_{10}S$ SAM before fabrication. The bright area that is 30 nm in diameter and 0.25 nm higher than the rest of the surface is due to a single atomic layer of Au(111) covered by the $C_{10}S$ SAM. (b) After fabrication of two parallel $C_{18}S$ nanolines with dimensions of 10 × 50 nm² and a pitch of 20 nm. (c) Erasure of the right line by scanning its area under a high imaging force in a $C_{10}SH$ solution. (d) Refabrication of the second line by scanning under a high imaging force in $C_{18}SH$ solution. The interline spacing was increased to 65 nm. Reprinted with permission from S. Xu, S. Miller, P. E. Laibinis and G.-Y. Liu, *Langmuir*, 1999, **15**, 7244. Copyright 1999 American Chemical Society.

radicals reacted with alkene molecules to form patterns of alkyl groups. The authors drew lines at different forces and observed lines as narrow as 30 nm, although the measured surface roughness under the liquid was too great to resolve anything narrower.[115] But hydrogen-terminated silicon is not very stable, especially in the presence of a reactive molecule, and allows molecules to react in undesired areas. If instead, hydrogen-terminated silicon is reacted with an alkene in advance, this alkyl layer can still be shaved with a higher force, but provides better protection against non-specific binding. Observed linewidths were greater than observed on hydrogen-terminated surfaces, but that is likely due to the higher forces required to break the Si–C bond.[78]

For nanografting on a truly insulating surface, silicon dioxide is preferred. Silane chemistry is the most likely candidate, but the Si–O bond is significantly stronger than the SiC bond.[71] Indeed, monolayers of a trifunctional silane on silicon dioxide were not able to be shaved from a silicon dioxide surface, likely because of the silane network and the combined strength of three Si–O

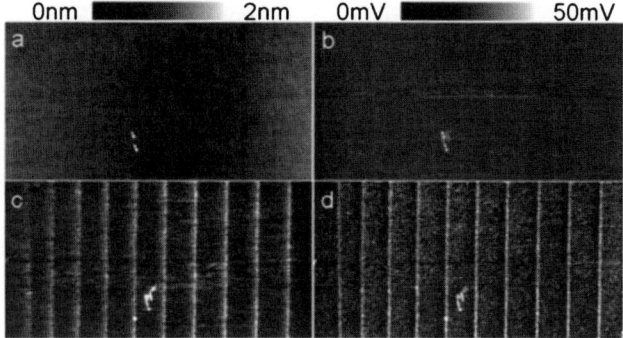

Figure 15.11 Lines of amine-terminated silane nanografted through a C_8-monochlorosilane monolayer on silicon dioxide. Panels a and c are AFM topography images; panels b and d are lateral force images. Each panel shows a 10 μm × 5 μm region. Panels a and b were obtained *in situ*, immediately after patterning and show a definite chemical change in panel b. Panels c and d were imaged in air subsequent to copper plating. After plating the lines are raised. Reprinted with permission from M. V. Lee, K. A. Nelson, L. Hutchins, H. A. Becerril, S. T. Cosby, J. C. Blood, D. R. Wheeler, R. C. Davis, A. T. Woolley, J. N. Harb and M. R. Linford, *Chem. Mater.*, 2007, **19**, 5052–5054. Copyright 2007 American Chemical Society.

linkages.[116] Nevertheless, although it required higher forces, a monolayer of monofunctional silane on silicon dioxide could be successfully nanoshaved, and an amine-terminated monolayer was nanografted in the pattern. Amine-terminated lines down to 50 nm wide were used to bind DNA molecules from an aqueous solution, or alternatively as a template for electroless plating of nanoscale metal lines (Figure 15.11).[117]

15.4.2.3 Dip-pen Nanolithography

Another method for SPM nanoscale patterning is dip-pen nanolithography (DPN).[118] DPN uses an AFM tip as a pen by dipping the tip in a chemical "ink" and then bringing the tip in contact with a surface where one would choose to write a pattern, which transfers the "ink" to the surface. DPN is versatile in the inks that are available and scalability of pattern creation. Small molecules are applied as a molecular ink, commonly including thiols and silanes, and transferred to the surface through dissolution in the water meniscus that forms naturally between the tip and surface.[119–121] Larger conductive molecules can also be used for DPN at higher loadings by directly flowing them to the surface, but feature resolution degrades. The multiple scales are shown in Figure 15.12.[122] Silane inks on oxide and thiol inks on gold will form covalent bonds, but generally the inks will physisorb on the surface and can also be erased.[123] Many excellent reviews have been published on the procedures, capability, and chemistry for successfully applying DPN to a virtually any

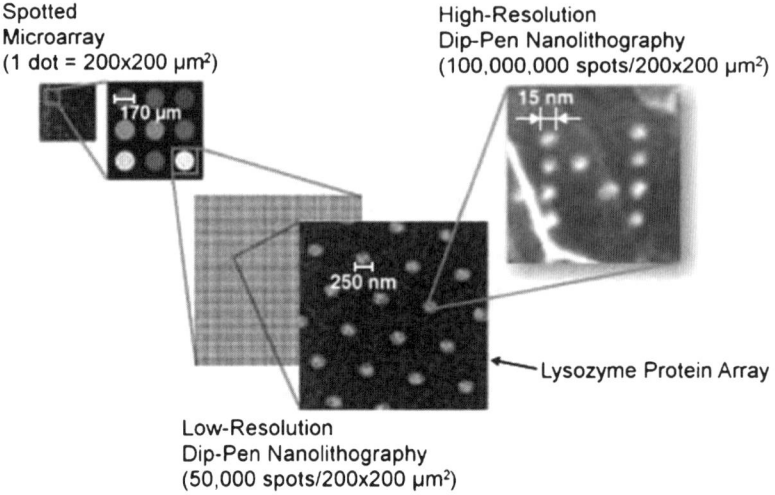

Figure 15.12 Schematic illustration of the range of scales possible by dip-pen nanolithography. Reprinted with permission D. S. Ginger, H. Zhang and C. A. Mirkin, *Angew. Chem. Int. Ed.*, 2004, **43**, 30–45. Copyright 2004 John Wiley and Sons.

surface for a variety of applications.[122] Here we will focus on the merits and challenges with regard to interconnects for molecular electronic devices.

DPN provides multiple advantages over the electrical methods and nanografting methods discussed earlier, but still has disadvantages. A wider variety of molecules are available to DPN; multiple inks can even be used on the same surface.[124] Multiple inks gives significant advantage over other methods. Additionally, the surface doesn't require a protective layer to prevent non-specific adsorption like nanografting, because DPN doesn't require the entire surface to be coated with the desired patterning agent. Instead the tip is coated with the desired ink and the ink is only transferred to the substrate where the tip touches the surface. Thus, DPN can use significantly less reagent. This is a great advantage in patterning for molecular electronics, since the molecules that will be interesting to molecular electronics will likely be expensive designer molecules. Additionally, this simplifies the process and also avoids much of the sample clean-up and reduces the total processing time. Still, however, the smallest features possible by DPN are on the order of 5–10 nanometres,[125,126] which is about an order of magnitude larger than single molecule manipulation. DPN would require assistance by self-assembly or another process that can provide finer alignment.

15.4.3 Vacuum Deposited Molecular Arrays

Adsorbed monolayer patterns may also provide a path to placement and alignment of individual molecules for molecular electronics. Atoms deposited on a clean metal surface that don't covalently bind will migrate across the surface and aggregate or align on surface features to form one or two dimensional patterns.[127] In similar fashion molecules can also physisorb onto very clean crystalline surfaces in well-defined monolayer patterns.[128–130] These patterns are inherently aligned to the crystallinity of the underlying substrate. In two dimensions, the arrangements are very regular and can often be represented by Penrose patterns. When stimulated or constrained, the molecules will change their conformation to fit the surface, often forming a

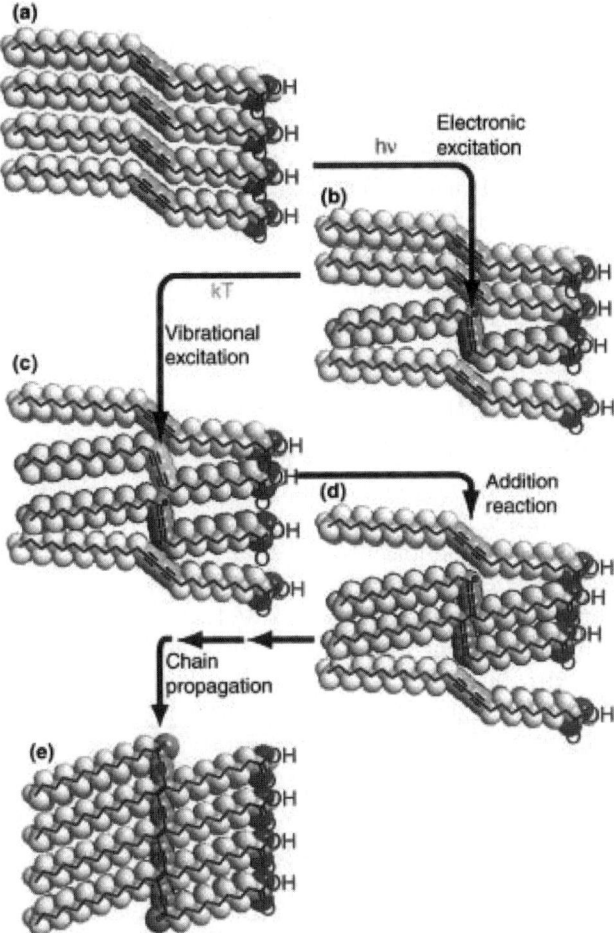

Figure 15.13 Illustration of the mechanism of chain polymerization on a surface. (a) Array of the monomer molecules. (b) Diradical formation by optical excitation. (c) Approach of a neighboring diacetylene moiety by vibrational excitation. (d) Radical dimer formation by an addition reaction. (e) Extended polymer formation by chain propagation reaction. Reprinted with permission from Y. Okawa and M. Aono, *J. Chem. Phys.*, 2001, **115**, 2317. Copyright 2001, American Institute of Physics.

different, but compatible tiling pattern.[131] If multiple components are adsorbed together in proportion, they can form multiple component patterns to accommodate the different shapes of the molecules.[132] If there are defects or surface features other than a perfect crystalline surface, molecules, *e.g.*, fullerenes, will align relative to the imperfect structure.[133] Because these patterns are regular over nanometre to tens of nanometre areas, they may provide an alternate path to patterning the desired molecule on a surface in the correct place.

One of the advantages of molecular arrays on a surface is that simple stimuli can cause a controllable reaction that rearranges the surface or changes its properties. One example of this is linear chain polymerization of 10,12-pentacosadiynoic acid self assembled on a graphite surface. The molecules arrange themselves into arrays of rows that can be 100s of nanometers long. When initiated by an STM tip, the acetyl subunits in the center of each molecule begin to polymerize. The molecules form a backbone structure similar to polyacetylene, which should be highly conductive (Figure 15.13).[134] Although the molecules are not bound to the surface by covalent bonds, the relative placement of molecules is precise to the atomic scale. Such precision is required for practical molecular electronics.

15.5 Scale Registration

Possibly the greatest challenge in nanoscience is reliably controlling the position and motion of molecules and atoms relative to a fixed point. In microelectronics, this idea is described as registration. Registration is required to fabricate an interface to molecular or atomic scale structures and is the key to making useful devices. Registration can be accomplished in many ways and some methods that are found in the literature are discussed below. These methods for registration include: direct positioning of molecules, designing self-assembling structures that accept and position the molecule of interest with respect to prefabricated structures, polymerization of molecules between electrodes that act as the interface, or alignment by external stimuli to an existing interface.

15.5.1 Direct Patterning

Direct patterning of both electrodes and active molecules has been attempted with limited success. One recent work prepared a phosphorus nanowire between electrodes on germanium with STM patterning as shown in Figure 15.14.[135] Current between the surface and an STM tip desorbed hydrogen atoms from a hydrogen-terminated germanium surface. Hydrogen atoms were removed to form the shape of electrodes and a fine wire between them. Dosing with phosphine gas provided phosphine molecules to bind to the bare germanium sites. This basic process has been practiced since hydrogen was first desorbed by an STM tip.[96,136] Single hydrogen desorption and

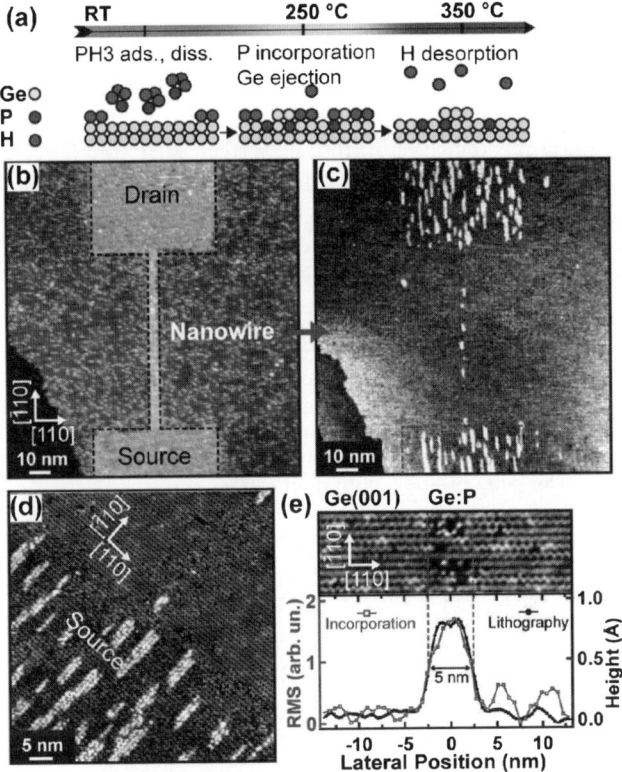

Figure 15.14 Preparing a phosphorus nanowire between electrodes on germanium. (a) Schematic of the process. Filled-state STM images of a 5 nm wide nanowire (b) before and (c) after P incorporation and H removal. (d) Close-up of the source contact of the nanowire. (e) Filled-state STM image of a section of the 5 nm wide nanowire after P incorporation plus line scans before and after annealing. The lines match which shows that lateral diffusion is low. Reprinted with permission from G. Scappucci, G. Capellini, B. Johnston, W. M. Klesse, J. A. Miwa, and M. Y. Simmons, *Nano Lett.*, 2011, **11**, 2272–2279. Copyright 2011 American Chemical Society.

replacement are a simple method to place single molecules on a surface in a repeatable and reproducible manner.

15.5.2 Self-assembly

The self-assembling power of DNA has been used to form the greatest complexity with the most controllability. DNA can be bound to organic molecules, which could be used to align an active molecule between electrodes.[137] But possibly the most impressive molecular design has been demonstrated with DNA building blocks in the form of DNA Origami. By

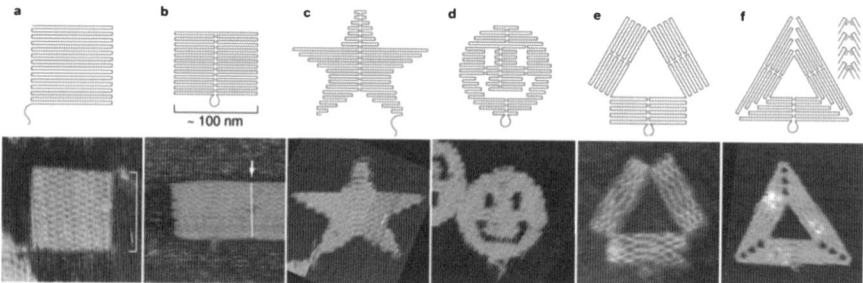

Figure 15.15 DNA origami shapes. Top row, folding paths. a, square; b, rectangle; c, star; d, disk with three holes; e, triangle with rectangular domains; f, sharp triangle with trapezoidal domains and bridges between them. Dangling curves and loops represent unfolded sequence. Bottom row, AFM images. All images and panels are the same size, 165nm × 165 nm. Adapted by permission from Macmillan Publishers Ltd: *Nature*, 2006, **440**, 297–302, copyright (2006).

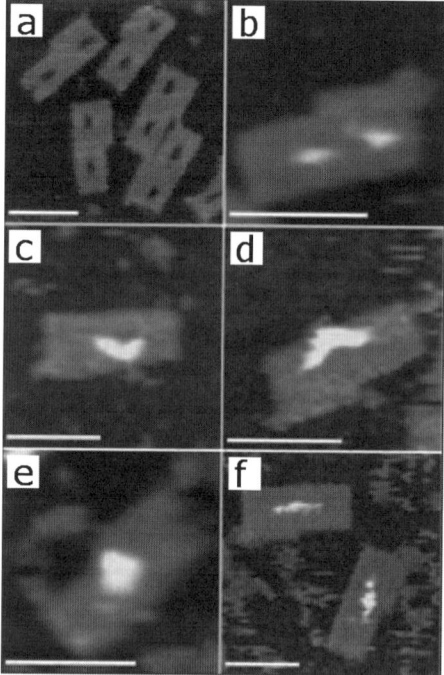

Figure 15.16 Origami arrays and capture molecules. Atomic force micrographs. (a) DNA origami structures lying on the surface with two empty regions to capture molecules. (b) two separate molecules trapped in the empty regions. (c–f) the two recognition molecules bridged by a down-pointing triangle, up pointing triangle, diamond, and straight-bridging molecule, respectively. The scale bars are all 100 nm. Reprinted by permission from Macmillan Publishers Ltd: *Nat. Nano.*, 2009, **4**, 245–248, copyright (2009).

designing complementary regions into strands of DNA, the folding of that strand can be directed such that complex shapes can be produced (Figure 15.15).[138] These shapes can be used to capture and direct the placement of a particle or an active molecule (Figure 15.16).[139] With complementary single strands of DNA on the Origami shape and on fixed points on a surface, the DNA Origami shapes can be aligned between patterned anchors, thus registering the molecular scale to the scale of cutting-edge patterning, as shown in Figure 15.17.[140] In Figure 15.17, the anchor points were prepared by a non-lithographic means, but surface probe lithography could also prepare these structures in a defined pattern.

Monolayer self-assembly can also form complex patterns that are regular over large areas and may be very useful for registration, although generally the molecules don't form stable bonds to the surface. One group found that in the

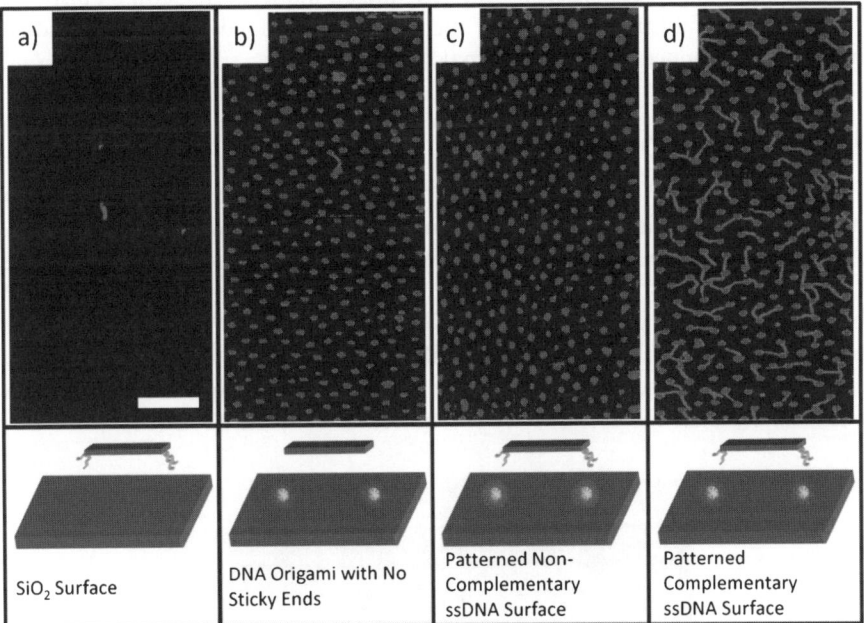

Figure 15.17 AFM images of control experiments were used to probe the mechanism of DNA origami attachment. (a) Sticky end modified DNA origami were placed on a clean SiO_2 surface. (b) Nonmodified DNA origami rectangles were placed on a patterned ssDNA surface. (c) Sticky end modified DNA origami were placed on a patterned noncomplementary ssDNA surface. (d) Modified DNA origami were placed a patterned complementary ssDNA surface. All reaction times were 4 h. The scale bar is 200 nm and applies to all images. The AFM height scale is 6 nm. Reprinted with permission from A. C. Pearson, E. Pound, A. T. Woolley, M. R. Linford, J. N. Harb and R. C. Davis, *Nano Lett.*, 2011, **11**, 1981–1987. Copyright 2011 American Chemical Society.

regular patterns that form in deposition of molecules at the graphite–liquid interface, the electrical characteristics vary according to the region of molecule being tested. They suggested that if this is the case, the phase separation that forms the regular monolayers may be an effective method for integrating such electrical function into nanoscale devices.[141] They then used this idea to demonstrate the field-effect in single molecules aligned on HOPG.[142] Such an arrangement would require vertical rather than lateral fabrication of devices, which is less advanced in microscale semiconductor fabrication, although research is intensifying. The alignment of molecules that is observed on the conducting surfaces also likely occurs on non-conducting surfaces. If molecules can be fixed in position after alignment by self-assembly, this could also be useful for lateral positioning of active molecules. Additionally, the differing electrical character at different points on the molecule is consistent with the quantum state of a molecule being stimulated and observing a response like a Hamiltonian operator. Self-assembled mixed-monolayers could also be potentially useful for registration if the position of individual molecules could be fixed and connected to an external interface.

15.5.3 Polymerization

Polymerization and other chain reactions have been used to bind and align molecules on a surface. This achieves much of what is needed for registration. An interesting example of this was demonstrated by alignment of polyaniline between electrodes. As shown in Figure 15.18, aniline was polymerized between electrodes to form a single channel or active element.[143] While the scale of the polyaniline molecule is still large compared to what is typically considered for molecular electronics, it illustrates well the principle of using a chemical reaction to reproducibly fix molecules between electrodes.

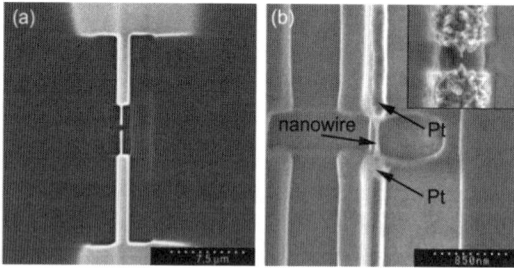

Figure 15.18 Polymerization of aniline between nanoscale electrodes. (a) FIB image of platinum deposited nanoelectrodes after fabrication. (b) SEM image of the same device after electrochemical deposition of a single nanowire between the nanoelectrodes. The inset, an SEM image of a device with incomplete nanowire growth, shows the two-nanowire approach to bridging the gap. Reprinted with permission from N. T. Kemp, D. McGrouther, J. W. Cochrane and R. Newbury, *Adv. Mater.*, 2007, **19**, 2634–2638. Copyright 2007 John Wiley and Sons.

This same principle can be used at the molecular scale. As others have demonstrated,[134] if diacetylene molecules form a monolayer on a surface, then stimulation by STM can initiate a radical-mediated polymerization. The reaction forms a polyacetylene backbone in an atomic scale straight line.[144] Since polyacetylene is conductive, the polymer chain could allow fabrication of a single molecular, conductive electronic channel between electrodes. If a single active molecule were placed in the center of the chain, it would then be aligned between the electrodes in a way that could be stimulated and observed.

Along the same line of reasoning molecular scale polymerization, multiple groups have applied the chain reaction of alkenes on hydrogen terminated

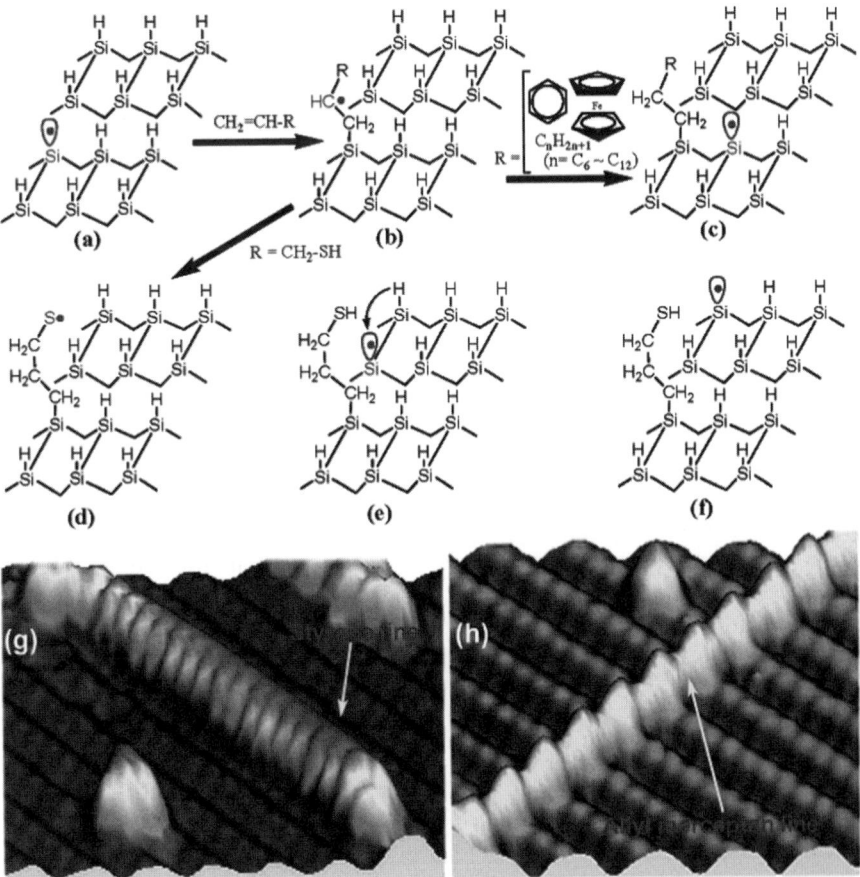

Figure 15.19 Chain polymerization on a Si(100) surface in orthogonal directions. (a–f) Schematic of the mechanism of radical chain reactions along and across the dimer rows on the H-terminated Si(100)−(2 × 1) surface. (g,h) STM images of molecular lines running along (styrene) and across (allyl mercaptan) the dimer rows. Reprinted with permission from M.Z. Hossain, H.S. Kato and M. Kawai, *J. Phys. Chem. B*, 2005, **109**, 23129. Copyright 2005 American Chemical Society.

Si(100) to create lines of styrenyl organic groups covalently bound in lines along the dimer rows on the silicon surface using either phenylethylene or phenylacetylene.[64,79,145] Simultaneous creation of lines in the perpendicular direction were demonstrated later with a different molecule.[146] The reaction parallel to the rows is kinetically controlled, which restricts the alkene molecules that can be used to form lines at a certain temperature; small molecule alkenes can only form stable line patterns significantly below room temperature.[147] The interaction between the mechanisms in the perpendicular directions allowed both directions to be combined on a single surface—even as a continuation of the chain reaction of the same radical (Figure 15.19).[148-150] Multi-directional control of patterning on the atomic scale is truly impressive.

15.5.4 Alignment to Prefabricated Electrodes

Many groups have worked to create nanoscale gaps which can trap a single molecule. Simply casting molecules onto a surface, *e.g.*, regioregular poly-3-hexylthiophene on electrode patterns and measuring the response can be effective for laboratory experiments, but it is not useful for practical devices.[151] One approach that has yielded good success is dielectrophoresis to align molecules between electrodes. This has been used primarily to align carbon nanotubes between electrodes; the method works better for metallic than semiconducting nanotubes.[152,153]

For molecules smaller than carbon nanotubes, the first significant challenge is just preparing an electrode gap of the right size. This can be accomplished by adjusting conditions for e-beam lithography and fine-tuning resist development,[154] electromigration,[155] mechanical distortion of the substrate,[38] or fabrication of electrodes by focused-ion beam.[156,157] Although the actual size of the gap cannot be accurately measured, functional molecules can bind in place and the electrical characteristics of individual molecules can be measured.[37] At single molecule scales, van der Waals forces can also be great enough to hold larger molecules in place, for instance symmetrical C_{60} molecules. C_{60} molecules that are trapped between electrodes have produced some interesting phenomena. For instance a nanomechanical oscillation response[158] or the Kondo effect at 50 K and coherence between the unpaired spin on the molecule the electrons at the electrodes.[159]

15.6 Concluding Remarks

Theories for molecular and atomic scale applications, including quantum electronics, have advanced about as far as possible without testing through experimental realization. The chemistry of attachment is also quite advanced. Realization of molecular electronics is still hindered by the difficulty forming controlled designs at resolutions below 10 nm. Since the solubility of conductive molecules is limited to about 2 nm, even AFM patterning is about an order of magnitude away from single molecules. Only direct manipulation

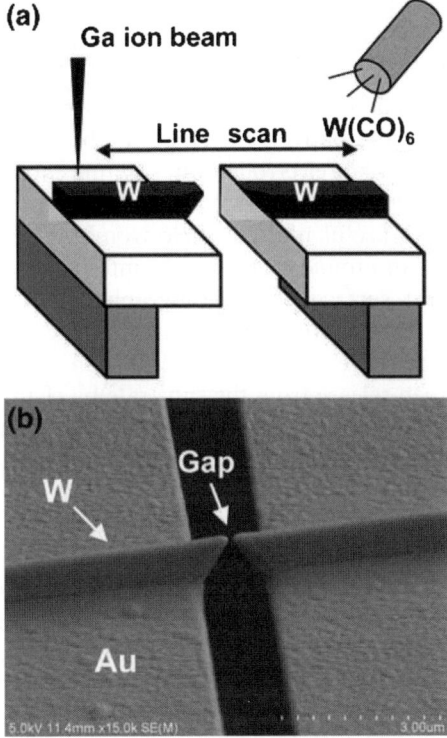

Figure 15.20 Fabrication of precise single nanometre electrode gap. (a) Schematic illustration of the gap fabrication process. Tungsten is deposited through the decomposition of a tungsten hexacarbonyl vapor onto the substrate where the Ga ion beam is irradiated. The tungsten is deposited not only on the substrate but also on the edge of the slit. Consequently, the tungsten electrodes grow from both sides of the slit. (b) SEM image obtained for a sample during the fabrication process. Reprinted from K. Shigeto, M. Kawamura, A. Y. Kasumov, K. Tsukagoshi, K. Kono and Y. Aoyagi, Reproducible formation of nanoscale-gap electrodes for single-molecule measurements by combination of FIB deposition and tunneling current detection, Microelectronic Engineering, 2006, **83**, 1471–1473, copyright (2006), with permission from Elsevier.

by STM or self-assembly processes have been able to reach to the scale of single molecules and atoms. These limits are not likely to be achieved by other methods directly; even if photolithography can continue to beat the diffraction limit and even if e-beam lithography develops further, these methods will eventually need to manipulate few individual molecules, which is completely different than working with a uniform film. Thus, creative ideas in nanoscale fabrication are certainly required.

Currently the most promising method of alignment by self-assembly that has been found in the literature is templating with DNA Origami structures. These structures can be aligned to anchor points on a substrate. The DNA structures can align single molecules relative to the fixed supports and hold active molecules in the desired position. Unfortunately, DNA is not inherently conductive, so additional conductive materials will need to by aligned to or plated on the DNA. Or alternatively the DNA can be used to fix the molecules in position between nanoscale electrodes, but still fine enough conducting leads are required to stimulate and observe states for quantum calculations, which will likely be the most useful molecular electronics. One likely strategy for successful creation of an atomic or molecular interface is to create nanoscale anchors by a covalent surface probe lithography method on an insulating substrate, and allow DNA structures to align to the pattern. If the DNA structures carry a conductive molecule that can span the gap, hold it in place while it reacts, and then release from the substrate by changing pH or heating, the self-assembling power could be used as a shuttle to align the molecules where other methods cannot. This could work as a parallel process and allow for unique molecules to be placed.

Other options are suggested by the recent introduction of graphene synthesis and processing.[160] Aromatic molecules even on the scale of metres can be processed. Since the aromatic carbon that graphene comprises is the principle building block of electronically active molecules, perhaps instead of aligning molecules to electrodes or electrodes to molecules, the active units for molecular electronics could be carved out of graphene monolayers that have been placed in position. This provides a simple solution to alignment. Carving, functionalizing or derivatizing the graphene by STM lithography and then patterning electrodes over graphene may provide a simple alternative to the alignment issue that has plagued molecular electronics.

Other strategies that include direct patterning by SPL methods as well as self-assembly could also be successful. As self-assembly becomes more controllable, it will complement the capabilities of surface probes, especially that of the atomic force microscope. The combination of research in these areas is critical to successful domestication of atoms and molecules for electronics, as well as mechanical, photonic, and thermal applications on this same scale.

References

1. L. A. Bumm, J. J. Arnold, M. T. Cygan, T. D. Dunbar, T. P. Burgin, L. Jones, D. L. Allara, J. M. Tour and P. S. Weiss, *Science*, 1996, **271**, 1705–1707.
2. A. Aviram and M. A. Ratner, *Chem. Phys. Lett.*, 1974, **29**, 277–283.
3. A. S. Martin, J. R. Sambles and G. J. Ashwell, *Phys. Rev. Lett.*, 1993, **70**, 218.
4. D. Q. Andrews, G. C. Solomon, R. P. Van Duyne and M. A. Ratner, *J. Am. Chem. Soc.*, 2008, **130**, 17309–17319.
5. S. M. Lindsay, *Jpn. J. Appl. Phys.*, 2002, **41**, 4867–4870.
6. K. Müllen and J. P. Rabe, *Acc. Chem. Res.*, 2008, **41**, 511–520.

7. A. P. H. J. Schenning and E. W. Meijer, *Chem. Commun.*, 2005, 3245.
8. Y. Xue and M. A. Ratner, *Int. J. Quantum Chem.*, 2005, **102**, 911–924.
9. C. Joachim and M. A. Ratner, *Proc. Natl. Acad. Sci. U. S. A.*, 2005, **102**, 8801–8808.
10. P. W. Shor, *SIAM Journal on Computing*, 1997, **26**, 1484.
11. L. M. K. Vandersypen, M. Steffen, G. Breyta, C. S. Yannoni, R. Cleve and I. L. Chuang, *Phys. Rev. Lett.*, 2000, **85**, 5452–5455.
12. L. M. K. Vandersypen, M. Steffen, G. Breyta, C. S. Yannoni, M. H. Sherwood and I. L. Chuang, *Nature*, 2001, **414**, 883–887.
13. T. D. Ladd, D. Maryenko, Y. Yamamoto, E. Abe and K. M. Itoh, *Phys. Rev. B*, 2005, **71**, 014401.
14. M. E. Hawley, G. W. Brown, M. Y. Simmons and R. G. Clark, *Los Alamos Sci.*. 2002, **72**, 302–315.
15. G. Feher, *Phys. Rev.*, 1956, **103**, 834–835.
16. A. M. Tyryshkin, J. J. L. Morton, S. C. Benjamin, A. Ardavan, G. A. D. Briggs, J. W. Ager and S. A. Lyon, *J. Phys.: Condens. Matter*, 2006, **18**, S783–S794.
17. L. Viola and S. Lloyd, *Phys. Rev. A*, 1998, **58**, 2733–2744.
18. K. Khodjasteh and D. A. Lidar, *Phys. Rev. Lett.*, 2005, **95**, 180501.
19. J. J. L. Morton, A. M. Tyryshkin, A. Ardavan, S. C. Benjamin, K. Porfyrakis, S. A. Lyon and G. A. D. Briggs, *Nat. Phys.*, 2006, **2**, 40–43.
20. S. Simmons, R. M. Brown, H. Riemann, N. V. Abrosimov, P. Becker, H.-J. Pohl, M. L. W. Thewalt, K. M. Itoh and J. J. L. Morton, *Nature*, 2011, **470**, 69–72.
21. N. Renaud and C. Joachim, *J. Phys. A: Math. Theor.*, 2011, **44**, 155302.
22. N. Renaud, M. Ito, W. Shangguan, M. Saeys, M. Hliwa and C. Joachim, *Chem. Phys. Lett.*, 2009, **472**, 74–79.
23. W.-H. Soe, C. Manzano, N. Renaud, P. de Mendoza, A. De Sarkar, F. Ample, M. Hliwa, A. M. Echavarren, N. Chandrasekhar and C. Joachim, *ACS Nano*, 2011, **5**, 1436–1440.
24. N. Jlidat, M. Hliwa and C. Joachim, *Chem. Phys. Lett.*, 2008, **451**, 270–275.
25. I. Duchemin, N. Renaud and C. Joachim, *Chem. Phys. Lett.*, 2008, **452**, 269–274.
26. W. C. Bigelow, D. L. Pickett and W. A. Zisman, *J. Coll. Sci.*, 1946, **1**, 513–538.
27. A. Ulman, *Chem. Rev.*, 1996, **96**, 1533–1554.
28. D. K. Schwartz, *Annu. Rev. Phys. Chem.*, 2001, **52**, 107–137.
29. N. L. Abbott, J. P. Folkers and G. M. Whitesides, *Science*, 1992, **257**, 1380–1382.
30. D. A. Offord, C. M. John and J. H. Griffin, *Langmuir*, 1994, **10**, 761–766.
31. F. P. Zamborini and R. M. Crooks, *Langmuir*, 1998, **14**, 3279–3286.
32. K. Bandyopadhyay and K. Vijayamohanan, *Langmuir*, 1998, **14**, 625–629.
33. R. G. Nuzzo and D. L. Allara, *J. Am. Chem. Soc.*, 1983, **105**, 4481–4483.

34. J. M. Seminario, A. G. Zacarias and J. M. Tour, *J. Am. Chem. Soc.*, 1999, **121**, 411–416.
35. C. D. Bain, E. B. Troughton, Y. T. Tao, J. Evall, G. M. Whitesides and R. G. Nuzzo, *J. Am. Chem. Soc.*, 1989, **111**, 321–335.
36. B. Lüssem, L. Müller-Meskamp, S. Karthäuser, M. Homberger, U. Simon and R. Waser, *J. Phys. Chem. C*, 2007, **111**, 6392–6397.
37. H. Song, Y. Kim, Y. H. Jang, H. Jeong, M. A. Reed and T. Lee, *Nature*, 2009, **462**, 1039–1043.
38. A. R. Champagne, A. N. Pasupathy and D. C. Ralph, *Nano Letters*, 2005, **5**, 305–308.
39. C. Chu, J.-S. Na and G. N. Parsons, *J. Am. Chem. Soc.*, 2007, **129**, 2287–2296.
40. H. W. Kim, M. Han, H.-J. Shin, S. Lim, Y. Oh, K. Tamada, M. Hara, Y. Kim, M. Kawai and Y. Kuk, *Phys. Rev. Lett.*, 2011, **106**, 146101.
41. R. A. van Delden, M. K. J. ter Wiel, M. M. Pollard, J. Vicario, N. Koumura and B. L. Feringa, *Nature*, 2005, **437**, 1337–1340.
42. Y.-K. Kim, J. P. Koo and J. S. Ha, *Appl. Surf. Sci.*, 2005, **249**, 7–11.
43. A. Chandekar, S. K. Sengupta and J. E. Whitten, *Appl. Surf. Sci.*, 2010, **256**, 2742–2749.
44. W. R. Wilcox and T. J. LaChapelle, *J. Appl. Phys.*, 1964, **35**, 240.
45. U. Gösele, W. Frank and A. Seeger, *Appl. Phys.*, 1980, **23**, 361–368.
46. J. Sagiv, *J. Am. Chem. Soc.*, 1980, **102**, 92–98.
47. K. Ariga and Y. Okahata, *J. Am. Chem. Soc.*, 1989, **111**, 5618–5622.
48. D. A. Outka, J. Stohr, R. J. Madix, H. H. Rotermund, B. Hermsmeier and J. Solomon, *Surf. Sci.*, 1987, **185**, 53–74.
49. B. Arkles, *CHEMTECH*, 1977, **7**, 766–778.
50. P. Guyot-Sionnest, R. Superfine, J. H. Hunt and Y. R. Shen, *Chem. Phys. Lett.*, 1988, **144**, 1–5.
51. X. Zhao and R. Kopelman, *J. Phys. Chem.*, 1996, **100**, 11014–11018.
52. E. T. Vandenberg, L. Bertilsson, B. Liedberg, K. Uvdal, R. Erlandsson, H. Elwing and I. Lundström, *J. Colloid Interface Sci.*, 1991, **147**, 103–118.
53. J. A. Howarter and J. P. Youngblood, *Langmuir*, 2006, **22**, 11142–11147.
54. I. Haller, *J. Am. Chem. Soc.*, 1978, **100**, 8050–8055.
55. A. Hozumi, S. Asakura, A. Fuwa, N. Shirahata and T. Kameyama, *Langmuir*, 2005, **21**, 8234–8242.
56. E. T. de Givenchy, S. Amigoni, C. Martin, G. Andrada, L. Caillier, S. Géribaldi and F. Guittard, *Langmuir*, 2009, **25**, 6448–6453.
57. R. Cruz-Silva, M. E. Nicho, M. C. Reséndiz, V. Agarwal, F. F. Castillón and M. H. Farías, *Thin Solid Films*, 2008, **516**, 4793–4802.
58. N. L. Jeon, R. G. Nuzzo, Y. Xia, M. Mrksich and G. M. Whitesides, *Langmuir*, 1995, **11**, 3024–3026.
59. A. Revzin, R. J. Russell, V. K. Yadavalli, W.-G. Koh, C. Deister, D. D. Hile, M. B. Mellott and M. V. Pishko, *Langmuir*, 2001, **17**, 5440–5447.

60. M. F. Calhoun, J. Sanchez, D. Olaya, M. E. Gershenson and V. Podzorov, *Nat. Mater.*, 2008, **7**, 84–89.
61. D. H. Dinh, L. Vellutini, B. Bennetau, C. Dejous, D. Rebieère, E. Pascal, D. Moynet, C. Belin, B. Desbat, C. Labrugeère and J.-P. Pillot, *Langmuir*, 2009, **25**, 5526–5535.
62. C. D. Lorenz, E. B. Webb, M. J. Stevens, M. Chandross and G. S. Grest, *Tribol. Lett.*, 2005, **19**, 93–98.
63. M. R. Linford and C. E. D. Chidsey, *J. Am. Chem. Soc.*, 1993, **115**, 12631–12632.
64. M. R. Linford, P. Fenter, P. M. Eisenberger and C. E. D. Chidsey, *J. Am. Chem. Soc.*, 1995, **117**, 3145–3155.
65. Y.-Y. Lua, W. J. J. Fillmore, L. Yang, M. V. Lee, P. B. Savage, M. C. Asplund and M. R. Linford, *Langmuir*, 2005, **21**, 2093–2097.
66. A. Arafat, M. Giesbers, M. Rosso, E. J. R. Sudholter, K. Schroen, R. G. White, L. Yang, M. R. Linford and H. Zuilhof, *Langmuir*, 2007, **23**, 6233–6244.
67. J. M. Buriak, *Chem. Rev.*, 2002, **102**, 1271–1308.
68. T. L. Niederhauser, G. Jiang, Y.-Y. Lua, M. J. Dorff, A. T. Woolley, M. C. Asplund, D. A. Berges and M. R. Linford, *Langmuir*, 2001, **17**, 5889–5900.
69. E. G. Robins, M. P. Stewart and J. M. Buriak, *Chem. Commun.*, 1999, **24**, 2479–2480.
70. F. Zhang, L. Pei, E. Bennion, G. Jiang, D. Connley, L. Yang, M. V. Lee, R. C. Davis, V. S. Smentkowski, G. Strossman, M. R. Linford and M. C. Asplund, *Langmuir*, 2006, **22**, 10859–10863.
71. D. R. Lide, editor, *CRC Handbook of Chemistry and Physics*, CRC Press, New York, 90th edn, 2009.
72. M. V. Lee, R. Scipioni, M. Boero, P. L. Silvestrelli and K. Ariga, *Phys. Chem. Chem. Phys.*, 2011, **13**, 4862–4867.
73. T. K. Mischki, G. P. Lopinski and D. D. M. Wayner, *Langmuir*, 2009, **25**, 5626–5630.
74. M. Woods, S. Carlsson, Q. Hong, S. N. Patole, L. H. Lie, A. Houlton and B. R. Horrocks, *J. Phys. Chem. B*, 2005, **109**, 24035–24045.
75. Q.-Y. Sun, L. C. P. M. de Smet, B. van Lagen, M. Giesbers, P. C. Thune, J. van Engelenburg, F. A. de Wolf, H. Zuilhof and E. J. R. Sudholter, *J. Am. Chem. Soc.*, 2005, **127**, 2514–2523.
76. G. Jiang, T. L. Niederhauser, S. A. Fleming, M. C. Asplund and M. R. Linford, *Langmuir*, 2004, **20**, 1772–1774.
77. R. Boukherroub, S. Morin, D. D. M. Wayner, F. Bensebaa, G. I. Sproule, J.-M. Baribeau and D. J. Lockwood, *Chem. Mater.*, 2001, **13**, 2002–2011.
78. M. V. Lee, M. T. Hoffman, K. Barnett, J.-M. Geiss, V. S. Smentkowski, M. R. Linford and R. C. Davis, *J. Nanosci. Nanotechnol.*, 2006, **6**, 1639–1643.
79. G. P. Lopinski, D. D. M. Wayner and R. A. Wolkow, *Nature*, 2000, **406**, 48–51.

80. M. V. Lee, J. R. I. Lee, D. E. Brehmer, M. R. Linford and T. M. Willey, *Langmuir*, 2010, **26**, 1512–1515.
81. G. Binnig, H. Rohrer, C. Gerber and E. Weibel, *Phys. Rev. Lett.*, 1983, **50**, 120.
82. J. Tersoff and D. R. Hamann, *Phys. Rev. B*, 1985, **31**, 805.
83. N. D. Lang, *Phys. Rev. Lett.*, 1986, **56**, 1164.
84. R. J. Hamers, R. M. Tromp and J. E. Demuth, *Phys. Rev. Lett.*, 1986, **56**, 1972.
85. P. K. Hansma and J. Tersoff, *J. Appl. Phys.*, 1987, **61**, R1.
86. J. A. Dagata, *J. Vac. Sci. Technol., A*, 1992, **10**, 2105.
87. F. Grey, *Adv. Mater.*, 1993, **5**, 704–710.
88. K. Sattler, *Jpn. J. Appl. Phys.*, 2003, **42**, 4825–4829.
89. M. A. McCord, *J. Vac. Sci. Technol., B*, 1986, **4**, 86.
90. L. Stockman, G. Neuttiens, C. Van Haesendonck and Y. Bruynseraede, *Appl. Phys. Lett.*, 1993, **62**, 2935.
91. C. R. K. Marrian and E. S. Snow, *Microelectron. Eng.*, 1996, **32**, 173–189.
92. R. S. Becker, J. A. Golovchenko and B. S. Swartzentruber, *Nature*, 1987, **325**, 419–421.
93. D. M. Eigler and E. K. Schweizer, *Nature*, 1990, **344**, 524–526.
94. I.-W. Lyo and P. Avouris, *Science*, 1991, **253**, 173–176.
95. T. C. Shen, C. Wang, G. C. Abeln, J. R. Tucker, J. W. Lyding, P. Avouris and R. E. Walkup, *Science*, 1995, **268**, 1590–1592.
96. J. W. Lyding, *J. Vac. Sci. Technol., B*, 1994, **12**, 3735.
97. J. W. Lyding, T.-C. Shen, J. S. Hubacek, J. R. Tucker and G. C. Abeln, *Appl. Phys. Lett.*, 1994, **64**, 2010.
98. J. W. Lyding, T. C. Shen, G. C. Abeln, C. Wang and J. R. Tucker, *Nanotechnology*, 1996, **7**, 128.
99. J. A. Dagata, J. Schneir, H. H. Harary, C. J. Evans, M. T. Postek and J. Bennett, *Appl. Phys. Lett.*, 1990, **56**, 2001.
100. S. C. Minne, *J. Vac. Sci. Technol., B*, 1996, **14**, 2456.
101. J. A. Dagata, W. Tseng, J. Schneir and R. M. Silver, *Nanotechnology*, 1993, **4**, 194–199.
102. A. de Lozanne, *Jpn. J. Appl. Phys.*, 1994, **33**, 7090–7093.
103. G. Binnig, C. F. Quate and C. Gerber, *Phys. Rev. Lett.*, 1986, **56**, 930.
104. A. Majumdar, P. I. Oden, J. P. Carrejo, L. A. Nagahara, J. J. Graham and J. Alexander, *Appl. Phys. Lett.*, 1992, **61**, 2293.
105. P. Vettiger, M. Despont, U. Drechsler, U. Durig, W. Haberle, M. I. Lutwyche, H. E. Rothuizen, R. Stutz, R. Widmer and G. K. Binnig, *IBM J. Res. Dev.*, 2000, **44**, 323–340.
106. E. S. Snow and P. M. Campbell, *Appl. Phys. Lett.*, 1994, **64**, 1932.
107. P. T. Hurley, A. E. R. Ribbe and J. M. Buriak, *J. Am. Chem. Soc.*, 2003, **125**, 11334–11339.
108. S. H. Lee, T. Ishizaki, N. Saito and O. Takai, *Surf. Sci.*, 2007, **601**, 4206–4211.

109. O. Schneegans, A. Moradpour, F. Houze, A. Angelova, C. Henry de Villeneuve, P. Allongue and P. Chretien, *J. Am. Chem. Soc.*, 2001, **123**, 11486–11487.
110. J. N. Ngunjiri, S. S. S. Vegunta and J. C. Flake, *J. Electrochem. Soc.*, 2009, **156**, H516.
111. S. S. S. Vegunta, J. N. Ngunjiri and J. C. Flake, *J. Electrochem. Soc.*, 2010, **157**, D509.
112. R. Maoz, S. Matlis, E. DiMasi, B. M. Ocko and J. Sagiv, *Nature*, 1996, **384**, 150–153.
113. R. Maoz, S. R. Cohen and J. Sagiv, *Adv. Mater.*, 1999, **11**, 55–61.
114. S. Xu, S. Miller, P. E. Laibinis and G.-yu Liu, *Langmuir*, 1999, **15**, 7244–7251.
115. B. A. Wacaser, M. J. Maughan, I. A. Mowat, T. L. Niederhauser, M. R. Linford and R. C. Davis, *Appl. Phys. Lett.*, 2003, **82**, 808–810.
116. J. E. Headrick, M. Armstrong, J. Cratty, S. Hammond, B. A. Sheriff and C. L. Berrie, *Langmuir*, 2005, **21**, 4117–4122.
117. M. V. Lee, K. A. Nelson, L. Hutchins, H. A. Becerril, S. T. Cosby, J. C. Blood, D. R. Wheeler, R. C. Davis, A. T. Woolley, J. N. Harb and M. R. Linford, *Chem. Mater.*, 2007, **19**, 5052–5054.
118. M. Jaschke and H.-J. Butt, *Langmuir*, 1995, **11**, 1061–1064.
119. R. D. Piner, J. Zhu, F. Xu, S. Hong and C. A. Mirkin, *Science*, 1999, **283**, 661–663.
120. A. Ivanisevic and C. A. Mirkin, *J. Am. Chem. Soc.*, 2001, **123**, 7887–7889.
121. H. Jung, R. Kulkarni and C. P. Collier, *J. Am. Chem. Soc.*, 2003, **125**, 12096–12097.
122. D. S. Ginger, H. Zhang and C. A. Mirkin, *Angew. Chem., Int. Ed.*, 2004, **43**, 30–45.
123. J.-W. Jang, D. Maspoch, T. Fujigaya and C. A. Mirkin, *Small*, 2007, **3**, 600–605.
124. S. Hong, J. Zhu and C. A. Mirkin, *Science*, 1999, **286**, 523–525.
125. H. Zhang, N. A. Amro, S. Disawal, R. Elghanian, R. Shile and J. Fragala, *Small*, 2006, **3**, 81–85.
126. A. J. Senesi, D. I. Rozkiewicz, D. N. Reinhoudt and C. A. Mirkin, *ACS Nano*, 2009, **3**, 2394–2402.
127. H. Roder, E. Hahn, H. Brune, J.-P. Bucher and K. Kern, *Nature*, 1993, **366**, 141–143.
128. T. Yokoyama, S. Yokoyama, T. Kamikado, Y. Okuno and S. Mashiko, *Nature*, 2001, **413**, 619–621.
129. K. W. Hipps, L. Scudiero, D. E. Barlow and M. P. Cooke, *J. Am. Chem. Soc.*, 2002, **124**, 2126–2127.
130. S. Hoger, K. Bonrad, A. Mourran, U. Beginn and M. Moller, *J. Am. Chem. Soc.*, 2001, **123**, 5651–5659.
131. J. P. Hill, Y. Wakayama and K. Ariga, *Phys. Chem. Chem. Phys.*, 2006, **8**, 5034–5037.

132. E. Barrena, D. G. de Oteyza, H. Dosch and Y. Wakayama, *ChemPhysChem*, 2007, **8**, 1915–1918.
133. D. S. Deak, F. Silly, K. Porfyrakis and M. R. Castell, *Nanotechnology*, 2007, **18**, 075301.
134. Y. Okawa and M. Aono, *J. Chem. Phys.*, 2001, **115**, 2317.
135. G. Scappucci, G. Capellini, B. Johnston, W. M. Klesse, J. A. Miwa and M. Y. Simmons, *Nano Lett.*, 2011, **11**, 2272–2279.
136. J. L. O'Brien, S. R. Schofield, M. Y. Simmons, R. G. Clark, A. S. Dzurak, N. J. Curson, B. E. Kane, N. S. McAlpine, M. E. Hawley and G. W. Brown, *Phys. Rev. B*, 2001, **64**, 161401.
137. J. K. Lee, Y. H. Jung, R. M. Stoltenberg, J. B.-H. Tok and Z. Bao, *J. Am. Chem. Soc.*, 2008, **130**, 12854–12855.
138. P. W. K. Rothemund, *Nature*, 2006, **440**, 297–302.
139. H. Gu, J. Chao, S.-J. Xiao and N. C. Seeman, *Nat. Nanotechnol.*, 2009, **4**, 245–248.
140. A. C. Pearson, E. Pound, A. T. Woolley, M. R. Linford, J. N. Harb and R. C. Davis, *Nano Lett.*, 2011, **11**, 1981–1987.
141. F. Jäckel, Z. Wang, M. D. Watson, K. Müllen and J. P. Rabe, *Chem. Phys. Lett.*, 2004, **387**, 372–376.
142. F. Jäckel, M. D. Watson, K. Müllen and J. P. Rabe, *Phys. Rev. Lett.*, 2004, **92**, 188303.
143. N. T. Kemp, D. McGrouther, J. W. Cochrane and R. Newbury, *Adv. Mater.*, 2007, **19**, 2634–2638.
144. Y. Okawa and M. Aono, *Nature*, 2001, **409**, 683–684.
145. M. A. Walsh, S. R. Walter, K. H. Bevan, F. M. Geiger and M. C. Hersam, *J. Am. Chem. Soc.*, 2010, **132**, 3013–3019.
146. M. Z. Hossain, H. S. Kato and M. Kawai, *J. Am. Chem. Soc.*, 2005, **127**, 15030–15031.
147. M. Z. Hossain, H. S. Kato and M. Kawai, *J. Am. Chem. Soc.*, 2007, **129**, 3328–3332.
148. M. Z. Hossain, H. S. Kato and M. Kawai, *J. Phys. Chem. B*, 2005, **109**, 23129–23133.
149. M. Z. Hossain, H. S. Kato and M. Kawai, *J. Am. Chem. Soc.*, 2007, **129**, 12304–12309.
150. M. Z. Hossain, H. S. Kato and M. Kawai, *J. Am. Chem. Soc.*, 2008, **130**, 11518–11523.
151. J. A. Merlo and C. D. Frisbie, *J. Polym. Sci., Part B: Polym. Phys.*, 2003, **41**, 2674–2680.
152. R. Krupke, F. Hennrich, H. v. Löhneysen and M. M. Kappes, *Science*, 2003, **301**, 344–347.
153. M. Duchamp, K. Lee, B. Dwir, J. W. Seo, E. Kapon, L. Forroó and A. Magrez, *ACS Nano*, 2010, **4**, 279–284.
154. M. S. M. Saifullah, T. Ondar uhu, D. K. Koltsov, C. Joachim and M. E. Welland, *Nanotechnology*, 2002, **13**, 659–662.

155. H. Park, A. K. L. Lim, A. P. Alivisatos, J. Park and P. L. McEuen, *Appl. Phys. Lett.*, 1999, **75**, 301.
156. K. Shigeto, M. Kawamura, A. Y. Kasumov, K. Tsukagoshi, K. Kono and Y. Aoyagi, *Microelectron. Eng.*, 2006, **83**, 1471–1473.
157. A. Y. Kasumov, K. Tsukagoshi, M. Kawamura, T. Kobayashi, Y. Aoyagi, K. Senba, T. Kodama, H. Nishikawa, I. Ikemoto, K. Kikuchi, V. T. Volkov, Y. A. Kasumov, R. Deblock, S. Guéron and H. Bouchiat, *Phys. Rev. B*, 2005, **72**, 033414.
158. H. Park, J. Park, A. K. L. Lim, E. H. Anderson, A. P. Alivisatos and P. L. McEuen, *Nature*, 2000, **407**, 57–60.
159. L. H. Yu and D. Natelson, *Nano Lett.*, 2004, **4**, 79–83.
160. K. S. Novoselov, A. K. Geim, S. V. Morozov, D. Jiang, Y. Zhang, S. V. Dubonos, I. V. Grigorieva and A. A. Firsov, *Science*, 2004, **306**, 666–669.

CHAPTER 16

Nanomechanical Sensors and Membrane-type Surface Stress Sensor (MSS) for Medical, Security and Environmental Applications

GENKI YOSHIKAWA

International Center for Materials Nanoarchitectonics (WPI-MANA), National Institute for Materials Science (NIMS), 1-1 Namiki, Tsukuba, Ibaraki 305-0044, Japan
E-mail: YOSHIKAWA.Genki@nims.go.jp

16.1 Introduction

16.1.1 Demands for New Sensors

The demands for new sensors are rapidly growing in various fields; including for medical, security and environmental applications.

In a conventional blood test, for example, at least several days are required to identify the infectious bacteria. Accordingly, medical doctors prescribe an antibiotic based on their experiences (empiric therapy) as the first choice, thereby running the risk of fostering antibiotic-resistant bacteria everyday all over the world. Moreover, recent studies indicate that such overuse of antibiotics could be fuelling the dramatic increase in conditions such as obesity, type 1 diabetes, inflammatory bowel disease, allergies and asthma, because antibiotics kill the beneficial bacteria that we do want, as well as those we do not, and our friendly flora has been found to be never fully recovered once the natural balance is upset.

In the environmental field, the pollution of air, water, and soil have been the long-standing issues. We have to keep monitoring any tiny indication to prevent new pollution from being widely spread. The sick house syndrome or various allergies could also be attributable to the pollution of our surrounding environment.

As for the security issues, many countries are still suffering from the tremendous numbers of land mines (the total number is estimated at more than 70 million). Detection of drugs and explosives is one of the critical requirements to oppose terrorism and antisocial forces.

Nanomechanical sensors have potential to contribute to these global grand challenges owing to their intrinsic versatility. As discussed later, nanomechanical sensors detect volume and/or mass of target molecules, while there are no substances that do not have volume and mass. Therefore, nanomechanical sensors can be used for detecting virtually any kind of substance.

16.1.2 Introduction to Nanomechanical Sensors

A nanomechanical sensor is a mechanical structure which transduces analyte-induced stimuli into a signal *via* its structural change with nanometre precision. The definition of a nanomechanical sensor can also cover a mechanical transducer which has a size on the nanometre scale. In either senses, a cantilever sensor is a representative example among various geometries. Thus, in this chapter, we will rather focus on a cantilever sensor, while the recently developed membrane-type surface stress sensor (MSS) will also be highlighted as an optimized platform for actual applications.[1]

As discussed in detail in the next section, there are two basic modes in the operation of cantilever sensors; "static mode" and "dynamic mode". These modes can be correlated with "volume" and "mass" in terms of analyte-induced surface stress and mass change, respectively. Since there are no molecules that do not have "volume" and "mass", direct measurements of these fundamental characteristics can realize label-free and real-time detection of virtually any kind of molecule. In addition to economic and time considerations, this strategy is getting more and more important, especially in biological applications, for reliable analysis rather than monitoring artificially added features, such as fluorescent or radioactive tags, which may perturb the relative abundance and stoichiometry of samples, potentially biasing the analysis. While the measurement of mass change can be performed by other sensors, *e.g.* quartz crystal microbalance (QCM), surface stress-based sensing is a unique feature of nanomechanical sensors. Since surface stress emerges not only with the adsorption of molecules but also with the nanomechanical structural changes of substances on the surface, the scope of the surface stress-based sensing can cover various phenomena,[2,3] leading to a new type of sensing platform.

At first glance, nanomechanical sensors would seem to be rather simple in both structural and mechanistic aspects. However, the development of nanomechanical sensors requires quite a wide spectrum of expertise, such as mechanics, electronics, fluidics, analytical calculations and simulations, materials science and engineering, surface science, and medical considerations, in addition to the basic knowledge and technology of physics, chemistry, and biology. Since nanomechanical sensors intrinsically possess numerous possibilities as a versatile platform, systematic optimization for each application will realize superior sensors with high performance which can surpass other types of sensors in various applications. To demonstrate its high potential, it should be important to find a "killer application" and develop a really useful product. For that purpose, we have to carefully consider both advantages and disadvantages of nanomechanical sensors for each application in terms of requirements for commercialization, namely, cost, size, stability, robustness, and integration, in addition to the basic specifications: sensitivity, selectivity, dynamic range, number of channels, and reproducibility.

16.2 Cantilever Sensors

16.2.1 A Brief History of Cantilever Sensors

While the use of a cantilever as a sensing tool was already reported in 1968 by Wilfinge *et al.* using centimetre-scale silicon beams,[4] a microcantilever structure became well-known as a promising tool in various research fields after the development of the atomic force microscope (AFM).[5] The successful demonstration of using a cantilever in AFM as a scanning tool triggered intensive exploration of further applications of a cantilever partially owing to the increase in commercial availability of microfabricated cantilevers. The first chemical sensing application was demonstrated by Gimzewski *et al.* in 1994, revealing that the static bending of a cantilever can be used as a calorimeter which can detect the catalytic reaction taking place on the surface of a cantilever.[6] In the same year, Thundat *et al.* demonstrated mass detection with picogram resolution based on the shifts in cantilever resonance frequency induced by the exposure of a metal-coated cantilever to humidity or vapors of mercury.[7] Then, the target of cantilever sensors expanded to various phenomena, such as the formation of self-assembled monolayers[8] and the hybridization of DNA.[9]

Nowadays, nanomechanical cantilever sensors have been attracting more and more attention as a key device in various fields because of their fascinating advantages, such as label-free and real-time detection of target molecules with high sensitivity. Various applications have been demonstrated so far, including an electrical nose[10,11] and chemical and biological detection.[2,12–22]

Meanwhile, lots of methodological improvements have been made on each element in a cantilever sensor system, including the cantilever itself, surface functionalization, read-out technique, *etc.*

The most commonly used material for a cantilever is silicon as it has good material properties with adequate stiffness and robustness and matured microfabrication technology to form intended geometries with high precision and reproducibility. On the other hand, Johansson et al. fabricated cantilever arrays made of epoxy-based photoresist, SU-8, taking advantage of its softer characteristics.[23–26] For the read-out technique, the optical read-out is one of the most commonly used techniques because of its high sensitivity owing to the low noise level. The know-how accumulated in the development of AFM made it possible to achieve very high sensitivity up to even sub-nm precision with rather flexible cantilevers having a dimension of $500 \times 100 \times 1$ μm^3. For the functionalization of cantilever surfaces, various methods have been proposed, such as the "capillary method",[12] "inkjet printing",[27] etc. The details of surface functionalization will be discussed in the next section. The functionalization is definitely one of the most important challenges in nanomechanical sensors for actual applications. As for the reading-out technique, the optical read-out method has been commonly used, while other methods, e.g. piezoresistive read-out, have been proposed to overcome the problems related with optical read-out for actual applications. After the comprehensive optimization of a piezoresistive cantilever sensor, we have recently developed MSS,[1] although it has no longer the shape of a "cantilever". Since the MSS possesses various practical capabilities in addition to comparable or even higher sensitivity than that of optical read-outs, it will be one of the most promising platforms for fulfilling recent increasing demands on the realization of practical sensors in various applications. The details of MSS will be discussed later.

16.2.2 Operation Modes of Cantilever Sensors

There are two basic operation modes in cantilever sensors; "static mode" and "dynamic mode" (Figure 16.1). In the case of static mode, the adsorbates-induced surface stress makes the cantilever bend due to volume-related effects, while in the dynamic mode the adsorbates change the resonance frequency of a cantilever due to mass loading. The advantages and disadvantages of the static and dynamic modes are summarized in Table 16.1.

It should be noted that the combination of static and dynamic modes can offer more advanced level of measurements, as demonstrated by Ghatkesar et al.[2] They succeeded in monitoring both mass and structural changes that occurred at the binding of the bee venom peptide melittin to the lipid vesicles. This approach allows a comprehensive discussion of molecular interactions.

16.2.2.1 Static Mode

The static mode operation is a unique feature of cantilever sensors because it measures surface stress, which is not easy to measure with other sensing techniques. The important advantage of the static mode is that a cantilever

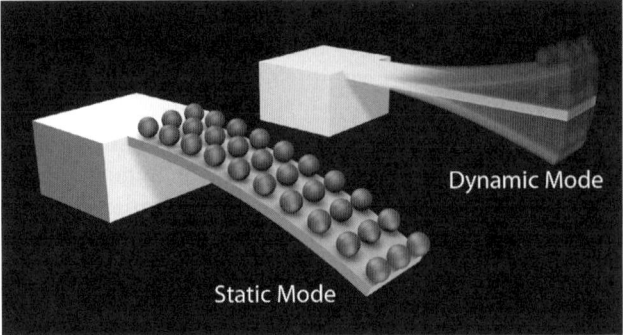

Figure 16.1 Schematic illustrations of static mode and dynamic mode operations. It is important to note that the bending of a cantilever is caused not by the gravity effect, but by the adsorbate-induced surface stress in the case of static mode.

does not suffer from damping effects because the bending motion caused by the analyte-induced surface stress is slow enough not to be affected by damping in most cases. Since the damping of liquid media severely decreases the sensitivity in dynamic mode, the static mode is a good option for the measurements in liquid environment. It also requires no complicated peripheral devices, such as actuators or the high frequency reading-out setup required for dynamic mode measurements.

There are several issues to be addressed in the static mode operation. The long-standing issue is the difficulties in the interpretation of obtained signals. Actually, there is still no well-accepted consensus on the origin of surface stress. Although it is roughly regarded as a result of an increase in electrostatic or steric interactions between the adsorbed analytes on the surface of a

Table 16.1

	Static mode	*Dynamic mode*
Principle	Surface stress	Mass
Advantages	No (or small) damping	Direct correlation with basic property: mass
	Simple set-up without an actuator	Higher sensitivity with smaller platform
Disadvantages and challenges	Difficulty in interpretation of signals (origin of surface stress, time dependence)	Large damping, especially in liquid
	Requirement of a certain amount of molecules to induce surface stress	Requirement of actuators
	Drift of baseline	Cross talk between channels
	Higher sensitivity with larger platform	Difficulty in interpretation of signals (position dependence, surface stress effect, *etc.*)

cantilever, a comprehensive model is still missing. Even for a model system of alkanethiols on a Au(111) surface, several explanations have been reported so far. Srinivasan *et al.* employed first-principle calculations to elucidate the origin of adsorption-induced surface stresses in alkanethiolate self-assembled monolayers on a Au(111) surface and proposed the mechanism in terms of a local rearrangement of surface Au atoms accompanying charge removal from the surface towards the Au–S bond.[28] Godin *et al.* compared the contributions from three different interactions; Lennard–Jones interactions (attractive [van der Waals] or repulsive [Pauli exclusion]), electrostatic interactions, and redistribution of the electronic structure of the substrate surface atoms, and concluded that the third effect induces up to ~ 2–4 orders of magnitude larger surface stress than those of the first and second effects.[29] The situation should be rather different in various systems, thereby making it difficult to establish a comprehensive model. Thus, the proper calibration should be performed for each application before the actual measurements.

For solid coating layers, we proposed a simple analytical model which can provide general reference values in terms of the strain induced in the coating layer.[41] It will help toward analyzing the static behavior of cantilever sensors and various nanomechanical sensors in conjunction with physical properties of coating films as well as optimizing the films for higher sensitivity. The details of the analytical model will be discussed later.

Another difficulty in the interpretation of a signal is the time-dependent complicated behavior, especially in the cases of polymer coatings[27] possibly due to the viscoelastic effects.[30]

16.2.2.2 Dynamic Mode

The concept of the dynamic mode operation is same as that of various resonators, such as the quartz crystal microbalance (QCM). In this mode, the resonance frequency shift due to the change in effective mass induced by the adsorption of analytes on a cantilever is measured. Since the obtained signals can be directly correlated to the basic property of adsorbate, *i.e.* mass, the dynamic mode is a useful and powerful technique for quantitative applications. As the sensitivity generally depends on the resonance frequency determined by the size of a cantilever, a nanometre scale cantilever operated at very high frequency (VHF) bands (~ 30–300 MHz) marked several milestones, such as ~ 7 zeptogram (10^{-21} g) resolution (equivalent to ~ 30 xenon atoms) in a cryogenically cooled, ultrahigh vacuum (below 10^{-10} Torr) apparatus,[31] and a mass resolution less than 1 attogram (10^{-18} g) in air at room temperature.[32]

The induced mass change (Δm) for a rectangular cantilever of length l, thickness t, width w, and Young's modulus E, can be calculated by the following equation:

$$\Delta m = \frac{k}{4\pi^2} \times \left(\frac{1}{f_1^2} - \frac{1}{f_0^2} \right) \quad (16.1)$$

where $k = Ewt^3/(4l^3)$ is a spring constant of the cantilever, and f_0 and f_1 are the eigenfrequencies before and after the mass change. This equation is derived by pure mechanics, assuming ideal conditions. Thus, in practical situations, there are several issues to be addressed. One of the most important issues is the damping effect induced by the surrounding media. The damping severely lowers the performance of a cantilever sensor in terms of a low quality factor Q, especially in a liquid environment where Q becomes several percent of that in air, resulting in low resolution for tracking the resonance frequency. Braun and Ghatkesar *et al.* proposed and demonstrated an elegant way to circumvent the damping effect in liquid environment using higher flexural modes.[15,33,34] They succeeded in detecting protein–ligand interactions in a physiological environment at a sensitivity of 2.5 pg/Hz,[15] and demonstrating a significant improvement in quality factor up to ~ 30 times with the 16th flexural mode.[34]

Other factors which affect the signals in dynamic mode are adsorption-induced effects, such as surface stress and position dependence, which can either stiffen or soften the cantilever, thereby varying the spring constant. The relation between the surface stress and stiffness of a cantilever has been intensively discussed.[35–37] Lee *et al.* visually demonstrated the dependence of resonance frequency on a pattern of gold layer on the surface of a cantilever.[38] In any cases, we have to be careful about these effects when we analyze the signals obtained with a dynamic mode.

16.2.2.3 Analytical Models for a Solid Coating Film

The commonly used analytical expression for the static mode is the Stoney's equation, which gives the deflection of the free-end of a cantilever (Δz) induced by surface stress (σ_{surf}):[39]

$$\Delta z = \frac{3(1-v_c)l_c^2}{E_c t_c^2} \sigma_{surf} \qquad (16.2)$$

where E_c and v_c are the Young's modulus and Poisson's ratio, and l_c and t_c correspond to the length and thickness of a cantilever, respectively. Sader proposed the improved asymptotic formula which includes the clamping effect and confirmed its significantly better accuracy by finite element analysis (FEA).[40] It should be noted that both of these models assume ideal surface stress and do not contain any parameters relating to a coating film which induces the surface stress in actual systems. It gives, on the one hand, a convenient feature to these models because they do not require any consideration of a coating film and surface stress can be correlated with the deflection only with the information on the properties of a cantilever. On the other hand, they can result in significant deviation from the reality in the presence of a solid coating film, especially with higher thickness. Moreover, as far as using these models is concerned, it is not possible to correlate the obtained signals (*i.e.* Δz) with physical properties of a coating film, preventing further consideration on the mechanism of the observed behavior of cantilever sensors or the origin of surface stress in conjunction with the physical properties of coating films.

In order to provide an analytical model which describes the actual static behavior of a cantilever with a solid coating film, we proposed a comprehensive model which contains all relevant physical parameters of both cantilever and coating film:[41]

$$\Delta z = \frac{3l^2(t_f+t_c)}{(A+4)t_f^2+(A^{-1}+4)t_c^2+t_f t_c} \quad (16.3)$$

where $A = (E_f w_f t_f (1 - v_c))/(E_c w_c t_c (1 - v_f))$, and ε_f represents externally induced strains to the coating film. E, v, l, w, and t correspond to Young's modulus, Poisson's ratio, length, width, and thickness of a cantilever (indicated by subscript c) and coating film (indicated by subscript f) (Figure 16.2 (a)). Since the present model contains all relevant parameters of both cantilever and coating film, it can be applied to cantilever sensors and coating films of virtually any material and dimension, verified by a good agreement with FEA (Figure 16.2 (b)).

This model can be used for translating obtained signals (Δz) into physical parameters (e.g. ε_f).[41] It can also provide the optimum coating thickness ($t_{f\text{-op}}$), which gives the largest signal (i.e. deflection) (Figure 16.2 (c)). The $t_{f\text{-op}}$ can be found by solving $d\Delta z/dt_f = 0$ for t_f, yielding

$$t_{f-op} = \frac{t_c}{d}\left(X^{1/3}+X^{-1/3}-1\right) \quad (16.4)$$

in the case of $w_c = w_f = w$, X is given as follows:

$$X = \frac{2U_c - U_f - 2\sqrt{U_c(U_c - U_f)}}{U_f} \quad (16.5)$$

where $U_c = E_c(1 - v_f)$ and $U_f = E_f(1 - v_c)$. Using eqn (16.3) and (16.4), one can readily find $t_{f\text{-op}}$, which turns out to be determined only by the Young's moduli (E_c and E_f), Poisson's ratios (v_c and v_f), and t_c, while the other parameters (length, width, and the amount of induced strain) determine the amount of deflection.

16.2.3 Surface Functionalization

To make a cantilever work as a sensor, the surface of a cantilever has to be functionalized by a receptor layer on which the target molecules adsorb or react. For the functionalization of cantilever surfaces, we have to take account of the following issues: (1) "one-side coating" to get reasonable deflection; (2) "individual coating" for simultaneous multiple detection of different targets,

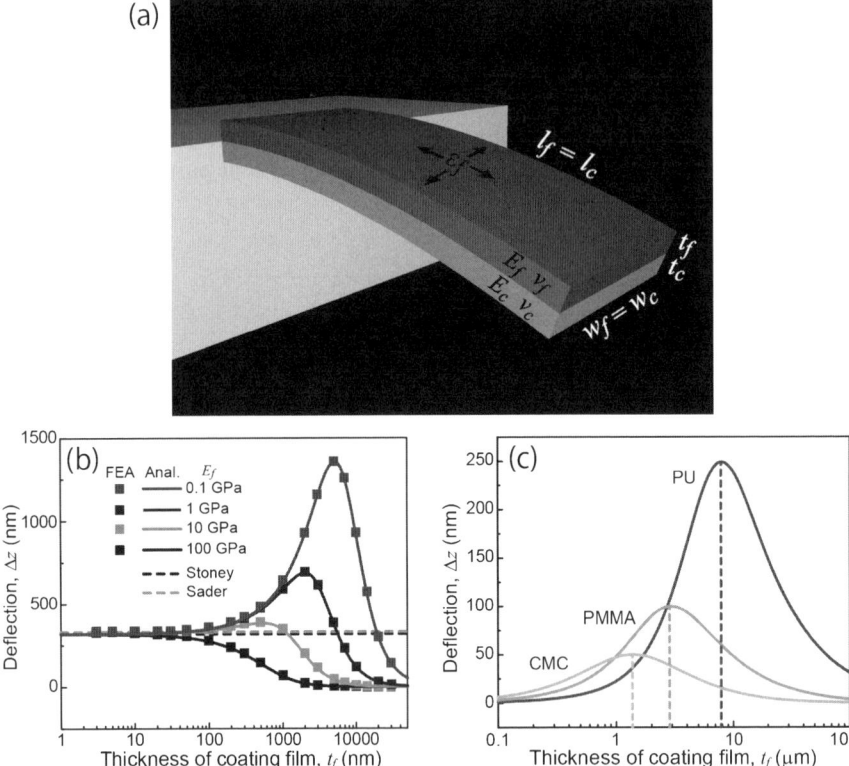

Figure 16.2 (a) Illustration of a cantilever covered by a coating film with isotropic internal strain. (b) Dependence of cantilever deflection on the thickness of coating films with various Young's moduli from 0.1 GPa to 100 GPa. The values calculated by finite element analysis (FEA) using COMSOL Multiphysics 3.5a are represented with filled squares. Black and gray dashed lines correspond to the cantilever deflection calculated by the Stoney's equation and Sader's model, respectively. $l = 500$ μm, $w_c = w_f = 100$ μm, $t_c = 1$ μm, $E_c = 170$ GPa, $v_c = 0.28$, $v_f = 0.30$, $\sigma_{surf} = 0.1$ N m^{-1}. (c) The dependence of cantilever deflection on coating film thickness at a specific strain. Dashed lines indicate the optimum thicknesses of PMMA, PU, and CMC for the highest sensitivity under the given conditions.[1]

avoiding cross contamination; (3) "quickness" to conserve the properties of the coating layer, especially for bio-molecules which can be easily denaturalized; (4) "reproducibility", "cost", and "easiness" for actual applications. In addition to these requirements, "reflection" has to be added for better signals in the case of the optical read-out technique and it is achieved by coating a thin gold layer, which also allows thiol chemistry for further functionalization. Although an ideal method which fulfills all the abovementioned requirements has not been established yet, several reasonable methods have been proposed and succeeded in demonstrating various measurements. Representative examples are the "capillary method", using arrays of dimension-matched capillaries,[12] and "ink-jet spotting".[21,22,27]

16.2.4 Read-out Methods

There are several methods for reading-out nanomechanical motion of a cantilever; we focus on the two major read-out methods; optical (laser) read-out and piezoresistive read-out. The comparison of these two methods is given in Table 16.2.

Table 16.2

Optical (laser) read-out	Piezoresistive read-out
Advantages:	Disadvantages:
High sensitivity (low noise)	Low sensitivity
No wiring required for each chip	Electrical wiring required for each chip
Disadvantages:	Advantages:
Bulky (large, complex), thereby difficult to miniaturize	Simple (small, easy), thereby easy to miniaturize
Expensive device	Low cost device
Laser alignment required	No laser alignment required
Impossible in opaque liquids	Possible in opaque liquids
Reflection layer required in some cases	No reflection layer required
	CMOS compatible (mass production)
	2D or 3D array

16.2.4.1 Optical (laser) Read-out

The most commonly used read-out method is so-called optical read-out, which utilizes laser light emitted from *e.g.* a vertical cavity surface emitting laser (VCSEL) and reflected on the surface of a cantilever and measured by a position sensitive detector (PSD) (Figure 16.3). The optical method is also frequently used in AFM. It gives high sensitivity in terms of signal-to-noise ratio because of its relatively low noise. Another practical advantage of this method is no requirement of wiring on a cantilever array sensor chip because both the source and detector of laser light are placed at remote positions.

However, the optical read-out has several drawbacks for actual applications. First of all, laser-related peripheries are bulky and expensive in most cases. For multiple cantilevers, the same number of laser sources must be prepared. While highly integrated VCSEL or multiple optical fibers system might be able to solve this problem, each laser light should be always aligned on each cantilever precisely. Thus, the optical read-out has low applicability for large one- or two-dimensional arrays. Another critical problem is the difficulty in performing measurements in opaque liquids, such as blood, because the optical signal is significantly attenuated due to low transmission or refractive index change in such a solution.

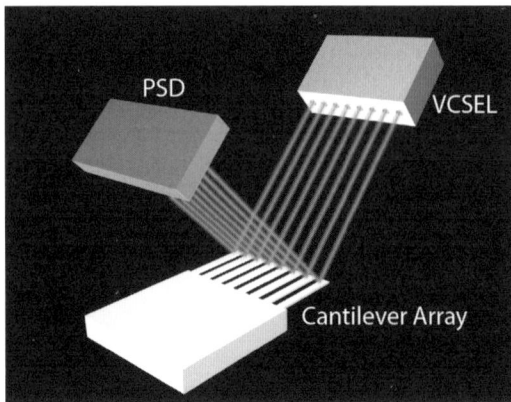

Figure 16.3 Typical setup for the optical (laser) read-out system. VCSEL is usually used as a source of multiple laser light. Each laser light reflected on the surface of each cantilever is measured by a PSD.

16.2.4.2 Piezoresistive Read-out

Piezoresistive read-out is based on a piezoresistive cantilever, in which a piezoresistor is embedded; thereby it is sometimes called as a "self-sensing" cantilever (Figure 16.4). It does not require any bulky and complex peripheries in contrast to an optical read-out because it electrically measures the signals in terms of the change in resistance of a piezoresistor (Figure 16.5). It can be also used for the detection in any opaque liquid with large multidimensional arrays. In addition, owing to its CMOS compatibility, the whole sensor unit including read-out parts can be integrated into common semiconductor devices, such as mobile phones. It is also technically feasible by mass production to produce inexpensive disposable chips, which are important for various applications, especially for medical diagnosis. Thus, the piezoresistive read-out has been regarded as one of the most promising approaches to overcome the problems related with the optical read-out. In spite of these inherent advantages, piezoresistive cantilevers have not been widely in use for sensing applications

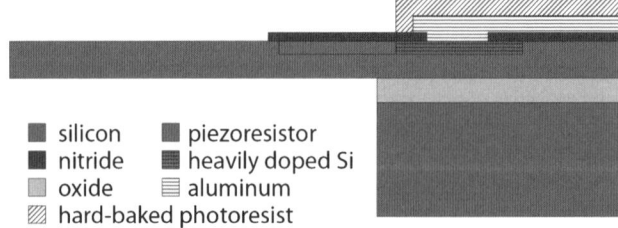

Figure 16.4 An example of the cross-section of a piezoresistive cantilever.[2] It is important to cover the piezoresistor by a passivation layer, such as silicon nitride, to prevent the leakage of current, especially in the case of measurements in a liquid environment.

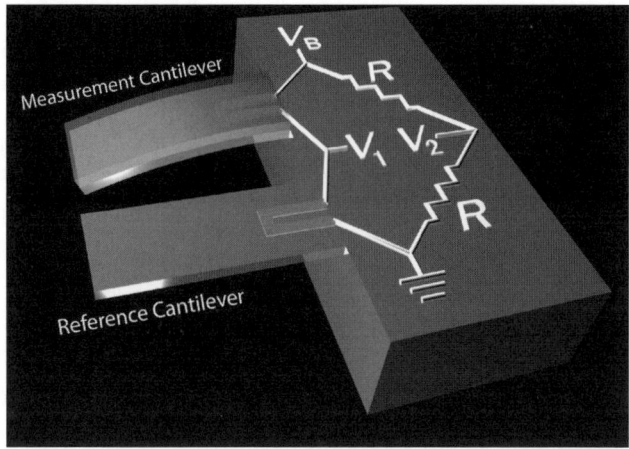

Figure 16.5 An example of electrical wiring of piezoresistive read-out. In this configuration, the differential signal is given as an output of the Wheatstone bridge, while it is also possible to place measurement and reference cantilevers in individual Wheatstone bridges and obtain differential signals by subtracting the output of the reference lever bridge from that of measurement lever bridge.

because of their critically low sensitivity. In other words, it will open the door to actual applications of nanomechanical sensors if the sensitivity of a piezoresistive cantilever can be significantly improved. Therefore, various trials have been made for more than a decade to improve the sensitivity of a piezoresistive cantilever for sensing applications. However, they have not yielded significant improvement in sensitivity to make piezoresistive detection comparable to the optical approach.

16.3 Membrane-type Surface Stress Sensor (MSS)

In this section, the motivation, strategy, and performance of the MSS will be discussed.

16.3.1 Strategy for the Improvement in Sensitivity; Towards Membrane-type Surface Stress Sensor (MSS)

Various strategies have been proposed for the improvement in the sensitivity of a piezoresistive cantilever for sensing applications by structural modification, such as making a through hole,[42,43] patterning of the cantilever surface,[44] or variation of geometrical parameters (*e.g.* length, width, and overall shapes).[45,46] Although these approaches gained some improvement in the sensitivity (typically, several tens of percent, while up to ~5 times in some cases[46]), they have not gained significant enhancement in sensitivity.

To address the appropriate scheme for the sensitivity enhancement of the piezoresistive cantilever sensors, it is important to note the basic properties of a piezoresistive cantilever for surface stress sensing.[1] Taking into account the advantage of their high piezocoefficient,[47–49] we focus on p-type piezoresistors created by boron diffusion onto a single-crystal Si with (100) surface. Plain stress (i.e. $\sigma_z = 0$) is assumed because of the intrinsically two dimensional feature of surface stress. In this case, relative resistance change can be described as follows:[49,50]

$$\frac{\Delta R}{R} \approx \frac{1}{2}\pi_{44}(\sigma_x - \sigma_y) \tag{16.6}$$

where π_{44} ($\sim 138.1 \times 10^{-11}$ Pa^{-1}) is one of the fundamental piezoresistance coefficients of the silicon crystal. σ_x, σ_y, and σ_z are stresses induced on the piezoresistor in [110], [1$\bar{1}$0], and [001] directions of the crystal, respectively. According to this equation, both enhancement of σ_x (σ_y) and suppression of σ_y (σ_x) are required to yield a substantial amount of $\Delta R/R$. However, in the case of surface stress sensing, the stress is basically isotropic; i.e. $\sigma_x \sim \sigma_y$, resulting in $\Delta R/R \sim 0$. Therefore, the piezoresistive signal is virtually zero on the whole surface irrespective of simple modifications, such as constrictions (Figure 16.6 (a) and (b)). Because of this intrinsic material property, it is difficult to gain significantly large improvement in the sensitivity by simple cantilever-type structures. It should be noted that the piezoresistive detection for force sensing applications, such as AFM, is very different from those for surface stress sensing. In such point-force loading cases (Figure. 16.6 (c) and (d)), the induced stress on the surface of a cantilever is uniaxial (i.e. $\sigma_x \gg \sigma_y$) and increases from zero at the free end, where the point-force is loaded in most cases, to maximum at the clamped end. The stress at the clamped end is readily amplified by a constriction.[51] Thus, the piezoresistors embedded at the constricted parts can yield a larger signal.

Our strategy to overcome this intrinsic problem is summarized in Figure 16.7. The first key to the improvement in sensitivity is the utilization of a double lever geometry (Figure 16.7 (b)),[52,53] which can transduce the whole analyte-induced surface stress on a so-called "adsorbate lever" to a "sensing lever," in which piezoresistors are embedded. In this case, the sensing lever is *pressed* by the adsorbate-lever, which is a similar situation to the point-force-loaded cantilever. Thus, the dominant stress applied on the sensing lever is in the x-direction (i.e. $\sigma_x \gg \sigma_y$) and it can be readily enhanced by a simple constriction. We made a comprehensive analysis of this geometry and found that sensitivity amplifications of up to a factor of 40 should be obtainable for conventional batch micro fabrication processing.[52]

Although the double lever geometry can already achieve significant improvement in sensitivity, this structure seems not so suitable for actual applications because of its asymmetric configuration, which also loses some amount of sensitivity at the connecting part. Actually, a much simpler

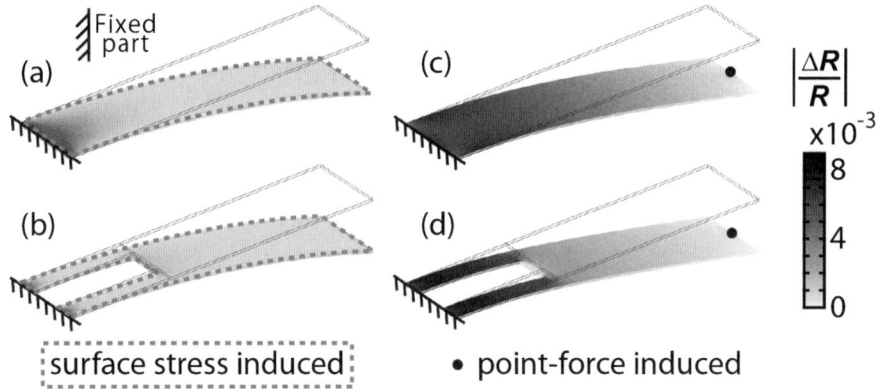

Figure 16.6 Distribution of $\Delta R/R$ on the surface of cantilevers with and without a constriction calculated by FEA using COMSOL Multiphysics 3.5a. The direction of current flow is assumed to be in the direction along the cantilever (x-direction; [110] of Si(100)). (a), (b): A compressive surface stress (-3.0 N m^{-1}) is applied uniformly on the whole surface surrounded by gray dashed lines. (c), (d): A point force is applied at the free end (filled circle position). Piezoresistors are supposed to be embedded inside the cantilever near to the top surface. Dimensions of all cantilevers are $135 \times 30 \times 1$ μm^3, while that of the constricted parts in (b) and (d) consist of two parts of $45 \times 8 \times 1$ μm^3. The deflection of the free ends of the cantilevers is the same (~ 700 nm, downward) in all cases.

structure like that presented in Figure 18.7 (c) can achieve a similar or even better performance. The adsorbate-covered part presses directly the piezoresistive sensing part. It should be noted that, at this stage, we departed from the concept of a "cantilever (*i.e.* one-side-clamped beam)."

We noticed the second key again in the basic property of single crystalline silicon. For p-type Si(100), the relative resistance change with a current flow in x-direction is given by eqn 16.6. In the case that all four resistors ($R_1 \sim R_4$) are practically equal and that the relative resistance changes are small with $\Delta R_i/R_i \ll 1$ ($i = 1$–4), the total output signal V_{out} of the Wheatstone bridge can be approximated by

$$V_{\text{out}} = \frac{V_B}{4}\left(\frac{\Delta R_1}{R_1} - \frac{\Delta R_2}{R_2} + \frac{\Delta R_3}{R_3} - \frac{\Delta R_4}{R_4}\right) \qquad (16.7)$$

where V_B is a bias voltage of the Wheatstone bridge. Thus, if the sign of the resistance changes ΔR_1 and ΔR_3 are opposite to that of ΔR_2 and ΔR_4, the full Wheatstone bridge yields an amplification of another factor of 4. If we configure the structure like Figure 16.7 (d), this condition is fulfilled because the dominant stresses induced by surface stress on the adsorbate membrane are σ_x in R_1, R_3 and σ_y in R_2, R_4, resulting in opposite signs for the relative resistance changes in each set of resistors, efficiently utilizing the whole induced surface stress. We call this structure a "membrane-type surface stress

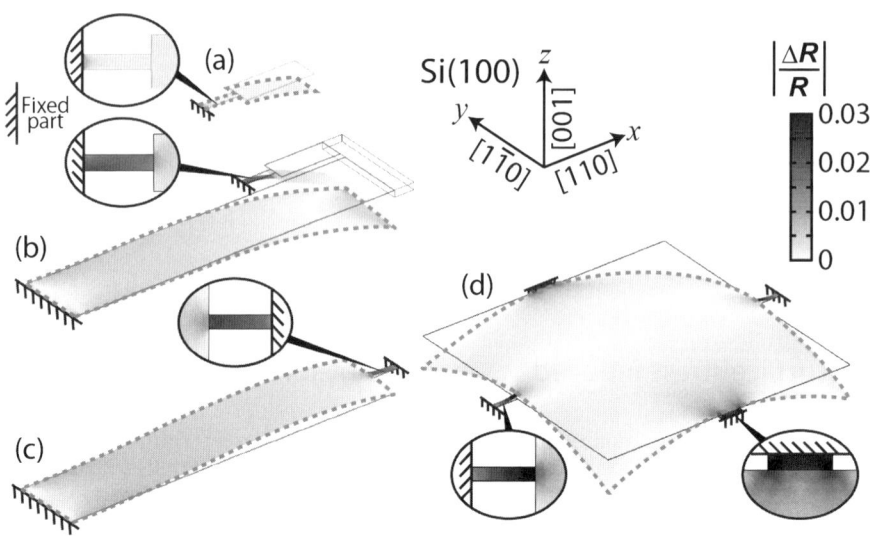

Figure 16.7 The FEA results of various geometries, assuming that all of the geometries are made of single-crystalline Si(100), which has a large piezocoefficient, and surface stress of 3.0 N m^{-1} is induced on the parts indicated by gray-dashed-lines. (a) Standard cantilever geometry, in which it is intrinsically difficult to gain signals even with some modifications, such as a constriction near to the clamped-end, because of the basic property of Si(100) (*i.e.* $\sigma_x \sim \sigma_y$, resulting in $\Delta R/R \sim 0$). (b) Double-lever geometry, in which the "adsorbate lever" and "sensing lever (with piezoresistors embedded)" are rigidly connected at their free-ends. Since the accumulated deflection caused by uniform surface stress on the adsorbate lever is transduced on the sensing lever in a similar form of point-force. Thus, the dominant stress induced on the sensing lever is along x-direction, thereby $\sigma_x \gg \sigma_y$, leading to $\Delta R/R \gg 0$. However, this asymmetric structure reduces some amount of signal at the connection part. (c) Both-side clamped geometry, in which both sensing and adsorbate levers are simply connected along x-direction. This structure is essentially same as that of double-lever geometry, while it is rather simpler, with a symmetric structure. It should be noted that this both-side clamped geometry and MSS are no longer a "cantilever-type" structure. (d) Membrane-type surface stress sensor (MSS) geometry. An "adsorbate membrane" is supported with four constrictions, "sensing beams", on which piezoresistors are embedded. The isotropic deformation, which is accumulated at the periphery of the membrane, is effectively converted into a force at the connection between the membrane and the small sensing beams. With this configuration, each sensing beam experiences the cumulative deformation of the membrane and, thus, piezoresistors embedded at these sensing beams can efficiently detect the whole surface stress applied on the adsorbate membrane.

sensor" (MSS, Figure 16.8 (a)).[1] It realizes fascinating properties for practical applications, such as unprecedentedly high sensitivity, self-compensating low-drift operation with full Wheatstone bridge, and stable operation without a free-end.

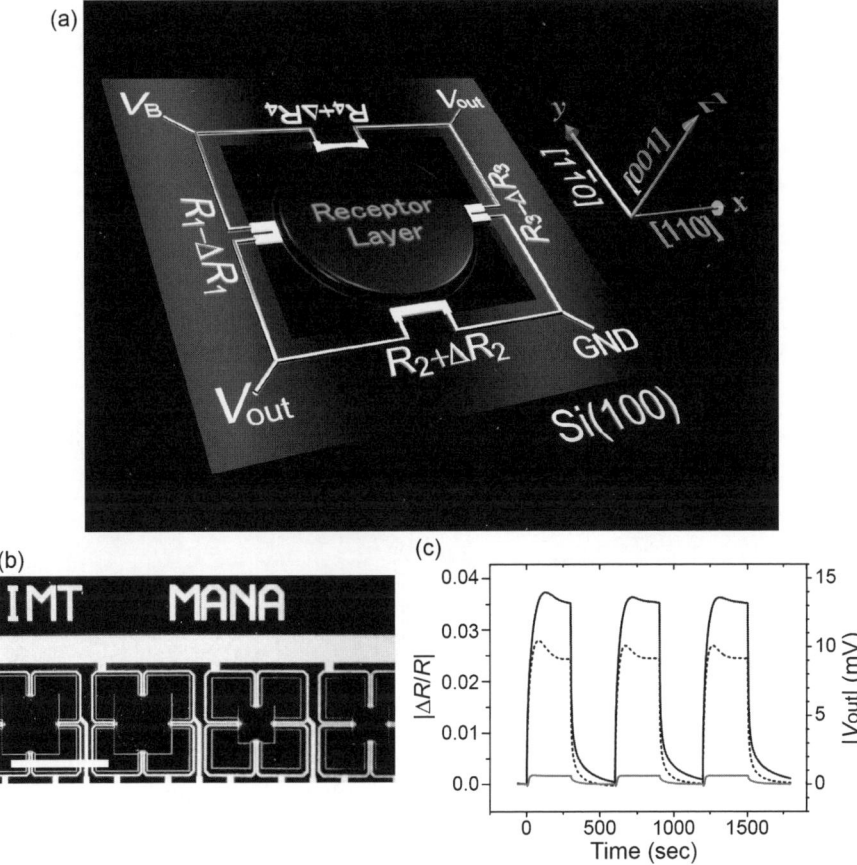

Figure 16.8 (a) Schematic illustration of MSS with the configuration of typical electrical wirings. The whole surface stress induced on the round center membrane is efficiently detected by piezoresistors (white colored parts) embedded in the constricted beams. (b) A photo of an actually fabricated MSS chip. White color bar corresponds to 1 mm. Membranes with diameters of 500 μm and 300 μm are fabricated in the same chip to examine the size dependence under the same condition. (c) Output signals (V_{out}) and corresponding values of $|\Delta R/R|$ obtained by the MSS having a diameters of 500 μm (solid black line) and 300 μm (dashed black line), and by a standard piezoresistive cantilever (gray line). Significant enhancement in sensitivity is confirmed in addition to the size dependence of MSS.

16.3.2 Performance of MSS

To verify the properties of MSS, we fabricated an array of MSSs (Figure 16.8 (b)). Two sizes of a membrane having diameters of 500 and 300 μm were fabricated in the same array. With simple gas measurements, more than 20 times higher sensitivity than that of a standard piezoresistive cantilever is

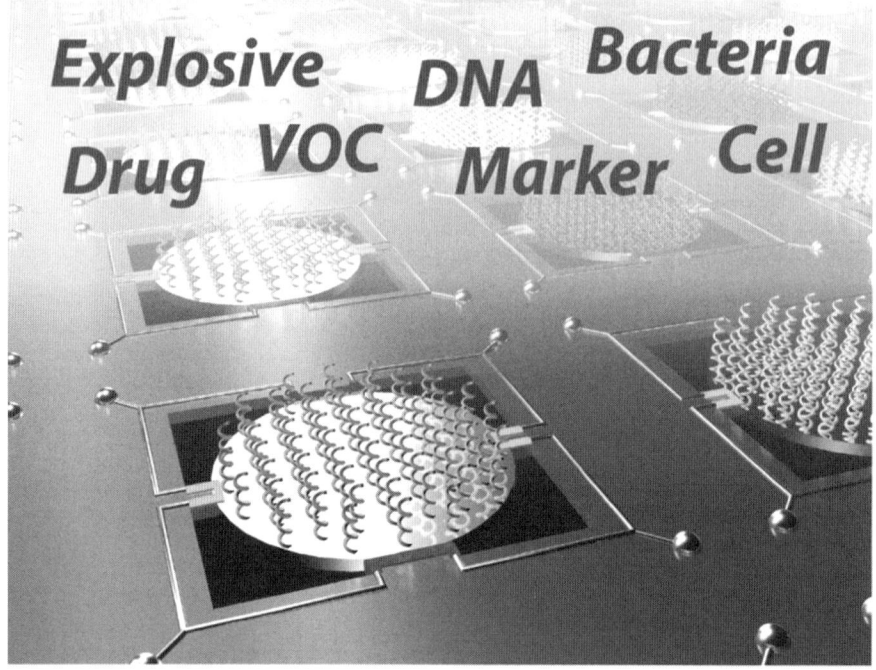

Figure 16.9 The schematic image of an array of MSS, and possible targets.

confirmed, as can be seen in Figure 16.8 (c).[1] We also quantitatively compared the performance of MSS with that of an optically read-out cantilever sensor in terms of signal-to-noise ratio (S/N). It is confirmed that the MSS has comparable sensitivity to the optical read-out methods with the experimentally demonstrated prototype chip shown in Figure 16.8 (b), while the standard piezoresistive cantilever is one or 2 orders of magnitude less sensitive than the optical read-out, as is consistent with the common observations. Moreover, the MSS structure has a potential for further enhancement in sensitivity up to several orders of magnitude by changing the dimensions of the adsorbate membrane and sensing beams, although the effective amplification would be somewhat reduced because of practical requirements, such as properties of piezoresistors or passivation layers on the sensing beams. It is also important to notice that the non-linearity of the bridge circuit or fracture stress of silicon will have non-negligible effects in some extreme cases.

Another important difference between MSS and a cantilever is the size dependence on a signal. Since the analyte induced surface stress can be efficiently transduced at the piezoresistive sensing beams in the MSS geometry, a larger adsorbate membrane or smaller sensing beams lead to a larger resistance change, that is, higher sensitivity. The larger aspect ratio between the beams and membrane can achieve a larger enhancement in MSS because of the constriction effects, whereas there is almost no dependence for the standard cantilever architecture.[1]

16.4 Summary and Future Prospects

Nanomechanical sensors intrinsically possess high performance for detecting minute amount of target molecules. In principle, it can detect any kind of molecule, ranging from gaseous to biological molecules, as already demonstrated in various systems. In order to make nanomechanical sensors practically useful in most actual applications, the sensing platforms must have various features, such as being small, simple, inexpensive, and capable of real-time and label-free measurements (without time-consuming preparation, such as laser alignment, labeling of samples with fluorescent tags, *etc.*), in addition to the basic specifications, such as high sensitivity.

One of the promising platforms to fulfill these practical requirements is MSS, which possesses all of these features in principle. In addition, it can be miniaturized with simple architectures and realize low cost chips owing to the CMOS compatibility for batch fabrication, and also can be used in opaque liquids, such as blood. Therefore, MSS is expected to be a highly sensitive versatile sensor for various fields, such as medical, biological, security, and environmental applications.

Acknowledgements

The author would like to sincerely thank Dr Heinrich Rohrer, Dr Terunobu Akiyama, Dr Sebastian Gautsch, and Dr Peter Vettiger for the indispensable contributions to realize MSS. The author expresses appreciation to Professor Masakazu Aono and Dr Tomonobu Nakayama from the International Center for Materials Nanoarchitectonics (MANA), National Institute for Materials Science (NIMS), Tsukuba, Japan, and Professor Nicolaas de Rooij from the Institute of Microengineering (IMT) of Ecole Polytechnique Federale de Lausanne (EPFL), Neuchatel, Switzerland, for their help and support. The author acknowledges the kind introduction to nanomechanical sensors, valuable discussions, and help and support by the members of the Cantilever Array Sensor Group, Department of Physics, University of Basel, Switzerland. The author thanks the staff of Comlab, the CSEM micro- and nanofabrication facility for the technical support. The financial support for the development of MSS were provided by the World Premier International Research Center (WPI) Initiative on Materials Nanoarchitronics (MANA) and by the Grant-in-Aid for Young Scientists (B) 21750083 (2009), MEXT, Japan.

References

1. G. Yoshikawa, T. Akiyama, S. Gautsch, P. Vettiger and H. Rohrer, *Nano Lett.*, 2011, **11**, 1044.
2. M. K. Ghatkesar, H. P. Lang, C. Gerber, M. Hegner and T. Braun, *PLoS One*, 2008, **3**, e3610.

3. R. Mukhopadhyay, V. V. Sumbayev, M. Lorentzen, J. Kjems, P. A. Andreasen and F. Besenbacher, *Nano Lett.*, 2005, **5**, 2385.
4. R. J. Wilfinge, P. H. Bardell and D. S. Chhabra, *IBM J. Res. Dev.*, 1968, **12**, 113.
5. G. Binnig, C. F. Quate and C. Gerber, *Phys. Rev. Lett.*, 1986, **56**, 930.
6. J. K. Gimzewski, C. Gerber, E. Meyer and R. R. Schlittler, *Chem. Phys. Lett.*, 1994, **217**, 589.
7. T. Thundat, R. J. Warmack, G. Y. Chen and D. P. Allison, *Appl. Phys. Lett.*, 1994, **64**, 2894.
8. R. Berger, E. Delamarche, H. P. Lang, C. Gerber, J. K. Gimzewski, E. Meyer and H. J. Guntherodt, *Science*, 1997, **276**, 2021.
9. J. Fritz, M. K. Baller, H. P. Lang, H. Rothuizen, P. Vettiger, E. Meyer, H. J. Guntherodt, C. Gerber and J. K. Gimzewski, *Science*, 2000, **288**, 316.
10. H. P. Lang, M. K. Baller, R. Berger, C. Gerber, J. K. Gimzewski, F. M. Battiston, P. Fornaro, J. P. Ramseyer, E. Meyer and H. J. Güntherodt, *Anal. Chim. Acta*, 1999, **393**, 59.
11. M. K. Baller, H. P. Lang, J. Fritz, C. Gerber, J. K. Gimzewski, U. Drechsler, H. Rothuizen, M. Despont, P. Vettiger, F. M. Battiston, J. P. Ramseyer, P. Fornaro, E. Meyer and H. J. Guntherodt, *Ultramicroscopy*, 2000, **82**, 1.
12. R. McKendry, J. Y. Zhang, Y. Arntz, T. Strunz, M. Hegner, H. P. Lang, M. K. Baller, U. Certa, E. Meyer, H. J. Güntherodt and C. Gerber, *Proc. Natl. Acad. Sci. U. S. A.*, 2002, **99**, 9783.
13. Y. Arntz, J. D. Seelig, H. P. Lang, J. Zhang, P. Hunziker, J. P. Ramseyer, E. Meyer, M. Hegner and C. Gerber, *Nanotechnology*, 2003, **14**, 86.
14. N. Backmann, C. Zahnd, F. Huber, A. Bietsch, A. Pluckthun, H. P. Lang, H. J. Güntherodt, M. Hegner and C. Gerber, *Proc. Natl. Acad. Sci. U. S. A.*, 2005, **102**, 14587.
15. T. Braun, V. Barwich, M. K. Ghatkesar, A. H. Bredekamp, C. Gerber, M. Hegner amd H. P. Lang, *Phys. Rev. E*, 2005, 72.
16. F. Huber, M. Hegner, C. Gerber, H. J. Guntherodt and H. P. Lang, *Biosens. Bioelectron.*, 2006, **21**, 1599.
17. J. Zhang, H. P. Lang, F. Huber, A. Bietsch, W. Grange, U. Certa, R. McKendry, H.-J. Güntherodt, M. Hegner and C. Gerber, *Nat. Nanotechnol.*, 2006, **1**, 214.
18. N. Nugaeva, K. Y. Gfeller, N. Backmann, M. Duggelin, H. P. Lang, H. J. Guntherodt and M. Hegner, *Microsc. Microanal.*, 2007, **13**, 13.
19. M. Watari, J. Galbraith, H. P. Lang, M. Sousa, M. Hegner, C. Gerber, M. A. Horton and R. A. McKendry, *J. Am. Chem. Soc.*, 2007, **129**, 601.
20. J. W. Ndieyira, M. Watari, A. D. Barrera, D. Zhou, M. Vogtli, M. Batchelor, M. A. Cooper, T. Strunz, M. A. Horton, C. Abell, T. Rayment, G. Aeppli and R. A. McKendry, *Nat. Nanotechnol.*, 2008, **3**, 691.
21. T. Braun, M. K. Ghatkesar, N. Backmann, W. Grange, P. Boulanger, L. Letellier, H. P. Lang, A. Bietsch, C. Gerber and M. Hegner, *Nat. Nanotechnol.*, 2009, **4**, 179.

22. G. Yoshikawa, H. P. Lang, T. Akiyama, L. Aeschimann, U. Staufer, P. Vettiger, M. Aono, T. Sakurai and C. Gerber, *Nanotechnology*, 2009, **20**, 015501.
23. A. Johansson, M. Calleja, P. A. Rasmussen and A. Boisen, *Sens. Actuators, A*, 2005, **123–24**, 111.
24. M. Calleja, M. Nordstrom, M. Alvarez, J. Tamayo, L. M. Lechuga and A. Boisen, *Ultramicroscopy*, 2005, **105**, 215.
25. L. Gammelgaard, P. A. Rasmussen, M. Calleja, P. Vettiger and A. Boisen, *Appl. Phys. Lett.*, 2006, **88**, 113508.
26. F. G. Bosco, E. T. Hwu, S. Keller, A. Greve and A. Boisen, *Microelectron. Eng.*, 2010, **87**, 708.
27. A. Bietsch, J. Y. Zhang, M. Hegner, H. P. Lang and C. Gerber, *Nanotechnology*, 2004, **15**, 873.
28. V. Srinivasan, G. Cicero and J. C. Grossman, *Phys. Rev. Lett.*, 2008, **101**, 185504.
29. M. Godin, V. Tabard-Cossa, Y. Miyahara, T. Monga, P. J. Williams, L. Y. Beaulieu, R. B. Lennox and P. Grutter, *Nanotechnology*, 2010, **21**, 075501.
30. S. M. Heinrich, M. J. Wenzel, F. Josse and I. Dufour, *J. Appl. Phys.*, 2009, 105.
31. Y. T. Yang, C. Callegari, X. L. Feng, K. L. Ekinci and M. L. Roukes, *Nano Lett.*, 2006, **6**, 583.
32. M. Li, H. X. Tang and M. L. Roukes, *Nat. Nanotechnol.*, 2007, **2**, 114.
33. M. K. Ghatkesar, V. Barwich, T. Braun, J. P. Ramseyer, C. Gerber, M. Hegner, H. P. Lang, U. Drechsler and M. Despont, *Nanotechnology*, 2007, **18**, 445502.
34. M. K. Ghatkesar, T. Braun, V. Barwich, J. P. Ramseyer, C. Gerber, M. Hegner and H. P. Lang, *Appl. Phys. Lett.*, 2008, 92.
35. A. W. McFarland, M. A. Poggi, M. J. Doyle, L. A. Bottomley and J. S. Colton, *Appl. Phys. Lett.*, 2005, 87.
36. M. J. Lachut and J. E. Sader, *Phys. Rev. Lett.*, 2007, 99.
37. M. J. Lachut and J. E. Sader, *Appl. Phys. Lett.*, 2009, 95.
38. C. Lee, S. Kim, N. Jung, T. Thundat and S. Jeon, *J. Appl. Phys.*, 2009, 106.
39. G. G. Stoney, *Proc. R. Soc. London, Ser. A*, 1909, **82**, 172.
40. J. E. Sader, *J. Appl. Phys.*, 2001, **89**, 2911.
41. G. Yoshikawa, *Appl. Phys. Lett.*, 2011, **98**, 173502.
42. J. H. He and Y. F. Li, *J. Phys.*, 2006, **34**, 429.
43. X. M. Yu, Y. Q. Tang, H. T. Zhang, T. Li and W. Wang, *IEEE Sens. J.*, 2007, **7**, 489.
44. N. L. Privorotskaya and W. P. King, *Microsyst. Technol.*, 2008, **15**, 333.
45. F. T. Goericke and W. P. King, *IEEE Sens. J.*, 2008, **8**, 1404.
46. A. Loui, F. T. Goericke, T. V. Ratto, J. Lee, B. R. Hart and W. P. King, *Sens. Actuators, A*, 2008, **147**, 516.
47. Y. Kanda, *IEEE Trans. Electron. Devices*, 1982, **29**, 64.

48. W. G. Pfann and R. N. Thurston, *J. Appl. Phys.*, 1961, **32**, 2008.
49. Y. Kanda, *Sens. Actuators, A*, 1991, **28**, 83.
50. P. A. Rasmussen, O. Hansen and A. Boisen, *Appl. Phys. Lett.*, 2005, **86**, 203502.
51. M. Tortonese, R. C. Barrett and C. F. Quate, *Appl. Phys. Lett.*, 1993, **62**, 834.
52. G. Yoshikawa and H. Rohrer, *7th International Workshop on Nanomechanical Cantilever Sensors*, Banff, 2010.
53. S. M. Yang, T. I. Yin and C. Chang, *Sens. Actuators, B*, 2007, **121**, 545.
54. L. Aeschimann, A. Meister, T. Akiyama, B. W. Chui, P. Niedermann, H. Heinzelmann, N. F. De Rooij, U. Staufer and P. Vettiger, *Microelectron. Eng.*, 2006, **83**, 1698.
55. G. Yoshikawa, *Appl. Phys. Lett.*, 2011, **98**, 173502.
56. L. Aeschimann, A. Meister, T. Akiyama, B. W. Chui, P. Niedermann, H. Heinzelmann, N. F. De Rooij, U. Staufer and P. Vettiger, *Microelectron. Eng.*, 2006, **83**, 1698–1701.

Subject Index

α-amino groups 233
α-diazo-β-hydroxy esters synthesis 138, 139, 142, 143
α particle emissions 261–2
α-sulfanyl-β-amino derivatives 143
β-amino acids 139, 143, 144
β particle emissions 261–2
β value 60–4
γ-ray emissions 261–2
π orbitals 359–60, 393
θ (contact angle) 31, 49, 73
ζ (zeta potential) 273, 283–4, 292

AAO (anodised aluminium oxide) 334, 335, 336
aaRSs (aminoacyl tRNA synthetases) 217, 218, 223–36
Aβ_{1-42} theranostics 196
acetic acid 125
acids
    acetic acid 125
    acidic porous clay heterostructures 92–3
    Brønsted acid sites 260, 264
    Lewis acid/base sites 135–6, 137, 140, 141, 142
    sorbent chemical stability assessment 272
active layer structures *see* conjugated polymer–fullerene organic solar cells
active sites *see* enzyme active site architectonics
adapted molecular conformations 13–14
additives, solvent boiling points 373
adenosine, carbon nanocages 117–18
adenosine triphosphate (ATP)
    aminoacyl tRNA synthetase active site 223–6, 228, 229, 230, 231–6
    binding pocket 224, 225, 230
    HisRS active site measurements 233, 234
adhesive interactions, thin films 33
administration routes 250
    see also drug delivery systems
adsorbate lever 440
adsorption
    carbon nanocages 115–19
    eicosyl alcohol on glass 397
    facet-controlled nanostructures 169
    ibuprofen on MCM-41 silica carrier 252–3
    layered titanates 104
    LbL mesoporous materials 124
    pillared clays 91
    porous clay heterostructures 94
AFM *see* atomic force microscopy (AFM)
agmatine 195
albumin 250, 251
aldol reactions 146–7
alginate 193–4, 250
alignment *see* substrate alignment
alkaline layered silicates 102–4
alkanedithiols 374
alkanethiol 433

alkenes 401–2, 403, 404, 407–8, 417–18
  see also individual alkenes
N-alkyl carbazole 20–1
alkyl chains in SAMs 304
alkynes 401
allosteric regulation 11–12
alpha amino groups 233
alpha electron emissions 261–2
alumina nanomembranes 334–6
alumina pillars, magadiite 102
aluminium incorporation 93, 96
aluminium oxide 96
aluminium polycations 90, 91
aluminium trioxide–silicon dioxide 96
Alzheimer's disease 196
amine group effects 399
amine-terminated silane 409
amino acids
  charge properties *in situ* modification 309–10
  chirality recognition 24
  enzyme active sites 214, 215, 216, 222–36
    binding pockets 224, 225, 230
    sequence conservation 217–19, 224, 229, 230, 234
    side chain variability in aaRS 227
  Mannich reaction 139, 143, 144
α-amino groups 233
aminoacyl tRNA synthetases (aaRSs) 217, 218, 223–36
aminopropylsilane (APS) 94, 327, 328
anatase-based DCPHs preparation 95–6
anchored molecules to substrates 397–402
aniline polymerisation 416
anion-complexation-induced charge separation enhancement 11–12
anisotropic wetting 74–8
anodic porous alumina nanomembranes 334–6
anodised aluminium oxide (AAO) 334, 335, 336
anti-bonding levels 175–6

Aono, Dr Masakazu 2, 112
APS (aminopropylsilane) 94, 327, 328
aptamer DNA bioconjugates 204, 205
aptamer-based FET biosensors 307
aromatic compounds 19–20, 144, 145, 146
  see also fullerenes
arrays
  block copolymer droplet wetting control 69–73
  gold nanoprisms hexagonal array on tungsten trioxide 180–1
  gold nanorod arrays image 332
  p/n heterojunction arrays 17
  porphyrin molecular 2D arrays 12–14
  silver nanoparticles NSL fabrication 343
  vacuum-deposited molecular arrays on patterned substrates 410–12
artificial atoms 175
artificial cells 206
N-arylation, heterocycles 45–6
as-transferred droplets 76
asbestos, CNT morphology similarity 252
ascorbic acid 327, 328, 329
aspect ratio 167, 168
assembly
  see also materials development; self-assembled monolayers (SAMs); self-assembly
  layered clays/nanoparticles 96–8
asthma, drug delivery 253
asymmetric ketones reduction 146
atomic force microscopy (AFM)
  development 430, 431
  electrical stimulation for patterning 405–7
  low-MW films spin-cast from chloroform/xylenes 370
  MDMO-PPV:PCBM composite films 365, 375
  multi-tip 405
  P3HT/PCBM blends 381, 382, 384

Subject Index

rr-P3HT:PCBM 1:1 blend films 372, 373
substrate alignment 393
surface patterning 402, 404–10
ATP *see* adenosine triphosphate (ATP)
Auger electron emissions 261–2
autophobic dewetting 30, 40
azido-terminated poly(*N*-isopropylacrylamide) (PNIPAM-N_3) 200

Bacillus spp.
  B. stearothermophilus (BS) 226, 229, 230, 232, 234, 235
  B. subtilis 344
bacterial alcohol dehydrogenase 221
bacterial detection 192, 196
bases
  Lewis acid/base sites 135–6, 137, 140, 141, 142
  sorbent chemical stability assessment 272
batch equilibrium method 274
Baylis–Hillman reaction 138, 139, 141, 142
Beckmann rearrangement 150
Becquerel, Henri 259
benzene 52
benzylation 144, 145
beta electron emissions 261–2
binary code/logic 9–10, 395, 396–7
binodal thin film instability 35
bioconjugate nanostructures 191–207
  biosensors, biomolecular charges at interfaces 202–11
  drug delivery systems 242–54
  nucleic acids 203–6
  phospholipids 195–9
  proteins 199–203
  saccharides 191–5
biomimetic mineralisation 134
biosensors
  biomolecular charges at interfaces 202–11
  DNA bioconjugates 204

lipid bioconjugates 196
materials fabrication 318–47
metal–DNA bioconjugate nanostructures 204–5
biothiols detection 204–5
bismuth vanadium oxide ($BiVO_4$) nanoplates 169, 170
block copolymers
  controlled dewetting 28–78
    experiments 41–2
    thin films 43–68
  controlled self-assembly 51–4
  controlled wetting 69–73
    anisotropic, droplets 74–8
  magnetic-core-gold-shell nanoparticles preparation 322–3
  nanowires 16–18
  P4VP-b-PS-b-P4VP 323, 324
  PEG–lipid bioconjugate nanostructures 197–8
  phase segregation 16–18
  solvent control 51–6
  stability 36–43
    experiments 39–43
    theory 36–9
  thermo-responsive drug delivery 245–6, 247
blood residence time 248
BNL (Brookhaven National Laboratory) 260, 264
boiling points, solvents 370, 373, 378
bond counting concept 176
bonding levels 175–6
Boolean truth tables 396, 397
Born repulsion 34
boron carbon nitride, mesoporous 123
boron nitride, mesoporous 121, 122, 123
borosilicate glass columns 274–5
bottom-up *see* wet chemistry (bottom-up) routes
boundary effect, pinned micelles 66, 67
boundary region molecular conformation adaptation 13–14
Bragg scattering 181, 182
BRCA-1 breast cancer gene 204

breakthrough point, sorption 274
Brønsted acid sites 136, 140
Brookhaven National Laboratory (BNL, USA) 260, 264
brush monolayers 30, 43, 68–73
Buckingham potential 34
Buckminster fullerenes *see* fullerenes
bulk heterojunction active layer structures *see* conjugated polymer–fullerene organic solar cells
bulk mechanical motion 22–4

C-SVA (controlled solvent vapour annealing) 378–9, 380
cadmium hydroxide nanosheet templates 178–9
cadmium sulfide 168, 176–7, 178–9
caffeine 115–17, 124
calcium phosphate–lipid bioconjugate nanostructures 198–9
cancer treatment
  bioconjugate nanostructures 195, 197, 198–9, 202, 203, 204
  carbon nanomaterials slow release 252
  carbon–protein bioconjugates 203
  MAbs delivery in liposomes 249
  molecular conjugate drug delivery 250–1
  PEG–lipid bioconjugate nanostructures 197, 199
  polysaccharide-based drug delivery 194–5
  temperature-triggered liposomes 249
cantilever sensors 430–9
  applications 430
  damping 432, 434
  double lever geometry 440, 441, 442
  finite element analysis 441, 442
  history 430–1
  operation modes 431–5
  optical (laser) readouts 431, 437–8, 444
  overview 428–30
  performance measurement 443–5
  piezoresistive readouts 431, 437, 438–9
  point-force loaded 440, 441
  read-out methods 431, 437–9
  sensing beams 442, 444
  surface functionalisation 435–7
capillary pressure/interaction 63, 68, 126, 127
capping agent removal damage 169
carbazole derivatives 20–1
carbohydrates *see* saccharides
carbon bonds, pi (π) orbitals 393
carbon capsules, mesoporous 125
carbon compound cyanation 142
carbon disulfide 379–80
carbon nanocages 114–19
  applications 114
  competitive adsorption 116–19
  DNA intercalation removal 119
  dyes adsorption 118–19
  nucleosides adsorption 117–18
  tea components separation 115–17
carbon nanomaterial drug delivery systems 251–2
carbon nanotubes (CNTs) 203, 252, 331–4
carbon nitride
  mesoporous
    elemental substitution method 121, 122, 123
    gold nanoparticle encapsulation 121, 122
    micelle assemblies 121
    preparation 121, 122
    synthesis 120–1, 122
carbon-60 molecules *see* fullerenes
carbon–protein nanostructures 202–3
carboxylic acids 9, 226–7, 233, 234
carcinoembryonic antigen (CEA) 203
carrier migration, semiconductors 166
catalysis
  see also enzyme active site architectonics
  enzyme action 216

Subject Index

metal oxides 129–55
pillared clays 91
porous clay heterostructures 94
silicon–hydrogen cleavage catalysis by oxygen 401
catechin 115–17, 124
cation exchange capacity (CAC) 89
cavity formation 23–4
CEA (carcinoembryonic antigen) 203
ceramics synthesis 154
CFoF1-ATPase biomimetic energy converter 206
chain polymerisation 411, 412, 417–18
chain transfer agent (CTA) 245
charge phenomena
  amino acids *in situ* modification 309–10
  anion-complexation-induced charge separation 11–12
  corona discharge coating technique 49
  FET biosensors 202–11, 307–8
  manipulation 308–11
  percolation pathways in OSCs 364, 365, 384
  polymer conjugate–fullerenes 384
  site-selective charge conversion of proteins 309, 310
  transport/collection in OSCs 360
chemical functionalisation 405–6, 407
chemical precipitation 131–2
chemical purity of radionuclides 276
chemical stability 267, 270
chemical structure
  conjugated polymer–fullerenes 366–9
  MDMO-PPV 362
  nanocrystalline zirconia sorbents 289–90
  P3HT 362
  PCBM 362
  polymer-embedded titania 280–1
chemically patterned substrates *see* patterned substrates

chemically-heterogeneous patterns 45–6
chemisorption 169
chip-based structures 324, 325, 326
chirality 8–9, 24, 219–23, 228
chitosan (CHI) 192–4, 250
chlorobenzene
  active layer film formation 363, 364, 365, 366
  conjugated polymer–fullerenes solubility 366, 367
  P3HT:PCBM OPV thin films 371
chloroform 370
1-chloronaphthalene 374
cholesterol 248–50
cholesterol-substituted triazocyclononane 24–5
chromophores 11–12
cilostazol 244
citrate-stabilised colloidal gold solutions 319, 320, 321
Claisen–Schmidt condensation 138, 139, 140
classification 2, 8, 213, 260–2, 296–7
clay material-based nanoarchitectures 87–104
  fibrous/tubular clays 98–102
  layered clays 88–98
  layered inorganic solids 102–4
click chemistry 196, 200
CMC (critical micelle concentration) 245
CMK-3 mesoporous carbon 115–19, 124, 125
CMOS (complementary metal oxide semiconductors) 303, 306, 307
co-catalysts 173
co-precipitation 132
coaxial one-dimensional hetero-nanostructures 168
coffee-stain effect 48
coherence times, nuclear spin 395
coiled conjugated polymer–fullerenes 367
colloidal crystal templates 182–3

colloidal lithographies 339–44
colloidal nanoparticles 319–31
colloidal plasmonic nanostructures
  core–shell nanoparticles 321–4
  lithographic methods 337–47
  lithography-free methods 331–6
  preparation 320–1
  surface nano-patterning
    periodic ordered nanostructures 329–31
    random-fashion structures 324–9
colloidal preparation routes 95, 99, 320–1
column chromatography
  radionuclides
    chemical stability 267
    generators 266–7, 268–75
    matrix radiation damage 266
    restricted sorption capacity 267
    retention affinity 266
competitive adsorption 115–19
complementary metal oxide semi-conductors (CMOS) 303, 306, 307
computational analysis of active sites 217, 218–19
concentric spread, brush monolayer 70, 71, 72, 76
conductive regions 400, 406, 407
confinement, active site 219–23
conformational changes, enzymes 13–14, 214, 216
conjugated polymer–fullerene organic solar cells 361–85
  active layer morphology tuning 362–74
    film drying times 369–72
    molecular structure/morphology 366–9
    ratio of polymer to PCBM 364–5
    solvent additives 372–4
    solvent effects 363–4
  basic principles 361–2
  ideal morphology 361

  post-film formation treatment 375–85
    solvent vapour annealing 378–81
    thermal annealing 375–8
    vertical phase separation 381–5
conservation of amino acid sequence 217–19, 224, 229, 230, 234
contact angle (θ) 31, 49, 73
controlled dewetting
  block copolymer films 50–68
    self-organised multiscale 28–78
  chemically heterogeneous patterns 45–6
  fundamentals 30–6
  imprinting lithography 46–7
  photolithography 44–5
  solvent evaporation 47–8
controlled release systems 126–7
controlled solvent vapour annealing (C-SVA) 378–9, 380
controlled wetting 48–50, 68–78
copper hydroxide 145–7
copper oxide 144–7
core–shell nanoparticles 321–4
corona discharge coating technique 49
costs 339, 404, 437
covalent bond instability 393
covalent self-assembled monolayers 397–9
crest/mesa regions 59–64
critical micelle concentration (CMC) 245
cross-conjugated molecules 393, 394
crystal growth, fullerenes 15, 16
crystalline fullerenes 361, 362, 367, 375
crystalline P3HT nanorods 383
crystallites, magnesium oxide 135–44
CTAB surfactant 327, 328, 329
cubic crystals 136, 170–1
curcumin derivatives 196
Curie, Marie & Pierre 259
cutinases 221
cyanation reaction 142
cyanohydrins synthesis 143

cyanosilylation of aldehydes 138, 139, 142–3
cyclens, *N*-substituted 24
cyclophane molecular machine 22–4
cyclotron-produced radioisotopes 262, 296
CYP450 2C9 221
cytochrome P450 221

damping, cantilever sensors 432, 434
daughter radionuclide activity 263, 264, 265, 266, 269, 272–3
Debye interaction 32
decay, radioactive 263, 264–5
delaminated porous clay heterostructures (DPCHs) 94–6
dendrimers, drug delivery 251
deposition masks, colloidal 339–44
designable guest selectivity 125
designer nanointerfaces 202–11
desorption 169
detection methods
 see also biosensors
 radiochemical impurities 276
dewetting
 fundamentals 30–6
 and partial layer inversion 46–7
 self-organised block copolymers 43–50
 controlled, multiscale 28–78
dextran (DEX) 195
dextro enantiomers 222–3
diabetes mellitus 250
diacetylene molecules 417
diagnostic radionuclides *see* medically-useful radioisotopes
dialectrophoresis alignment 418
3,6-diaminoacridine hydrochloride 118
α-diazo-β-hydroxy esters 138, 139, 142, 143
diblock copolymers 245–6, 247, 322–3
diblock lamellae 37–9
1,2-dichlorobenzene (DCB) 378
differential equations 264–5
1,8-diiodooctane 374

dip-coating 47–8
dip-pen nanolithography (DPN) 46, 409–10
direct asymmetric aldol reactions 146
direct patterning scale registration 412–13
directional solvent evaporation 55
discs, HCL nanofabrication 341, 342
disorder transition temperature (T_{ODT}) 70, 72
distribution ratio (K_d) 269, 272–3
DMAEAm (poly(*N*,*N*-dimethylaminoethylacrylamide)) 322–3
DNA
 binders/intercalators 308–9
 DNA-based bioconjugate nanostructures 203–6
 intercalator removal by carbon nanocages 119
 origami shapes 413–15, 419–20
dodecanethiol 321
domain–domain period control 65
donor/acceptor segregation 15–18, 360
dot-shaped polymeric patterns 48
double hydroxides, layered 104
double lever geometry 440, 441, 442
DPCHs (delaminated porous clay heterostructures) 94–6
DPN (dip-pen nanolithography) 46, 409–10
droplet behaviour 29–30, 69–73, 74–8
drug delivery systems 242–54
 carbon nanomaterials 251–2
 carbon–protein bioconjugates 203
 chitosan bioconjugates 192–4
 dendrimers 251
 DNA bioconjugates 205
 emulsions/microemulsions 246–8
 inhalable particles 253
 inorganic materials 252–3
 lipid bioconjugates 195–7
 liposomes 248–50
 molecular conjugates 250–1
 nanocrystals 243–4

PEG–lipid bioconjugate nanostructures 197–8
polymer nanogels 250
polymer/surfactant micelles 244–6
requirements/advantages 253–4
site-selective charge conversion of proteins 310
drying times, P3HT films 369–72, 373
dyes 118–19, 168, 169, 170, 171, 183
dynamic coupling of enzyme regions 213–14
dynamic interfaces 22–4
dynamic mode, stress sensors 429, 432, 433–4, 436
dynamic self-assembly 15–18

EBL (electron beam lithography) 337–8
EC categories of enzyme-catalysed reactions 213
ECC-LbL (electrochemical coupling layer-by-layer) 20–1
EDC/NHS coupling reaction 200, 201
ee (enantiomeric excess) 8–9
eicosyl alcohol 397
electrical charge see charge phenomena
electrical stimulation, AFM patterning 405–7
electrochemical coupling layer-by-layer (ECC-LbL) 20–1
electrochemical metal oxide nanoparticle formation 133–4
electrode gaps, precise 418, 419
electrode scale registration methods 416, 418
electron beam lithography (EBL) 337–8
β-electron emissions 261–2
electron tomography 383
electronic coupling assembly 175–9
electronic devices 9–10
see also molecular electronics
electron–hole pairs 166
electrostatics
    interactions 228–9
    interfaces 202–11
    layer-by-layer assembly 19–22
    potential 226, 234
electro-wetting control method 49
elemental substitution method 121, 122, 123
elliptical nanostructures 342
elution of radionuclides 275
emulsions, drug delivery 246–8
enantiomeric excess (ee) 8–9
enantioselectivity 219–23
energy band reconstruction 175–9
energy converters, biomimetic 206
enhanced permeability and retention (EPR) effect 245, 252
enoyl reductase 218
environmental nanomechanical sensors 429
enzyme active site architectonics 213–37
    amino acid sequence conservation 217–19, 224, 229, 230, 234
    chiral discrimination/specificity 219–23
    confinement 219–23
    definition/characteristics 214
    influence on enzymatic reactions 216–19
    nanodimensions 219–23
    overview 213–16
enzyme–substrate complexes 214
epoxide hydrolase, human soluble 218
EPR (enhanced permeability and retention) effect 245, 252
equilibrium state, emulsions 246
erythromycin polyketide synthetase 218
Escherichia coli (EC)
    HisRS quantum mechanical study 231–6
    histidyl tRNA synthetase 225, 226, 227, 228, 229, 230
    monosaccharide bioconjugate nanostructures 192, 196
etching 337, 338, 339–44

Subject Index

evaporative block copolymer dewetting control 52, 53, 54–6
evolution, active sites 214
excitons 360, 361, 364, 365
exfoliation–restacking approach 103, 104

fabrication *see* materials development
facet-controlled photocatalytic nanostructures 169–73
FE-TEM (field emission transmission electron microscopy) 371
FEA (finite element analysis) 434, 436, 441, 442
ferric oxide nanoparticles 323
ferrite, fibrous/tubular clays 100
fertilizers, nitrogen-based 129–30
FET *see* field-effect transistor (FET) biosensors
Feynman, Richard 130
FIB (focused ion beam) lithography 337, 338, 339
fibrous clays 98–102
field emission transmission electron microscopy (FE-TEM) 371
field-effect transistor (FET) biosensors
  biomolecular charges 202–11
  charge manipulation 308–11
  gate insulator 303, 304
  label-free biosensing 303–8
  nucleic acids sensing 305–7
  principles of action 302–3
  protein sensing 307–8
  site-selective charge conversion of proteins 309–11
  types 304
film instability 35
finger patterns 48, 54, 361
finite element analysis (FEA) 434, 436, 441, 442
fission molybdenum-99 278
flavanones synthesis 138, 139, 140, 141
fluorescence 9–10, 201–2, 204–5
fluoride-writable memory systems 10

focused ion beam (FIB) lithography 337, 338, 339
free energy, wetting/dewetting 31–9
free flow sorbents 269–70
Friedel–Crafts alkylation of indoles 148–9, 150
frontier layer 29, 68
fullerenes
  see also conjugated polymer–fullerene organic solar cells
  carbon-60 nanorods 379–80
  carbon-71-PCBM 373–4
  scale registration 418
  shape-shifting crystal development 15, 16
Fusarium solani pisi lipase 221

gallium arsenide 403
gallium nitride 168
gamma-ray emissions, therapeutic 261–2
gate insulators 303, 304
gate surfaces, FET biosensors 202–11
gelatin 250
gels *see* polymer nanogels
gene therapy 195, 197–9
generator-produced radioisotopes *see* radionuclide generators
genetic analysis 305–7
germanium 412–13
GI-XRD (grazing incidence X-ray diffraction) 372
Gibbs–Thomson relation 45
GISAXS (grazing-incidence small-angle X-ray scattering) 43
glass substrates
  chip-based nanorod structures for biosensors 327, 328
  colloidal silver nanoparticles 325, 327
  eicosyl alcohol adsorption 397
  periodic ordered colloidal plasmonic structures 329–31
glucose oxidase (GOD) 200–1, 206
glycine achirality 228
glycosides 249

GNPs *see* gold nanoparticles (GNPs)
gold EBL-fabricated heptamers 338
gold nanoparticles (GNPs)
  bioconjugates
    BRCA-1 breast cancer gene 204
    gold–protein 200–1
  biosensor materials 319–21
  drug delivery 252
  encapsulation in mesoporous carbon nitride 121, 122
  hybrid nanostructure synthesis 331–4
  titanium dioxide films 180, 181
gold nanoprisms 180–1
gold substrates
  alkanethiol self-assembled on gold surface 433
  colloidal
    ordered nanoplasmonic pores 340
    surface nano-patterning methods 324–9
    surface nano-patterning of periodic ordered nanostructures 329–31
  controlled anisotropic wetting of block copolymer droplets 75, 76
  covalent self-assembled monolayers 397–9
  gold/triphenylene interaction 397
  nanografted thiols 407, 408
  thiol group interaction 304
gold–carbon hybrid materials 331–4
gold–protein nanostructures 200–1
grafted monolayers 407–9
granular radionuclide sorbent matrices 269
graphene sheet/ionic liquid (GS-IL) 19–20
graphene synthesis/processing 420
graphene–DNA nanostructures 205–6
graphite surfaces 411, 412
gratings, periodic ordered 329–31
grazing incidence X-ray diffraction (GI-XRD) 372

grazing-incidence small-angle X-ray scattering (GISAXS) 43
green fluorescent protein (GFP) 195
Greene, Margaret 260, 264
GS-IL (graphene sheet/ionic liquid) 19–20
guanosine 117–18
guest reagents 11–12, 19–20, 125

Haber–Bosch process 129
halloysite-based fibrous/tubular clays 101–2
Hamaker theory/constant 33, 38
Hamiltonian equation/computing system 396
hand-operated nanotechnology 24
HCL (hole-mask colloidal lithography) 339, 340–3
HDT (hexadecanethiol) 56, 75–8
heating, localised 50
Heck reaction 138
heterogeneous catalysis 129–30
heterotropic allosteric regulation 11–12
hexadecanethiol (HDT) 56, 75–8
hexagonal patterns/structures
  gold nanoprisms on tungsten trioxide 180–1
  magnesium oxide nanocrystals 136
  mesoporous carbon nitride 120, 121
  non-circular block copolymer droplets 74, 76
  two-dimensional, substituted porphyrins 14–15
hierarchical structures 56–8, 123, 167, 173–5
high resolution transmission electron microscopy (HRTEM)
  gold–carbon nanotube hybrid structures 333
  mesoporous boron carbon nitride 123
  mesoporous carbon nitride 120
  P3HT:PCBM blend films 381, 382, 384

Subject Index

histidyl tRNA synthetase (HisRS) 225, 226, 227, 228, 229, 230, 232, 234, 235
historical aspects 112–13, 129–30, 263–4
hole and island formation 30, 38
hole mobilities 376
hole-mask colloidal lithography (HCL) 339, 340–3
holes in metal films 338–9
HOMO shifts 397
horseradish peroxidase (HRP) 202–3
host–guest chemistry 22–5
HPC (2-[2-hydroxypropylthio]-ethane-sulfonate) 221
HRTEM *see* high resolution transmission electron microscopy (HRTEM)
human T-lymphotropic virus type I (HTLV-I) 204
hyaluronan (HA) 194–5
hybrid materials 331–4
hydrazine ceramics synthesis method 154
hydro-/solvo-thermal metal oxide nanoparticle formation 132–3
hydrofluoric acid capping agent 169
hydrogen bonding 11–12, 14–15
hydrogen transfer 402, 407–8
hydrogen-terminated silicon alkenes
    AFM-mediated hydrogen atom displacement 407–8
    chain polymerisation on 417–18
    chemical functionalisation 405–6, 407
    self-assembled monolayers 400–2
    shaved alkyl layers 408
hydrolase active site residues 218–19
hydrophilicity 75–8, 401
hydrophobicity 56, 75–8, 401
hydroquinones 11
hydroxyapatite 100
(2-[2-hydroxypropylthio]-ethane-sulfonate) (HPC) 221
hypocrellin B 193–4, 197

ibuprofen 252–3
ICP-AES (inductively coupled plasma atomic emission spectroscopy) 272
immunosensors 202–3
imogolite-based fibrous/tubular clays 101–2
imprinting lithography 46–7, 337–8, 339
    see also nanoimprint lithography (NIL)
impurity levels
    chemicals 297
    nanomaterial sorbents 271, 272
    radionuclides 296
    generators 275–6
in situ substrate modification 407–9
indium tin oxide (ITO) 325, 329–31
indoles 148–9, 150
induced fit model 214, 216
inductively coupled plasma atomic emission spectroscopy (ICP-AES) 272
inhalable particles 253
inorganic drug delivery systems 243, 252–3
inorganic solids, layered 102–4
inorganic–inorganic nanocomposites 94–6
instability of films 35
insulating regions 406, 407
insulin delivery 250
intelligent polymers 200
interaction energy 231–6
interaction types
    alkanethiol on gold 433
    Born repulsion 34
    Debye interaction 32
    electrostatic 19–22, 226, 228–9, 234, 302–17
    hydrogen bonding 11–12, 14–15
    Keesom reaction 32
    London dispersive 32
    thin film cohesion 32, 33
    van der Waals forces 31–4

intercalation processes 88–90, 119, 308–9
interdigitation 361
interfaces 22–4, 31–6, 202–11
internal molecular strains 369, 371, 436
International Commission on Radio Protection (ICRP) 253
intervening regions in active sites 224, 225, 226, 227, 230
inverse opal photonic crystals 182–3
ion beams, focused 337, 338, 339
ion exchange properties 88, 89, 92, 94, 95
ION-loaded Cy5.5-conjugated oleyl chitosan 194
ion-sensitive field-effect transistors (ISFETs) 306, 307
iron (III) oleate 323
iron oxides 97, 153–6, 202, 252
ISFETs (ion-sensitive field-effect transistors) 306, 307
isotopes *see* medically-useful radio-isotopes
ITO (indium tin oxide) 325, 329–31

Kagomé lattice 14–15
K_d (distribution ratio) 269, 272–3
Keesom interaction 32
ketones asymmetric reduction 146
(2-(2-ketopropylthio) ethanesulfonate) (2-KPC) 221
kinetically-driven dewetting 47–8
kinetics of sorption/sorbents 270
KIT-5 mesoporous silica templates 114, 117, 118
Kondo effect 418
2-KPC (2-(2-ketopropylthio) ethane-sulfonate) 221

label-free detection
  see also membrane-type surface stress sensors (MSS)
  biosensing 303–8
labelling efficiency, radionuclides 276–7

lamellar structures 36–9, 55
lasers 437, 438
  Doppler velocimetry 273
layer-by-layer (LbL) assembly 18–22
  chitosan bioconjugates synthesis 193
  layered clays/nanoparticles 96–8
  lipid bioconjugate nanostructures 196
  mesoporous nanomaterials 123–6
  multifunctional core–shell microspheres 323
layered materials
  clays 88–98
  double hydroxides (LDHs) 100, 104
  hierarchic structures 123–7
  phosphates 104
  titanates 103
LbL *see* layer-by-layer (LbL) assembly
LCST (lower critical solution temperature) 245–6, 247
LDHs (layered double hydroxides) 100, 104
Lennard–Jones potential 34
Lewis acid/base sites 135–6, 137, 140, 141, 142
ligases 217, 218, 223–36
lipases 221
lipids 195–9, 246–50
liposomal drug delivery 248–50
Lippmann–Young equation 49
liquid ribbons 56
lithographic methods
  colloidal lithographies 339–44
  colloidal plasmonic nanostructures 337–47
  controlled dewetting 44–7
  nanoprint lithography 344–7
  scanning beam lithographies 337–9
lithography-free methods 331–6
localised heating 50
localised surface plasmon resonance (LSPR) 179, 180, 320, 321
lock-and-key principle 8–9, 214
London dispersive interaction 32

Subject Index

low temperature nanocrystalline metal oxide synthesis 131–4
lower critical solution temperature (LCST) 245–6, 247
LSPR (localised surface plasmon resonance) 179, 180, 320, 321
lubrication, stable thin films 29

magadiite pillars 102
magnesium oxide (NAP-MgO) 135–44
magnetism
  composites recycling 174
  nanoparticles 96, 101, 252, 321–4
  separation method 154
Mannich reaction 139, 143, 144
manufacturing methods
  materials for biosensor applications 318–47
  medically-useful radioisotopes 262
  nanocrystals for drug delivery 244
  pillared clays 90
  polymer nanogels 250
  radioisotopes for nuclear medicine 262
master stamps, EBL preparation 338
materials development
  biosensors 318–47
  clay material-based nanoarchitectures 87–104
  controlled multiscale dewetting of self-organised block copolymers 28–78
  intentional assembly for functional structures 18–22
  mechanical tuning 22–5
  mesoporous nanoarchitectonics 112–27
  metal oxides in catalysis 129–55
  photocatalytic materials 165–83
  self-assembly for functional structures 15–18
  supramolecular materials 7–26
MCM-41 silica carrier 252–3
MDMO-PPV
  chemical structure 362

MDMO-PPV:PCBM composite films 363–6, 375
  atomic force microscopy 365, 375
  scanning electron micrographs 366
  thermal annealing post-film formation 375–8
mechanical patterning 407–9
mechanical sensors *see* nanomechanical sensors
mechanical tuning 22–5
mechanically-driven fields 50
medical nanomechanical sensors 428–9
medically-useful radioisotopes 259–97
  classifications 260–2
  diagnostic radionuclides 260–1
  overview 259–60
  production 262
  radionuclide generators
    column chromatographic 266–7, 268–72
    history/development 263–4
    isotope quality control 275–7
    nanomaterials-based sorbents 278–95
    radioactive decay equations 264–5
    separation methods 266–8
    shelf-life 277
    sorbent material evaluation 272–5
  terminology 296–7
  therapeutic radionuclides 261–2
membrane-type surface stress sensors (MSS) 428–45
  dynamic mode 429, 432, 433–4, 436
  overview 428–30
  performance 443–5
  sensitivity improvements 439–43
  static mode 429, 431–3
membranous anodic porous alumina 334–6
memory systems, fluorescence 9–10
mercaptohexadecanoic acid (MHA) 75–8
mercaptopropyltrimethoxysilane (MPTMS) 94

mesa/crest regions 59–64
mesoporous nanomaterials 112–27
  applications 113, 114
  boron nitride 119–23
  carbon nanocage functions 114–19
  carbon nitride 119–23
  definition 113
  layered hierarchic structure 123–7
  overview 112–14
metal oxides in catalysis 129–55
  applications 129–30
  catalysis overview 134–5
  design/synthesis 130–1
  typical examples/applications 135–54
    copper oxide 144–7
    iron oxides 153–6
    magnesium oxide 135–44
    titania 147–50
    zinc oxide 150–3
  wet-chemical/low temperature synthesis routes 131–4
metals
  see also individual metals; metal oxides in catalysis
  metal oxides in clays 96, 99
  metal–DNA nanostructures 204–5
  pillared clays 90
  plasmonic colloidal nanoparticles 319–31
  scanning beam lithographies 337, 338
  solvent annealing combination for block copolymer dewetting control 54
methyl orange (MO) 168, 171
methyl violet 118–19
methyl-isobutyl ketone (MIBK) 104
MHA (mercaptohexadecanoic acid) 75–8
MIBK (methyl-isobutyl ketone) 104
micelles
  drug delivery 244–6
  mesoporous nanomaterials preparation 113
  pinned on topographic surface 64–8
  polyionic 310

porous clay heterostructure preparation 92, 93
Michael addition 149
micro-imprinting 46–7, 56, 57
microemulsions for drug delivery 246–8
microfibrous clay structures 88
micropatterned PS-*b*-PMMA copolymer film 55–6
microwave irradiation 90, 134
miniaturisation for portable technologies 7
mirror symmetric surfaces 176–7
MK-0869 nanocrystals 244
molecular capture, dynamic interfaces 23–4
molecular conformational change 13–14, 214, 216
molecular conjugates 250–1
  see also conjugated polymer–fullerene organic solar cells
molecular electronics
  coupling assembly 175–9
  ECC-LbL 20–1
  electro-wetting 48–50
  electron–hole pairs 166
  patterning by STM 404
  silicon–carbon bond 400
molecular inks 409, 410
molecular internal strains 369, 371, 436
molecular machines 10–11
molecular recognition 8–9, 214
molecular structures see chemical structures
molecular weight change 369
molecular-level supramolecular complexes 8–15
molybdenum-99/technetium-99m generators 260–96
monoclinic bismuth vanadate (m-BiVO$_4$) nanoplates 169, 170
monoclonal antibodies (MAbs) 249
monodispersed metallic nanoparticles 98
monolayers

Subject Index 463

see also self-assembled monolayers
brush monolayers 30, 43, 68–73
compression/expansion 22–4
monolayer–bilayer patterns 406, 407
mononuclear phagocyte system (MPS) 248
monosaccharides 191–2
morphology
   conjugated polymer–fullerenes 366–9, 372, 373
   metal oxides in catalysis 135
   photocatalytic materials 167–75, 183
MPS (mononuclear phagocyte system) 248
MPTMS (mercaptopropyltrimethoxysilane) 94
MSS see membrane-type surface stress sensors (MSS)
multi-tip atomic force microscopy lithography 405
multicomponent coupling reactions (MCR) 146, 147
multifunctional core–shell microspheres 323
multiple component patterns 412
multiwall carbon nanotubes (MWCNTs) 332
mutation studies 217, 218–19

N-arylation 45–6
nanobelts 167, 168
nanoblock moulds 345–6
nanocages 179
nanocomposites
   hierarchical composite nanostructures 173–5
   inorganic–inorganic nanocomposites 94–6
   MDMO-PPV:PCBM composite films 365, 375
   pillared clays 88–90
nanocrystals
   drug delivery systems 243–4
   magnesium oxide 135–44

molybdenum-98/tungsten-186 recovery from spent columns 295
molybdenum-99/technetium-99m generator development 294–5
quality control of technetium-99m/rhenium-188 295
sorbents for radionuclide separation 279, 288–95
synthesis 288–9
tungsten-188/rhenium-188 generator development 294–5
zeta potential determination 292
zirconia, applications 291–5
nanodiscs 341, 342
nanoimprint lithography (NIL) 344–7
nanointerfaces see interfaces
nanomechanical sensors 428–45
   cantilever sensors 430–9
   demand for 428–30
   membrane-type surface stress sensors 439–44
nanomembranes
   see also membrane-type surface stress sensors (MSS)
   anodic porous alumina 334–6
nanoparticles (NPs)
   see also gold nanoparticles (GNPs)
   colloidal plasmonic 319–31
      core–shell nanoparticles 321–4
      noble metal nanostructures 319–20
      NSL fabrication 343
   definition 242
   delaminated porous clay heterostructures 94, 95, 96
   fibrous/tubular clays 98–101
   hybrid gold–carbon 331–4
   size for biomedical use 242–3
nanoprint lithography 344–7
nanorings 60, 339–40
nanorods
   carbon-60 nanorods 379–80
   chip-based structures 325, 327–9
   crystalline P3HT nanorods 383
   gold nanorod arrays image 332

oblique angle deposition method 331, 332
nanosheets 19–20, 87–8, 169–70, 178–9
nanosphere lithography (NSL) 339, 343–4
nanostructuring metal oxide effects 135
nanotubes 101–2, 167, 179
nanowires (NW)
  cadmium sulfide assemblies 179
  NW-FETs 304, 305
  periodic ordered colloidal plasmonic structures 329–31
  phosphorus NW between electrodes on germanium 412–13
  photocatalytic materials 167–8
  trigeminal porphyrins 15–18
NAP magnesium oxide 135–6, 137, 140, 141, 142
neocarzinostatin 251
nickel nanoparticles 101
NIL (nanoimprint lithography) 344–7
NIPAAm pH/temperature-sensitive copolymer 249–50
nitrendipine 247–8
nitrobenzene 374
nitrogen-based fertilisers 129–30
noble metal nanostructures 319–20
nonchiral–chiral solvating agents 9–10
non-circular block copolymer droplets 74–8
non-conserved residues 230
nonequilibrium states 369
nonvolatile memory systems 9–10
NOR gate 397
NSL (nanosphere lithography) 339, 343–4
nuclear medicine *see* medically-useful radioisotopes
nuclear reactors 278
nuclear spin coherence times 395
nucleation and growth mechanism 29, 35, 36, 39, 44
nucleic acids
  bioconjugate nanostructures 203–6
  DNA

  binders/intercalators 308–9
  intercalator removal 119
  origami shapes 413–15, 419–20
  field-effect transistor biosensors 305–7
nucleosides 117–18, 139, 143, 144

oblique angle deposition method (OAD) 331, 332
octahedral sheets, layered clays 87–8
1,8-octanedithiol 374
OD (optical density) 382
ODT (order-disorder transition temperature) 40
oleic acid 323
oleyl-chitosan bioconjugates 194
oleylamine 323
on-wire-lithography (OWL) 335–6
one-dimensional photocatalytic nanostructures 167–8
opaque liquids 437
open-mouthed capsules 21–2
optical biosensors 324, 325, 326, 340
optical coupling assembly 181–3
optical density (OD) 382
optical (laser) readouts 431, 437–8, 444
oral absorption 243, 244, 247–8, 250
order-disorder transition temperature (ODT) 40
ordered nanoplasmonic pores 341
ordered semicrystalline domains, P3HT 368
organic carbonates synthesis 138, 139, 140, 141
organic photovoltaic devices 16–18
organic solar cells (OSCs) *see* conjugated polymer–fullerene organic solar cells
organosilanes 304
oriented elliptical nanostructures 342
origami DNA shapes 413–15, 419–20
orotidine monophospphate (OMP) 215
OSCs (organic solar cells) *see* conjugated polymer–fullerene organic solar cells

Subject Index

osmium tetroxide nanoparticles 102
Ostwald–Freudlich equation 243
oversized PCBM aggregates 376, 377, 378
OWL (on-wire-lithography) 335–6
oxidative silicon/gallium arsenide patterns 403
oxide nanoparticles 94, 95, 96
oxide substrates 399–400
oxidoreductases, active site residues replacement 218
oxoporphyrinogen 11–12
oxygen 169, 170, 172, 401

P3HT:PCBM films 381–5
P3HT (poly(3-hexylthiophene)) 362, 367, 368, 369–72, 373
$p4mm$ symmetry 65–6, 67
P4VP-b-PS-b-P4Vp (triblock copolymer polystyrene/poly(4-vinyl pyridine)) 323, 324
paclitaxel 195, 250–1
PALS (phase analysis light scattering) 273
palygorskite-based fibrous/tubular clays 98–101
PAMAM (polyamidoamine) 251
papain 194
parallel lamellae stability 38–9
parent–daughter radionuclides 263, 264, 265, 266, 269, 272–4
partial dewetting theory 35
passivating medium effects 177
patterned substrates
  block copolymer arrays 69–73
  colloidal plasmonic nanostructures 324–31
  dip-pen nanolithography 46, 409–10
  hydrophilic to hydrophobic surface properties change 401
  mechanical patterning 407–9
  nanoshaved/nanografted 407–9
  solvent controlled dewetting of block copolymer thin films 56–8

vacuum-deposited molecular arrays 410–12
PCBM (6,6-phenyl C_{61} butyric acid methyl ester)
  chemical structure 362
  MDMO-PPV:PCBM thin films 363–6, 375
  oversized aggregates 376, 377, 378
  PEDOT:PSS-coated fused silica 384
  ratio of polymer to PCBM in organic solar cells 364–5
PCHs (porous clay heterostructures) 91–4
PCPDTBT low band gap polymer 373–4
PDI (polydispersity index) 368
PDMS (poly(dimethylsiloxane)) 69–73
PEDOT:PSS-coated fused silica 384
PEG–lipid bioconjugate nanostructures 197–8, 248
Penrose patterns 411
10,12-pentacosadiynoic acid 411, 412
peptide bond formation 222–3
peptidyl transferase centre (PTC) 222–3
periodic mesoporous organosilicates (PMO) 113–14
periodic ordered colloidal plasmonic structures 329–31
PET (positron emission tomography) 260–1
pH
  nanocrystalline zirconia zeta potential 292
  pH-responsive liposomes 249–50
  pH-sensitive dewetting 50
  polymer-embedded titania 283–4
phase analysis light scattering (PALS) 273
phase separation 16–18, 361, 362, 366, 375
6,6-phenyl C_{61} butyric acid methyl ester see PCBM (6,6-phenyl C_{61} butyric acid methyl ester)
phosphates, layered 104

phospholipids 195–9, 246–50
phosphorus atoms 395
phosphorus nanowires 412–13
photo-driven molecular machines 10–11
photocatalytic materials 165–83
　facet-controlled nanostructures 169–73
　fibrous/tubular clays 100
　hierarchical composite nanostructures 173–5
　layered clays/nanoparticles assembly 96
　layered titanates 103
　morphology control 167–75, 183
　nano-assembly 175–83
　optical coupling assembly 181–3
　plasmon–exciton coupling assemblies 179–81
photolithography 44–5
photonic band gap 181
photonic crystals 181–3
photoresist (PR) gratings 329–30
photothermal therapy 197, 199
photovoltaic switching response 20–1
physical events regulation 11–12
physical properties
　diagnostic radionuclides 260, 261
　magnesium oxide crystallites 135
　nanomaterial sorbents for radionuclide separation 271, 272
　therapeutic radionuclides 261, 262
pi (π) orbitals, carbon bonds 359–60, 393
piezoresistive readouts 431, 437, 438–9
pillared clays (PILCs) 88–91
pillars 88–90, 102, 103
Pinnavaia's method 92, 96
plasma clearance time 251
plasmonic nanomaterials 319–31
　see also gold...; silver
plasmon–exciton coupling assemblies 179–81
platinum microcapsules 21–2
PMO (periodic mesoporous organosilicates) 113–14

p–n junctions/heterojunctions 17, 20, 393, 394
PNIPAAm (poly(n-isopropylacrylamide)) 245–6, 247
PNIPAM-N_3 (azido-terminated poly(N-isopropylacrylamide)) 200
point-force loaded cantilevers 440, 441
Poisson's ratio 434–5
poly[2-methoxy-5-(3′,7′-dimethyloctyloxy)-1,4-phenylene vinylene] *see* MDMO-PPV
poly(3-hexylthiophene) (P3HT) 367, 368, 369–72, 373
polyacetylene 360, 417
polyamidoamine (PAMAM) 251
polyaniline 416
poly(dimethylsiloxane) (PDMS) 69–73
polydispersity index (PDI) 368
poly(ethylene glycol) 243–4
polyfluorene copolymers 382
polygons, polymer thin films 57
polyhedral structures 136, 170–3
polyionic micelles 310
polymer-embedded titania (TiP)
　application 282–3
　molybdenum-98/tungsten-186 recovery from spent columns 288
　radionuclide generator development using 286–7
　rhenium-188/technetium-99m quality control 287–8
　sorption capacity determination 284–6
　structural characterisation 280–1
　synthesis 280
　zeta potential determination 283–4
polymers
　see also individual polymers
　aniline 416
　conjugated *see* conjugated polymer–fullerene organic solar cells
　micelles 244–6
　nanogels 250

Subject Index

polymer–protein nanosystems 199–203
thin films 28–78
poly(n-isopropylacrylamide) (PNIPAAm) 245–6, 247
poly(N,N-dimethylamineoethylacrylamide) (DMAEAm) 322–3
poly(p-phenylene) 360
polypyrrole 360
polysaccharides 192–5
polystyrene spheres 339, 340, 341
poly(styrene-*block*-4-hydroxystyrene) (PS-*b*-PHOST) 60–8
poly(styrene-*block*-ethylene oxide) (PS-*b*-PEO) 51–4, 69–73
polystyrene-*block*-poly(methyl methylacrylate) (PS-*b*-PMMA) 59–60
poly(styrene-co-maleic acid) nanogels 251
polythiophene 360
poly(vinylpyridine) (PVP) 325
poorly soluble drugs 243, 247–8
porous clay heterostructures (PCHs) 91–6
porous materials
  alumina nanomembranes 334–6
  ordered nanoplasmonic 341
  pillared clays 88–90
  size of pores 182, 334–5
porphine macrocycles 9–10
porphyrin 9–18
porphysomes 197, 198, 199
portable technologies 7
positron emission tomography (PET) 260–1
potassium cadmium chloride nanowires 178–9
powders 145, 151–3
PR (photoresist) gratings 329–30
precipitation 131–2
precise electrode gaps 418, 419
prefabricated electrodes 418
production *see* manufacturing methods
propargylamines 121

proteins
  bioconjugate nanostructures 199–203
  biosynthesis 223–36
  cages, biomimetic mineralisation 134
  protein-based nanosystems 199–203
  proteomic analysis 307–8
  quantum dot–protein nanostructures 201–2
  ribosomes 218–19, 222–3
  sensing 307–8
  site-selective charge conversion 309–11
PS-*b*-PEO (poly(styrene-*block*-ethylene oxide)) 51–4, 69–73
PS-*b*-PHOST (poly(styrene-*block*-4-hydroxystyrene)) 60–8
PS-*b*-PMMA (polystyrene-*block*-poly(methyl methylacrylate)) 59–60
PTC (peptidyl transfer centre) 222–3
pulmonary drug delivery 242–3, 253
purines 117–18
purity, radionuclides 271–2
PVP (poly(vinylpyridine)) 325
pyridine derivatives 139, 143, 144
pyrimidines 117–18
pyrogenicity, radionuclides 277, 297

QCM electrodes/resonators 124–6
quality control, isotopes 275–7, 295
quality factor (Q) 434
quantum dots 201–2, 252
quantum effects 231–6, 392, 393–7
quantum qubits 395

radiation stability 270
radical-mediated polymerisation 417
radioactive decay equations 264–5
radiochemical purity 276
radioisotopes *see* medically-useful radioisotopes
radionuclide generators
  column chromatographic 266–7, 268–75
  concept 263–4

daughter radionuclide activity 263, 264, 265, 266, 269, 272–3
elution efficiency 275
history/development 263–4
mathematical equations 264–5
nanomaterial sorbents 270–2
parent nucleotide activity 263, 264, 265, 266, 269, 272–4
preparation 266–8
quality control of isotopes 275–7
radioactive decay equations 264–5
radiochemical purity 276
radionuclide purity 275–6
separation methods 266–8
shelf-life 277
solvent extraction 267–8
sorbents 266–7, 269–75, 278–95
RAFT (reversible addition–fragmentation chain transfer) 200
random colloidal plasmonic structures 324–9
Re (rhenium) 278–95
reaction chemistry 215, 216
reaction rates, active sites 214, 215
reactive ion etching (RIE) 340, 341
reactor-produced radioisotopes 262
read-out, cantilever sensors 437–9
real-time detection *see* membrane-type surface stress sensors (MSS)
recognition proteins 200
recycling of nanostructures 174
redox reactions 166
reducing agents 320, 321, 322, 323
regioregular conjugated polymers 367, 368, 369–72, 373
registration, substrates 412–13
relative resistance change 440, 441
relaxed model structures 176
RES (reticuloendothelial system) 245
resistance, cantilever sensors 440, 441
resolution, DPN range of scales 409–10
retention affinity 266
reticuloendothelial system (RES) 245
reusable masks 338

reversible addition–fragmentation chain transfer (RAFT) 200
RhB (rhodamine B) dye 169, 170, 171
rhenium-188 (^{188}Rh) 278–95
rhodamine B (RhB) 169, 170, 171
rhombic dodecahedrons 170–3
ribosomes 218–19, 222–3
RIE (reactive ion etching) 340, 341
rod-like twisted structures 114
  see also nanorods
rr (regioregular) conjugated polymers 367, 368, 369–72, 373

saccharides 191–5
Saccharomyces cerevisiae (SC) 226, 229, 230, 232, 234, 235
sacrificial resist layers 340–1, 342
Sader's model 436
SAMs *see* self-assembled monolayers (SAMs)
SBA-15 mesoporous silica template 115, 116, 118, 119, 120
scale registration
  alignment to prefabricated electrodes 418
  direct patterning method 412–13
  polymerisation method 416–18
  self-assembly method 413–16
  substrate alignment 412–18
scales resolution 409–10
scanning beam lithographies 337–9
scanning electron microscopy (SEM) 332, 338, 366
scanning tunnelling microscopy (STM)
  graphene synthesis/processing 420
  low cost/low damage to substrate 404
  molecular conformation adaptation 13
  organic self-assembled monolayers on gold 398, 399
  phosphorus nanowires between electrodes on germanium 412–13
  substrate alignment 393, 394, 397, 398, 399
  surface patterning 402–4

Subject Index

Schiff's bases 193
Schottky junctions 181
SCL (sparse colloidal lithography) 339–40
scribing (mechanical patterning) 401
security applications 429
selective sorption/sorbents 270
self-assembled monolayers (SAMs)
  alkyl chains 304
  controlled dewetting 45
  eicosyl alcohol adsorption on glass 397
  gold substrate 397–9
  hexadecanethiol hydrophobic 56, 75–8
  mercaptohexadecanoic acid hydrophilic 75–8
  oxide substrate 399–400
  silicon substrate 400–2
self-assembly
  block copolymers, controlled 51–4
  colloidal particles 339–44
  conjugated polymer–fullerenes, manipulation 359–85
  functional structures 15–18
  layered clays/nanoparticles 97
  regioregular conjugated polymers 367
  scale registration 413–16
self-microemulsifying formulations 247–8
self-organisational film drying 369–72, 373
self-organised block copolymers 28–78
SEM (scanning electron microscopy) 332, 338, 366
semiconducting nanoparticles 98
  photocatalytic *see* photocatalytic materials
semiconductors 173, 438
  p–n junctions 17, 20, 393, 394
sensing beams, cantilevers 442, 444
sensors *see* biosensors
separation methods 259–97

sepiolite-based fibrous/tubular clays 98–101
sepiolite–metal oxide nanoparticles 98, 99
sequence conservation 217–19, 224, 229, 230, 234
shape change 15, 16, 76–8
  see also morphology
shaved nanostructures 407–9
shelf-life, radionuclide generators 277
short interfering RNA (siRNA) 198
side chain length variation 367
silanes 399, 409
silica
  capsules 126–7
  pillars 93, 102
  silica-based DCPHs preparation 95
  sources 92
  templates 112–27
silicon
  see also hydrogen-terminated silicon
  cantilevers 431
  hydrogen transfer, 1-alkene monolayers on silicon 402
  isotopically-pure 395
  radicals/alkenes reaction 403, 404
  silicon crystal gates 395
  silicon dioxide 96, 408–9
  silicon nitride 340–1
  silicon–carbon bond 400, 401
  silicon–hydrogen bond 401
  substrates 400–2
  wafers 340–1
silver
  chip-based nanorod structures for biosensors 327, 328
  drug delivery 252
  halloysite-based heterostructures 101
  nanoparticles for biosensor materials 319–21, 325, 327
  nanorods 331, 332
  NSL array fabrication 343
  silver nitrate 169, 170, 327, 328, 329
  silver phosphate 170–3

silver/silver chloride core-shell
    nanowires 168
silver–DNA bioconjugates 204–5
single atom/molecule calculations
    393, 394, 395, 396–7
single nucleotide polymorphisms
    (SNPs) 308
single-walled carbon nanotubes
    (SWNT) FET biosensors
    304, 305
siRNA (short interfering RNA) 198
site-directed mutagenesis 217, 218–19
size aspects
  catalytic chemistry, metal oxides
    129–55
  particles for biomedical use 242–3, 252
  pinned micelles on topographic
    surface 65
  pore size 182, 334–5
  solubility relationship 243
slip–stick motion 54
slow drying (solvent annealing) 52,
    53–4, 69–73
slow-light effect 182
SMAD (solvated metal atom
    dispersion method) 332
small interfering RNA (siRNA) 195
smart adjustment 64–8
smectites 94
sodium hydroxide 327, 328, 329
soft lithography 44–6
solar cells
  see also conjugated polymer–
    fullerene organic solar cells
  dye-sensitised 183
solid coating films 434–5
solid-state (top-down) routes 131, 244
  see also lithographic methods
solubility
  conjugated polymer–fullerenes
    366–9
  micelle delivery systems 247–8
  nanocrystal delivery systems 243
  paclitaxel 251
  particle size relationship 243

solution preparation 362–74
solvated metal atom dispersion
    method (SMAD) 332
solvent vapour annealing (SVA) 378–81
solvents
  active layer formation 363–4
  block copolymers
    nanostructures control 51–6
    thin film dewetting control 47–8,
      369–70
  boiling points 373
  solvent annealing 52, 53–4, 69–73
  solvent extraction 267–8
solvo-thermal method 132–3
sol–gel process 95, 99, 132
Sonagashira reaction 138
sonochemical method 133
sorbent matrices for radionuclide
    generators 266–7, 269–75
  chemical stability 270
  evaluation 272–5
  free flow characteristics 269–70
  granularity 269
  nanomaterials-based 270–2
    nanocrystalline zirconia-based
      288–95
    polymer embedded nanocrystal-
      line titania 280–8
  radiation stability 270
  selectivity of sorption/elution 270
  sorption kinetics 270
  sorption capacity determination 274,
    284–6, 292–4
source–drain channels 304
sparse colloidal lithography (SCL)
    339–40
species similarities/variations 217
specific activity, radionuclides 296
specificity, enzyme action 216, 219–23
SPECT imaging 260, 261
spin coating
  active layer film formation 370, 371
  organic solar cells 361, 362, 363
  block copolymer dewetting control
    51–4, 62

Subject Index

spinel ferrites 153
spinodal dewetting 29, 35
SPM *see* surface probe microscopy (SPM)
stabilising agents 321
stability, thin films 36–43
Staphylococcus spp.
   S. aureus (SA) 226, 229, 230, 231–6
   S. hyicus lipase 221
static mode, stress sensors 429, 431–3
static sorption capacity 274
stent coatings 250
stereospecific active sites 219–23
sterility, radionuclides 277, 297
steroid cyclophane molecular machine 22–4
stick–slip motion 48
stimuli-responsive liposomes 249–50
Stoney's equation 434, 436
strains, internal molecular 369, 371
strand-displacement DNA polymerisation 206
Strecker reaction 138, 139, 141, 142
stressed surfaces *see* membrane-type surface stress sensors (MSS)
stripe-shaped polymeric patterns 48
strong segregation theory 37
structure *see* chemical structure
styrene
   lines growth on hydrogen-terminated silicon 403, 404, 417–18
   polystyrene spheres 339, 340, 341
   poly(styrene-*block*-4-hydroxystyrene) 60–8
   poly(styrene-*block*-ethylene oxide) 51–4, 69–73
   polystyrene-*block*-poly(methyl methylacrylate) 59–60
   poly(styrene-co-maleic acid) nanogels 251
   triblock copolymer polystyrene/poly(4-vinyl pyridine) 323, 324
5-substituted 1*H*-tetrazoles 153
substituted pyridine derivatives 139, 143, 144

substrate alignment
   scanning probe lithography 392–420
      molecule anchoring to substrates 397–402
      overview 392–3
      scale registration 412–18
      single molecule/atom calculations 393–7
      surface patterning 402–12
succinic anhydride (SA) 309, 310
sucrose 114, 115
sugar rings, chirality 222–3
α-sulfanyl-β-amino derivatives 143
sulfhydryls 398, 399
supported lipid bilayers 195–6
supramolecular materials 7–26
surface probe microscopy (SPM) 392–420
surfaces
   crystal surface energy 172
   functionalisation 435–7
   patterning 324–31, 402–10
      see also patterned substrates
   sorbent surface area 271, 272
   stress sensors *see* membrane-type surface stress sensors (MSS)
   two-dimensional molecular arrays formation 12–14
surfactants
   chip-based nanorod structures for biosensors 327, 328, 329
   fibrous/tubular clay preparation 99
   mesoporous nanomaterials ynthesis 114
   micelles, drug delivery 244–6
   porous clay heterostructure preparation 92
SVA (solvent vapour annealing) 378–81

Tanner's law 73
tannic acid 115–17, 124
targeted drugs 245
technetium-99m radioisotope generators 260–96

TEM (transmission electron microscopy) 379
temperature-triggered liposomes 249
TEOS (tetraethylorthosilicate) 93
terminology 2, 8, 213, 260–2, 296–7
terraced droplets 40, 43, 69–73
terraced hierarchical structures 29, 68
tetraethylorthosilicate (TEOS) 93
tetragonal symmetries 67, 68
tetrahedral DNA bioconjugates 204
tetrahedral sheets 87–8
tetramethyl-ammonium hydroxide (TMAH) 340
tetrazoles 153
therapeutic radionuclides 261–2
   see also medically-useful radioisotopes
thermal annealing 375–8
thermolysis 133
thermoresponsivity 192, 200, 245–6, 247
Thermus thermophilus (TT) 226, 229, 230, 231–6
thick homopolymer films 32
thiols
   alkanethiol 433
   biothiols detection 204–5
   dodecanethiol 321
   gold interaction 304
   hexadecanethiol 56, 75–8
   molecular inks 409
   nanografted on gold substrates 407, 408
   1,8-octanedithiol 374
   organic self-assembled monolayers 398, 399
threshold voltage (V_T) 308, 309
thymidine 117–18
thymine 24–5
TiO_2 nanoparticles 177–8
TiP (polymer-embedded titania) 279, 280–8
TIPOTI (titanium *tetra*-isopropoxide) 93
titania
   applications 147–8
   nanocrystalline for catalysis 147–50
   physical/chemical properties of nanocrystals 149–50
   polymer-embedded (TiP) 279, 280–8
   rutile nanotitania properties 150
titania–silica pillars 93
titanium dioxide films 180, 181
titanium oxide 169, 177–8
titanium *tetra*-isopropoxide (TIPOTI) 93
TMAH (tetramethyl-ammonium hydroxide) 340
TNS (6-(*p*-toluidino)naphthalene-2-sulfonate) 24
T_{ODT} (disorder transition temperature) 70, 72
toluene
   active layer film formation 363–4, 365, 366
   benzylation 144, 145
   conjugated polymer–fullerenes solubility 366
   rr-P3HT thin films 369
(6-(*p*-toluidino)naphthalene-2-sulfonate) (TNS) 24
top-down *see* solid-state (top-down) routes
topographic prepatterns 45, 46, 47, 59–68
transmission electron microscopy (TEM) 379
triazocyclononane 24–5
triblock copolymer polystyrene/poly(4-vinyl pyridine) (P4VP-b-PS-b-P4VP) 323, 324
trigeminal porphyrin molecule 15–18
trigeminal structural motif 15–18
triglycerides 246–8
triphenylene 397
triphosphate groups 226
truth tables, Boolean 396, 397
tubular clays 98–102
Tucker, Walter 260, 264
tungsten trioxide (WO_3) 173, 174, 180–1, 182–3

tungsten-186 (^{186}W) 288, 295
tungsten-188 (^{188}W) 278–95
tuning
   anodic porous alumina nanomembranes 334
   liposomal drug release 249–50
   photonic crystals 181–2
   ultra-thin tunable masks 335, 336
twisted rod-like structures 114
twisting motions 24
two-dimensional hexagonal structures 14–15
two-dimensional molecular arrays 12–14
type I/II/III/IV thin film systems 35–6

ultra-thin alumina masks (UTAMs) 335, 336
ultrasound method 133
unexpected properties 2
uracil 24–5
uridine monophosphate (UMP) 215
UTAMs (ultra-thin tunable masks) 335, 336
UV-lithography 340–1

VA (vapour annealing) 378–81
vacuum-deposited molecular arrays 410–12
van der Waals forces 31–4
vapour annealing (VA) 378–81
vapour detection 19–20
vapour pressure 378–9
variable-angle spectroscopic ellipsometry (VASE) 381, 382, 384
VCSEL (vertical cavity surface emitting laser) 437, 438
vermiculite 95
vertical cavity surface emitting laser (VCSEL) 437, 438
vertical phase separation 381–5
V_T (threshold voltage) 308, 309

W *see* tungsten...
Wadsworth–Emmons reaction 138, 139
warfarin 221
water photolysis 168

water release 126–7
web-like nanostructures 55
wet chemistry (bottom-up) routes
   fibrous/tubular clay-based heteroarchitectures 98–102
   metal oxide nanoparticle formation 131–4
      biomimetic mineralisation 134
      electrochemical 133–4
      hydro-/solvo-thermal 132–3
      hydrolysis/chemical precipitation 131–2
      microwave synthesis 134
      sol–gel synthesis 132
      sonochemical 133
      thermolysis 133
   nanocrystals for drug delivery 244
wet-philic SAMs 75–8
wet-phobic SAMs 56, 75–8
wetting 30–6, 68–78
Wheatstone bridges 439, 441, 442
Wittig reaction 138, 139, 140
Wyckoff notation 66, 68

X-ray photoelectron spectroscopy (XPS) 121
xylene 370

Young's equation 31
Young's modulus 433, 434, 435, 436

Z-scheme systems 173
zeolite–sepiolite structures 100–1
zeta potential (ζ) 273, 283–4, 292
zinc blende cadmium sulfide crystals 176–7
zinc cations 228
zinc oxide
   layered clays/nanoparticles assembly 96, 97
   nanocrystalline for catalysis 150–3
   one-dimensional photocatalytic nanostructures 168
   powder
      heat treatment/activation 152–3
      synthesis 151–2
zirconia sorbents 279, 288–95